ALGEBRA FOR COLLEGE STUDENTS

REVISED AND EXPANDED EDITION

BERNARD KOLMAN

Drexel University

ARNOLD SHAPIRO

Pennsylvania State University,
Ogontz Campus

ACADEMIC PRESS

A Subsidiary of Harcourt Brace Jovanovich, Publishers
New York / London
Paris / San Diego / San Francisco
São Paulo / Sydney / Tokyo / Toronto

To Our Wives, Lillie and Reba

Cover art: *Arcturus II* by Vasarely
courtesy Hirshhorn Museum and Sculpture Garden,
Smithsonian Institution

Academic Press, Inc.
111 Fifth Avenue, New York, New York 10003

United Kingdom Edition published by
Academic Press, Inc. (London) Ltd.
24/28 Oval Road, London, NW1 7DX

ISBN: 0-12-417875-8
Library of Congress Catalog Card Number: 79-50492

Printed in the United States of America

CONTENTS

Preface   vii
Acknowledgments   ix
To the Student   xi

CHAPTER ONE
THE REAL NUMBER SYSTEM   1
1.1 The Real Number System *1*
1.2 Arithmetic Operations: Fractions *6*
1.3 Algebraic Expressions *16*
1.4 Operating with Signed Numbers *19*
1.5 Properties of Real Numbers *24*
1.6 Absolute Value and Inequalities *28*

CHAPTER TWO
POLYNOMIALS   41
2.1 Polynomials *41*
2.2 Addition and Subtraction of Polynomials *48*
2.3 Multiplication of Polynomials *52*
2.4 Factoring *57*
2.5 Special Factors *63*

CHAPTER THREE
LINEAR EQUATIONS AND INEQUALITIES   69
3.1 Linear Equations in One Variable *69*
3.2 Applications *73*
3.3 Linear Inequalities *79*
3.4 Absolute Value in Equations and Inequalities *84*

CHAPTER FOUR
WORD PROBLEMS   91
4.1 From Words to Algebra *91*
4.2 Coin Problems *95*
4.3 Investment Problems *99*
4.4 Distance (Uniform Motion) Problems *102*
4.5 Mixture Problems *106*

CHAPTER FIVE
ALGEBRAIC FRACTIONS   113
5.1 Multiplication and Division of Fractions *114*
5.2 Addition and Subtraction of Fractions *119*
5.3 Complex Fractions *125*
5.4 Ratio and Proportion *128*
5.5 Equations and Inequalities with Fractions *132*
5.6 Applications; Work Problems *136*

CHAPTER SIX
FUNCTIONS   145
6.1 Rectangular Coordinate Systems *145*
6.2 Functions and Function Notation *155*
6.3 Graphs of Functions *166*
6.4 Increasing and Decreasing Functions *173*
6.5 Direct and Inverse Variation *176*

CHAPTER SEVEN
THE STRAIGHT LINE   187
7.1 Slope of the Straight Line *187*
7.2 Equations of the Straight Line *192*
7.3 Further Properties of the Straight Line *199*
7.4 Linear Inequalities in Two Variables *207*

CHAPTER EIGHT
EXPONENTS, RADICALS, AND COMPLEX NUMBERS   215
8.1 Positive Integer Exponents *215*
8.2 Integer Exponents *220*
8.3 Rational Exponents and Radicals *224*
8.4 Evaluating and Simplifying Radicals *230*
8.5 Operations with Radicals *235*
8.6 Complex Numbers *238*

CHAPTER NINE
SECOND-DEGREE EQUATIONS AND INEQUALITIES   249
9.1 Solving Quadratic Equations *249*
9.2 The Quadratic Formula *255*
9.3 Roots of a Quadratic Equation: The Discriminant *259*
9.4 Applications *262*
9.5 Forms Leading to Quadratics *266*
9.6 Second-Degree Inequalities *270*

CHAPTER TEN
ROOTS OF POLYNOMIALS   279
10.1 Polynomial Division and Synthetic Division *280*
10.2 The Remainder and Factor Theorems *284*
10.3 Factors and Roots *288*
10.4 Real and Rational Roots *296*

CHAPTER ELEVEN
EXPONENTIAL AND LOGARITHMIC FUNCTIONS   307
11.1 Combining Functions; Inverse Functions *308*
11.2 Exponential Functions *319*
11.3 Logarithmic Functions *327*
11.4 Properties of Logarithms *334*
11.5 Computing with Logarithms *340*
11.6 Exponential and Logarithmic Equations *346*

CHAPTER TWELVE
ANALYTIC GEOMETRY: THE CONIC SECTIONS   353
12.1 Distance and Midpoint Formulas *353*
12.2 Symmetry *359*
12.3 The Circle *362*
12.4 The Parabola *366*
12.5 The Ellipse and Hyperbola *372*
12.6 Identifying the Conic Sections *380*

CHAPTER THIRTEEN
SYSTEMS OF EQUATIONS   387
13.1 Systems of Linear Equations *387*
13.2 Solving by Elimination *395*
13.3 Applications *399*
13.4 Systems of Linear Equations in Three Unknowns *408*
13.5 Systems Involving Nonlinear Equations *411*

CHAPTER FOURTEEN
MATRICES AND DETERMINANTS   419
14.1 Matrices and Linear Systems *419*
14.2 Determinants *428*
14.3 Cramer's Rule *433*

CHAPTER FIFTEEN
TOPICS IN ALGEBRA   441
15.1 Arithmetic Progressions *441*
15.2 Geometric Progressions *449*
15.3 The Binomial Theorem *456*
15.4 Counting: Permutations and Combinations *461*

APPENDIX/TABLES   T-1
ANSWERS TO ODD-NUMBERED EXERCISES AND
PROGRESS TESTS   A-1
INDEX   I-1

PREFACE

This book is a complete and self-contained presentation of the fundamentals of algebra which has been designed for use *by the student*.

The authors believe that it is almost impossible to oversimplify an idea in the fundamentals of mathematics. Thus, we have adopted an informal, supportive style to encourage the student to read the text and to develop confidence under its guidance. Concepts are introduced gradually with accompanying diagrams and illustrations which aid the student to grasp intuitively the "reasonableness" of results. The mathematical technique or result is immediately reinforced by one or more fully worked examples. Only then is the student asked to tackle parallel problems, called **Progress Checks,** to which answers are provided immediately.

The following features have been included to encourage the student to make use of the book and to assure that the student masters the material.

☐ Split-screen presentation of algebraic techniques. The steps of an algorithm or procedure are listed on one side of a page and an illustrative example is worked out in parallel on the other side of the page.

☐ Fully worked examples accompanying the text

☐ Chapter summaries that include
Terms and Symbols with appropriate page references
Key Ideas for Review to stress the concepts as well as the techniques
Common Errors
Progress Tests with answers in the back of the book

Most students have previously encountered much of the material covered in the early chapters of this book and many have developed incorrect practices. To help eliminate these misconceptions and bad mathematical habits, we have inserted numerous **Warnings** that point out those incorrect practices most commonly found in homework and exam papers.

An algebra course at this level must develop competence in two areas, mechanical and conceptual. The large number of carefully graded exercises at the end of each section provide reinforcement in the mechanical aspects of algebra. A unique emphasis on applied problems throughout the book utilizes each new technique and develops the conceptual aspects of algebra. In addition, Chapter Four is devoted exclusively to the solution of word problems and begins with an attack on a common obstacle to student progress: translating from words to algebraic expressions. A variety of problem types are discussed and analyzed, and an effective "chart" approach is used to help with the mathematical formulation of a given problem.

The book also demonstrates that elementary mathematics is adequate for understanding contemporary concepts in modern economics such as "break-even analysis" and "supply-demand curves." Mathematics is thus seen to be not just an abstract set of rules and procedures but an integral part of the real world.

This book is at the level of an intermediate algebra text with sufficient material for use in many courses in college algebra. The "review" material is handled at the slower pace consistent with courses in intermediate algebra, but the function concept is introduced at the same relative point as in most college algebra books. Thus, the language and notation of functions can be used in the discussion of first- and second-degree equations in two variables.

We have chosen to deemphasize the set-theoretic approach to algebra. Basic concepts of sets and set notation are reviewed in Chapter One and are used sparingly throughout the book. More importantly, the book provides an introduction to matrices and to their application in solving linear systems. Also, we have struck a compromise in the presentation of the conic sections. We present the geometric definition of each conic, but omit the derivation of the equations of the ellipse and hyperbola.

[Although gratified by the wide acceptance of *Algebra for College Students* by our colleagues, we have noted repeated requests for the following items: (a) inclusion of material on the theory of polynomial equations (b) inclusion of the formulas for factoring a sum and a difference of cubes The current edition responds to these requests and provides us with an opportunity to eliminate typographical errors and several incorrect answers in the answer section.

Chapter 10, *Roots of Polynomials*, provides an introduction to the theory of polynomial equations appropriate to this level. The chapter is a logical concluding unit of the material in Chapters 8 and 9. For those instructors choosing to skip this material, we suggest inclusion of Section 10.1 on polynomial division and synthetic division.

The material on factoring occupies two sections in this edition. It has been expanded to introduce more examples, factoring by grouping, and factoring of a sum and a difference of cubes.]

A number of exercises calling for the use of a hand calculator have been included. These are identified by the calculator symbol shown in the margin. Numerical answers to all odd-numbered exercises are given in the back of the book. A Solutions Manual, providing sample tests and the answers to all even-numbered exercises, is available from the publisher (gratis) to instructors.

Bernard Kolman
Arnold Shapiro

ACKNOWLEDGMENTS

We thank the following for their review of the manuscript and their helpful comments: Professor Stanley Lukawecki at Clemson University; Professor Norman Mittman at Northeastern Illinois University; Professor Maurice L. Monahan at South Dakota State University; Professor Richard Spangler at Tacoma Community College; Professor Monty J. Strauss at Texas Tech University; Professor Dell Swearingen at Linn-Benton Community College; and Professor Donald W. Bellairs of Grossmont College. A special thanks is due Dr. Charles G. Denlinger of Millersville State College for his review of various stages of the manuscript. His comments were especially useful in perfecting the presentation of material and in maintaining a balanced approach between detail and general overview of the algebraic concepts.

Our thanks to Erica Shapiro for her excellent typing of the manuscript. Her patience and suggestions were an unanticipated bonus. Another extraordinary contributor was our copyeditor and proofreader, Harry Spector. His unflagging efforts resulted in galleys that were remarkably free of error. Another meaningful contributor was Mr. James Frock of Keystone Computer Associates whose technical assistance is much appreciated.

We would also like to thank the following for solving the exercises, proofreading, and providing sketches of the graphs for the answer section: Jeffrey Eisen, Harold E. Schwalm, Jr., Stephen Yankovich, Jacqueline Shapiro, Lisa Kolman, and Thomas Coates.

TO THE STUDENT

This book was written for you. It gives you every possible chance to succeed—if you use it properly.

We would like to have you think of mathematics as a challenging game—but not as a spectator sport. Which leads to our primary rule: *Read this textbook with pencil and paper handy*. Every new idea or technique is illustrated by fully worked examples. When you feel you understand the material, do the **Progress Check** that follows each example. The key to success in a math course is working problems, and the **Progress Check** is there to provide immediate practice with the material you have just learned.

Your instructor will assign homework from the extensive selection of exercises that follow each section in the book. *Do the assignments regularly, thoroughly, and independently*. By doing lots of problems you will develop the necessary skills in algebra and your confidence will grow. Since algebraic techniques and concepts build upon previous results, you can't afford to skip any of the work.

To help you eliminate improper habits and to help you avoid those errors that we see each semester as we grade papers, we have interspersed **Warnings** throughout each chapter. The **Warnings** point out common errors and show you the proper method. They are summarized at the end of the chapter under the heading **Common Errors.**

There is other important review material at the end of each chapter. The **Terms and Symbols** should all be familiar by the time you reach them. If your understanding of a term or symbol is hazy, use the page reference to find the place in the text where it is introduced. Go back and read the definition.

It is possible to become so involved with the details of techniques that you may lose track of the broader concepts. The list of **Key Ideas for Review** at the end of each chapter will help you focus on the principal ideas.

After you have reviewed the material of a chapter, try *Progress Test A*. You will soon pinpoint your weak spots and can go back for further review and more exercises in those areas. Then, and only then, should you proceed to *Progress Test B*.

The authors believe that the eventual "payoff" in studying mathematics is an improved ability to tackle practical problems in your chosen field of interest. To that end, this book places special emphasis on word problems, which recent surveys show are often troublesome to students. Since algebra is the basic language of most mathematical techniques as used in virtually all fields, the mastery of algebra is well worth your effort.

CHAPTER ONE
THE REAL NUMBER SYSTEM

Arithmetic teaches us that "two plus two equals four" is independent of the object to which the rule applies. Since it doesn't matter whether the objects are apples or ants, countries or cars, the idea evolved to study the properties of numbers in an abstract sense, that is, to find those properties which apply to *all* numbers.

We will be dealing in much of our work with the *real numbers* and our studies begin with a review of the *real number system*. We will then introduce symbols to denote arbitrary numbers, a practice characteristic of algebra. The remainder of the chapter will be devoted to explaining some of the fundamental properties of the real number system.

1.1
THE REAL NUMBER SYSTEM

Although this text will not stress the set approach to algebra, the concept and notation of sets will at times be useful.

A **set** is a collection of objects or numbers which are called the **elements** or **members** of the set. The elements of a set are written within braces so that

$$A = \{4, 5, 6\}$$

tells us that the set A consists of the numbers 4, 5, and 6. The set

$$B = \{\text{Exxon, Ford, Honeywell}\}$$

consists of the names of these three corporations. We also write $4 \in A$ which we read as "4 is a member of the set A." Similarly, Ford $\in B$ is read as "Ford is a member of the set B," and I.B.M. $\notin B$ is read as "I.B.M. is not a member of the set B."

If every element of a set A is also a member of a set B, then A is a **subset** of B. For example, the set of all robins is a subset of the set of all birds.

EXAMPLE 1

The set C consists of the names of all coins whose denomination is less than 50 cents. We may write C in set notation as

$$C = \{\text{penny, nickel, dime, quarter}\}$$

We see that dime $\in C$ but half-dollar $\notin C$. Further, the set $H = \{\text{nickel, dime}\}$ is a subset of C.

PROGRESS CHECK 1

The set V consists of the vowels in the English alphabet.

(a) Write V in set notation.
(b) Is the letter k a member of V?
(c) Is the letter u a member of V?
(d) List the subsets of V having four elements.

Answers
(a) $V = \{a, e, i, o, u\}$ (b) *No*   (c) *Yes*   (d) $\{a, e, i, o\}$, $\{e, i, o, u\}$, $\{a, i, o, u\}$, $\{a, e, o, u\}$, $\{a, e, i, u\}$

Much of our work in algebra deals with the set of real numbers. Let's review the composition of this number system.

The numbers 1, 2, 3, . . . used for counting form the set of **natural numbers.** If we had only these numbers to use to show the profit earned by a company, we would have no way to indicate that the company has no profit or has a loss. To indicate losses we need to introduce negative numbers. The numbers

$$\ldots, -2, -1, 0, 1, 2, \ldots$$

form the set of **integers.** Thus, every natural number is an integer. However not every integer is a natural number.

When we try to divide two apples equally among four people we find no number in the set of integers that will express how many apples each person should get. We need to introduce the **rational numbers,** which are numbers that can be written as a ratio of two integers,

$$\frac{p}{q} \quad \text{with } q \text{ not equal to zero}$$

Examples of rational numbers are

$$0, \quad \frac{2}{3}, \quad -4, \quad \frac{7}{5}, \quad \frac{-3}{4}$$

Thus, when we divide two apples equally among four people, each person gets $\frac{1}{2}$ apple. Since every integer n can be written as $n/1$, we see that the integers are all rational numbers. The number 1.3 is also a rational number since $1.3 = \frac{13}{10}$.

We have now seen three fundamental number systems: the natural number system, the system of integers, and the rational number system. Each later system includes the previous system(s) and is more complicated. However, the rational number system is still inadequate for mature uses of mathematics since there exist numbers which are not rational, that is, these numbers cannot be written as the ratio of two integers. It is easy to show that $\sqrt{2}$ is such a number. The number π, which is the ratio of the circumference of a circle to its diameter, is also such a number. These are called **irrational numbers.**

The rational and irrational numbers together comprise the **real number system** (Figure 1.1).

Real numbers

Rational numbers

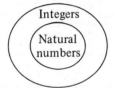

Irrational numbers

FIGURE 1.1

The **decimal form** of a rational number will either terminate, as

$$\frac{3}{4} = 0.75; \qquad -\frac{4}{5} = -0.8$$

or will form a repeating pattern, as

$$\frac{2}{3} = 0.\underset{\smile}{666}\ldots; \qquad \frac{1}{11} = 0.\underset{\smile}{09090}\ldots; \qquad \frac{1}{7} = 0.\underset{\smile}{1428571}\ldots$$

Remarkably, the decimal form of an irrational number *never* forms a repeating pattern. Although we sometimes write $\pi = 3.14$, this is only an approximation, as is

$$\pi = 3.14159\ 26536\ldots$$

There is a very simple and very useful geometric interpretation of the real number system. Draw a horizontal straight line, which we will call the **real number line.** Pick a point, label it with the number 0, and call it the **origin.** Choose the **positive direction** to the right of the origin and the **negative direction** to the left of the origin.

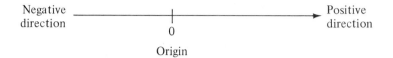

Now select a unit of length for measuring distance. With each positive real number r we associate the point that is r units to the right of the origin, and with each negative number $-r$ we associate the point that is r units to the left of the origin. We can now show some points on the real number line.

The numbers to the right of zero are called **positive;** the numbers to the left of zero are called **negative.** The positive numbers and zero together are called the **nonnegative** numbers, while the combination of zero and the negative numbers yields the **nonpositive** numbers. Thus, the real numbers are identified with all possible points on a straight line. For every point on the line there is a real number and vice versa.

We will frequently turn to the real number line to help us picture the results of algebraic computations.

EXAMPLE 2

Draw a real number line and plot the following points: $-\dfrac{3}{2}$, 2, $\dfrac{13}{4}$.

Solution

PROGRESS CHECK 2

Determine the real numbers denoted on the real number line as A, B, C, and D.

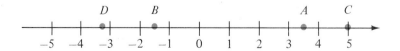

Answers

$A:$ $\dfrac{7}{2}$ $B:$ $-\dfrac{3}{2}$ $C:$ 5 $D:$ $-\dfrac{13}{4}$

EXERCISE SET 1.1

In the following exercises, choose the correct answer(s) from the following:
(a) rational number   (b) natural number   (c) real number   (d) integer   (e) irrational number.

1. The number 2 is
2. The number -3 is
3. The number $-\dfrac{2}{3}$ is
4. The number 0.8 is
5. The number 3 is
6. The numbers $-1, -2, -3$ are
7. The numbers 0, 1, 2 are
8. The numbers $0, \dfrac{1}{2}, 1, \dfrac{2}{3}, -\dfrac{4}{5}$ are
9. The numbers $\sqrt{2}, \sqrt{3}, \pi$ are
10. The numbers 0.5, 0.8 are
11. The numbers $\dfrac{\pi}{3}, 2\pi$ are
12. The numbers $0, \dfrac{1}{2}, \sqrt{2}, \pi, 4, -4$ are

In Exercises 13–25 determine whether the given statement is true (T) or false (F).

13. -14 is a natural number.
14. $-\dfrac{4}{5}$ is a rational number.
15. $\dfrac{\pi}{3}$ is a rational number.
16. $\sqrt{9}$ is an irrational number.
17. -1207 is an integer.
18. 0.75 is an irrational number.
19. $\dfrac{4}{5}$ is a real number.
20. 3 is a rational number.
21. $\sqrt{5}$ is a real number.
22. The sum of two rational numbers is always a rational number.
23. The sum of two irrational numbers is always an irrational number.
24. The product of two rational numbers is always a rational number.
25. The product of two irrational numbers is always an irrational number.
26. Draw a real number line and plot the following points.

    (a) 4   (b) -2 (c) $\dfrac{5}{2}$ (d) -3.5 (e) 0

27. Draw a real number line and plot the following points.

    (a) -5 (b) 4   (c) -3.5 (d) $\dfrac{7}{2}$ (e) $-\pi$

28. On the following real number line give the real number associated with the points A, B, C, D, O, and E.

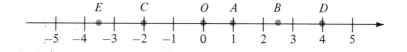

29. Represent the following by real numbers.
    (a) a profit of $10
    (b) a loss of $20
    (c) a temperature of 20° above zero
    (d) a temperature of 5° below zero

In the following exercises indicate which of the two given numbers appears first, viewed from left to right, on the real number line.

30. 4, 6   31. 2, 0   32. −2, 3

33. 0, −4   34. −5, −2   35. 4, −5

In the following exercises, indicate which of the two given numbers appears second, viewed from left to right, on the real number line.

36. 9, 8   37. 0, 3   38. −4, 2

39. −4, 0   40. −3, −4   41. −2, 5

In the following exercises indicate the given sets of numbers on the real number line.

42. The natural numbers less than 8
43. The natural numbers greater than 4 and less than 10
44. The integers that are greater than 2 and less than 7

1.2
ARITHMETIC OPERATIONS: FRACTIONS
TERMINOLOGY AND ORDER OF OPERATIONS

The vocabulary of arithmetic will carry over to our study of algebra. In multiplying two real numbers, each of the numbers is called a **factor** and the result is called the **product.**

$$6 \cdot 5 = 30$$

factors   product

In this text we will indicate multiplication by a dot, as in the example just given, or by parentheses.

$$(6)(5) = 30$$

We will avoid use of the multiplication sign $\times$ since it may be confused with other algebraic symbols.

Let's look at the terminology of division.

$$30 \div 6 = 5$$

dividend   divisor   quotient

Most of the time we will write division this way.

$$\frac{30}{6} = 5$$

Numbers to be multiplied have the same name (factors), whereas numbers to be divided have distinct roles (dividend, divisor). This suggests that we can interchange the factors in multiplication without altering the product, but interchanging the dividend and divisor in division will alter the quotient. In general:

> Addition can be performed in any order.
> Multiplication can be performed in any order.
> Subtraction must be performed in the given order.
> Division must be performed in the given order.

What happens when more than one operation appears in a problem? To avoid ambiguity, we adopt this simple rule.

> Always do multiplication and division
> before addition and subtraction.

EXAMPLE 1
Perform the indicated operations.

(a) $2 + (3)(4) - 5 = 2 + 12 - 5 = 9$

(b) $\dfrac{10}{5} + 7 + 3 \cdot 6 = 2 + 7 + 18 = 27$

PROGRESS CHECK 1
Perform the indicated operations.

(a) $2 \cdot 4 - 6 + 3$ (b) $(3)(4) - 2 + \dfrac{9}{3}$

Answers
(a) 5   (b) 13

FRACTIONS

It is important to master the *arithmetic* of fractions since this serves as background for the *algebra* of fractions. We prefer to write $10 \div 5$ in the form $\frac{10}{5}$ which we call a **fraction.** The number above the line is called the **numerator** while that below the line is called the **denominator.**

$$\text{numerator} \diagdown \frac{10}{5} \diagup$$
$$\text{denominator}$$

Multiplication of fractions is straightforward.

Multiplication of Fractions

Step 1. Multiply the numerators of the given fractions to find the numerator of the product.

Step 2. Multiply the denominators of the given fractions to find the denominator of the product.

EXAMPLE 2

Multiply.

(a) $\dfrac{3}{5} \cdot \dfrac{7}{2} = \dfrac{3 \cdot 7}{5 \cdot 2} = \dfrac{21}{10}$

(b) $\dfrac{2}{9} \cdot \dfrac{5}{3} \cdot 4 = \dfrac{2}{9} \cdot \dfrac{5}{3} \cdot \dfrac{4}{1} = \dfrac{2 \cdot 5 \cdot 4}{9 \cdot 3 \cdot 1} = \dfrac{40}{27}$

PROGRESS CHECK 2

Multiply.

(a) $\dfrac{4}{3} \cdot \dfrac{7}{3}$ (b) $\dfrac{5}{12} \cdot \dfrac{7}{3} \cdot \dfrac{1}{2}$

Answers

(a) $\dfrac{28}{9}$ (b) $\dfrac{35}{72}$

We **invert** the fraction $\frac{3}{4}$ by forming the fraction $\frac{4}{3}$ which is called the **reciprocal** of the fraction $\frac{3}{4}$. Note that the product of a real number and its reciprocal is always equal to 1. The number 0 does not have a reciprocal since the product of 0 and any number is 0.

Division of fractions can always be converted into multiplication problems by forming the reciprocal.

Division of Fractions

Step 1. Invert the denominator.

Step 2. Multiply the resulting fractions.

Since 0 has no reciprocal, division by 0 is not defined.

EXAMPLE 3

Divide.

(a) $\dfrac{\dfrac{2}{3}}{\dfrac{5}{7}} = \dfrac{2}{3} \cdot \dfrac{7}{5} = \dfrac{2 \cdot 7}{3 \cdot 5} = \dfrac{14}{15}$

(b) $\dfrac{4}{9} \div \dfrac{3}{5} = \dfrac{\dfrac{4}{9}}{\dfrac{3}{5}} = \dfrac{4}{9} \cdot \dfrac{5}{3} = \dfrac{4 \cdot 5}{9 \cdot 3} = \dfrac{20}{27}$

PROGRESS CHECK 3

Divide.

(a) $\dfrac{8}{7} \div \dfrac{3}{2}$ (b) $\dfrac{\frac{1}{2}}{3}$ (c) $\dfrac{2}{11} \div \dfrac{5}{3}$

Answers

(a) $\dfrac{16}{21}$ (b) $\dfrac{1}{6}$ (c) $\dfrac{6}{55}$

The same fractional value can be written in many ways. Thus,

$$\frac{3}{2} = \frac{6}{4} = \frac{18}{12} = \frac{72}{48}$$

are **equivalent fractions.**

Equivalent Fractions

The value of a fraction is not changed by multiplying or dividing *both* the numerator and denominator by the same number (other than 0). The result is called an equivalent fraction.

If we multiply a fraction, say $\frac{5}{2}$ by $\frac{3}{3}$, we are really multiplying by a "disguised" or equivalent form of 1 since $\frac{3}{3} = 1$. Thus,

$$\frac{5}{2} = \frac{5}{2} \cdot \frac{3}{3} = \frac{15}{6}$$

EXAMPLE 4

Find the equivalent fraction.

$$\frac{7}{3} = \frac{?}{12}$$

Since $12 = 3 \cdot 4$, we multiply the original denominator by 4 to obtain the new denominator. Then we must also multiply the numerator by 4.

$$\frac{7}{3} \cdot \frac{4}{4} = \frac{7 \cdot 4}{3 \cdot 4} = \frac{28}{12}$$

PROGRESS CHECK 4

Find the equivalent fraction.

(a) $\dfrac{5}{4} = \dfrac{?}{20}$ (b) $6 = \dfrac{?}{4}$ (c) $\dfrac{2}{3} = \dfrac{8}{?}$

Answers

(a) $\dfrac{25}{20}$ (b) $\dfrac{24}{4}$ (c) $\dfrac{8}{12}$

Now let's reverse the process. We saw that $\frac{7}{3}$ can be written as

$$\frac{7}{3} \cdot \frac{4}{4} = \frac{28}{12}$$

which is an equivalent fraction. If we begin with $\frac{28}{12}$ we can write the numerator and denominator as a product of factors and obtain

$$\frac{28}{12} = \frac{7 \cdot 4}{3 \cdot 4} = \frac{7}{3} \cdot \frac{4}{4} = \frac{7}{3} \cdot 1 = \frac{7}{3}$$

We say that $\frac{7}{3}$ is the **reduced form** of $\frac{28}{12}$. This is called the **cancellation principle.**

Cancellation Principle

Common factors appearing in both the numerator and denominator of a fraction can be canceled without changing the value of the fraction. When a fraction has no common factors in its numerator and denominator, it is said to be in reduced form.

EXAMPLE 5

Write $\dfrac{15}{27}$ in reduced form.

Solution

$$\frac{15}{27} = \frac{5 \cdot 3}{9 \cdot 3} = \frac{5}{9} \cdot \frac{3}{3} = \frac{5}{9} \cdot 1 = \frac{5}{9}$$

PROGRESS CHECK 5

Write in reduced form.

(a) $\dfrac{22}{60}$ (b) $\dfrac{90}{15}$ (c) $\dfrac{32}{12}$

Answers

(a) $\dfrac{11}{30}$ (b) 6 (c) $\dfrac{8}{3}$

 WARNING

Only multiplicative factors common to both the *entire* numerator and the *entire* denominator can be canceled. *Don't* write

$$\frac{6 + 5}{3} = \frac{\overset{2}{\cancel{6}} + 5}{\cancel{3}} = \frac{7}{1} = 7$$

Since 3 is not a multiplicative factor common to the *entire* numerator, we may not cancel.

The rule for the addition or subtraction of fractions is:

We can add or subtract fractions directly only if they have the same denominator.

When fractions do have the same denominator, the process is easy: add or subtract the numerators and keep the common denominator. Thus,

$$\frac{3}{4} + \frac{15}{4} = \frac{3 + 15}{4} = \frac{18}{4} = \frac{9}{2}$$

If the fractions we wish to add or subtract do not have the same denominator, we must rewrite them as equivalent fractions which do have the same denominator. There are easy ways to find the **least common denominator (L.C.D.)** of two or more fractions, that is, the smallest number which is divisible by each of the given denominators.

To find the L.C.D. of two or more fractions, say $\frac{1}{2}$, $\frac{5}{6}$, and $\frac{4}{9}$, we first write each denominator as a product of prime numbers. (A **prime number** is a natural number greater than 1 whose only factors are itself and 1.) Thus,

$$2 = 2$$
$$6 = (2)(3)$$
$$9 = (3)(3)$$

We then form a product in which each distinct prime factor appears the greatest number of times that it occurs in any single denominator. This product is the L.C.D. In our example, the prime factor 2 appears at most once in any denominator, while the prime factor 3 appears twice in a denominator. Thus, the L.C.D. is $(2)(3)(3) = 18$.

The L.C.D. is the tool we need to add or subtract fractions with different denominators. Here is the process.

TABLE 1.1

Addition and Subtraction of Fractions	Example
	$\frac{2}{5} + \frac{3}{4} - \frac{2}{3}$
Step 1. Find the L.C.D. of the fractions.	L.C.D. = 60
Step 2. Convert each fraction to an equivalent fraction with the L.C.D. as its denominator.	$\frac{2}{5} = \frac{2}{5} \cdot \frac{12}{12} = \frac{24}{60}$ $\frac{3}{4} = \frac{3}{4} \cdot \frac{15}{15} = \frac{45}{60}$ $\frac{2}{3} = \frac{2}{3} \cdot \frac{20}{20} = \frac{40}{60}$
Step 3. The fractions now have the same denominator. Add and subtract all the numerators indicated.	$\frac{2}{5} + \frac{3}{4} - \frac{2}{3}$ $= \frac{24}{60} + \frac{45}{60} - \frac{40}{60}$ $= \frac{24 + 45 - 40}{60}$ $= \frac{29}{60}$
Step 4. Write the answer in reduced form.	Answer: $\frac{29}{60}$

EXAMPLE 6

Perform the indicated operations and simplify.

(a) $\dfrac{1}{2} - \dfrac{1}{6} + \dfrac{1}{3}$

We see that the L.C.D. is 6 and

$$\frac{1}{2} = \frac{1}{2} \cdot \frac{3}{3} = \frac{3}{6}$$

$$\frac{1}{6} = \frac{1}{6} \cdot \frac{1}{1} = \frac{1}{6}$$

$$\frac{1}{3} = \frac{1}{3} \cdot \frac{2}{2} = \frac{2}{6}$$

Thus,

$$\frac{1}{2} - \frac{1}{6} + \frac{1}{3} = \frac{3}{6} - \frac{1}{6} + \frac{2}{6} = \frac{3 - 1 + 2}{6} = \frac{4}{6} = \frac{2}{3}$$

(b) $\dfrac{\dfrac{4}{3} - \dfrac{1}{2}}{\dfrac{1}{3} + \dfrac{3}{4}}$

The L.C.D. of the fractions in the numerator is 6 and hence the numerator is

$$\frac{4}{3} - \frac{1}{2} = \frac{4}{3} \cdot \frac{2}{2} - \frac{1}{2} \cdot \frac{3}{3} = \frac{8}{6} - \frac{3}{6} = \frac{5}{6}$$

The L.C.D. of the fractions in the denominator is 12 and hence the denominator is

$$\frac{1}{3} + \frac{3}{4} = \frac{1}{3} \cdot \frac{4}{4} + \frac{3}{4} \cdot \frac{3}{3} = \frac{4}{12} + \frac{9}{12} = \frac{13}{12}$$

Then the given fraction is

$$\frac{\dfrac{5}{6}}{\dfrac{13}{12}} = \frac{5}{\cancel{6}} \cdot \frac{\cancel{12}^{\,2}}{13} = \frac{10}{13}$$

PROGRESS CHECK 6

Perform the indicated operations and simplify.

(a) $\dfrac{3}{2} + \dfrac{5}{9} - \dfrac{1}{3}$ (b) $\dfrac{\dfrac{2}{3} + \dfrac{5}{6}}{\dfrac{1}{2} + \dfrac{2}{3}}$

Answers

(a) $\dfrac{31}{18}$ (b) $\dfrac{9}{7}$

PERCENT

Percent is a way of writing a fraction whose denominator is 100. The percent sign % means "place the number over 100." Thus, 7% means $\frac{7}{100}$.

A fraction whose denominator is 100 is converted to decimal form by moving the decimal point in the numerator two places to the left and eliminating the denominator.

$$\frac{65}{100} = 0.65$$

Since a percent is understood to mean a fraction whose denominator is 100, we see that

$$7\% = 7.0\% = 0.07$$

Similarly, we change a decimal to a percent by moving the decimal point two places to the right.

$$0.065 = 6.5\%$$

To write a fraction as a percent, we first find the equivalent fraction whose denominator is 100. For example, to write $\frac{1}{20}$ as a percent we write

$$\frac{1}{20} = \frac{5}{100}$$

from which we see that $\frac{1}{20} = 5\%$.

EXAMPLE 7

Write each percent as a decimal and fraction, and each decimal and fraction as a percent.

(a) $25\% = 0.25$; $25\% = \dfrac{25}{100} = \dfrac{1}{4}$

(b) $142\% = 1.42$; $142\% = \dfrac{142}{100} = \dfrac{71}{50}$

(c) $0.06 = 6\%$ (d) $2.1 = 210\%$

(e) $\dfrac{3}{4} = \dfrac{75}{100} = 75\%$ (f) $\dfrac{21}{5} = \dfrac{420}{100} = 420\%$

PROGRESS CHECK 7

Write each percent as a decimal and fraction, and each decimal and fraction as a percent.

(a) 62.5% (b) $\dfrac{1}{2}\%$ (c) 0.26 (d) 3.475 (e) $\dfrac{1}{8}$ (f) $\dfrac{5}{2}$

Answers

(a) $0.625, \dfrac{5}{8}$ (b) $0.005, \dfrac{1}{200}$ (c) 26% (d) 347.5% (e) 12.5%

(f) 250%

It is common to state business problems in terms of percent. You have heard and read statements such as:

Ms. Smith was promised an 8% salary increase.

The Best Savers Bank pays 5.75% interest per year.

Automobile prices will increase 4.62% on July 1.

During the sale period, all merchandise is reduced by 20%.

To find the **percent of a number,** we must convert the percent to a decimal or fraction and then *multiply* by the number.

EXAMPLE 8
(a)   What is 30% of 15? $30\% = 0.3$ and $(0.3)(15) = 4.5$

(b)   What is 5% of 400? $5\% = 0.05$ and $(0.05)(400) = 20$

(c)   The price of a refrigerator selling at $600 is to be reduced by 20%. What is the sale price?

$20\% = 0.2$ and $(0.2)(\$600) = \$120$ discount

Sale price $= \$600 - \$120 = \$480$

(d)   A bank pays 6.75% interest per year. What will the annual interest be on a deposit of $500?

$6.75\% = 0.0675$ and $(0.0675)(\$500) = \$33.75$

PROGRESS CHECK 8
(a)   What is 40% of 60?

(b)   What is 2% of 1200?

(c)   How much interest will be earned during one year on a deposit of $2500 at 7.5% per year?

(d)   The price of an automobile selling at $6800 will be increased by 4%. What is the new price?

Answers

(a)   *24*   (b)   *24*   (c)   *$187.50*   (d)   *$7072*

EXERCISE SET 1.2
Perform the indicated operations in each of the following.

1. $\dfrac{2(6 + 2)}{4}$

2. $\dfrac{(4 + 5)6}{18}$

3. $\dfrac{6(3 + 1)}{2} + 3 \cdot 5$

4. $\dfrac{8(6 - 1)}{4} - 3 \cdot 2$

5. $\dfrac{(4 + 5)(2 + 3)}{3} - 3 \cdot 5$

6. $\dfrac{(7 - 2)(8 - 2)}{3} + 7 \cdot 4$

7. $\dfrac{2}{11} \cdot \dfrac{10}{3} \cdot \dfrac{2}{5}$

8. $\dfrac{7}{5} \cdot \dfrac{4}{3} \cdot 2$

9. $\dfrac{\frac{2}{3}}{\frac{1}{5}}$

10. $\dfrac{\frac{3}{4}}{\frac{4}{3}}$

11. $\dfrac{\frac{2}{5}}{\frac{3}{10}}$

12. $\dfrac{\dfrac{1}{2}}{\dfrac{5}{6}}$

13. $\dfrac{2}{3} \div \dfrac{4}{9}$

14. $\dfrac{3}{5} \div \dfrac{9}{25}$

Find the equivalent fraction in each of the following.

15. $\dfrac{4}{3} = \dfrac{?}{9}$

16. $\dfrac{3}{4} = \dfrac{15}{?}$

17. $1 = \dfrac{7}{?}$

18. $2 = \dfrac{?}{14}$

19. $\dfrac{5}{4} = \dfrac{?}{20}$

20. $\dfrac{4}{7} = \dfrac{12}{?}$

Perform the indicated operations and simplify your answer.

21. $\dfrac{3}{4} + \dfrac{2}{3}$

22. $\dfrac{2}{3} + \dfrac{5}{6}$

23. $\dfrac{1}{4} + \dfrac{2}{3} - \dfrac{1}{2}$

24. $\dfrac{1}{5} - \dfrac{1}{2} + \dfrac{1}{3}$

25. $\dfrac{\dfrac{1}{6} + \dfrac{1}{2}}{\dfrac{5}{4} + \dfrac{2}{3}}$

26. $\dfrac{\dfrac{1}{2} - \dfrac{3}{8}}{\dfrac{1}{3} + \dfrac{1}{4}}$

27. $\dfrac{2 - \dfrac{1}{3}}{3 + \dfrac{1}{4}}$

28. $\dfrac{\dfrac{3}{5} - \dfrac{1}{10}}{1 + \dfrac{1}{2}}$

Change the given percent to fraction and decimal form in each of the following.

29. 20%

30. 60%

31. 65.5%

32. 32.5%

33. 4.8%

34. 5.5%

35. 120%

36. 160%

37. $\dfrac{1}{5}\%$

38. $\dfrac{1}{10}\%$

Convert each number to percent.

39. 0.05

40. 0.03

41. 0.425

42. 0.345

43. 6.28

44. 7.341

45. $\dfrac{3}{5}$

46. $\dfrac{5}{6}$

47. $\dfrac{9}{4}$

48. $\dfrac{6}{5}$

49. $\dfrac{2}{7}$

50. $\dfrac{4}{3}$

51. What is 35% of 60?
52. What is 60% of 80?
53. What is 140% of 30?
54. What is 160% of 50?

55. What is $\dfrac{1}{2}\%$ of 40?

56. What is $\dfrac{2}{5}\%$ of 20?

57. A bank pays 7.25% interest per year. If a depositor has $800 in a savings account, how much interest will the bank pay at the end of one year?

58. Suppose that you buy a $5000 General Motors bond that pays 9.3% interest per year. What will be the amount of the dividend when mailed to you by G.M. at the end of the year?

59. A savings bank pays 8% interest per year. If a depositor has $6000 in a savings account, what will be the amount in the account at the end of one year, assuming that no withdrawals are made?

60. An $800 stereo system will be sold at a 25% reduction. What will be the new price?

61. In order to cope with rising costs, an oil producer plans to raise prices by 15%. If a barrel of oil now sells for $13.20, what will be the new price?

62. A student has borrowed $3000 at a rate of 4% per year. How much interest is owed to the bank at the end of one year?

63. A record store embarks on an advertising campaign to raise its profits by 20%. If this year's profits were $96,000, what should next year's profits be if the campaign is to succeed?

64. A boat that originally sold for $600 is now on sale for $540. What is the percent of discount?

65. The holder of an $8000 savings certificate gets a $640 check at the end of the year. What is the annual rate of simple interest?

1.3
ALGEBRAIC EXPRESSIONS

A rational number is one that can be written as p/q where p and q are integers (and q is not zero). These symbols can take on more than one distinct value. For example, when $p = 5$ and $q = 7$, we have the rational number $\frac{5}{7}$; when $p = -3$ and $q = 2$, we have the rational number $\frac{-3}{2}$. The symbols p and q are called **variables** since various values can be assigned to them.

If we invest P dollars at an interest rate of 6% per year we will have $P + 0.06P$ dollars at the end of the year. We call $P + 0.06P$ an **algebraic expression.** Note that an algebraic expression involves **variables** (in our case, P), **constants** (such as 0.06), and **algebraic operations** (such as $+, -, \times, \div$). Virtually everything we do in algebra involves algebraic expressions, sometimes as simple as our example and sometimes very complicated.

When we assign a value to each variable in an algebraic expression, we are "evaluating" the expression.

EXAMPLE 1
Evaluate.

(a) $2x + 5$ when $x = 3$

Substituting 3 for x we have

$2(3) + 5 = 6 + 5 = 11$

(b) $\dfrac{3m + 4n}{m + n}$ when $m = 3, n = 2$

Substituting, we have

$$\frac{3(3) + 4(2)}{3 + 2} = \frac{9 + 8}{3 + 2} = \frac{17}{5}$$

PROGRESS CHECK 1

Evaluate.

(a) $\dfrac{r + 2s}{s - r}$ when $r = 1$, $s = 3$

(b) $4x + 2y - z$ when $x = 1$, $y = 4$, $z = 2$

Answers

(*a*) $\dfrac{7}{2}$ (*b*) 10

If we want to evaluate

$$3(x + 2)(y - 4) + \frac{6y}{2(y - 2)} + x \quad \text{when } x = 1 \quad \text{and} \quad y = 5$$

in what order should we perform the operations? Here's an easy procedure to use when the order of operations is not clear.

TABLE 1.2

Hierarchy of Operations	Example
	$3(x + 2)(y - 4) + \dfrac{6y}{2(y - 2)} + x$
	when $x = 1$, $y = 5$
Step 1. Substitute the given values.	$3(1 + 2)(5 - 4) + \dfrac{6 \cdot 5}{2(5 - 2)} + 1$
Step 2. Perform all operations within parentheses.	$= 3(3)(1) + \dfrac{6 \cdot 5}{2 \cdot 3} + 1$
Step 3. Perform multiplication and division.	$= 9 + \dfrac{30}{6} + 1 = 9 + 5 + 1$
Step 4. Perform addition and subtraction.	$= 15$

EXAMPLE 2

Evaluate each expression.

(a) $\dfrac{3(x - 1)}{y + 2} + (x - y)(x + y)$ when $x = 5$, $y = 2$

Substituting, we have

$$\frac{3(5 - 1)}{(2 + 2)} + (5 - 2)(5 + 2)$$

$$= \frac{3(4)}{4} + (3)(7) \qquad \text{Operations in parentheses}$$

$$= 3 + 21 \qquad \text{Multiplication and division}$$

$$= 24 \qquad \text{Addition}$$

(b) $(x + 1)(x + 2)(x + 3) + (x - 1)(x - 2)$ when $x = 3$

Substituting $x = 3$,

$(3 + 1)(3 + 2)(3 + 3) + (3 - 1)(3 - 2)$

$= 4 \cdot 5 \cdot 6 + 2 \cdot 1 = 120 + 2 = 122$

PROGRESS CHECK 2
Evaluate.

(a) $\dfrac{2(x - 1)}{(x + 1) + (x + 3)}$ when $x = 2$

(b) $1 + (2 - x) + \dfrac{y}{y - 1}$ when $x = 1, y = 2$

Answers

(a) $\dfrac{1}{4}$ (b)   4

EXERCISE SET 1.3
In the following exercises determine whether the given statement is true (T) or false (F).

1. $3x + 2 = 8$ when $x = 2$

2. $5x - 1 = 11$ when $x = 2$

3. $2xy = 12$ when $x = 2, y = 3$

4. $\dfrac{2x + y}{y} = 7$ when $x = 3, y = 1$

In the following exercises evaluate the given expression when $x = 4$.

5. $2x + 3$ 6. $3x - 2$ 7. $\dfrac{1}{2}x$

8. $3(x - 1)$ 9. $(2x)(2x)$ 10. $(2x + 1)x$

11. $\dfrac{1}{2x + 3}$ 12. $\dfrac{x}{2x - 4}$

In the following exercises evaluate the given expression when $a = 3, b = 4$.

13. $2a + b$ 14. $3a - b$ 15. $2b - a$

16. ab 17. $3(a + 2b)$ 18. $\dfrac{1}{3a + b}$

19. $\dfrac{2a - b}{b}$ 20. $\dfrac{a + b}{b - a}$

Evaluate the given expression when $r = 2, s = 3, t = 4$ in the following exercises.

21. $r + 2s + t$ 22. rst 23. $\dfrac{rst}{r + s + t}$

24. $(r + s)t$ 25. $\dfrac{r + s}{t}$ 26. $\dfrac{r + s + t}{t}$

27. $\dfrac{t - r}{rs}$ 28. $\dfrac{3(r + s + t)}{s}$

29. Evaluate $2\pi r$ when $r = 3$ (remember that π is approximately 3.14).

30. Evaluate $\dfrac{9}{5}C + 32$ when $C = 37$.

31. Evaluate $0.02r + 0.314st + 2.25t$ when $r = 2.5$, $s = 3.4$, and $t = 2.81$.

32. Evaluate $10.421x + 0.821y + 2.34xyz$ when $x = 3.21$, $y = 2.42$, and $z = 1.23$.

33. If P dollars are invested at a simple interest rate of r percent per year for t years, the amount on hand at the end of t years is $P + Prt$. Suppose \$2000 is invested at 8% per year ($r = 0.08$). How much money is on hand after
     (a) one year?
     (b) three years?
     (c) half a year?
     (d) eight months? (*Hint:* Express eight months as a fraction of a year.)

34. The perimeter of a rectangle is given by the formula $P = 2(L + W)$, where L is the length and W is the width of the rectangle. Find the perimeter if
     (a) $L = 2$ feet, $W = 3$ feet

     (b) $L = \dfrac{1}{2}$ meter, $W = \dfrac{1}{4}$ meter

     (c) $L = 13$ inches, $W = 15$ inches

1.4
OPERATING WITH SIGNED NUMBERS

Let's review the rules for operating with signed numbers before using them in more complicated problems.

We say that $-a$ is the **opposite** of a if a and $-a$ are equidistant from the origin and lie on opposite sides of the origin.

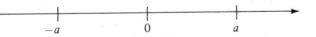

Since $-a$ is opposite in direction to a, then $-(-a)$ must have the same direction as a. We therefore see that

$$-(-a) = a$$

We also refer to $-a$ as the **negative** of a. Of course, the negative of a number need not be negative: $-(-5) = 5$.

Here are the rules for operating with signed numbers.

Addition with Like Signs

Ignoring the signs, add the numbers. The sign of the answer is the same as the sign of the original numbers.

$$5 + 2 = 7$$
$$(-5) + (-2) = -7$$

Addition with Unlike Signs

Ignoring the signs, subtract the smaller number from the larger. The sign of the answer is the sign of the larger number.

$6 + (-4) = 2$

$(-7) + 3 = -4$

$3 + (-5) = -2$

$(-5) + 8 = 3$

EXAMPLE 1

(a) $2 + 4 = 6$

(b) $2 + (-4) = -2$

(c) $(-2) + (-4) = -6$

(d) $(-2) + 4 = 2$

(e) $3 + [(-5) + (-2)] = 3 - 7 = -4$

(f) $[(-4) + (-7)] + 3 = -11 + 3 = -8$

(g) $8 + 0 = 8$

PROGRESS CHECK 1
Add.

(a) $(-3) + (-7)$ (b) $(-5) + 1$ (c) $2 + (-6)$

(d) $-2 + [5 + (-4)]$ (e) $0 + (-8)$ (f) $[3 + (-6)] + (-1)$

Answers

(a)   −10   (b)   −4   (c)   −4   (d)   −1   (e)   −8   (f)   −4

Subtraction problems can be converted into addition of signed numbers.

Subtraction

Change

$a - b$

to

$a + (-b)$

$5 - 2 = 5 + (-2) = 3$

$-3 - 4 = -3 + (-4) = -7$

$2 - (-6) = 2 + (+6) = 8$

$-6 - (-5) = -6 + (+5) = -1$

EXAMPLE 2

(a) $7 - 4 = 7 + (-4) = 3$

(b) $10 - (-6) = 10 + (+6) = 16$

(c) $-8 - 2 = -8 + (-2) = -10$

(d) $-5 - (-4) = -5 + (+4) = -1$

(e) $-7 - (-7) = -7 + (+7) = 0$

(f) $[-2 - 5] + 6 = [-2 + (-5)] + 6 = -1$

PROGRESS CHECK 2

Perform the operations.

(a) $3 - 8$ (b) $-6 - 7$ (c) $-9 - (-5)$ (d) $16 - (-9)$

(e) $(14 - 5) - 4$ (f) $(-11 + 2) - 4$ (g) $(-6 - 4) - 2$

Answers

(a) -5 (b) -13 (c) -4 (d) 25 (e) 5

(f) -13 (g) -12

The rules for determining the sign in multiplication and division are straightforward.

Multiplication and Division

If both numbers have the same sign, the result is positive. If the numbers have opposite signs, the result is negative.

$$3 \cdot 4 = 12 \qquad \frac{6}{3} = 2$$

$$(-2)(-5) = 10 \qquad \frac{-8}{-4} = 2$$

$$(-4)(6) = -24 \qquad \frac{-10}{2} = -5$$

$$(7)(-3) = -21 \qquad \frac{12}{-3} = -4$$

$$(-1)(4) = -4 \qquad \frac{-4}{-1} = 4$$

EXAMPLE 3

(a) $4 \cdot \left(-\dfrac{1}{5}\right) = -\dfrac{4}{5}$ (b) $\left(-\dfrac{2}{3}\right)(-3) = 2$ (c) $\dfrac{-4}{8} = -\dfrac{1}{2}$

(d) $\dfrac{-16}{-24} = \dfrac{2}{3}$ (e) $(-5) \cdot \dfrac{1}{4} = -\dfrac{5}{4}$ (f) $\dfrac{18}{-2} = -9$

PROGRESS CHECK 3

(a) $(-3)\left(\dfrac{2}{-7}\right)$ (b) $\left(\dfrac{2}{3}\right)\left(\dfrac{-3}{4}\right)$ (c) $\dfrac{20}{-6 - 4}$ (d) $\dfrac{4 - 5}{3 - 6}$

(e) $(-4)\left(\dfrac{2 - 3}{4}\right)$ (f) $\left(\dfrac{1}{5}\right)\left(\dfrac{5 - 6}{4}\right)$

Answers

(a) $\dfrac{6}{7}$ (b) $-\dfrac{1}{2}$ (c) -2 (d) $\dfrac{1}{3}$ (e) 1 (f) $-\dfrac{1}{20}$

Let's apply the rules to the evaluation of algebraic expressions.

EXAMPLE 4

Evaluate the given expression when $x = -1, y = -1$.

(a) $2x + \dfrac{x-1}{y+2}$

Substituting, we have

$$2(-1) + \frac{(-1-1)}{-1+2} = -2 + \frac{-2}{1} = -2 + (-2) = -4$$

(b) $-3(2x - y) + (-4)(2y - x)$

Substituting,

$$-3[2(-1) - (-1)] + (-4)[2(-1) - (-1)]$$
$$= -3(-2 + 1) + (-4)(-2 + 1)$$
$$= -3(-1) + (-4)(-1) = 3 + 4 = 7$$

PROGRESS CHECK 4

Evaluate the given expression when $x = 2, y = -1$.

(a) $-(-y)$ (b) $2 - 3x + y$ (c) $\dfrac{2 - 2x}{2 - 2y}$ (d) $\dfrac{x + y}{x - y}$

Answers

(a) -1 (b) -5 (c) $-\dfrac{1}{2}$ (d) $\dfrac{1}{3}$

EXERCISE SET 1.4

Simplify.

1. $3 + 5$ 2. $-2 + (-3)$ 3. $(-3) + (-4)$

4. $2 + (-3)$ 5. $4 + (-2)$ 6. $-4 + 6$

7. $-4 + 2$ 8. $0 + (-2)$ 9. $3 + 0$

10. $3 + (-3)$ 11. $5 - 3$ 12. $5 - 8$

13. $5 - (-3)$ 14. $4 - (-4)$ 15. $-8 - (-3)$

16. $-6 - (-7)$ 17. $2 + [(3 - 4)]$ 18. $[(1 - 6)] + 2$

19. $[(-2 - 3)] + 7$ 20. $[(-3 - 1)] - 4$ 21. $[(-4 - 2)] - (-3)$

22. $2\left(\dfrac{3}{4}\right)$ 23. $(-2)(-5)$ 24. $(-3)\left(-\dfrac{8}{6}\right)$

25. $\left(-\dfrac{5}{6}\right)\left(\dfrac{9}{15}\right)$ 26. $\left(\dfrac{3}{5}\right)\left(-\dfrac{10}{4}\right)$ 27. $\dfrac{8}{2}$

28. $\dfrac{-10}{-2}$ 29. $\dfrac{-15}{5}$ 30. $\dfrac{20}{-4}$

31. $\dfrac{-15}{25}$ 32. $\dfrac{15}{-\dfrac{3}{4}}$ 33. $\dfrac{-12}{-\dfrac{2}{3}}$

34. $(-4)\left(-\dfrac{5}{2}\right)$ 35. $\dfrac{3}{5}\left(-\dfrac{15}{2}\right)$ 36. $\left(-\dfrac{3}{4}\right)0$

37. $\left(-\dfrac{4}{5}\right)\left(-\dfrac{15}{2}\right)$ 38. $-(-2)$ 39. $\dfrac{4-4}{2}$

40. $\dfrac{14}{2+5}$ 41. $\dfrac{5+(-5)}{3}$ 42. $\dfrac{-18}{-3-6}$

43. $\dfrac{15}{7-2}$ 44. $\dfrac{24}{2-8}$ 45. $\dfrac{-8-4}{3}$

46. $-5(2-4)$ 47. $-4(4-1)$ 48. $\dfrac{3(-5+1)}{-4(2-6)}$

49. $-(-2x+3y)$ 50. $(-x)(-y)$ 51. $\dfrac{-x}{-y}$

52. $\dfrac{-x}{\frac{1}{2}}$ 53. $\dfrac{2}{\frac{x}{-2}}$ 54. $\dfrac{-a}{(-b)(-c)}$

In the following exercises evaluate the given expression when $x = -2$.

55. $x - 5$ 56. $-2x$ 57. $\dfrac{x}{x-1}$

In the following exercises evaluate the given expression when $x = -3$, $y = -2$.

58. $x + 2y$ 59. $x - 2y$ 60. $\dfrac{4x-y}{y}$

61. Subtract 3 from -5.

62. Subtract -3 from -4.

63. Subtract -5 from -2.

64. Subtract -2 from 8.

65. At 2 P.M. the temperature is $10°C$ above zero and at 11 P.M. it is $2°C$ below zero. How many degrees has the temperature dropped?

66. Repeat Exercise 65 if the temperature at 2 P.M. is $8°C$ below zero and it is $-4°C$ at 11 P.M.

67. A stationery store had a loss of $400 for its first year of operation and a loss of $800 for its second year. How much money did the store lose for the first two years of its existence?

68. A bicycle repair shop had a profit of $150 for the month of July and a loss of $200 for the month of August. How much money did the shop gain or lose over the two-month period?

69. E. & E. Fabrics had a loss of x dollars during its first business year and a profit of y dollars its second year. Write an expression for the net profit or loss after two years.

70. S. & S. Hardware had a profit of x dollars followed by a loss which exceeded twice the profit by $200. Write an expression for the loss.

1.5
PROPERTIES OF REAL NUMBERS

The real numbers obey laws that enable us to manipulate algebraic expressions with ease. We know that

$$3 + 4 = 7 \qquad \text{and} \qquad 3 \cdot 4 = 12$$
$$4 + 3 = 7 \qquad\qquad\qquad 4 \cdot 3 = 12$$

That is, we may *add or multiply real numbers in any order*. Writing this in algebraic symbols, we have

$a + b = b + a$	**Commutative Law of Addition**
$a \cdot b = b \cdot a$	**Commutative Law of Multiplication**

EXAMPLE 1

(a) $5 + 7 = 7 + 5$; $5 \cdot 7 = 7 \cdot 5$

(b) $3 + (-6) = -6 + 3$; $3 \cdot (-6) = (-6) \cdot 3$

(c) $3x + 4y = 4y + 3x$; $(3x)(4y) = (4y)(3x)$

PROGRESS CHECK 1

Use the commutative laws to write the following in another form.

(a) $(-3) + 6$ (b) $(-4) \cdot 5$ (c) $-2x + 6y$ (d) $\left(\frac{3}{2}x\right)\left(\frac{1}{2}y\right)$

Answers

(a) $6 + (-3)$ (b) $5 \cdot (-4)$ (c) $6y + (-2x)$ (d) $\left(\frac{1}{2}y\right)\left(\frac{3}{2}x\right)$

When we add $2 + 3 + 4$, does it matter in what order we group the numbers? No. We see that

$$(2 + 3) + 4 = 5 + 4 = 9$$
and
$$2 + (3 + 4) = 2 + 7 = 9$$

Similarly, for multiplication of $2 \cdot 3 \cdot 4$ we have

$$(2 \cdot 3) \cdot 4 = 6 \cdot 4 = 24$$
and
$$2 \cdot (3 \cdot 4) = 2 \cdot 12 = 24$$

Clearly, when *adding or multiplying real numbers we may group in any order*. Translating into algebraic symbols, we have

$a + (b + c) = (a + b) + c$	**Associative Law of Addition**
$a(bc) = (ab)c$	**Associative Law of Multiplication**

EXAMPLE 2

(a) $5 + (2 + 3) = (5 + 2) + 3 = 10$

(b) $5 \cdot (2 \cdot 3) = (5 \cdot 2) \cdot 3 = 30$

(c) $(3x + 2y) + 4z = 3x + (2y + 4z)$

(d) $3(4y) = (3 \cdot 4)y = 12y$

(e) $(-2)[(-5)(-x)] = [(-2)(-5)](-x) = 10(-x) = -10x$

PROGRESS CHECK 2

Use the associative laws to simplify.

(a) $3 + (2 + x)$ (b) $6 \cdot 2xy$

Answers

(a) $5 + x$ (b) *12xy*

We can combine the commutative and associative laws to simplify algebraic expressions.

EXAMPLE 3

Use the commutative and associative laws to simplify.

(a) $(3 + x) + 5 = (x + 3) + 5$ Commutative law of addition

$\qquad\qquad\qquad = x + (3 + 5)$ Associative law of addition

$\qquad\qquad\qquad = x + 8$

(b) $\left(\frac{2}{3}y\right)\left(\frac{3}{4}\right) = \frac{3}{4} \cdot \left(\frac{2}{3}y\right)$ Commutative law of multiplication

$\qquad\qquad\quad = \left(\frac{3}{4} \cdot \frac{2}{3}\right)y$ Associative law of multiplication

$\qquad\qquad\quad = \frac{1}{2}y$

(c) $(2x - 4) + 7 = [2x + (-4)] + 7$

$\qquad\qquad\qquad = 2x + [(-4) + 7]$ Associative law of addition

$\qquad\qquad\qquad = 2x + 3$

PROGRESS CHECK 3

Use the commutative and associative laws to simplify.

(a) $4 + (2x + 2)$ (b) $\left(\frac{4}{5}x\right)\left(\frac{10}{2}\right)$ (c) $(5 - 3x) + 6$

Answers

(a) $6 + 2x$ (b) *4x*   (c) *11 − 3x*

The **distributive laws** deal with both addition and multiplication. For instance,

$$2(3 + 4) = 2(7) = 14$$

and

$$(1 + 2)5 = (3)5 = 15$$

We may notice that

$$2(3) + 2(4) = 6 + 8 = 14$$

and

$$1(5) + 2(5) = 5 + 10 = 15$$

produce the same results. The distributive laws tell us that this is not an accident but rather is a rule that we may always use.

Distributive Laws

$$a(b + c) = ab + ac$$

$$(a + b)c = ac + bc$$

The distributive laws can be extended to factors that are a sum of more than two terms. Thus,

$$3(5x + 2y - 4z) = 3(5x) + 3(2y) + 3(-4z)$$
$$= 15x + 6y - 12z$$

EXAMPLE 4

(a) $4(2x + 3) = 4(2x) + 4(3) = 8x + 12$

(b) $(4x + 2)6 = (4x)(6) + (2)(6) = 24x + 12$

(c) $2(x + 5y + 2z) = 2x + 2(5y) + 2(2z)$
$$= 2x + 10y + 4z$$

PROGRESS CHECK 4

Simplify, using the distributive laws.

(a) $5(3x + 4)$ (b) $(x + 3)7$ (c) $-2(3a - b + c)$

Answers

(a) $15x + 20$ (b) $7x + 21$ (c) $-6a + 2b - 2c$

It is easy to show that the commutative and associative laws *do not* hold for subtraction and division. For example,

$$2 - 5 = -3 \quad \text{but} \quad 5 - 2 = 3$$

and, in general,

$$a - b = -(b - a) \neq b - a$$

Similarly, $12 \div 3 \neq 3 \div 12$ shows that the commutative law does not hold for division.

The student is encouraged to provide counterexamples to show that the associative law does not hold for subtraction and division (see Exercise 19).

EXERCISE SET 1.5

In the following exercises justify the given equation by using one or more properties of real numbers.

1. $2 + 5 = 5 + 2$
2. $(3 \cdot 2)(-4) = 3[(2)(-4)]$
3. $-2 \cdot 5 = 5(-2)$
4. $2(4 + 5) = 2 \cdot 4 + 2 \cdot 5$
5. $(4 + 3)2 = 4 \cdot 2 + 3 \cdot 2$
6. $3(2 - 4) = 3 \cdot 2 - 3 \cdot 4$
7. $-2(4 - 5) = (-2)(4) - 2(-5)$
8. $(3 + 2) + 4 = 3 + (2 + 4)$
9. $-3 + (2 + 5) = (-3 + 2) + 5$
10. $(2 - 5) + 8 = 2 + (-5 + 8)$
11. $3 + a = a + 3$
12. $2(x + 2) = 2x + 4$
13. $2(ab) = (2a)b$
14. $5(a + b) = 5(b + a)$
15. $2(xy) = x(2y)$
16. $4(a + b) = 4b + 4a$
17. $5 + (a + 2) = 2 + (5 + a)$
18. $(5x)y = 5(yx)$
19. Give examples showing that the operation of subtraction does not satisfy the commutative and associative laws.

Find and correct the mistake.

20. $a + 2a = 2a^2$
21. $2(a + 2) = 2a + 2$
22. $3(x - 2) = x - 6$
23. $(a - b)2 = 2a - b$
24. $3(ab) = (3a)(3b)$
25. $(2a + 3) + a = 3(a + 2)$

Simplify the following.

26. $(2 + x) + 4$
27. $(2 - x) + 2$
28. $(x - 3) - 4$
29. $(x - 5) - 2x$
30. $(2x)(-5)$
31. $(-3x)(-4)$
32. $2(-3x)$
33. $a(2b)(3c)$
34. $4\left(\dfrac{3}{2}a\right)$
35. $\dfrac{2}{3}(9 + 12a - 6b)$
36. $\dfrac{4x}{-2}$
37. $\dfrac{-2}{4x}$
38. $\dfrac{-8x}{-4}$
39. $(3x)\left(-\dfrac{4}{9}y\right)$
40. $\dfrac{1}{4}(4a)$
41. $\dfrac{1}{5}(10\,ab)$
42. $\dfrac{4(b + 2b)}{3}$
43. $\dfrac{3(5x - x)}{12}$
44. $6x + \dfrac{(x - 1)4}{2}$
45. $3a + \dfrac{(a - 3a)}{2}(-5)$
46. $3(a - 2) - 2(b + 4)$
47. $4(x + y) - 2(x - 2y)$
48. $3\left[\dfrac{2u + v}{6}\right] + \dfrac{1}{2}(4u - v)$
49. $2\left(\dfrac{x}{2} + y - 2z\right) - (-2x + 4y + z)$
50. $4\left(\dfrac{a - 2b + 4c}{2}\right) + \dfrac{1}{2}(2a - 6b + 2c)$

51. $-3.65\left(\dfrac{0.47 - 2.79}{6.44}\right)$

52. $\dfrac{6.92}{4.7}\left(\dfrac{2.01}{1.64 - 3.53}\right)$

53. $0.40\left(\dfrac{17.52 - 6.48 + 2.97}{3.60}\right) - 0.25(-4.75 + 2.92)$

54. $16.33\left(\dfrac{14.94}{3.87} - \dfrac{2.22 + 7.46}{2.96}\right)$

1.6
ABSOLUTE VALUE AND INEQUALITIES

When we introduced the real number line we pointed out that positive numbers lie to the right of the origin and negative numbers lie to the left of the origin.

Negative direction ← ─────────────── → Positive direction
$-4 \quad -3 \quad -2 \quad -1 \quad 0 \quad 1 \quad 2 \quad 3 \quad 4$

Suppose we are interested in the *distance* between the origin and the points labeled 4 and -4. Each of these points is four units from the origin, that is, the *distance is independent of the direction.*

When we are interested in the size or magnitude of a number, a, and don't care about the direction or sign, we use the notation of **absolute value** which we write as $|a|$. Thus,

$$|4| = 4$$
$$|-4| = 4$$

EXAMPLE 1
(a) $|-6| = 6$ (b) $|17.4| = 17.4$ (c) $|0| = 0$

(d) $|4 - 9| = |-5| = 5$ (e) $\left|\dfrac{2}{5} - \dfrac{4}{5}\right| = \left|-\dfrac{2}{5}\right| = \dfrac{2}{5}$

PROGRESS CHECK 1
Find the values.

(a) $|22|$ (b) $\left|-\dfrac{2}{7}\right|$ (c) $|4 - 4|$ (d) $|6 - 8|$

(e) $\left|\dfrac{1}{7} - \dfrac{3}{7}\right|$

Answers

(a) 22   (b) $\dfrac{2}{7}$ (c) 0   (d) 2   (e) $\dfrac{2}{7}$

EXAMPLE 2
(a) $|-3| + |-6| = 3 + 6 = 9$

(b) $|3 - 5| - |8 - 6| = |-2| - |2| = 2 - 2 = 0$

(c) $\dfrac{|4-7|}{|-6|} = \dfrac{|-3|}{|-6|} = \dfrac{3}{6} = \dfrac{1}{2}$

(d) $\left|\dfrac{2-8}{3}\right| = \left|\dfrac{-6}{3}\right| = |-2| = 2$

PROGRESS CHECK 2
Find the values.

(a) $|-2| - |-4|$ (b) $\dfrac{|2-5|}{-3}$ (c) $\left|\dfrac{1-5}{2-8}\right|$

(d) $\dfrac{|-3| - |-6|}{4 - |-10|}$

Answers

(a) -2 (b) -1 (c) $\dfrac{2}{3}$ (d) $\dfrac{1}{2}$

The absolute value of a number is, then, always nonnegative. But what can we do with the absolute value of a variable, say $|x|$? We don't know whether x is positive or negative, so we can't write $|x| = x$. For instance, when $x = -4$ we have

$$|x| = |-4| = 4 \neq x$$

and when $x = 4$

$$|x| = |4| = 4 = x$$

We must define absolute value so that it works for both positive and negative values of a variable.

$$|x| = \begin{cases} x & \text{if } x \text{ is } 0 \text{ or positive} \\ -x & \text{if } x \text{ is negative} \end{cases}$$

When x is positive, say $x = 4$, the absolute value is the number itself; when x is negative, say $x = -4$, the absolute value is the negative of x or $+4$. Thus, the absolute value is always nonnegative.

INEQUALITIES

Other important relations involving real numbers are those of **less than** and **greater than.** Here is how these relations are written with respect to the relative positions on the real number line.

$a < b$	a is less than b	a is to the left of b
$a > b$	a is greater than b	a is to the right of b
$a \leq b$	a is less than or equal to b	a coincides with b or is to the left of b
$a \geq b$	a is greater than or equal to b	a coincides with b or is to the right of b

If we think of the symbols $>$ and $<$ as pointers, then they always point to the lesser of the two numbers.

EXAMPLE 3

(a) $2 < 5$ (b) $-1 < 3$ (c) $6 > 4$

PROGRESS CHECK 3

Replace the square by the symbol $<$ or $>$ to make a true statement.

(a) $7 \square 10$ (b) $16 \square 8$ (c) $4 \square -2$

Answers
(a) $<$ *(b)* $>$ *(c)* $>$

We can use the real number line to illustrate the relations $<$ and $>$. For example, the inequality $x < 3$ is satisfied by *all* points to the left of 3.

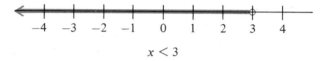

$$x < 3$$

Similarly, for $x \geq -1$ we mean *all* points to the right of (and including) -1.

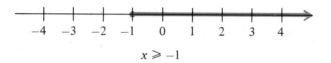

$$x \geq -1$$

For $x < 3$, the point labeled 3 does not satisfy the inequality; we indicate this by an open circle

3

For $x \geq -1$, the point labeled -1 does satisfy the inequality; we indicate this by a solid circle

-1

In the two number lines above, the shading indicates the "set" of *all x* which satisfy the given inequality. This set is called the **solution set** of the inequality, and we are said to have *sketched* or *marked* the solution set of the inequality.

EXAMPLE 4
In the figure below,

(a) $a > b$ since a is to the right of b.

(b) $c < a$ since c is to the left of a.

(c) $b < 0$ since b is to the left of 0.

(d) $d > a$ since d is to the right of a.

PROGRESS CHECK 4
For the figure of Example 4, replace each square by the symbol $<$ or $>$ to make a true statement.

(a) $b \square d$ (b) $a \square c$ (c) $d \square 0$ (d) $b \square a$

Answers

(a) $<$ (b) $>$ (c) $>$ (d) $<$

EXAMPLE 5
Show the inequality on a real number line.

(a) $x \geq 2$

(b) $x < 2$

We sometimes refer to this as **graphing the inequality.**

PROGRESS CHECK 5
Graph the inequality on a real number line.

(a) $x < 0$ (b) $x \geq -1$ (c) $x < -2$

Answers

(a)

(b)

(c)

We also write double inequalities such as

$$-1 \leq x < 2$$

The solution set to this inequality consists of all real numbers which satisfy

$$-1 \leq x \quad \text{and} \quad x < 2$$

that is, all numbers between -1 and 2 and including -1 itself. We can easily graph the solution set on a real number line.

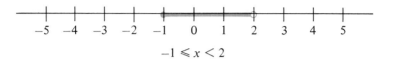

$$-1 \leq x < 2$$

EXAMPLE 6

Graph $-3 < x < -1$, x a real number.

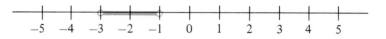

PROGRESS CHECK 6

Graph $-2 \leq x \leq 3$, x a real number.

Answer

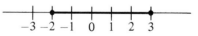

We can also graph the solution set to the inequality

$$-1 \leq x < 2, \quad x \text{ an integer}$$

that is, the set of integers greater than or equal to -1 and less than 2. The solution set is $\{-1, 0, 1\}$.

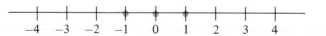

EXAMPLE 7

Graph the given inequality on a real number line.

(a) $-5 < x < 4$, x a natural number

(b) $-5 < x < 4,$ x an integer

(c) $-5 < x < 4,$ x a real number

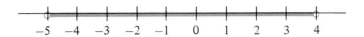

PROGRESS CHECK 7

Graph the given inequality on a real number line.

(a) $-2 \le x < 3,$ x a natural number

(b) $-3 \le x \le 2,$ x an integer

(c) $-4 \le x \le 0,$ x a real number

Answers

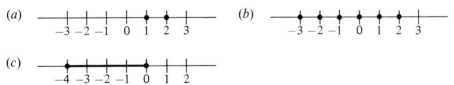

It is sometimes convenient to use set-builder notation as a way of writing statements such as "*A* is the set of integers between -3 and 2." If we let *I* represent the set of integers, we write

$$A = \{x \in I \mid -3 < x < 2\}$$

Each part of this symbolic expression has an explicit meaning.

$\{ \quad \}$	*read*	the set of all
$x \in I$	*read*	integers x
$\mid$	*read*	such that
$-3 < x < 2$	*read*	x is between -3 and 2

EXAMPLE 8

If *N* is the set of natural numbers, the statement "the set of all natural numbers less than 17" is written in set-builder notation as

$$\{x \in N \mid x < 17\}$$

PROGRESS CHECK 8

If *N* is the set of natural numbers, write the statement "*A* is the set of all odd natural numbers less than 12" in set-builder notation. List the elements of *A*.

Answer

$A = \{x \in N \mid x < 12 \text{ and } x \text{ is odd}\} = \{1, 3, 5, 7, 9, 11\}$

EXERCISE SET 1.6

Find the values of the following.

1. $|2|$

2. $\left|-\dfrac{2}{3}\right|$

3. $|1.5|$

4. $|-0.8|$

5. $-|2|$

6. $-\left|-\dfrac{2}{5}\right|$

7. $|2 - 3|$

8. $|2 - 2|$

9. $|2 - (-2)|$

10. $|2| + |-3|$

11. $\dfrac{|14 - 8|}{|-3|}$

12. $\dfrac{|2 - 12|}{|1 - 6|}$

13. $\dfrac{|3| - |2|}{|3| + |2|}$

14. $\dfrac{|4| - 2|4||3|}{|4 - 3|}$

15. $|x| - |y|$ when $x = -1$, $y = -2$

16. $|x| - |x \cdot y|$ when $x = -3$, $y = 4$

17. $|x + y| + |x - y|$ when $x = -3$, $y = 2$

18. $\dfrac{|a - 2b|}{2a}$ when $a = 1$, $b = 2$

19. $\dfrac{|x| + |y|}{|x| - |y|}$ when $x = -3$, $y = 4$

20. $\dfrac{-|a - 2b|}{|a + b|}$ when $a = -2$, $b = -1$

21. $\dfrac{|x - y| - |x + y|}{|xy|}$ when $x = -3$, $y = -2$

22. $\dfrac{|-|2a + b||}{|a - b|}$ when $a = -3$, $b = 2$

23. $\dfrac{|-|3a - 2b| + c|}{|a - b|}$ when $a = 1$, $b = 3$, $c = 1$

24. $\dfrac{|a - b| - 2|c - a|}{|a - b + c|}$ when $a = -2$, $b = 3$, $c = -5$

25. $\dfrac{|2a - b| - |c + a|}{a|a + b - 2c|}$ when $a = 1.69$, $b = -7.43$, $c = 2.98$

26. $\dfrac{|-b|c - a||}{c|b - a|}$ when $a = 12.44$, $b = 4.74$, $c = -5.83$

Write each of the following statements using the symbols $<$, $>$, $\leq$, $\geq$.

27. 4 is greater than 1

28. -2 is less than -1

29. 2 is not greater than 3

30. 3 is not less than 1

31. 3 is nonnegative

32. -2 is nonpositive

In each of the following, replace the square by the symbol $<$ or $>$ to make a true statement.

33. $3 \square 5$

34. $8 \square 2$

35. $4 \square -3$

36. $4 \square -6$

37. $-3 \square -2$

38. $-5 \square -4$

39. $-\dfrac{1}{2} \square \dfrac{1}{3}$ 40. $\dfrac{1}{2} \square -\dfrac{1}{4}$ 41. $-\dfrac{1}{5} \square -\dfrac{1}{3}$

42. $|-3| \square |5|$ 43. $-|3| \square |4|$ 44. $|-4| \square |-3|$

45. $|-2| \square 1$ 46. $|4| \square 0$ 47. $-|4| \square 0$

48. $-3 \square |3|$ 49. $-|2.1| \square 2.1$ 50. $|3.2| \square -|-3.2|$

Using the number line below, replace the square in the following exercises by the symbol $<$ or $>$ to make a true statement.

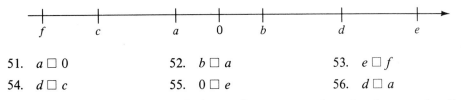

51. $a \square 0$ 52. $b \square a$ 53. $e \square f$

54. $d \square c$ 55. $0 \square e$ 56. $d \square a$

In the following exercises state the inequality represented on the given number line.

57.

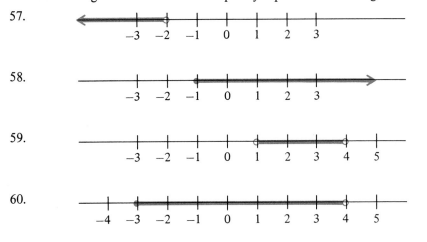

58.

59.

60.

Graph the given inequalities.

61. $x \le -2$ 62. $x \ge -3$ 63. $x < 4$

64. $x > 1$ 65. $x > 0$ 66. $x \le 3$

67. $-3 \le x \le 2$ 68. $-4 < x < -2$ 69. $-2 < x < 0$

70. $1 < x < 3$ 71. $0 < x \le 4$ 72. $1 \le x \le 5$

73. $-2 < x \le 3$, x an integer   74. $1 < x < 5$, x a natural number

75. $-3 \le x \le -1$, x an integer   76. $-3 \le x < 2$, x a natural number

If I is the set of integers, N is the set of natural numbers, and R is the set of real numbers, write each of the following in set-builder notation.

77. The set of natural numbers between -3 and 4, including -3
78. The set of integers between 2 and 4, including 4
79. The set of negative integers
80. The set of even natural numbers less than 6
81. The set of real numbers between -1.6 and 2.5
82. The set of real numbers between $-\dfrac{3}{2}$ and 6.2

If I is the set of integers and N is the set of natural numbers, list the elements of each of the following.

83. $\{x \in I \mid -5 < x \leq -1\}$ 84. $\{x \in I \mid -4 \leq x < 0\}$

85. $\{x \in N \mid -2 \leq x < 4\}$ 86. $\{x \in N \mid 0 \leq x \leq 6\}$

87. $\{x \in I \mid -4 \leq x \leq 10 \text{ and } x \text{ is even}\}$

88. $\{x \in N \mid 1 < x \leq 6 \text{ and } x \text{ is odd}\}$

TERMS AND SYMBOLS

set (p. 1)

element (p. 1)

member (p. 1)

$\in$ (p. 2)

$\notin$ (p. 2)

$\{ \quad \}$ (p. 2)

subset (p. 2)

natural number (p. 2)

integer (p. 2)

rational number (p. 2)

irrational number (p. 3)

real number system (p. 3)

real number line (p. 4)

origin (p. 4)

positive (p. 4)

negative (p. 4)

nonnegative (p. 4)

nonpositive (p. 4)

factor (p. 6)

product (p. 6)

dividend (p. 6)

divisor (p. 6)

quotient (p. 6)

fraction (p. 7)

numerator (p. 7)

denominator (p. 7)

reciprocal (p. 8)

equivalent fraction (p. 9)

reduced form (p. 10)

cancellation principle
  (p. 10)

common factor (p. 10)

L.C.D. (p. 11)

prime number (p. 11)

percent (p. 13)

% (p. 13)

variable (p. 16)

algebraic expression
  (p. 16)

constant (p. 16)

commutative laws (p. 24)

associative laws (p. 24)

distributive laws (p. 26)

absolute value (p. 28)

$| \; |$ (p. 28)

$<$ (p. 29)

$>$ (p. 29)

$\leq$ (p. 29)

$\geq$ (p. 29)

solution set (p. 30)

graphing an inequality
  (p. 31)

set-builder notation
  (p. 33)

KEY IDEAS FOR REVIEW

☐ Numbers and algebraic expressions may be added in any order.

☐ Numbers and algebraic expressions may be multiplied in any order.

☐ Subtraction and division must be performed in the given order.

☐ Division of fractions is handled by multiplying the numerator by the reciprocal of the denominator.

☐ Common multiplicative factors in the numerator and denominator of a fraction can be canceled.

☐ To add or subtract fractions with the same denominator, add or subtract the numerators and keep the same denominator.

☐ To add or subtract fractions with different denominators, find the L.C.D. and convert each fraction to an equivalent fraction with the L.C.D. as the denominator.

☐ Percent is a fraction whose denominator is 100.

☐ Fractions, decimals, and percents can be converted from any form to any other form.

☐ A rational number can be written as p/q, where p and q are both integers and $q \neq 0$.

☐ The real number system consists of the rational numbers and the irrational numbers.

☐ The decimal form of a rational number either terminates or forms a repeating pattern.

☐ The decimal form of an irrational number never forms a repeating pattern.

☐ Evaluating an algebraic expression means substituting numbers for the variables.

☐ Operations within parentheses should be done *before* multiplication and division; addition and subtraction are done last.

☐ Distributive laws: $a(b + c) = ab + ac$; $(a + b)c = ac + bc$

☐ Absolute value represents distance and is always nonnegative.

☐ Inequalities can be graphed on the real number line.

☐ $|x| = \begin{cases} x & \text{if } x \geq 0 \\ -x & \text{if } x < 0 \end{cases}$

COMMON ERRORS

1. $2x - y \neq 2(x - y)$. Don't assume grouping where it isn't indicated.
2. $3(a - b) = 3a - 3b$. *Don't* write $3(a - b) = 3a - b$.
3. $(-2)(-3)(-4) = -24$, not $+24$. When the number of negative factors is odd, the product is negative.
4. To evaluate an expression such as

$$\frac{3x + 2y}{x + 3y} \quad \text{when } x = 2 \quad \text{and} \quad y = 1$$

work independently on the numerator and denominator before dividing.

$$\frac{3(2) + 2(1)}{2 + 3(1)} = \frac{6 + 2}{2 + 3} = \frac{8}{5}$$

Don't write

$$\frac{3(\cancel{2}) + \cancel{2}(1)}{\cancel{2} + 3(1)} = \frac{3 + 1}{1 + 3} = \frac{4}{4} = 1$$

5. The absolute value bars act as grouping symbols. We must work *inside* these grouping symbols before we can remove them.
6. The number π is irrational; therefore, we cannot say $\pi = \frac{22}{7}$ or $\pi = 3.14$. These are approximations for computational use only. We write $\pi \approx 3.14$ where the symbol $\approx$ is read as "approximately equal to."

PROGRESS TEST 1A

1. The numbers 3, $-\frac{2}{3}$, 0.72 are   (a)   natural numbers   (b)   rational numbers   (c)   irrational numbers   (d)   none of these.

2. The numbers $-\frac{\pi}{2}$, $\sqrt{3}$, $-\frac{4}{5}$ are   (a)   irrational numbers   (b)   rational numbers   (c)   real numbers   (d)   none of these.

3. On a real number line, indicate the integers that are greater than -3 and less than 4.

4. Evaluate $\dfrac{3a - 4b}{2a - b}$ when $a = 3$, $b = 2$.

5. Evaluate $\dfrac{2x - 6y}{x - y}$ when $x = -1$, $y = 1$.

6. Evaluate $\dfrac{3[(a + 2b) - (2b - a)]}{c}$ when $a = -2$, $b = -1$, $c = -6$.

7. Simplify $5(x - y) - 3(2x - y)$.

8. Simplify $2\left(\dfrac{a + b}{4}\right) + \left(\dfrac{b - a}{2}\right)$.

9. Evaluate $\dfrac{|2 - 8|}{|2| + |-8|}$.

10. Evaluate $3|x| - 2|2y|$ when $x = -2$, $y = -3$.

11. Evaluate $\left|\dfrac{-2|3x| + 3|-y|}{|x + y|}\right|$ when $x = -2$, $y = 1$.

12. Evaluate $|x| \cdot |y| - 2|x \cdot y|$ when $x = -1$, $y = 2$.

13. Graph the inequality $-1 \le x < 3$.

14. Graph the inequality $-4 \le x \le 1$, x an integer.

15. Give the inequality represented on the following number line.

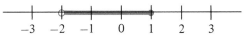

PROGRESS TEST 1B

1. The numbers -2, 0.45, $\dfrac{7}{9}$ are   (a)   integers   (b)   irrational numbers   (c)   rational numbers   (d)   none of these.

2. The numbers $-\sqrt{7}$, 2π, -0.49 are   (a)   natural numbers   (b)   irrational numbers   (c)   real numbers   (d)   none of these.

3. On a real number line, indicate the integers that are greater than -5 and less than -1.

4. Evaluate $\dfrac{2m - 5n}{3m - n}$ when $m = 2$, $n = 4$.

5. Evaluate $\dfrac{x + 2y}{2x - y}$ when $x = 3$, $y = -5$.

6. Evaluate $\dfrac{-2[(p - 2q) - 2r - p)]}{p \cdot q}$ when $p = 2$, $q = -3$, $r = \dfrac{1}{2}$.

7. Simplify $7(x + 2y) - 2(3x - y)$.

8. Simplify $3\left(\dfrac{a - 2b}{6}\right) - \left(\dfrac{b - 2a}{2}\right)$.

9. Evaluate $\dfrac{|-3| - |4 - 7|}{|-4| + |-2|}$.

10. Evaluate $\dfrac{|x|}{2} - 3|3y|$ when $x = -4$, $y = -5$.

11. Evaluate $\dfrac{3|2y| - 3|x|}{-|x - 2y|}$ when $x = -2$, $y = 1$.

12. Evaluate $\left|\dfrac{x}{3} \cdot \dfrac{y}{2}\right| - 3|2x \cdot y|$ when $x = -6$, $y = 4$.

13. Graph the inequality $-2 < x \le 5$.

14. Graph the inequality $-5 \le x \le -1$, x an integer.

15. Give the inequality represented on the following number line.

CHAPTER TWO
POLYNOMIALS

Polynomials play an important role in the study of algebra since many word problems translate into equations or inequalities that involve polynomials. We first study the manipulative and mechanical aspects of polynomials; this will serve as background for dealing with their applications in later chapters.

2.1
POLYNOMIALS

We use the notation of exponents to indicate repeated multiplication of a number by itself. For instance

$$a^1 = a$$
$$a^2 = a \cdot a$$
$$a^3 = a \cdot a \cdot a$$
$$\vdots \qquad \vdots$$
$$a^n = \underbrace{a \cdot a \dots a}_{n \text{ factors}}$$

where n is a natural number and a is a real number. We call a the **base** and n the **exponent,** and say that a^n is the nth **power** of a. When $n = 1$, we simply write a rather than a^1.

EXAMPLE 1

(a) $\left(\frac{1}{2}\right)^3 = \frac{1}{2} \cdot \frac{1}{2} \cdot \frac{1}{2} = \frac{1}{8}$ Base is $\frac{1}{2}$, exponent is 3.

(b) $x^4 = x \cdot x \cdot x \cdot x$ Base is x, exponent is 4.

(c) $2x^3 = 2 \cdot x \cdot x \cdot x$

(d) $-3x^2y^3 = -3 \cdot x \cdot x \cdot y \cdot y \cdot y$

PROGRESS CHECK 1
Write without using exponents.

(a) 2^4 (b) $\left(\frac{1}{3}\right)^2$ (c) x^3y (d) $\frac{1}{2}xy^3$

Answers

(a) $2 \cdot 2 \cdot 2 \cdot 2$ (b) $\frac{1}{3} \cdot \frac{1}{3}$ (c) $x \cdot x \cdot x \cdot y$ (d) $\frac{1}{2} \cdot x \cdot y \cdot y \cdot y$

WARNING

Note the difference between

and
$$(-2)^4 = (-2)(-2)(-2)(-2) = 16$$
$$-2^4 = -(2 \cdot 2 \cdot 2 \cdot 2) = -16$$

There is a rule of exponents that we will need later in this chapter. We see that

$$a^2 \cdot a^3 = \underbrace{(a \cdot a)}_{2 \text{ factors}} \cdot \underbrace{(a \cdot a \cdot a)}_{3 \text{ factors}}$$

$$= \underbrace{(a \cdot a \cdot a \cdot a \cdot a)}_{2 + 3 = 5 \text{ factors}} = a^5$$

and, in general, if m and n are any natural numbers and a is any real number,

$$a^m \cdot a^n = a^{m+n}$$

EXAMPLE 2
(a) $x^2 \cdot x^3 = x^{2+3} = x^5$

(b) $(3x)(4x^4) = 3 \cdot 4 \cdot x \cdot x^4 = 12x^{1+4} = 12x^5$

PROGRESS CHECK 2
Multiply.

(a) $x^5 \cdot x^2$ (b) $(2x^6)(-2x^4)$

Answers

(a) x^7 (b) $-4x^{10}$

When we combine exponent forms in one or more variables, as in

$$2x^2 + 3xy - x + 4$$

each part connected by addition or subtraction is called a **term** and the entire expression is called a **polynomial.** The constants 2, 3, -1, and 4 are also given a special name: **coefficients.** Thus,

$2x^2$	$+ \ 3x$	$- \quad x$	$+ \ 4$	Polynomial
$2x^2,$	$3x,$	$-x,$	4	Terms
$2,$	$3,$	$-1,$	4	Coefficients

Note that for polynomials we insist that the exponents of the variables be nonnegative integers.

EXAMPLE 3

Find the terms and coefficients.

(a) $x^2y - y^2 + 3xy$

Terms: $x^2y, \ -y^2, \ 3xy$

Coefficients: $1, \ -1, \ 3$

(b) $\dfrac{1}{2}x^3 - \dfrac{2}{3}y^3$

Terms: $\dfrac{1}{2}x^3, \ -\dfrac{2}{3}y^3$

Coefficients: $\dfrac{1}{2}, \ -\dfrac{2}{3}$

PROGRESS CHECK 3

Find the terms and coefficients.

(a) $\dfrac{1}{4}x^7$ (b) $2x^2y^2 + 4xy - y^2$

Answers

(a) *Terms:* $\dfrac{1}{4}x^7$ *Coefficients:* $\dfrac{1}{4}$

(b) *Terms:* $2x^2y^2, \ 4xy, \ -y^2$ *Coefficients:* $2, \ 4, \ -1$

Here are some examples of polynomials.

$$2x \qquad \frac{1}{2}x^3 \qquad xy^2 \qquad 3x - 2 \qquad x^3 + 6x^2$$

$$4x^2 - 2x + 1 \qquad 3x^2 + 4x^2y - xy^2 - 5y^3$$

We will later see that the product of two polynomials, such as

$$(2x + 1)(x^2 - 2x + 1)$$

is also a polynomial. However, the quotient of polynomials, such as

$$\frac{2x + 1}{x^2 - 2x + 1}$$

is not always a polynomial.

EXAMPLE 4
Which of the following are not polynomials?

(a) $3x^{1/2} + xy^2 + 2y$ (b) $2x^2y - 5$ (c) $x^5 - x^{-1} + 2$

Solution
Every exponent in a polynomial must be a natural number. Thus, (a) is not a polynomial since it contains a fractional exponent, $x^{1/2}$; (c) is not a polynomial since it contains a negative exponent, x^{-1}.

PROGRESS CHECK 4
Which of the following are not polynomials?

(a) $-2xy + 3x - 3y$ (b) $xy^{2/5} - 2x^2$ (c) $-3xy + 3x^{-3}y$

Answer
(b) and (c)

The **degree of a term** of a polynomial is found by adding the exponents of each variable in that term. (The degree of a constant term is zero.) For instance, the terms of

$$2x^3 - 3xy^2 + 5x^2y^2 + xy - 7$$

have the following degrees:

$2x^3$	is of degree 3
$-3xy^2 = -3x^1y^2$	is of degree $1 + 2 = 3$
$5x^2y^2$	is of degree $2 + 2 = 4$
$xy = x^1y^1$	is of degree $1 + 1 = 2$
-7	is of degree 0 and is often called the **constant term**

The **degree of a polynomial** is the degree of the term with nonzero coefficient that has the highest degree in the polynomial. The polynomial $2x^3 - 3xy^2 + 5x^2y^2 + xy - 7$ is of degree 4 since the highest degree term, $5x^2y^2$, is of degree 4.

EXAMPLE 5
Find the degree of each term and of the polynomial $4x^5 - 2x^3y + x^2y^2 - 3$.

Solution

$4x^5$	degree 5
$-2x^3y$	degree 4 (since $3 + 1 = 4$)

x^2y^2 degree 4 (since $2 + 2 = 4$)

-3 degree 0

Degree of the polynomial $= 5$ (highest degree term).

PROGRESS CHECK 5

Find the degree of each term and of the polynomial $2x^6y - x^3y^2 + 7xy^2 - 12$.

Answers
Degree of each term, in sequence: 7, 5, 3, 0
Degree of the polynomial: 7

APPLICATIONS

Polynomials occur in many applications. We will now look at several simple examples; many others will occur later throughout the book.

EXAMPLE 6

Consider the square shown in Figure 2.1. Each side is of length x. The polynomial $4x$ gives the perimeter of the square. The polynomial x^2 gives the area of the square.

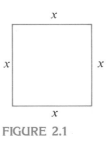

FIGURE 2.1

PROGRESS CHECK 6

Consider the rectangle in Figure 2.2, whose sides are x and y.
(a)   Write the polynomial representing the perimeter of the rectangle.
(b)   Write the polynomial representing the area of the rectangle.

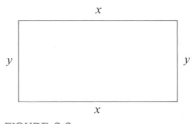

FIGURE 2.2

Answers
(*a*)   *2x + 2y*   (*b*)   *xy*

EXAMPLE 7

A grocery bag contains x apples, each costing 12 cents, and y pears, each costing 10 cents. What does the polynomial $12x + 10y$ represent?

Solution

The term $12x$ gives the total cost (in cents) of the apples in the bag and the term $10y$ gives the total cost (in cents) of the pears in the bag. Thus, the polynomial $12x + 10y$ represents the total cost of the contents of the grocery bag.

PROGRESS CHECK 7

If a car travels at the rate of r miles per hour for t hours, what does the polynomial rt represent?

Answer
the distance traveled in t hours

EXERCISE SET 2.1

Identify the base(s) and exponent(s) in each of the following.

1. 2^5 2. $(-2)^4$ 3. t^4 4. w^6

5. $3y^5$ 6. $-2t^3$ 7. $3x^2y^3$ 8. $-4u^3v^4$

Write the given expression using exponents.

9. $3 \cdot 3 \cdot 3$ 10. $(-5)(-5)(-5)(-5)$

11. $\left(\frac{1}{3}\right)\left(\frac{1}{3}\right)\left(\frac{1}{3}\right)\left(\frac{1}{3}\right)$ 12. $x \cdot x \cdot x \cdot x \cdot x \cdot x$

13. $3 \cdot y \cdot y \cdot y \cdot y$ 14. $-2 \cdot \frac{1}{p} \cdot \frac{1}{p} \cdot \frac{1}{p}$

15. In the expression $\left(\frac{2}{3}\right)^4$, the base is (a) 2   (b) 3   (c) 4   (d) $\frac{2}{3}$

   (e) none of these.

16. In the expression $\left(-\frac{3}{4}\right)^5$ the base is (a) 3   (b) $\frac{3}{4}$ (c) $-\frac{3}{4}$ (d) 5

   (e) none of these.

Carry out the indicated operations in each of the following.

17. $b^5 \cdot b^2$ 18. $x^3 \cdot x^5$

19. $(3x^2)(2x^4)$ 20. $(6x^3)(5x)$

21. $(4y^3)(-5y^6)$ 22. $(-6x^4)(-4x^7)$

23. $\left(\frac{4}{3}v^3\right)\left(\frac{5}{7}v^5\right)$ 24. $\left(\frac{4}{3}w^2\right)\left(\frac{5}{2}w^4\right)$

25. $\left(\frac{3}{2}x^3\right)(-2x)$ 26. $\left(-\frac{5}{3}x^6\right)\left(-\frac{3}{10}x^3\right)$

27. Which of the following expressions are *not* polynomials?

  (a) $-3x^2 + 2x + 5$ (b) $-3x^2y$ (c) $-3x^{2/3} + 2xy + 5$

  (d) $-2x^{-4} + 2xy^3 + 5$

28. Which of the following expressions are not polynomials?

  (a) $4x^5 - x^{1/2} + 6$ (b) $\frac{2}{5}x^3 + \frac{4}{3}x - 2$ (c) $4x^5y$

  (d) $x^{4/3}y + 2x - 3$

Give the terms and coefficients for each given polynomial.

29. $4x^4 - 2x^2 + x - 3$

30. $\frac{1}{3}x^2 + 2x - 5$

31. $\frac{2}{3}x^3y + \frac{1}{2}xy - y + 2$

32. $2.5x^3y - 3xy^2 + 4x^2 + 8$

33. $\frac{1}{3}x^3 + \frac{1}{2}x^2y - 2x + y + 7$

34. $-4x^4 + 3x^3y - y^3 + 12$

Find the degree of each term in the following polynomials.

35. $3x^3 - 2x^2 + 3$

36. $4x^2 + 2x - y + 3$

37. $4x^4 - 5x^3 + 2x^2 - 5x + 1$

38. $5x^5 + 2x^2y^2 + xy^3 + 3y$

39. $\frac{3}{2}x^4 + 2xy^2 + y^3 - y + 2$

40. $3x^8 - 3y^5 + 4x + 2$

Find the degree of each of the following polynomials.

41. $2x^3 + 3x^2 - 5$

42. $4x^5 - 8x^3 + x + 5$

43. $3x^2y + 2x^2 - y^2 + 2$

44. $4xy^3 + xy^2 + 4y^2 - y$

45. $\frac{3}{5}x^4 + 2x^2 - x^2y + 4$

46. $4x^5y^2 + x^3y - 2xy^2 + 7$

47. The degree of the polynomial $\frac{3}{5}x^4 + 2x^2 + 3x - 2$ is (a) $\frac{3}{5}$ (b) 4

  (c) 1   (d) none of these.

48. The degree of the polynomial $-2x^3y + y^3 + x^2 + 3$ is (a) -2 (b) 3
  (c) 4   (d) none of these.

49. Find the value of the polynomial $2x^2 - 2x + 1$ when $x = 3$.

50. Find the value of the polynomial $2x^3 + x^2 - x + 4$ when $x = -2$.

51. Find the value of the polynomial $3x^2y^2 + 2xy - x + 2y + 7$ when $x = 2$, $y = -1$.

52. Find the value of the polynomial $0.02x^2 + 0.3x - 0.5$ when $x = 0.3$.

53. Find the value of the polynomial $2.1x^3 + 3.3x^2 - 4.1x - 7.2$ when $x = 4.1$.

54. Find the value of the polynomial $0.3x^2y^2 - 0.5xy + 0.4x - 0.6y + 0.8$ when $x = 0.4$, $y = 0.25$.

55. Write a polynomial giving the area of a circle of radius r.

56. Write a polynomial giving the area of a triangle of base b and height h.

57. Consider the field shown in Figure 2.3. What does each of the following polynomials represent?

(a) $x^2 + xy$

(b) $2x + 2y$

(c) $4x$

(d) $4x + 2y$

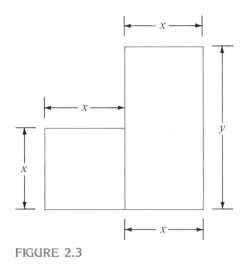

FIGURE 2.3

58. An investor buys x shares of G.E. stock at \$55 per share, y shares of Exxon stock at \$45 per share, and z shares of A.T.T. stock at \$60 per share. What does the polynomial $55x + 45y + 60z$ represent?

2.2
ADDITION AND SUBTRACTION OF POLYNOMIALS

Those terms of a polynomial which differ only in their coefficients are called **like terms.** Here are some examples of like terms:

$$4x^2 \quad \text{and} \quad -3x^2$$
$$-5xy^3 \quad \text{and} \quad 17xy^3$$
$$2x^2y^2 \quad \text{and} \quad -2x^2y^2$$

We add polynomials by adding the coefficients of like terms. It is often helpful to regroup the terms before adding. For example,

$$
\begin{array}{ccc}
x^2 \text{ terms} & x \text{ terms} & \text{constant terms}
\end{array}
$$
$$(3x^2 - 2x + 5) + (x^2 + 4x - 9)$$
$$= 3x^2 + x^2 - 2x + 4x + 5 - 9$$
$$= 4x^2 + 2x - 4$$

EXAMPLE 1
Add.

(a) $4x^2 - 3xy + 2y^2$ and $2x^2 - xy - y^2$

Grouping like terms, we have

$4x^2 + 2x^2 - 3xy - xy + 2y^2 - y^2 = 6x^2 - 4xy + y^2$

(b) $x^3 - 2x^2 + 6x$, $x^2 - 4$, and $2x^3 - x + 6$

Grouping like terms, we have

$x^3 + 2x^3 - 2x^2 + x^2 + 6x - x - 4 + 6 = 3x^3 - x^2 + 5x + 2$

PROGRESS CHECK 1
Simplify by combining like terms.

(a) $2x^2 + x - 3 + 4x^2 - 5x - 8$

(b) $-2x^3 + 2x^2y^2 - 4y^2 + 4x^3 + 2y^2 + xy - 7$

Answers
(*a*) $6x^2 - 4x - 11$ (*b*) $2x^3 + 2x^2y^2 - 2y^2 + xy - 7$

Sometimes we must remove parentheses before we can combine terms. The key to this is the distributive law. For example,

$$2(x - 3y) - 3(2x + 4y)$$
$$= 2x - 6y - 6x - 12y \qquad \text{Distributive law}$$
$$= -4x - 18y$$

The same idea permits us to subtract polynomials. For example, to subtract $x^2 - 3x + 1$ from $3x^2 - x - 5$ we have

$$(3x^2 - x - 5) - (x^2 - 3x + 1)$$
$$= 3x^2 - x - 5 - x^2 + 3x - 1$$
$$= 2x^2 + 2x - 6$$

WARNING
Don't write

$$(x + 5) - (x + 2) = x + 5 - x + 2 = 7$$

The coefficient -1 must multiply each term in the parentheses:

$$(x + 5) - (x + 2) = x + 5 - x - 2 = 3$$

EXAMPLE 2
Simplify.

(a) $3(x^2 - 2xy + \frac{1}{3}y^2) - 2(2y^2 + x^2 - \frac{1}{2}xy)$

$$= 3x^2 - 6xy + y^2 - 4y^2 - 2x^2 + xy$$
$$= 3x^2 - 2x^2 - 6xy + xy + y^2 - 4y^2$$
$$= x^2 - 5xy - 3y^2$$

(b) $2x(x - 5) + 4(x - 3)$

$= 2x^2 - 10x + 4x - 12$

$= 2x^2 - 6x - 12$

PROGRESS CHECK 2
Simplify.

(a) $6(r^2 + 2rs - 1) - 4(-rs - 2 + 2r^2)$

(b) $4\left(\frac{1}{2}x^2 + \frac{1}{4}x + 1\right) + 5\left(2x^2 - \frac{2}{5}x - 1\right)$

Answers
(a) $-2r^2 + 16rs + 2$ (b) $12x^2 - x - 1$

EXERCISE SET 2.2
Add the given polynomials in each of the following.

1. $5x;\quad 2x$ 2. $5y;\quad -3y$

3. $3x^3;\quad -6x^2$ 4. $-3x^2;\quad -5x^2$

5. $x^2 - 3x + 1;\quad 3x^2 + 2x + 3$ 6. $2x^2 + \frac{5}{2}x + 2;\quad -3x^2 - \frac{5}{3}x + 7$

7. $2x^3 + 2x^2 - x + 1;\quad -2x^3 + 5x^2 + x + 2$

8. $3xy;\quad 4xy$ 9. $2rs;\quad -5rs$

10. $2x^2y^2 - xy + 2x + 3y + 3;\quad x^2y^2 + 3xy + 2x + 7$

11. $\frac{2}{5}rs^3 + 4r^2s^2 + 2r^2 + 2;\quad \frac{4}{5}rs^3 - 6r^2s^2 - r^2s + 5$

Find the mistake(s) in each of the following. Obtain the correct answer.

12. $(x + 3) - (x + 5) = 8$

13. $(x^2 + 2x + 4) - 2(x^2 + 3x - 5) = -x^2 + 5x - 1$

14. $(x^2y^2 + 2x^2 + y) - (3x^2y^2 + x^2 - y + 2) = -x^2y^2 + x^2 - 2$

15. $(y^2 + xy + y) - 2(x^2 + xy - 3y) = y^2 - 2x^2 - xy - 2y$

Subtract the second polynomial from the first.

16. $8x;\quad 3x$ 17. $18y;\quad -6y$

18. $3x^2;\quad 4x^2$ 19. $3x^2 + 2x - 5;\quad -3x^2 + 2x - 2$

20. $\frac{3}{2}x^3 + 2x^2 + 5;\quad \frac{5}{2}x^3 - x^2 - \frac{1}{2}x + 3$

21. $3x^2y^2 + 2xy - y;\quad 2x^2y^2 - xy + x + 2y + 3$

Subtract the first polynomial from the second.

22. $4x^2 + 2x - 5;\quad 3x^2 - 3x + 5$

23. $\dfrac{5}{2}x^3 - 2x^2 + x - 2;\quad \dfrac{3}{5}x^3 + x^2 - 4$

24. $2xy^2 + xy + x - 3;\quad 2x^2y - xy^2 + y + x - 2$

25. $3rs^3 - 2rs^2 + rs + 3;\quad -2rs^2 + 3rs - r + s$

Simplify by combining like terms.

26. $(4x^2 + 3x + 2) + (3x^2 - 2x - 5)$

27. $5x^2 + 2x + 7 - 3x^2 - 8x + 2$

28. $(2x^2 + 3x + 8) - (5 - 2x + 2x^2)$

29. $3xy + 2x + 3y + 2 + (1 - y - 2x + xy)$

30. $4xy^2 + 2xy + 2x + 3 - (-2xy^2 + xy - y + 2)$

31. $(3r^2s^2 + rs^2 - rs + r) + (2r^2s^2 - r^2s + s + 1)$

32. $3a^2b - ba^2 + 2a - b + 2a^2b - 3ba^2 + b + 1$

33. $(2s^2t^3 - st^2 + st - s + t) - (3s^2t^2 - 2s^2t - 4st^2 - t + 3)$

34. $3xy^2z - 4x^2yz + xy + 3 - (2xy^2z + x^2yz - yz + x - 2)$

35. $a^2bc + ab^2c + 2ab^3 - 3a^2bc - 4ab^3 + 3$

36. On Monday morning an investor buys x shares of A.T.T. stock at $60 per share and y shares of Exxon stock at $50 per share. On Monday afternoon the same investor buys x shares of G.E. stock at $55 per share and y shares of Bethlehem Steel stock at $20 per share.
     (a) How much money was invested during the morning transactions?
     (b) How much money was invested during the afternoon transactions?
     (c) How much money was invested by the end of Monday?

37. An investor buys x shares of I.B.M. stock at $260 per share at Thursday's opening of the stock market. Later in the day, he sells y shares of Gulf and Western stock at $13 per share and z shares of Holiday Inn stock at $17 per share. Write a polynomial which expresses the net of his transactions for the day.

38. An artist takes a rectangular piece of cardboard whose sides are x and y and cuts out a square of side $x/2$ (see Figure 2.4) to obtain a mat for a painting. Write a polynomial giving the area of the mat; that is, what is the area of the remaining figure?

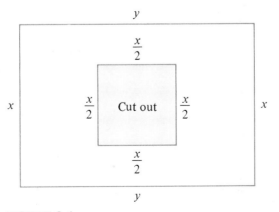

FIGURE 2.4

2.3
MULTIPLICATION OF POLYNOMIALS

We have already dealt with multiplication of some simple polynomial forms such as

$$3x^2 \cdot x^4 = 3x^6$$

and

$$2x(x + 4) = 2x^2 + 8x$$

By using the rule for exponents

$$a^m \cdot a^n = a^{m+n}$$

and the distributive laws

$$a(b + c) = ab + ac$$
$$(a + b)c = ac + bc$$

we can handle the product of any two polynomials.

EXAMPLE 1
Multiply.

(a) $3x^3(2x^3 - 6x^2 + 5)$
$$= (3x^3)(2x^3) + (3x^3)(-6x^2) + (3x^3)(5)$$
$$= (3 \cdot 2)x^{3+3} + 3(-6)x^{3+2} + 3 \cdot 5x^3$$
$$= 6x^6 - 18x^5 + 15x^3$$

(b) $-x^2y(x^3 - 4xy^2 + 6y^3)$
$$= (-x^2y)(x^3) + (-x^2y)(-4xy^2) + (-x^2y)(6y^3)$$
$$= -x^5y + 4x^3y^3 - 6x^2y^4$$

PROGRESS CHECK 1
Multiply.

(a) $2x^2\left(\dfrac{1}{2}x^2 - 3x - 4\right)$ (b) $-xy(x^2 - 2x^2y - 3xy^2)$

Answers
(a) $x^4 - 6x^3 - 8x^2$ (b) $-x^3y + 2x^3y^2 + 3x^2y^3$

Let's try to find the product

$$(x + 2)(3x^2 - x + 5)$$

The key here is to "rename" terms and groups of terms to fit into the alternate form of the distributive law $(a + b)c = ac + bc$. Here is how we can do it.

$$\underbrace{(x + 2)}_{(a + b)}\underbrace{(3x^2 - x + 5)}_{c} = x\underbrace{(3x^2 - x + 5)}_{ac} + 2\underbrace{(3x^2 - x + 5)}_{bc}$$
$$= 3x^3 - x^2 + 5x + 6x^2 - 2x + 10$$
$$= 3x^3 + 5x^2 + 3x + 10$$

In general, we can multiply two polynomials by multiplying one polynomial by each term of the other polynomial and adding the resulting products.

EXAMPLE 2

Multiply.

(a) $(3x - 2)(x + 4)$

$= 3x(x + 4) - 2(x + 4)$

$= 3x^2 + 12x - 2x - 8$

$= 3x^2 + 10x - 8$

(b) $(2x - 3)(-4x^2 + x + 5)$

$= 2x(-4x^2 + x + 5) - 3(-4x^2 + x + 5)$

$= -8x^3 + 2x^2 + 10x + 12x^2 - 3x - 15$

$= -8x^3 + 2x^2 + 12x^2 + 10x - 3x - 15$

$= -8x^3 + 14x^2 + 7x - 15$

PROGRESS CHECK 2

Multiply.

(a) $(3x - 1)(x - 3)$ (b) $(x^2 + 2)(x^2 - 3x + 1)$

Answers

(*a*) $3x^2 - 10x + 3$ (*b*) $x^4 - 3x^3 + 3x^2 - 6x + 2$

The work can be arranged in a "long multiplication" format. Here is an example.

EXAMPLE 3

Find the product of $2x - 3$ and $-4x^2 + x + 5$.

Solution

We arrange the work as follows:

$$
\begin{array}{r}
-4x^2 + x + 5 \\
2x - 3 \\
\hline
-8x^3 + 2x^2 + 10x \qquad [= 2x(-4x^2 + x + 5)] \\
12x^2 - 3x - 15 \qquad [= -3(-4x^2 + x + 5)] \\
\hline
-8x^3 + 14x^2 + 7x - 15
\end{array}
$$

PROGRESS CHECK 3

Repeat Progress Check 2(b) using long multiplication.

Products of the form $(2x + 3)(5x - 2)$ are so important that we must learn to handle them mentally. Let's work through this problem:

$(2x + 3)(5x - 2) = 2x(5x - 2) + 3(5x - 2)$

$= (2x)(5x) + 2x(-2) + 3(5x) + 3(-2)$

$= 10x^2 - 4x + 15x - 6$

We have stopped just short of the last step because we want to show the relationships between the factors and the products. If we take the product of the first term of each expression

$$(2x + 3)(5x - 2)$$
$$10x^2$$

we have the term containing x^2. Similarly, taking the product of the last term of each expression

$$(2x + 3)(5x - 2)$$
$$-6$$

we have the constant term. The term containing x can be found by adding the product of the "inners" and the product of the "outers":

$$(2x + 3)(5x - 2)$$
$$15x$$
$$-4x$$
$$\text{Sum} = 11x$$

Thus, $(2x + 3)(5x - 2) = 10x^2 + 11x - 6$.

EXAMPLE 4
Multiply.

(a) $(x - 1)(2x + 3)$

We diagram the process so that you can learn to do these mentally.

$$(x - 1)(2x + 3) \qquad (x - 1)(2x + 3) \qquad (x - 1)(2x + 3)$$
$$2x^2 \qquad\qquad -2x \qquad\qquad -3$$
$$3x$$
$$\text{Sum} = +x$$

Thus, $(x - 1)(2x + 3) = 2x^2 + x - 3$.

(b) $(2x + 2)(2x - 2)$

Once more, we have

$$(2x + 2)(2x - 2) \qquad (2x + 2)(2x - 2) \qquad (2x + 2)(2x - 2)$$
$$4x^2 \qquad\qquad 4x \qquad\qquad -4$$
$$-4x$$
$$\text{Sum} = +0x$$

Thus, $(2x + 2)(2x - 2) = 4x^2 - 4$.

PROGRESS CHECK 4
Multiply mentally.

(a) $(x + 2)(x + 1)$ (b) $(t - 2)(2t + 3)$

(c) $(2x - 3)(3x - 2)$ (d) $(3x + 2)(3x - 2)$

Answers

(a) $x^2 + 3x + 2$ (b) $2t^2 - t - 6$ (c) $6x^2 - 13x + 6$ (d) $9x^2 - 4$

Here are three forms that occur frequently and are worthy of special attention.

$$(x + y)^2 = (x + y)(x + y) = x^2 + 2xy + y^2$$
$$(x - y)^2 = (x - y)(x - y) = x^2 - 2xy + y^2$$
$$(x + y)(x - y) = x^2 - y^2$$

EXAMPLE 5

Multiply mentally.

(a) $(x + 2)^2$

$= (x + 2)(x + 2) = x^2 + 4x + 4$

(b) $(x - 3)^2$

$= (x - 3)(x - 3) = x^2 - 6x + 9$

(c) $(x + 4)(x - 4)$

$= x^2 - 16$

PROGRESS CHECK 5

Multiply mentally.

(a) $(x - 4)^2$ (b) $(x + 1)^2$ (c) $(2x - 3)^2$ (d) $(2x + 3)(2x - 3)$

Answers

(a) $x^2 - 8x + 16$ (b) $x^2 + 2x + 1$ (c) $4x^2 - 12x + 9$

(d) $4x^2 - 9$

EXERCISE SET 2.3

Perform the indicated multiplication in each of the following.

1. $(2x^3)(3x^2)$ 2. $(6x^2)(-5x^4)$

3. $(3ab^2)(2ab)$ 4. $(-3s^2t)(4s)$

5. $2x(x^2 + 3x - 5)$ 6. $6x^2(2x^3 - 2x^2 + 5)$

7. $-2s^3(2st^2 - 2st + 6)$ 8. $a^3(-3a^2 - a + 2)$

9. $4a^2b^2(2a^2 + ab - b^2)$ 10. $4y(2y^3 - 3y + 3)$

11. $(x + 2)(x - 3)$ 12. $(x - 1)(x + 4)$

13. $(y + 5)(y + 2)$ 14. $(a - 3)(a - 4)$

15. $(x + 3)^2$ 16. $(y - 2)(y - 2)$

Perform the indicated multiplication mentally in each of the following.

17. $(s + 3)(s - 3)$ 18. $(t + 6)(t - 6)$

19. $(3x + 2)(x - 1)$ 20. $(2x - 3)(x + 1)$

21. $(a - 2)(2a + 5)$ 22. $(a + 3)(3a - 2)$

23. $(2y + 3)(3y + 2)$ 24. $(3x - 2)(2x + 3)$

25. $(2a + 3)(2a + 3)$

26. $(3x + 2)(3x + 2)$

27. $(2y + 5)(2y - 5)$

28. $(4t + 3)(4t - 3)$

29. $(3x - 4)(3x + 4)$

30. $(5b - 2)(5b + 2)$

Perform the indicated multiplication in the following. Use long multiplication where convenient.

31. $(x^2 + 2)(x^2 + 2)$

32. $(y^2 - 3)(y^2 - 3)$

33. $(x^2 - 2)(x^2 + 2)$

34. $(2x^2 - 5)(2x^2 + 5)$

35. $(x + 1)(x^2 + 2x - 3)$

36. $(x - 2)(2x^3 + x - 2)$

37. $(2s - 3)(s^3 - s + 2)$

38. $(-3s + 2)(-2s^2 - s + 3)$

39. $(a + 2)(3a^2 - a + 5)$

40. $(b + 3)(-3b^2 + 2b + 4)$

41. $(x^2 + 3)(2x^2 - x + 2)$

42. $(2y^2 + y)(-2y^3 + y - 3)$

43. $(x^2 + 2x - 1)(2x^2 - 3x + 2)$

44. $(a^2 - 4a + 3)(4a^3 + 2a + 5)$

45. $(3x^2 - 2x + 2)(2x^3 - 4x + 2)$

46. $(-3y^3 + 3y - 4)(2y^2 - 2y + 3)$

47. $(2a^2 + ab + b^2)(3a - b^2 + 1)$

48. $(-3a + ab + b^2)(3b^2 + 2b + 2)$

49. $5(2x - 3)^2$

50. $2(3x - 2)(2x + 3)$

51. $x(2x - 1)(x + 2)$

52. $3x(2x + 1)^2$

53. $(x - 1)(x + 2)(x + 3)$

54. $(3x + 1)(2x - 4)(3x + 2)$

55. In the product $(x - 1)(x^2 - 2x + 3)$ give the coefficient of (a) x^2 (b) x.

56. In the product $(3x - 2)(2x^2 + 3x - 4)$ give the coefficient of (a) x^2 (b) x.

57. In the product $(x^2 - 2x + 1)^2$ give the coefficient of (a) x^4 (b) x^2.

58. In the product $(x^2 - 2x + 3)(x^2 - 3x - 5)$ give the coefficient of (a) x^3 (b) x^2.

Compute each of the following.

59. $3x(yx^2 + xy) + xy(x^2 - x)$

60. $(x - y)(x + y) - x(x + y)$

61. $(x - 1)(x + 3) - x^2$

62. $(2x - 1)(3x + 2) - (x - 2)(x + 3)$

63. $2x(x - 3) - 4(x^2 - 4)$

64. $x(-x - 3) + (-x + 4)^2$

65. $(x + 4)(x - 4) - (x - 2)^2$

66. $(x - 2)^2 - x(x - 1)$

67. When a polynomial of degree 3 is multiplied by a polynomial of degree 4, the degree of the product is (a) 12   (b) 1   (c) 7   (d) $\frac{4}{3}$ (e) none of these.

68. When a polynomial of degree 6 is multiplied by a polynomial of degree 3, the degree of the product is (a) 18   (b) 3   (c) 2   (d) 9   (e) none of these.

Compute.

69. $(1.25x - 3.67)^2$

70. $(-3.74 + 7.39y)^2$

71. $(5.74y^2 - 2.82)(3.96y^2 + 1.15)$

72. $2.62x(4.78x - 16.42)(3.76x + 4.91)$

73. $(x - 0.04)(3.25x - 2.00)(6.67x + 3.48)$

74. $(6.94 - 10.01x^2)(4.72 + 9.97x^2)$

2.4
FACTORING

Now that we can find the product of two polynomials, let's consider the reverse problem: Given a polynomial, can we find factors whose product will yield the given polynomial? This process is known as **factoring.** We will approach factoring by learning to recognize the situations in which factoring is possible.

COMMON FACTORS

Look at the polynomial

$$x^2 + x$$

Is there some factor common to *each* term? Yes—each term contains the variable x. If we remove x and write

$$x^2 + x = x(\quad + \quad)$$

we can see that we must have

$$x^2 + x = x(x + 1)$$

EXAMPLE 1
Factor.

(a) $15x^3 - 10x^2$

Both 5 and x^2 are common to *both* terms.

$15x^3 - 10x^2 = 5x^2(3x - 2)$

(b) $4x^2y - 8xy^2 + 6xy$

Here, we see that 2, x, and y are common to *each* term.

$4x^2y - 8xy^2 + 6xy = 2xy(2x - 4y + 3)$

PROGRESS CHECK 1
Factor.

(a) $4x^2 - x$ (b) $3x^4 - 9x^2$

Answers

(a) $x(4x - 1)$ (b) $3x^2(x^2 - 3)$

EXAMPLE 2
Factor.

(a) $2ab - 8bc$

We see that both 2 and b are found in each term. Don't be misled by the position of b to the right in the first term and to the left in the second term—remember, multiplication is commutative!

$$2ab - 8bc = 2b(a - 4c)$$

(b) $2x(x + y) - 5y(x + y)$

Here, $(x + y)$ is found in both terms. Factoring,

$$2x(x + y) - 5y(x + y) = (x + y)(2x - 5y)$$

PROGRESS CHECK 2
Factor.

(a) $3r^2t - 15t^2u + 6st^3$ (b) $3m(2x - 3y) - n(2x - 3y)$

Answers

(a) $3t(r^2 - 5tu + 2st^2)$ (b) $(2x - 3y)(3m - n)$

FACTORING BY GROUPING

It is sometimes possible to discover common factors by first grouping terms. The best way to learn the method is by studying some examples.

EXAMPLE 3
Factor.

(a) $2ab + b + 2ac + c$

Begin by grouping those terms containing b and those terms containing c.

$2ab + b + 2ac + c = (2ab + b) + (2ac + c)$	Grouping
$= b(2a + 1) + c(2a + 1)$	Common factors b, c
$= (2a + 1)(b + c)$	Common factor $2a + 1$

(b) $2x - 4x^2y - 3y + 6xy^2$

$= (2x - 4x^2y) - (3y - 6xy^2)$	Grouping, with sign change
$= 2x(1 - 2xy) - 3y(1 - 2xy)$	Common factors $2x$, $3y$
$= (1 - 2xy)(2x - 3y)$	Common factor $1 - 2xy$

PROGRESS CHECK 3
Factor.

(a) $2m^3n + m^2 + 2mn^2 + n$ (b) $2a^2 - 4ab^2 - ab + 2b^3$

Answers

(a) $(2mn + 1)(m^2 + n)$ (b) $(a - 2b^2)(2a - b)$

FACTORING SECOND-DEGREE POLYNOMIALS

Another type of factoring involves second-degree polynomials. We now know that

$$(x + 2)(x + 3) = x^2 + 5x + 6$$

and can do the multiplication mentally. We will need these mental gymnastics to allow us to reverse the process.

Let's look at

$$x^2 + 5x + 6$$

and think of this in the form

$$x^2 + 5x + 6 = (\qquad)(\qquad)$$

If we restrict ourselves to integer coefficients, then the term x^2 can only have come from $x \cdot x$ so we can write

$$x^2 + 5x + 6 = (x \qquad)(x \qquad)$$

The constant $+6$ can be the product of either two positive numbers or two negative numbers. (If we choose one positive factor and one negative factor we can't produce a positive product.) But the middle term is positive and results from adding two terms. Then the signs must both be positive.

$$x^2 + 5x + 6 = (x + \qquad)(x + \quad)$$

Finally, the number 6 can be written as the product of two integers in just two ways:

$$1 \cdot 6 \quad \text{or} \quad 2 \cdot 3$$

If you try each pair, you will quickly see that the pair $1 \cdot 6$ produces a middle term of $7x$—which is not what we want. But the pair $2 \cdot 3$ does work.

$$x^2 + 5x + 6 = (x + 2)(x + 3)$$

EXAMPLE 4
Factor.

(a) $x^2 - 7x + 10$

Since the constant term is positive and the middle term is negative, we must have

$$x^2 - 7x + 10 = (x - \qquad)(x - \quad)$$

The possible integer factors of 10 are

$$1 \cdot 10 \quad \text{and} \quad 2 \cdot 5$$

The middle term results from the use of $2 \cdot 5$ so that

$$x^2 - 7x + 10 = (x - 2)(x - 5)$$

(b) $x^2 - 3x - 4$

Since the constant term is negative, we must have

$$x^2 - 3x - 4 = (x + \qquad)(x - \quad)$$

The integer factors of 4 are

$$1 \cdot 4 \quad \text{and} \quad 2 \cdot 2$$

We find that using $1 \cdot 4$ will produce the correct middle term of $-3x$ if we associate the larger factor, 4, with the negative sign.

$$x^2 - 3x - 4 = (x + 1)(x - 4)$$

PROGRESS CHECK 4

Factor.

(a) $x^2 - 11x + 24$ (b) $x^2 + 6x + 9$ (c) $x^2 - 2x - 8$

Answers

(*a*) $(x - 3)(x - 8)$ (*b*) $(x + 3)(x + 3)$ (*c*) $(x + 2)(x - 4)$

Before you get the impression that every second-degree polynomial can be factored as a product of polynomials of lower degree with integer coefficients, try your hand at factoring

$$x^2 + x + 1$$

There simply are no polynomials with integer coefficients that allow us to factor $x^2 + x + 1$.

Now we will try something a bit more difficult, for example,

$$2x^2 - x - 6$$

When the coefficient of x^2 is a number other than 1, we can use the same approach but the number of possible combinations increases.

First, we see that $2x^2$ can only result from the factors $2x$ and x. Thus, we write

$$2x^2 - x - 6 = (2x \quad)(x \quad)$$

The constant -6 can result from

$$(-1)(6) \quad (1)(-6) \quad (6)(-1) \quad (-6)(1)$$
$$(-2)(3) \quad (2)(-3) \quad (3)(-2) \quad (-3)(2)$$

That's a long list for trial-and-error—which is why it is important to be able to handle the multiplication mentally. We find that the correct middle term $-x$ is obtained by using $(3)(-2)$, and thus

$$2x^2 - x - 6 = (2x + 3)(x - 2)$$

EXAMPLE 5

Factor.

(a) $3x^2 + 7x + 4$

We start with

$$3x^2 + 7x + 4 = (3x + \quad)(x + \quad) \quad \text{(Why?)}$$

The possible integer factors of 4 are

$$1 \cdot 4 \quad 4 \cdot 1 \quad 2 \cdot 2$$

and, by trial-and-error, we have

$$3x^2 + 7x + 4 = (3x + 4)(x + 1)$$

(b) $6x^2 + 5x - 4$

We list the possible integer factors of $+6$ and -4.

+6	−4	
(1)(6)	(−1)(4)	(4)(−1)
(6)(1)	(−2)(2)	(2)(−2)
(2)(3)	(−4)(1)	(1)(−4)
(3)(2)		

Since this is a massive trial-and-error problem, we had better organize our search. Lock in a pair of factors of $+6$, say $(1)(6)$, and then try each pair of factors of -4. If none will do, proceed to the next pair of factors of $+6$ and repeat. You will find that $6x^2 + 5x - 4 = (x \quad)(6x \quad)$ cannot be made to work, but eventually you will find that

$$6x^2 + 5x - 4 = (2x - 1)(3x + 4)$$

PROGRESS CHECK 5
Factor.

(a) $3x^2 - 16x + 21$ (b) $2x^2 + 3x - 9$ (c) $4x^2 + 12x + 5$

Answers

(a) $(3x - 7)(x - 3)$ (b) $(2x - 3)(x + 3)$ (c) $(2x + 1)(2x + 5)$

COMBINING METHODS

We conclude with problems that combine the various methods of factoring that we have studied. Here is a good rule to follow:

Always remove common factors before attempting any other factoring techniques.

EXAMPLE 6
Factor.

(a) $x^3 - 6x^2 + 8x$

Following our rule, we first remove the common factor:

$$x^3 - 6x^2 + 8x = x(x^2 - 6x + 8)$$
$$= x(x - 2)(x - 4)$$

(b) $2x^3 + 4x^2 - 30x$

Removing $2x$ as a common factor,

$$2x^3 + 4x^2 - 30x = 2x(x^2 + 2x - 15)$$
$$= 2x(x - 3)(x + 5)$$

(c) $3y(y + 3) + 2(y + 3)(y^2 - 1)$

Removing the common factor $y + 3$,

$$3y(y + 3) + 2(y + 3)(y^2 - 1) = (y + 3)[3y + 2(y^2 - 1)]$$
$$= (y + 3)(3y + 2y^2 - 2)$$
$$= (y + 3)(2y^2 + 3y - 2)$$
$$= (y + 3)(2y - 1)(y + 2)$$

PROGRESS CHECK 6

Factor.

(a) $x^3 + 5x^2 - 6x$ (b) $2x^3 - 2x^2y - 4xy^2$

(c) $-3x(x + 1) + (x + 1)(2x^2 + 1)$

Answers

(a) $x(x + 6)(x - 1)$ (b) $2x(x + y)(x - 2y)$ (c) $(x + 1)(2x - 1)(x - 1)$

EXERCISE SET 2.4

Factor completely.

1. $2x + 6$

2. $5x - 15$

3. $3x - 9y$

4. $\frac{1}{2}x + \frac{1}{4}y$

5. $-2x - 8y$

6. $3x + 6y + 15$

7. $4x^2 + 8y - 6$

8. $3a + 4ab$

9. $5bc + 25b$

10. $2x^2 - x$

11. $y - 3y^3$

12. $2x^4 + x^2$

13. $-3y^2 - 4y^5$

14. $-\frac{1}{2}y^2 + \frac{1}{8}y^3$

15. $3abc + 12bc$

16. $3x^2 + 6x^2y - 9x^2z$

17. $5r^3s^4 - 40r^4s^3t$

18. $9a^3b^3 + 12a^2b - 15ab^2$

19. $8a^3b^5 - 12a^5b^2 + 16$

20. $7x^2y^3z^4 - 21x^4yz^5 + 49x^5y^2z^3$

21. $x^2 + 4x + 3$

22. $x^2 + 2x - 8$

23. $y^2 - 8y + 15$

24. $y^2 + 7y - 8$

25. $a^2 - 7ab + 12b^2$

26. $x^2 - 14x + 49$

27. $y^2 + 6y + 9$

28. $a^2 - 7a + 10$

29. $25 - 10x + x^2$

30. $4b^2 - a^2$

31. $x^2 - 5x - 14$

32. $x^2 - 9$

33. $4 - y^2$

34. $a^2 + ab - 6b^2$

35. $x^2 - 6x + 9$

36. $a^2 - 4ab + 4b^2$

37. $x^2 - 12x + 20$

38. $x^2 - 8x - 20$

39. $x^2 + 11x + 24$

40. $y^2 + 4y + 3$

41. $2x^2 - 3x - 2$

42. $2x^2 + 7x + 6$

43. $3a^2 - 11a + 6$

44. $4x^2 - 9x + 2$

45. $6x^2 + 13x + 6$

46. $4y^2 + 4y - 3$

47. $8m^2 - 6m - 9$

48. $9x^2 + 24x + 16$

49. $10x^2 - 13x - 3$

50. $6a^2 + ab - 2b^2$

51. $6a^2 - 5ab - 6b^2$

52. $4x^2 + 20x + 25$

53. $10r^2s^2 + 9rst + 2t^2$

54. $16 - 24xy + 9x^2y^2$

55. $6 + 5x - 4x^2$

56. $8n^2 - 18n - 5$

57. $25r^2 + 4s^2$

58. $15 + 4x - 4x^2$

59. $2x^2 - 2x - 12$

60. $3y^2 + 6y - 45$

61. $30x^2 + 28x - 16$

62. $30x^2 - 35x + 10$

63. $12x^2b^2 + 2xb^2 - 24b^2$

64. $x^4y^4 + x^2$

65. $18x^2m + 35xm + 9m$

66. $8x^3 + 14x^2 - 15x$

67. $25m^2n^3 - 5m^2n$

68. $12x^2 - 22x^3 - 20x^4$

69. $xy + \dfrac{1}{4}x^3y^3$
70. $10r^2 - 5rs - 15s^2$
71. $x^4 + 2x^2y^2 + y^4$
72. $a^4 - 8a^2 + 16$
73. $b^4 + 2b^2 - 8$
74. $4b^4 + 20b^2 + 25$
75. $6b^4 + 7b^2 - 3$
76. $4(x + 1)(y + 2) - 8(y + 2)$
77. $2(x + 1)(x - 1) + 5(x - 1)$
78. $3(x + 2)^2(x - 1) - 4(x + 2)^2(2x + 7)$
79. $4(2x - 1)^2(x + 2)^3(x + 1) - 3(2x - 1)^5(x + 2)^2(x + 3)$
80. $5(x - 1)^2(y - 1)^3(x + 2) - (3x - 1)(x - 1)^3(y - 1)$
81. $(7 - 2x)^3(2)(5x)(5) + (5x)^2(3)(7 - 2x)^2(-2)$
82. $3(4x)^2(4)(7x - 2)^2 + (4x)^3(2)(7x - 2)(6)$

2.5
SPECIAL FACTORS

There is a special case of the second-degree polynomial that occurs frequently and factors easily. Given the polynomial $x^2 - 9$, we see that each term is a perfect square. You may easily verify that

$$x^2 - 9 = (x + 3)(x - 3)$$

In general,

Difference of Two Squares

$$a^2 - b^2 = (a + b)(a - b)$$

is a rule that works whenever we are dealing with a difference of two squares.

EXAMPLE 1
Factor.

(a) $x^2 - 16$
$x^2 - 16 = (x + 4)(x - 4)$

(b) $4x^2 - 25$.   With $a = 2x$ and $b = 5$,
$4x^2 - 25 = (2x + 5)(2x - 5)$

(c) $9x^2 - 16y^2$.   With $a = 3x$ and $b = 4y$,
$9x^2 - 16y^2 = (3x + 4y)(3x - 4y)$

PROGRESS CHECK 1
Factor.

(a) $x^2 - 49$ (b) $16x^2 - 9$ (c) $25x^2 - y^2$

Answers
(a) $(x + 7)(x - 7)$ (b) $(4x + 3)(4x - 3)$ (c) $(5x + y)(5x - y)$

WARNING

Don't confuse a *difference* of two squares, such as $4x^2 - 9$, and a *sum* of two squares, such as $x^2 + 25$. In the case of a difference of two squares,

$$4x^2 - 9 = (2x + 3)(2x - 3)$$

But a sum of two squares such as $x^2 + 25$ cannot be factored.

The formulas for a sum of two cubes and a difference of two cubes can be verified by multiplying the factors on the right-hand sides of the following equations.

Sum and Difference of Two Cubes

$$a^3 + b^3 = (a + b)(a^2 - ab + b^2)$$
$$a^3 - b^3 = (a - b)(a^2 + ab + b^2)$$

These formulas provide a direct means for factoring a sum and difference of two cubes and are used in the same way as the formula for a difference of two squares. Be careful as to the placement of plus and minus signs when using these formulas.

EXAMPLE 2

Factor each of the following.

(a) $x^3 + 1$ (b) $27m^3 - 64n^3$

Solution

(a) $x^3 + 1$. When $a = x$ and $b = 1$, the formula for a sum of two cubes yields

$$x^3 + 1 = (x + 1)(x^2 - x + 1)$$

(b) $27m^3 - 64n^3$. Note that $27m^3 - 64n^3 = (3m)^3 - (4n)^3$. We then use the formula for a difference of two cubes with $a = 3m$ and $b = 4n$.

$$27m^3 - 64n^3 = (3m - 4n)(9m^2 + 12mn + 16n^2)$$

PROGRESS CHECK 2

Factor.

(a) $8x^3 + y^3$ (b) $8s^3 - 27t^3$

Answers

(a) $(2x + y)(4x^2 - 2xy + y^2)$ (b) $(2s - 3t)(4s^2 + 6st + 9t^2)$

EXAMPLE 3

Factor each of the following.

(a) $\dfrac{1}{27}u^3 - 8v^3$ (b) $125x^6 + 8y^3$

Solution

(a) $\frac{1}{27}u^3 - 8v^3$. Since

$$\frac{1}{27}u^3 - 8v^3 = \left(\frac{u}{3}\right)^3 - (2v)^3$$

we may use the formula for a difference of two cubes with $a = \frac{u}{3}$ and $b = 2v$.

$$\frac{1}{27}u^3 - 8v^3 = \left(\frac{u}{3} - 2v\right)\left(\frac{u^2}{9} + \frac{2}{3}uv + 4v^2\right)$$

(b) $125x^6 + 8y^3$. Rewrite the polynomial as $(5x^2)^3 + (2y)^3$. With $a = 5x^2$ and $b = 2y$, the formula for a sum of two cubes tells us that

$$125x^6 + 8y^3 = (5x^2 + 2y)(25x^4 - 10x^2y + 4y^2)$$

PROGRESS CHECK 3

Factor each of the following.

(a) $125r^3 + \frac{1}{125}s^3$ (b) $27a^6 - 64b^6$

Answers

(a) $\left(5r + \frac{s}{5}\right)\left(25r^2 - rs + \frac{s^2}{25}\right)$ (b) $(3a^2 - 4b^2)(9a^4 + 12a^2b^2 + 16b^4)$

EXERCISE SET 2.5

Use the sum of cubes and difference of cubes formulas to find the following products.

1. $(2x + y)(4x^2 - 2xy + y^2)$ 2. $(x + 3y)(x^2 - 3xy + 9y^2)$

3. $(x - 2y)(x^2 + 2xy + 4y^2)$ 4. $(4x - y)(16x^2 + 4xy + y^2)$

5. $(3r + 2s)(9r^2 - 6rs + 4s^2)$ 6. $(2a - 3b)(4a^2 + 6ab + 9b^2)$

7. $(2m - 5n)(4m^2 + 10mn + 25n^2)$ 8. $(4a + 3b)(16a^2 - 12ab + 9b^2)$

9. $\left(\frac{x}{2} - 2y\right)\left(\frac{1}{4}x^2 + xy + 4y^2\right)$ 10. $\left(3x - \frac{y}{3}\right)\left(9x^2 + xy + \frac{1}{9}y^2\right)$

Factor each of the following.

11. $x^3 + 27y^3$ 12. $8x^3 + 125y^3$ 13. $27x^3 - y^3$

14. $64x^3 - 27y^3$ 15. $a^3 + 8$ 16. $8r^3 - 27$

17. $\frac{1}{8}m^3 - 8n^3$ 18. $8a^3 - \frac{1}{64}b^3$ 19. $(x + y)^3 - 8$

20. $27 + (x + y)^3$ 21. $8x^6 - 125y^6$ 22. $a^6 + 27b^6$

23. $27r^6 + 8s^6$ 24. $64m^6 - \frac{1}{8}n^6$ 25. $\frac{1}{64}x^6 + 125y^6$

TERMS AND SYMBOLS

base (p. 41)   coefficient (p. 43)   like terms (p. 48)

exponent (p. 41)   degree of a term (p. 44)   factoring (p. 57)

power (p. 41)   degree of a polynomial

polynomial (p. 43)   (p. 44)

term (p. 43)   constant term (p. 44)

KEY IDEAS FOR REVIEW

☐ In a^n, a is the base and n is the exponent.

☐ $a^m \cdot a^n = a^{m+n}$

☐ A polynomial is a sum of terms of the form $3x^2$, $-5x^2y$, and so on.

☐ The degree of a term is the sum of the exponents of each variable.

☐ The degree of a polynomial is the degree of the term of highest degree.

☐ To add or subtract two polynomials, combine like terms.

☐ To multiply two polynomials, multiply one polynomial by each term of the other and add the products.

☐ $(a + b)^2 = a^2 + 2ab + b^2$
$(a - b)^2 = a^2 - 2ab + b^2$
$(a + b)(a - b) = a^2 - b^2$

☐ Factoring a polynomial means that we write it as a product of polynomials of lower degree.

☐ $a^2 - b^2 = (a + b)(a - b)$
$a^3 + b^3 = (a + b)(a^2 - ab + b^2)$
$a^3 - b^3 = (a - b)(a^2 + ab + b^2)$

COMMON ERRORS

1. *Don't* write

$$(2x + 3) - (x + 3) = 2x + 3 - x \; \oplus \; 3 = x + 6.$$

This is probably the most common and persistent error made by algebra students! You must use the distributive law to give

$$(2x + 3) - (x + 3) = 2x + 3 - x \; \ominus \; 3 = x$$

2. The notation $a \div b$ is read "a divided by b" and is equivalent to $\dfrac{a}{b}$, *not* $\dfrac{b}{a}$.

PROGRESS TEST 2A

1. In $\left(-\dfrac{1}{5}\right)^4$, what is the base? What is the exponent?

2. Find the degree of the polynomial $-3x^5 + 2x^3 + xy^2 - 2$.

3. Find the value of the polynomial $3x^2y + 2xy - 3x + 1$ when $x = -2, y = 3$.

4. Find a polynomial giving the volume of a cube of side s.

5. Add the polynomials $2x^3 + 3x^2 + 1$ and $-5x^3 + x - 3$.

6. Simplify $x - 3x^2y + 2y + 1 + 3y - 2x + 5x^2y + y - 3$.

7. Find $(2x^3 + x^2 - 2x + 1) - (3x^3 + 4x - 2)$.

8. Subtract $2xy^2 + x^2 - y + 2$ from $3x^2y + 5xy^2 + 3y$.

9. Multiply mentally $(2x - 5y)^2$.

10. Multiply $(x^2 + 2)(3x^2 - 2x + 5)$.

11. Factor $x^2 + 4x - 12$.

12. Factor completely $2x^2y - 8xy + 6y$.

13. Factor $4a^2 - 49$.

14. Factor $3x^2 - 13xy - 10y^2$.

15. Factor $\dfrac{x^3}{125} - 125y^3$.

PROGRESS TEST 2B

1. State the base and exponent in $\left(-\dfrac{2}{3}\right)^5$.

2. Find the degree of the polynomial $2x^4 - x^3y^2 + 4y^3 + 7$.
3. Evaluate the polynomial $-x^3 + 2x^2y - y + 1$ when $x = -2, y = -1$.
4. Find a polynomial giving the area of a square of side s, less the area of an isosceles right triangle of side s.
5. Add the polynomials $6x^5 - x^3 + x^2$ and $4x^4 + 3x^3 - 2x^2 + 1$.
6. Simplify $2x^2 - 3x^2y + y^2 - 7 + 4x^2y - 4y^2 + 4x^2 - 3$.
7. Find $(-3x^3 - 2x^2 + x - 3) - (2x^3 - 4x^2 - 2x + 5)$.
8. Subtract $2x^2y^2 - 3xy^2 - 2$ from $6x^2y^2 - 4xy^2 + xy$.
9. Multiply mentally $(3x - 4)^2$.
10. Multiply $(x + y)(2x - 3y - 2)$.
11. Factor $r^2 + 9r + 14$.
12. Factor completely $3x^2 + 6x + 3$.
13. Factor $(16y^2 - 64x^2)$.
14. Factor $3x^2 - 17x + 10$.

15. Factor $8a^3 + \dfrac{b^3}{8}$.

CHAPTER THREE
LINEAR EQUATIONS AND INEQUALITIES

Finding solutions to equations has long been a major concern of algebra. Recent work in inequalities, much of it post World War II, has elevated the importance of solving inequalities as well. Oil refining and steel producing are among the major industries using computers daily to solve problems involving thousands of inequalities. The solutions enable these companies to optimize their "product mix" and their profitability.

In this chapter we will learn to solve the most basic forms of equations and inequalities. But even this rudimentary capability will prove adequate to allow us to tackle a wide range of applications in both this and the following chapter.

LINEAR EQUATIONS IN ONE VARIABLE

Here are some examples of equations in the variable x.

$$x - 2 = 0 \qquad x^2 - 9 = 0$$

$$3(2x - 5) = 3 \qquad 2x + 5 = x - 7$$

$$\frac{1}{2x + 3} = 5 \qquad x^3 - 3x^2 = 32$$

An **equation** states that two algebraic expressions are equal. The expression to the left of the equal sign is called the **left-hand side** of the equation while the expression to the right of the equal sign is called the **right-hand side.**

Our task is to find values of the variable for which the equation holds true. These values are called **solutions** or **roots** of the equation and the set of all solutions is called the **solution set.** For example, the equation

$$x - 5 = 3$$

is a true statement only when $x = 8$. Then 8 is a solution of the equation and $S = \{8\}$ is the solution set.

When we say that we want to "solve an equation" we mean that we want to find all the solutions or roots. If we can replace an equation by another, simpler, equation that has the same roots, we will have an approach to solving equations. Equations having the same roots are called **equivalent equations.** There are two important rules that allow us to replace an equation by an equivalent equation.

Simplifying Equations

The solutions of a given equation are not affected by the following operations:

Addition or subtraction of a number or expression to both sides of the equation.

Multiplication or division of both sides of the equation by a number other than 0.

Let's apply these rules to the equation

$$2x + 5 = 13$$

Since we want to isolate x, it seems reasonable to get rid of the $+5$. This we can do by *subtracting* $+5$ *from both sides* of the equation.

$$2x + 5 - 5 = 13 - 5$$
$$2x + 0 = 8$$
$$2x = 8$$

We can find x by eliminating the 2 in $2x$. Since the 2 and x are tied by multiplication, we can get rid of the 2 by *dividing both sides* by 2.

$$\frac{2x}{2} = \frac{8}{2}$$
$$x = 4$$

We have arrived at the solution: $x = 4$. It's a good idea to check that 4 does indeed satisfy the original equation.

$$2x + 5 = 13$$
$$2(4) + 5 \overset{?}{=} 13$$
$$13 \overset{\checkmark}{=} 13$$

Then 4 is a root or solution of the equation $2x + 5 = 13$.

EXAMPLE 1

Solve $3x - 1 = x + 9$.

We gather the terms involving x on one side of the equation and the constant terms on the other. Here are the steps.

$$3x - 1 = x + 9$$
$$3x - 1 + 1 = x + 9 + 1 \qquad \text{Add 1 to both sides.}$$
$$3x = x + 10 \qquad \text{Combine like terms.}$$
$$3x - x = x + 10 - x \qquad \text{Subtract } x \text{ from both sides.}$$
$$2x = 10 \qquad \text{Combine like terms.}$$

Now it's easy to solve for x.

$$\frac{2x}{2} = \frac{10}{2} \qquad \text{Divide both sides by 2.}$$
$$x = 5$$

Check:

$$3(5) - 1 \overset{?}{=} 5 + 9$$
$$15 - 1 \overset{?}{=} 5 + 9$$
$$14 \overset{\checkmark}{=} 14$$

PROGRESS CHECK 1

Solve and check.

(a) $4x + 7 = 3$ (b) $x - 6 = 5x - 26$

Answers

(a) -1 (b) 5

EXAMPLE 2

Solve $\dfrac{5}{6}x - \dfrac{4}{3} = \dfrac{3}{5}$.

The L.C.D. of all fractions appearing in the equation is 30. We multiply both sides of the equation by the L.C.D. to clear the equation of fractions.

$$30\left(\frac{5}{6}x - \frac{4}{3}\right) = 30\left(\frac{3}{5}\right)$$
$$30\left(\frac{5}{6}x\right) + 30\left(-\frac{4}{3}\right) = 30\left(\frac{3}{5}\right) \qquad \text{Distributive law}$$
$$25x - 40 = 18$$
$$25x = 58 \qquad \text{Add 40 to both sides.}$$
$$x = \frac{58}{25} \qquad \text{Divide both sides by 25.}$$

Verify that $\dfrac{58}{25}$ is a solution of the original equation!

PROGRESS CHECK 2

Solve and check.

(a) $-\frac{2}{3}(x - 5) = \frac{3}{2}(x + 1)$ (b) $\frac{1}{3}x + 2 - 3\left(\frac{x}{2} + 4\right) = 2\left(\frac{x}{4} - 1\right)$

Answers

(a) $\frac{11}{13}$ (b) $-\frac{24}{5}$

The equations we have solved are all of the first degree and involve only one variable. Such equations are called **first-degree equations in one variable,** or more simply, **linear equations.** By rearranging and collecting like terms, any such equation can be put into the general form

$$ax + b = 0$$

where a and b are any real numbers and $a \neq 0$. Let's see how we would solve this equation.

$$ax + b = 0$$
$$ax + b - b = 0 - b \qquad \text{Subtract } b \text{ from both sides.}$$
$$ax = -b$$
$$\frac{ax}{a} = \frac{-b}{a} \qquad \text{Divide both sides by } a.$$
$$x = -\frac{b}{a}$$

What is interesting to us about this result is that it demonstrates the following principle:

Roots of a Linear Equation

The linear equation $ax + b = 0$, $a \neq 0$, has exactly one solution: $-\dfrac{b}{a}$.

EXERCISE SET 3.1

In the following exercises determine whether the given statement is true (T) or false (F).

1. $x = 2$ is a solution to $3x = 6$.

2. $x = -2$ is a solution to $4x = -6$.

3. $x = 3$ is a solution to $3x - 1 = 10$.

4. $x = -5$ is a solution to $2x + 3 = -7$.

5. $x = \frac{3}{2}$ is a solution to $2x + 1 = 4$.

6. $x = \frac{5}{2}$ is a solution to $3x - 4 = \frac{5}{2}$.

7. $x = \dfrac{6}{4-k}$ is a solution to $kx + 6 = 4x$.

8. $x = \dfrac{7}{3k}$ is a solution to $2kx + 7 = 5x$.

Solve the given linear equation and check your answer in the following exercises.

9. $2x = 8$

10. $3x = -6$

11. $2x = -\dfrac{5}{2}$

12. $2x + 3 = 7$

13. $3x + 5 = -1$

14. $5r + 10 = 0$

15. $3s - 1 = 2$

16. $4 - x = 2$

17. $2 - 3a = 6$

18. $2 = 3x + 4$

19. $3 = 2x - 1$

20. $\dfrac{1}{2}s + 2 = 4$

21. $\dfrac{3}{2}t - 2 = 7$

22. $-\dfrac{2}{3}x + 3 = 2$

23. $-1 = -\dfrac{2}{3}x + 1$

24. $0 = -\dfrac{1}{2}a - \dfrac{2}{3}$

25. $2x + 2 = x + 6$

26. $4r + 3 = 3r - 2$

27. $-5x + 8 = 3x - 4$

28. $2x - 1 = 3x + 2$

29. $-2x + 6 = -5x - 4$

30. $6x + 4 = -3x - 5$

31. $2(3b + 1) = 3b - 4$

32. $-3(2x + 1) = -8x + 1$

33. $4(2x - 1) = 5x + 5$

34. $-3(3x - 1) = -4x - 2$

35. $4(x - 1) = 2(x + 3)$

36. $-3(x - 2) = 2(x + 4)$

37. $2(x + 4) - 1 = 0$

38. $3a + 2 - 2(a - 1) = 3(2a + 3)$

39. $-2(x - 1) + 3(x - 1) = 4(x + 5)$

40. $2(y - 1) + 3(y + 2) = 8$

41. $-4(2x + 1) - (x - 2) = -11$

42. $3(a + 2) - 2(a - 3) = 0$

Solve for x.

43. $kx + 8 = 5x$

44. $8 - 2kx = -3x$

45. $2 - k + 5(x - 1) = 3$

46. $3(2 + 3k) + 4(x - 2) = 5$

3.2
APPLICATIONS

Many applied problems lead to linear equations that must be solved. In this section we will apply our technique for solving a linear equation to a number of applied word problems. If these word problems prove somewhat troublesome at this point, you will find the following chapter, which is devoted exclusively to the analysis of word problems, to be helpful.

EXAMPLE 1
In planning a party for a class of 30 students it is calculated that the total expense will be $75. How much must each student contribute to the expense fund?

Solution

Although this problem can easily be solved mentally, we want to set it up using algebra. The first question to ask is: What do I want to find? It's reasonable to represent the unknown by a variable, so let

$$x = \text{the contribution of each student (in dollars)}$$

Next, we must seek a relationship involving this variable. Since

$$\text{total income} = (\text{number of students}) \cdot (\text{contribution per student})$$

we can write

$$75 = 30x$$

But this is an equation we know how to solve.

$$\frac{75}{30} = \frac{30x}{30}$$

$$2.5 = x$$

Each student must contribute $2.50.

PROGRESS CHECK 1

A family allocates $25 of its budget for purchasing ground beef. If ground beef sells for $1.25 per pound, how many pounds can be purchased?

Answer
20 pounds

EXAMPLE 2

30 is 40% of what number?

Solution

We begin by defining the unknown quantity.

$$\text{Let} \quad x = \text{the desired number}$$

From our work on percent, we know that the phrase "40% of a number" means we are to multiply the number by 40% or 0.4. Then

$$30 = 0.4x$$

$$\frac{30}{0.4} = \frac{0.4x}{0.4}$$

$$75 = x$$

PROGRESS CHECK 2

If you pay $63 for a car radio after receiving a 30% discount, what was the price of the radio before the discount?

Answer
$90

EXAMPLE 3

Find three consecutive integers whose sum is 54.

Solution
Let's have the unknown represent the first of the three consecutive integers. Then

$$n = \text{the first integer}$$

$$n + 1 = \text{the second integer}$$

and

$$n + 2 = \text{the third integer}$$

But the sum of these three integers is to be 54.

$$n + (n + 1) + (n + 2) = 54$$
$$3n + 3 = 54$$
$$3n = 51$$
$$n = 17$$

Then 17, 18, and 19 are the three consecutive integers we seek.

PROGRESS CHECK 3
Find three consecutive integers whose sum is 78.

Answer
25, 26, 27

WARNING
Don't be fooled into thinking that every problem has a solution. If you try to find three consecutive integers whose sum is 23 you will have a very frustrating experience—there simply aren't any.

EXAMPLE 4
John has taken two quizzes in algebra and has received scores of 90 and 96. What must he receive as a grade on his third quiz to achieve an average of 92?

Solution
We let the unknown represent the third quiz score.

$$\text{Let } x = \text{score on the third quiz}$$

Since

$$\text{average} = \frac{\text{sum of scores}}{\text{number of scores}}$$

we must have

$$92 = \frac{90 + 96 + x}{3}$$

We then multiply both sides of the equation by 3, the denominator of the fraction. This will clear the equation of fractions.

$$(92)(3) = 90 + 96 + x$$
$$276 = 186 + x$$
$$90 = x$$

John must achieve a grade of 90 on his third quiz.

PROGRESS CHECK 4

A golf pro has an average of 71 in the last four tournaments. If the scores were 68, 70, and 72 in three of the events, what was the score in the fourth tournament?

Answer
74

EXAMPLE 5

A right triangle whose area is 12 square feet has a base which measures 3 feet. What is the length of the other side?

Solution

It's a good idea to draw a figure for problems of this type (see Figure 3.1). We have labeled the figure to indicate that the base measures 3 feet and the height is our unknown x. Since

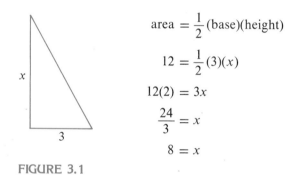

$$\text{area} = \frac{1}{2}(\text{base})(\text{height})$$

$$12 = \frac{1}{2}(3)(x)$$

$$12(2) = 3x$$

$$\frac{24}{3} = x$$

$$8 = x$$

FIGURE 3.1

Thus the length of the other side is 8 feet.

PROGRESS CHECK 5

A carpenter is instructed to build a rectangular room with a perimeter of 18 meters. If one side of the room measures 5 meters, what are the dimensions of the room?

Answer
4 meters by 5 meters

LITERAL EQUATIONS

The circumference C of a circle is given by the formula

$$C = 2\pi r$$

where r is the radius of the circle. For every value of r, the formula gives us a value of C. If $r = 20$, we have

$$C = 2\pi(20) = 40\pi$$

It is sometimes convenient to be able to turn a formula around in order to solve for a different variable. For example, if we want to express the radius of a circle in terms of the circumference, we have

$$C = 2\pi r$$

$$\frac{C}{2\pi} = \frac{2\pi r}{2\pi}$$

$$\frac{C}{2\pi} = r$$

Now, given a value of C we can determine the value of r.

EXAMPLE 6

If an amount P is invested at the simple annual interest rate r, then the amount A available at the end of t years is

$$A = P + Prt$$

Solve for P.

Solution

$$A = P + Prt$$

$A = P(1 + rt)$ Common factor P

$\dfrac{A}{1 + rt} = \dfrac{P(1 + rt)}{1 + rt}$ Divide both sides by $1 + rt$.

$\dfrac{A}{1 + rt} = P$ Cancel factor $1 + rt$.

PROGRESS CHECK 6

Solve the equation $d = v + 0.5at$ for t.

Answer

$$t = \frac{2d - 2v}{a}$$

EXERCISE SET 3.2

1. A vacation club charters an airplane to carry its 200 members to Rome. If the airline charges the club $50,000 for the round trip, how much will each member pay for the trip?
2. Suppose that a camp director needs to provide each camper with a small portable radio that sells for $4. The director has been given $900 for the purchase of the radios. How many campers will receive a radio?
3. How many $7 transistor radios can a dealer buy with $350?
4. Suppose that 12.5 gallons of special fuel are required to fill the tank of a tractor. If it costs $7.50 to fill the tank, what is the price per gallon?
5. If a dozen rolls cost $1.02, what is the price per roll?
6. A school district tries to give each of 110 typewriters a routine maintenance check, which costs $12 per typewriter once a year. This year the school has $960 on hand for the typewriter maintenance program. Will all typewriters be checked? If not, how many will not be checked?

7. 78 is 30% of what number?
8. 24 is 60% of what number?
9. 96 is 120% of what number?
10. 2 is $\frac{1}{2}$% of what number?
11. A car dealer advertises a $5400 sedan at a "30% discount" for $4000. Is the dealer telling the truth?
12. The local discount store sells a camera for $180, which is a 25% discount from the suggested retail price. What is the suggested retail price?
13. A stationery store sells a dozen ballpoint pens for $3.84, which represents a 20% discount from the price charged when less than a dozen pens are bought at any one time. How much does it cost to buy three pens?
14. A copying service advertises as follows: " cents per copy for 4 or fewer copies per page; 12 cents per copy, a 20% reduction, for 5 or more copies per page." Since the typesetter forgot to set the original prediscount price per copy, calculate it.
15. The sum of a certain number and 4 is 12. Find the number.
16. The difference of a certain number and 8 is 16. Find the number.
17. The difference of 24 and a certain number is 14. Find the number.
18. Find a number which when subtracted from 4 times itself yields 36.
19. Find a number which when added to 3 times itself gives 24.
20. Find three consecutive integers whose sum is 21.
21. A certain number is 3 more than another number. If their sum is 21, find the numbers.
22. A certain number is 5 less than another number. If their sum is 11, find the numbers.
23. A certain number is 2 more than 3 times another. If their sum is 14, find the numbers.
24. A certain number is 5 less than twice another. If their sum is 19, find the numbers.
25. A resort guarantees that the average temperature over the period Friday, Saturday, and Sunday will be exactly 80°F, or else each guest only pays half price for the facilities. If the temperatures on Friday and Saturday were 90°F and 82°F, respectively, what must the temperature be on Sunday so that the resort does not lose half of its revenue?
26. A patient's temperature was taken at 6 A.M., 12 noon, 3 P.M. and 8 P.M. The first, third, and fourth readings were 102.5, 101.5, and 102°F, respectively. The nurse forgot to write down the second reading but remembered that the average of the four readings was 101.5°F. What was the second temperature reading?
27. Suppose that an investor buys 100 shares of stock on each of four successive days at $10 per share on the first day, $10.50 per share on the second day, and $12 per share on the fourth day. If the average price of the stock is $11.20 per share, what was the price per share on the third day?
28. In an election for president of a local volunteer organization there were 84 votes cast. If candidate A received 24 votes more than candidate B, how many votes did each candidate receive?
29. A 12-meter-long steel beam is to be cut into two pieces so that one piece will be 4 meters longer than the other. How long will each piece be?
30. A rectangular grazing field whose length is 10 meters longer than its width is to be enclosed with 100 meters of fencing material. What are the dimensions of the field?
31. The perimeter of a rectangle is 36 meters. If the width is 3 times the length, find the dimensions.

32. The length of a rectangle is 4 meters more than 3 times its width. If the perimeter is 24 meters, find the dimensions.
33. A triangle whose area is 36 square centimeters has an altitude that measures 8 centimeters. What is the length of the base?
34. The perimeter of an isosceles triangle is 32 centimeters. The two equal sides are each 2 centimeters shorter than the base. Find the dimensions of each side.
35. In an isosceles triangle, the two equal angles are each 15° more than the third angle. Find the measure of each angle. (Recall that the sum of the angles of a triangle is 180°.)
36. An investor invested $4000 at 5% per year. How much additional money should be invested at 8% per year so that the total invested will pay 6% per year?
37. Solve the equation $A = bt + c$ for t.
38. Solve the equation $S = 3abt + 2ab$ for b.
39. Solve the equation $5Av + 3bvt + 2kt = 0$ for v.

3.3
LINEAR INEQUALITIES

To solve an inequality such as

$$2x + 5 > x - 3$$

means to find its solution set, that is, to find *all* values of x that make it true. We need to know what operations we can perform on inequalities to simplify the expressions and allow us to isolate the variable.

Let's see if we can deduce the rules for inequalities. If we begin with

$$8 > 3$$

and add a positive number, say $+12$, to both sides, we have

$$8 + 12 \overset{?}{>} 3 + 12$$

$$20 > 15$$

Similarly, if we add a negative number, say -4, to both sides, we have

$$8 + (-4) \overset{?}{>} 3 + (-4)$$

$$4 > -1$$

We can say that

Any number can be added to or subtracted from both sides of an inequality without affecting the inequality.

Now let's see what happens when we multiply both sides of the inequality

$$8 > 3$$

by a positive number, say $+6$. We see that

$$8 \cdot 6 \overset{?}{>} 3 \cdot 6$$

$$48 > 18$$

If we multiply both sides by a negative number, say -4, we have

$$(8)(-4) \overset{?}{>} (3)(-4)$$

$$-32 < -12$$

Look at what happened! Multiplication by a negative number changed the direction of the inequality sign. The rules we must follow are:

Multiplying or dividing both sides of an inequality by a positive number doesn't affect the inequality.

Multiplying or dividing both sides of an inequality by a negative number reverses the direction of the inequality.

We can summarize the rules for handling inequalities in this way:

The same operations can be performed with inequalities as with equations, except that multiplication or division by a *negative* number reverses the inequality.

EXAMPLE 1

Solve the inequality and graph the solution set.

$$2x + 5 \geq x - 3$$

Solution

We will perform addition and subtraction to collect terms in x just as we did for equations.

$$2x + 5 \geq x - 3$$
$$2x + 5 - 5 \geq x - 3 - 5$$
$$2x \geq x - 8$$
$$2x - x \geq x - 8 - x$$
$$x \geq -8$$

The graph of the solution set consists of -8 and all points to the right of -8.

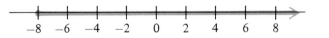

PROGRESS CHECK 1

Solve the inequality and graph the solution set.

$$3x - 2 \geq 2x + 4$$

Answer

$x \geq 6$

EXAMPLE 2
Solve the inequality $5x < 2(x - 1)$ and graph the solution set.

Solution
We proceed just as if we were dealing with an equation.

$$5x < 2(x - 1)$$
$$5x < 2x - 2$$
$$3x < -2$$

To solve for x we must divide by $+3$. Our rules say we may divide by a positive number without affecting the direction of the inequality.

$$\frac{3x}{3} < \frac{-2}{3}$$
$$x < -\frac{2}{3}$$

The graph of the solution set looks like this:

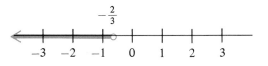

PROGRESS CHECK 2
Solve the inequality $8x + 2 \le 3(x - 1)$ and graph the solution set.

Answer

$x \le -1$

WARNING
Don't write

$$-2x \ge -4$$
$$x \le -2$$

Division by a negative number changes the direction of the inequality. But the signs obey the usual rules of algebra. Thus,

$$\text{if} \quad -2x \ge -4$$
$$\text{then} \quad x \le 2$$

Word problems can also result in a linear inequality. Here is an example.

EXAMPLE 3

A taxpayer may choose to pay a 20% tax on the gross income or to pay a 25% tax on the gross income less $4000. Above what income level should the taxpayer elect to pay at the 20% rate?

Solution

If we let x = gross income, then the choice available to the taxpayer is
(a)   pay at the 20% rate on the gross income, that is, pay $0.20x$, or
(b)   pay at the 25% rate on the gross income less $4000, that is, pay $0.25(x - 4000)$.
To determine when (a) produces a lower tax than (b), we must solve

$$0.20x \le 0.25(x - 4000)$$
$$0.20x \le 0.25x - 1000$$
$$-0.05x \le -1000$$

This time we must divide by -0.05. Our rule says that division by a negative number will change the direction of the inequality so that $\le$ becomes $\ge$. Thus,

$$\frac{-0.05x}{-0.05} \ge \frac{-1000}{-0.05}$$
$$x \ge 20,000$$

The taxpayer should choose to pay at the 20% rate if the income exceeds $20,000.

PROGRESS CHECK 3

A customer is offered the following choice of telephone services: unlimited local calls at a $20 monthly charge or a base rate of $8 per month plus 6 cents per message unit. When does it cost less to choose the unlimited service?

Answer
When the anticipated use exceeds 200 units.

We can solve double inequalities such as

$$1 < 3x - 2 \le 7$$

by operating on both inequalities at the same time.

$$3 < 3x \le 9 \qquad \text{Add } +2 \text{ to each member.}$$
$$1 < x \le 3 \qquad \text{Divide each member by 3.}$$

EXAMPLE 4

Solve the inequality.

$$-3 \le 1 - 2x < 6$$

Operating on both inequalities we have

$$-4 \le -2x < 5 \qquad \text{Add } -1 \text{ to each member.}$$
$$2 \ge x > -\frac{5}{2} \qquad \text{Divide each member by } -2.$$

PROGRESS CHECK 4

Solve the inequality $-5 < 2 - 3x < -1$.

Answer

$$\frac{7}{3} > x > 1$$

EXERCISE SET 3.3

Select the values of x that satisfy the given inequality in the following exercises.

1. $x < 3$ (a) 4   (b) 5   (c) -2 (d) 0   (e) 1.2
2. $x > 4$ (a) 8   (b) 4   (c) 6   (d) -3 (e) 9.1
3. $x \le 5$ (a) 3   (b) 7   (c) 5   (d) 4.3   (e) -5
4. $x \ge -1$ (a) 0   (b) -4 (c) 1   (d) -2 (e) -1

Solve the given inequality and graph the result in the following exercises.

5. $x + 4 < 8$
6. $x + 5 < 4$
7. $x + 3 < -3$
8. $x - 2 \le 5$
9. $x - 3 \ge 2$
10. $x + 5 \ge -1$
11. $2 < a + 3$
12. $-5 > b - 3$
13. $2y < -1$
14. $3x < 6$
15. $2x \ge 0$
16. $-\frac{1}{2}y \ge 4$
17. $2r + 5 < 9$
18. $3x - 2 > 4$
19. $3x - 1 \ge 2$
20. $4x + 3 \le 11$
21. $\frac{1}{2}y - 2 \le 2$
22. $\frac{3}{2}x + 1 \ge 4$
23. $3 \le 2x + 1$
24. $4 \ge 3b - 2$
25. $-3x - 2 \le 4$
26. $-5x + 2 > -8$
27. $4(2x + 1) < 16$
28. $3(3r - 4) \ge 15$
29. $2(x - 3) < 3(x + 2)$
30. $4(x - 3) \ge 3(x - 2)$
31. $3(2a - 1) > 4(2a - 3)$
32. $2(3x - 1) + 4 < 3(x + 2) - 8$
33. $3(x + 1) + 6 \ge 2(2x - 1) + 4$
34. $4(3x + 2) - 1 \le -2(x - 3) + 15$
35. $-2 < 4x \le 5$
36. $3 \le 6x < 12$
37. $4 < -3x < 10$
38. $-5 < -2x < 9$
39. $-4 \le 2x + 2 \le -2$
40. $5 \le 3x - 1 \le 11$
41. $3 \le 1 - 2x < 7$
42. $5 < 2 - 3x \le 11$
43. $-8 < 2 - 5x \le 7$
44. $-10 < 5 - 2x < -5$

45. You can rent a compact car from firm A for $160 per week with no charge for mileage, or from firm B for $100 per week plus 20 cents for each mile driven. At what mileages does it cost less to rent from firm A?

46. An appliance salesperson is paid $30 per day plus $25 for each appliance sold. How many appliances must be sold for the salesperson's income to exceed $130 per day?

47. A pension trust invests $6000 in a bond that pays 5% interest per year. It also wishes to invest additional funds in a more speculative bond paying 9% interest per year so that the return on the total investment will be at least 6%. What is the minimum amount that must be invested in the more speculative bond?

48. A book publisher spends $19,000 on editorial expenses and $6 per book for manufacturing and other expenses in the course of publishing a psychology textbook. If the book sells for $12.50, how many copies must be sold to show a profit?

3.4
ABSOLUTE VALUE IN EQUATIONS AND INEQUALITIES

We repeat the definition of absolute value given in Chapter 1.

$$|x| = \begin{cases} x & \text{if } x \geq 0 \\ -x & \text{if } x < 0 \end{cases}$$

For example,

$$|5| = 5$$
$$|-5| = 5$$
$$|0| = 0$$

Let's apply the definition of absolute value to solving the equation

$$|x - 3| = 5$$

We have two cases.

Case 1 $x - 3 \geq 0$

Then, using the definition,

$|x - 3| = x - 3 = 5$

$\qquad\qquad x = 8$

Case 2 $x - 3 < 0$

Then, using the definition,

$|x - 3| = -(x - 3) = 5$

$\qquad -x + 3 = 5$

$\qquad\qquad\qquad x = -2$

It's a good idea to check the answers by substituting.

$|8 - 3| \overset{?}{=} 5$

$|5| \overset{?}{=} 5$

$5 \overset{\checkmark}{=} 5$

$|-2 - 3| \overset{?}{=} 5$

$|-5| \overset{?}{=} 5$

$5 \overset{\checkmark}{=} 5$

EXAMPLE 1
Solve the equation $|2x - 7| = 11$.

Solution
We have to solve two equations.

$2x - 7 = 11 \qquad \text{and} \qquad -(2x - 7) = 11$

$\qquad 2x = 18 \qquad\qquad\qquad -2x + 7 = 11$

$\qquad\quad x = 9 \qquad\qquad\qquad\qquad x = -2$

PROGRESS CHECK 1
Solve and check.

(a) $|x + 8| = 9$ (b) $|3x - 4| = 7$

Answers

(a)   *1, −17*   (b) $\dfrac{11}{3}, -1$

To solve inequalities involving absolute value, we recall that $|x|$ is the distance between the origin and x on the real number line. We can then easily graph the solution set for each of the inequalities $|x| < a$ and $|x| > a$.

$|x| < a$

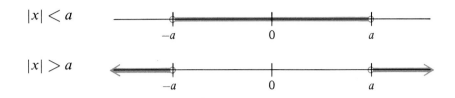

$|x| > a$

We can summarize the result this way:

For a given positive number a,

$|x| < a$ is equivalent to $-a < x < a$

$|x| > a$ is equivalent to $x > a$ or $x < -a$

EXAMPLE 2
Solve $|2x - 5| \leq 7$ and graph the solution set.

Solution
We must solve the equivalent double inequality.

$$-7 \leq 2x - 5 \leq 7$$

$$-2 \leq 2x \leq 12 \qquad \text{Add } +5 \text{ to each member.}$$

$$-1 \leq x \leq 6 \qquad \text{Divide each member by 2.}$$

The graph of the solution set is then

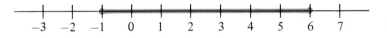

PROGRESS CHECK 2
Solve, and graph the solution set.

(a) $|x| < 3$ (b) $|3x - 1| \leq 8$ (c) $|x| < -2$

Answers

(a) $-3 < x < 3$

(b) $-\dfrac{7}{3} \leq x \leq 3$

(c) *No solution. Since $|x|$ is always nonnegative, $|x|$ cannot be less than -2.*

EXAMPLE 3

Solve $|2x - 6| > 4$ and graph the solution set.

Solution

We must solve the equivalent inequalities.

$$2x - 6 > 4 \qquad \text{or} \qquad 2x - 6 < -4$$
$$2x > 10 \qquad\qquad\qquad 2x < 2$$
$$x > 5 \qquad\qquad\qquad\quad x < 1$$

The graph of the solution set is then

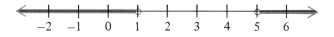

PROGRESS CHECK 3

Solve, and graph the solution set.

(a) $|5x - 6| > 9$ (b) $|2x - 2| \geq 8$

Answers

(a) $x < -\dfrac{3}{5}, \quad x > 3$

(b) $x \leq -3, \quad x \geq 5$

WARNING

Don't write

$$1 > x > 5$$

When written this way, the notation requires that x be simultaneously less than 1 and greater than 5, which is impossible. Write this as

$$x < 1 \quad \text{or} \quad x > 5$$

The answer must always be written this way when the graph consists of disjoint segments.

EXERCISE SET 3.4

1. Which of the following are solutions to $|x - 3| = 5$?

    (a) -8 (b) 8   (c) 2   (d) -2 (e) none of these

2. Which of the following are solutions to $|2x + 5| = 6$?

    (a) $\dfrac{1}{2}$ (b) $-\dfrac{1}{2}$ (c) $-\dfrac{11}{2}$ (d) $\dfrac{11}{2}$ (e) none of these

3. Which of the following are solutions to $|3a + 5| < 20$?

(a) 5   (b) 4   (c) −10   (d) −8   (e) none of these

4. Which of the following are solutions to $|2b − 3| \leq 6$?

(a) −1   (b) 0   (c) 5   (d) 4   (e) −2

5. Which of the following are solutions to $|−3x + 2| > 11$?

(a) −3   (b) −4   (c) 5   (d) 2   (e) 4

6. Which of the following are solutions to $|3x − 5| \geq 7$?

(a) $-\dfrac{1}{2}$ (b) 4   (c) 1   (d) 6   (e) −1

7. Which of the following are solutions to $|x − 3| < −5$?

(a) −2   (b) 1   (c) 8   (d) 0   (e) none of these

8. Which of the following are solutions to $|3x + 1| > −2$?

(a) $\dfrac{1}{3}$ (b) −1   (c) 0   (d) 1   (e) all of these

Solve and check in the following exercises.

9. $|x + 2| = 3$ 10. $|x − 3| = 5$ 11. $|r − 5| = \dfrac{1}{2}$

12. $|2r − 4| = 2$ 13. $|2x + 1| = 3$ 14. $|3x − 1| = 5$

15. $|3y − 2| = 4$ 16. $|5y + 1| = 11$ 17. $|−3x + 1| = 5$

18. $|−4x − 3| = 9$ 19. $|2t + 2| = 3$ 20. $|2t + 2| = 0$

Solve and graph in the following exercises.

21. $|x| < 5$ 22. $|x| \leq 3$ 23. $|x| > 4$

24. $|x| \geq 8$ 25. $|x| > −3$ 26. $|x| > 0$

27. $|x + 3| < 5$ 28. $|x − 2| \leq 4$ 29. $|x + 1| > 3$

30. $|x + 2| > −3$ 31. $|x − 3| \geq 4$ 32. $|2x + 1| < 5$

33. $|3x + 6| \leq 12$ 34. $|4x − 1| > 3$ 35. $|3x + 2| \geq −1$

36. $|2x + 3| \geq 7$ 37. $|1 − 2x| \leq 3$ 38. $\left|\dfrac{1}{3} − x\right| < \dfrac{2}{3}$

39. $|1 − 3x| > 4$ 40. $|1 + 2x| > 0$ 41. $\left|\dfrac{1}{2} + x\right| > \dfrac{1}{2}$

42. $|1 − 2x| < 0$ 43. $\left|\dfrac{x − 1}{2}\right| < 3$ 44. $\dfrac{|2x + 1|}{3} < 0$

45. $\dfrac{|2x − 1|}{4} < 2$ 46. $\dfrac{|3x + 2|}{2} < 4$

47. A machine that packages 100 vitamin pills per bottle can make an error of 2 pills per bottle. If x is the number of pills in a bottle, write an inequality using absolute value that indicates a maximum error of 2 pills per bottle. Solve the inequality.

48. The weekly income of a worker in a manufacturing plant differs from $300 by no more than $50. If x is the weekly income, write an inequality using absolute value that expresses this relationship. Solve the inequality.

TERMS AND SYMBOLS

equation (p. 69)	equivalent equation	literal equation (p. 76)
left-hand side (p. 70)	(p. 70)	linear inequality (p. 79)
right-hand side (p. 70)	linear equation (p. 72)	absolute value (p. 84)
solution (p. 70)	first-degree equation	
root (p. 70)	(p. 72)	
solution set (p. 70)	formula (p. 76)	

KEY IDEAS FOR REVIEW

☐ Solutions of an equation are found by changing the equation into a succession of simpler equivalent equations which have the same roots.

☐ The linear equation $ax + b = 0$, $a \neq 0$, has exactly one solution.

☐ Linear inequalities are solved in a manner very similar to linear equations except that multiplication or division by a negative number reverses the direction of the inequality.

☐ Linear equations and inequalities involving absolute value can be solved using the definition of absolute value.

☐ Assuming $a > 0$, the solution set to the inequality $|x| < a$ is the interval $-a < x < a$, whereas the solution set to the inequality $|x| > a$ consists of two disjoint intervals.

COMMON ERRORS

1. When multiplying an equation or inequality by a constant, remember to multiply both sides by the constant. This requires that each term of each side be multiplied by the constant, and the constant must never be zero.

2. When multiplying or dividing an inequality by a negative number, remember to change the direction of the inequality.
   Write

$$-3x \leq 6$$
$$x \geq -2$$

   Don't write

$$-3x \leq 6 \quad \text{or} \quad -3x \leq 6$$
$$x \geq 2 \quad\quad\quad\quad x \leq -2$$

   Both of these are wrong!

3. Inequalities of the form

$$|x - 4| \geq 9$$

   will result in the two disjoint segments, $x \leq -5$ or $x \geq 13$. This result must be written as shown; *don't* write $-5 \geq x \geq 13$, since this notation makes no sense.

PROGRESS TEST 3A

1. Solve and check: $4x - 6 = 9$.
2. Solve and check: $2(x - 2) = 3(2x + 4)$.
3. True or false: -1 is a root of $2x - 1 = 3x + 1$.
4. Solve for h: $V = \pi r^2 h$.
5. Solve for x: $-2x + 3 = 4 + kx$.
6. 28 is 40% of what number?

7. 8 is $\frac{1}{4}$% of what number?

8. The length of a rectangle is 3 meters longer than its width. If the perimeter is 36 meters, find the dimensions of the rectangle.

9. Find three consecutive even integers whose sum is 48.

10. Part of a $5000 trust fund is invested in a mutual fund yielding 6% per year in dividends, and the balance in a corporate bond yielding 7% interest per year. If the total annual interest is $320, how much is invested in each?

11. Solve and graph: $3(2 - x) < 12$.

12. Solve and graph: $5(3x - 2) \geq 2(4 - 5x) + 7$.

13. Solve: $|2x - 2| = 5$.

14. Solve and graph: $|2 - x| \leq 12$.

15. Solve and graph: $|2x + 5| > 7$.

PROGRESS TEST 3B

1. Solve and check: $5x + 4 = -6$.

2. Solve and check: $-(x + 3) = 4(x - 7)$.

3. True or false: -2 is a root of $-3x - 5 = 2x - 3$.

4. Solve for b: $A = \frac{1}{2}h(b + c)$.

5. Solve for x: $2(3 - kx) = (5 - 2x)$.

6. 44 is 110% of what number?

7. 56 is 70% of what number?

8. The width of a rectangle is 1 centimeter less than twice its length. If the perimeter is 22 centimeters, find the dimensions of the rectangle.

9. A certain number is 4 less than another number. If their sum is 46, find the numbers.

10. An $8000 pension fund is invested in two parts yielding 5% and 8% interest per year, respectively. If the total annual interest is $520, how much is invested in each part?

11. Solve and graph: $5(4 - 2x) > 45$.

12. Solve and graph: $4(x + 2) \leq 3(2 - 3x) - 11$.

13. Solve: $|3x - 4| = 5$.

14. Solve and graph: $|3 - 2x| \geq 12$.

15. Solve and graph: $|3x - 1| < 2$.

CHAPTER FOUR
WORD PROBLEMS

Most students have had previous exposure to word problems. Many students have had some difficulty in handling such problems. This entire chapter is devoted to methods of attacking word problems and is designed to develop the ability to analyze words and to heighten confidence in solving word problems.

We begin by outlining procedures for converting words to algebra and applying these procedures to some problems. We will then examine some types of word problems that require special attention.

FROM WORDS TO ALGEBRA

The process of solving word problems is not unlike the role of a detective in solving a crime. The clues are there. Properly interpreted, they will lead to a solution. For the detective, the clues point to the criminal; for us, the clues point to an algebraic expression that we can solve.

These are the typical steps used in solving word problems.

Step 1. Read the problem until you understand what is required.

Step 2. Isolate what is known and what is to be found.

Step 3. In many problems, the unknown quantity answers questions such as "how much" or "how many." Let an algebraic symbol, say x, represent the unknown.

Step 4. Represent other quantities in the problem in terms of *x*.

Step 5. Find the relationship in the problem that lets you write an equation (or an inequality).

Step 6. Solve. Check your answer to see that it (a) satisfies the original question and (b) satisfies the equation (or inequality).

Some students have trouble with word problems because they are unfamiliar with the mathematical interpretation of certain words and phrases. Practice will, of course, help; we also suggest that you read the problem very carefully. Table 4.1 has a list of words and phrases you will come across, with examples of how they are used, which may be helpful.

TABLE 4.1

Word or phrase	Algebraic symbol	Example	Algebraic expression
Sum	$+$	Sum of two numbers	$a + b$
Difference	$-$	Difference of two numbers	$a - b$
		Difference of a number and 3	$x - 3$
Product	$\times$ or $\cdot$	Product of two numbers	$a \cdot b$
Quotient	$\div$ or $/$	Quotient of two numbers	$\dfrac{a}{b}$ or a/b
Exceeds		a exceeds b by 3	$a = b + 3$
More than		a is 3 more than b	or
More of		There are 3 more of a than of b	$a - 3 = b$
Twice		Twice a number	$2x$
		Twice the difference of x and 3	$2(x - 3)$
		3 more than twice a number	$2x + 3$
		3 less than twice a number	$2x - 3$
Is or equals	$=$	The sum of a number and 3 is 15.	$x + 3 = 15$

Let's apply our steps for analyzing word problems.

EXAMPLE 1

The sum of the ages of a man and his daughter is 40 years. Nineteen years from now, the man will be twice as old as his daughter will be then. Find the present ages of the man and his daughter.

Solution

After reading the problem, it is clear that we may choose the unknown to be the age of either the man or his daughter. If we

then

$\quad$ let $\;n =$ the current age of the man $\qquad$ Step 3

$\quad 40 - n =$ the current age of the daughter $\qquad$ Step 4

since the sum of their ages is 40. We have now represented all quantities in the problem in terms of *n*. The situation 19 years from now leads to an equation that is the relationship we seek.

$$\text{man's current age} + 19 = 2(\text{daughter's current age} + 19)$$
$$n + 19 = 2(40 - n + 19) \qquad \text{Step 5}$$
$$n + 19 = 2(59 - n)$$
$$n + 19 = 118 - 2n$$
$$3n = 99$$
$$n = 33 = \text{man's current age} \qquad \text{Step 6}$$
$$40 - n = 7 = \text{daughter's current age}$$

Now check that this solution satisfies the problem.

PROGRESS CHECK 1
Loren is 3 times as old as Jody. Ten years from now, Loren's age will exceed twice Jody's age by 2 years. What are the present ages of Loren and Jody?

Answer
Jody is 12 and Loren is 36.

EXAMPLE 2
The larger of two numbers is 1 more than the smaller. Five times the larger exceeds four times the smaller by 12. Find the numbers.

Solution
The unknown may represent either the larger or smaller number. If we

let $n = $ the smaller number

then

$$n + 1 = \text{the larger number}$$

The equation we need is

$$5 \cdot \text{larger} = 4 \cdot \text{smaller} + 12$$
$$5(n + 1) = 4n + 12$$
$$5n + 5 = 4n + 12$$
$$n = 7 = \text{the smaller number}$$
$$n + 1 = 8 = \text{the larger number}$$

Verify that the answer is correct.

PROGRESS CHECK 2
Write the number 30 as the sum of two numbers such that twice the larger is 3 less than 7 times the smaller.

Answer
7, 23

EXAMPLE 3
The length of a rectangle is 2 feet more than twice its width. If the perimeter is 22 feet, find the dimensions of the rectangle.

Solution

Since the length is expressed in terms of the width, we let (see Figure 4.1)

$$w = \text{the width of the rectangle}$$

Then we see that

$$\text{length} = 2(\text{width}) + 2 = 2w + 2$$

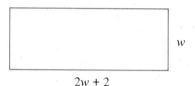

$$2w + 2$$

FIGURE 4.1

Since

$$\text{perimeter} = 2(\text{length} + \text{width})$$
$$22 = 2[(2w + 2) + w]$$
$$22 = 2(3w + 2)$$
$$22 = 6w + 4$$
$$18 = 6w$$
$$w = 3$$

and

$$l = 2(3) + 2 = 8$$

The rectangle has the dimensions 3 feet by 8 feet.

PROGRESS CHECK 3

A side of a triangle is 2 cm longer than the second side and 1 cm more than twice the third side. If the perimeter of the triangle is 20 cm, find the dimensions of the three sides.

Answer
4, 7, and 9 cm

EXERCISE SET 4.1

Translate from words to algebra in Exercises 1–10.

1. The sum of John's age and Mary's age is 39.
2. The difference in height between a large maple and a smaller maple is 4 meters.
3. The cost of a bag of n pencils if each pencil costs 8 cents.
4. The number of rolls of film you can buy with $12.40 if one roll of film costs x cents.
5. The average speed of the train is 20 miles per hour more than the average speed of the car.
6. The amount invested in a stock is twice the amount invested in a bond.
7. The number of blue chips is 3 more than twice the number of red chips.
8. The number of sedans on a parking lot is 20 fewer than 3 times the number of station wagons.
9. The sum of a certain number and twice that number is 18.
10. Five less than 6 times a number is 37.
11. A young man is 3 years older than his brother. Thirty years from now the sum of their ages will be 111. Find the current ages of the brothers.

12. An elderly man is 22 years older than his daughter. Fifty years ago, the sum of their ages was 34. Find the current age of the man.

13. Joan is 3 times as old as Anne. Fifteen years from now Joan will be twice as old as Anne will be then. How old is each now?

14. The sum of the ages of a woman and her son is 36 years. Six years from now the woman will be twice as old as her son. How old is each now?

15. John is presently 12 years older than Joseph. Four years ago John was twice as old as Joseph. How old is each now?

16. At the present time Albert is 20 years old and Steven is 16 years old. How many years ago was Albert $1\frac{1}{2}$ times as old as Steven?

17. At the present time Lisa is 24 years old and Erica is 16 years old. How many years ago was Lisa twice as old as Erica?

18. A certain number is 3 more than twice another. If their sum is increased by 8, the result is 41. Find the numbers.

19. The larger of two numbers is 3 more than twice the smaller. If their sum is 18, find the numbers.

20. Separate 36 into two parts so that 4 times the smaller minus 3 times the larger is 11.

21. The length of a rectangle is 5 feet more than twice its width. If the perimeter is 40 feet, find the dimensions of the rectangle.

22. The length of a rectangle is 3 cm less than four times its width. If the perimeter is 34 cm, find the dimensions of the rectangle.

23. A farmer plans to enclose a rectangular field, whose length is 16 meters more than its width, with 140 meters of chain-link fencing. What are the dimensions of the field?

24. Suppose that one angle of a triangle is 20° larger than the smallest angle, while the third angle is 10° larger than the smallest angle. Find the number of degrees in each angle.

25. One side of a triangle is 1 meter more than twice the shortest side, while the third side is 3 meters more than the shortest side. If the perimeter is 24 meters, what is the length of each side?

4.2
COIN PROBLEMS

Here is the key to the solution of coin problems.

> You must distinguish between the *number* of coins and the *value* of the coins.
>
> n nickels have a value of $5n$ cents.
>
> n dimes have a value of $10n$ cents.
>
> n quarters have a value of $25n$ cents.
>
> $\vdots$

If you have 8 quarters, what is their value? You find the answer by using this relationship.

Number of coins	×	Number of cents in each coin	=	Value in cents

Since each quarter has a value of 25 cents, the total value of the quarters is

$$8 \times 25 = 200 \text{ cents}$$

EXAMPLE 1

A purse contains $3.20 in quarters and dimes. If there are 3 more quarters than dimes, how many coins of each type are there?

Solution

In this problem, we may let the unknown represent the number of either quarters or dimes. We make a choice.

then

$$\text{Let} \quad n = \text{number of quarters}$$

$$n - 3 = \text{number of dimes}$$

since "there are 3 more quarters than dimes."

	Number of coins	$\times$ Number of cents in each coin	$=$ Value in cents
Quarters	n	25	$25n$
Dimes	$n - 3$	10	$10(n - 3)$

We know that

$$\text{total value} = (\text{value of quarters}) + (\text{value of dimes})$$

$$320 = 25n + 10(n - 3)$$

$$320 = 25n + 10n - 30$$

$$350 = 35n$$

$$10 = n$$

Then

$$n = \text{number of quarters} = 10$$

$$n - 3 = \text{number of dimes} = 7$$

Now verify that the value is $3.20

PROGRESS CHECK 1

(a) Solve Example 1, letting the unknown n represent the number of dimes.
(b) A class collected $3.90 in nickels and dimes. If there were 6 more nickels than dimes, how many coins were there of each type?

Answers
(a) *10 quarters, 7 dimes*   (b) *24 dimes, 30 nickels*

EXAMPLE 2

A jar contains 25 coins worth $3.05. If the jar contains only nickels and quarters, how many coins are there of each type?

Solution

Let $n = $ number of nickels. Since there are a total of 25 coins, we see that

$$25 - n = \text{number of quarters}$$

	Number of coins ×	Number of cents in each coin =	Value in cents
Nickels	n	5	$5n$
Quarters	$25 - n$	25	$25(25 - n)$

We know that

$$\text{total value} = (\text{value of nickels}) + (\text{value of quarters})$$
$$305 = 5n + 25(25 - n)$$
$$305 = 5n + 625 - 25n$$
$$-320 = -20n$$
$$n = 16 = \text{number of nickels}$$
$$25 - n = 9 = \text{number of quarters}$$

Verify that the coins have a total value of $3.05.

PROGRESS CHECK 2
A pile of coins worth $10 consisting of quarters and half-dollars is lying on a desk. If there are 20 quarters, how many half-dollars are there?

Answer
10

EXAMPLE 3
A man purchased 10¢, 15¢, and 20¢ stamps with a total value of $8.40. If the number of 15¢ stamps is 8 more than the number of 10¢ stamps and there are 10 more of the 20¢ stamps than of the 15¢ stamps, how many of each did he receive?

Solution
This problem points out two things: (a) it is possible to phrase coin problems in terms of stamps or other objects and (b) a "wordy" word problem can be attacked by the same type of analysis.
We let n be the number of 15¢ stamps (since the 10¢ and 20¢ stamps are specified in terms of the 15¢ stamps).

	Number of stamps ×	Denomination of each stamp =	Value in cents
10¢	$n - 8$	10	$10(n - 8)$
15¢	n	15	$15n$
20¢	$n + 10$	20	$20(n + 10)$

Thus

$$n = \text{number of 15¢ stamps}$$

Since

$$\text{total value} = \left(\begin{array}{c}\text{value of}\\10\text{¢ stamps}\end{array}\right) + \left(\begin{array}{c}\text{value of}\\15\text{¢ stamps}\end{array}\right) + \left(\begin{array}{c}\text{value of}\\20\text{¢ stamps}\end{array}\right)$$
$$840 = 10(n - 8) + 15n + 20(n + 10)$$

$$840 = 10n - 80 + 15n + 20n + 200$$
$$840 = 45n + 120$$
$$720 = 45n$$
$$16 = n$$

Thus
$$n = \text{number of 15¢ stamps} = 16$$
$$n - 8 = \text{number of 10¢ stamps} = 8$$
$$n + 10 = \text{number of 20¢ stamps} = 26$$

Verify that the total value is $8.40.

PROGRESS CHECK 3

The pretzel vendor finds that her coin-changer contains $8.75 in nickels, dimes, and quarters. If there are twice as many dimes as nickels and 10 fewer quarters than dimes, how many of each kind of coin are there?

Answer
15 nickels, 30 dimes, and 20 quarters

EXERCISE SET 4.2

1. A soda machine contains $3.00 in nickels and dimes. If the number of dimes is 5 more than twice the number of nickels, how many coins of each type are there?
2. A donation box has $8.50 in nickels, dimes, and quarters. If there are twice as many dimes as nickels, and 4 more quarters than dimes, how many coins of each type are there?
3. A wallet has $460 in $5, $10, and $20 bills. The number of $5 bills exceeds twice the number of $10 bills by 4, while the number of $20 bills is 6 fewer than the number of $10 bills. How many bills of each type are there?
4. A traveler buys $990 in traveler's checks, in $10, $20, and $50 denominations. The number of $20 checks is 3 less than twice the number of $10 checks, while the number of $50 checks is 5 less than the number of $10 checks. How many traveler's checks were bought in each denomination?
5. A movie theater charges $3 admission for an adult and $1.50 for a child. If 700 tickets were sold and the total revenue received was $1650, how many tickets of each type were sold?
6. At a gambling casino a red chip is worth $5, a green one $2, and a blue one $1. A gambler buys $27 worth of chips. The number of green chips is 2 more than 3 times the number of red ones, while the number of blue chips is 3 less than twice the number of red ones. How many chips of each type did the gambler get?
7. A student buys 5¢, 10¢, and 15¢ stamps, with a total value of $6.70. If the number of 5¢ stamps is 2 more than the number of 10¢ stamps, while the number of 15¢ stamps is 5 more than one half the number of 10¢ stamps, how many stamps of each denomination did the student obtain?
8. A railroad car, designed to carry containerized cargo, handles crates that weigh $1, \frac{1}{2}$, and $\frac{1}{4}$ ton. On a certain day, the railroad car carries 17 tons of cargo. If the number of $\frac{1}{2}$-ton containers is twice the number of 1-ton containers, while the number of $\frac{1}{4}$-ton containers is 8 more than 4 times the number of 1-ton containers, how many containers of each type are in the car?
9. An amateur theater group is converting a large classroom into an auditorium for a forthcoming play. The group will sell $3, $5, and $6 tickets. The group wants

to receive exactly $503 from the sale of the tickets. If the number of $5 tickets is twice the number of $6 tickets, and the number of $3 tickets is 1 more than 3 times the number of $6 tickets, how many tickets of each type are there?

10. An amusement park sells 10¢, 25¢, and 50¢ tickets. A teacher purchases $15 worth of tickets and a student remarks that there are twice as many 25¢ tickets as there are 50¢ tickets and that the number of 10¢ tickets is 30 more than the number of 25¢ tickets. How many tickets of each type are there?

4.3

INVESTMENT PROBLEMS

If $500 is invested at an annual interest rate of 6%, then the simple interest at year's end will be

$$I = (0.06)(500) = \$30$$

In general,

Simple annual interest = Principal × Annual rate

or

$$I = P \cdot r$$

This formula will be used in all investment problems.

EXAMPLE 1

A part of $7000 is invested at 6% annual interest and the remainder at 8%. If the total amount of annual interest is $460, how much was invested at each rate?

Solution

Let n = amount invested at 6%

then

$7000 - n$ = amount invested at 8%

since the total amount is $7000. Displaying the information, we have:

	Amount invested	×	Rate	=	Interest
6% portion	n		0.06		$0.06n$
8% portion	$7000 - n$		0.08		$0.08(7000 - n)$

Since the total interest is the sum of the interest from the two parts,

$$460 = 0.06n + 0.08(7000 - n)$$

$$460 = 0.06n + 560 - 0.08n$$

$$0.02n = 100$$

$$n = \$5000 = \text{portion invested at } 6\%$$

$$7000 - n = \$2000 = \text{portion invested at } 8\%$$

PROGRESS CHECK 1

A club decides to invest a part of $4600 in stocks earning $4\frac{1}{2}$% annual dividends and the remainder in bonds paying $7\frac{1}{2}$%. How much must the club invest in each to obtain a net return of 5.4%?

Answer
$3220 in stocks, $1380 in bonds

EXAMPLE 2

A part of $12,000 is invested at 5% annual interest and the remainder at 9%. The annual income on the 9% investment is $100 more than the annual income on the 5% investment. How much was invested at each rate?

Solution

then

Let $n =$ amount invested at 5%

$12,000 - n =$ amount invested at 9%

	Amount invested	$\times$	Rate	=	Interest
5% investment	n		0.05		$0.05n$
9% investment	$12,000 - n$		0.09		$0.09(12,000 - n)$

Since the interest on the 9% portion is $100 more than the interest on the 5% portion,

$$0.09(12,000 - n) = 0.05n + 100$$
$$1080 - 0.09n = 0.05n + 100$$
$$980 = 0.14n$$
$$n = 7000$$

Thus, $7000 is invested at 5% and $5000 at 9%.

PROGRESS CHECK 2

$7500 is invested in two parts yielding 5% and 15% annual interest. If the interest earned on the 15% portion is twice that on the 5% portion, how much is invested in each?

Answer
$4500 at 5%, $3000 at 15%

EXAMPLE 3

A shoe store owner had $6000 invested in inventory. The profit on women's shoes was 35%, while the profit on men's shoes was 25%. If the profit on the entire stock was 28%, how much was invested in each type of shoe?

Solution

then

Let $n =$ amount invested in women's shoes

$6000 - n =$ amount invested in men's shoes

	Amount invested	×	Rate	=	Profit
Women's shoes	n		0.35		$0.35n$
Men's shoes	$6000 - n$		0.25		$0.25(6000 - n)$
Combined stock	6000		0.28		$0.28(6000)$

The profit on the entire stock was equal to the sum of the profits on each portion:

$$0.28(6000) = 0.35n + 0.25(6000 - n)$$
$$1680 = 0.35n + 1500 - 0.25n$$
$$180 = 0.1n$$
$$n = 1800$$

The store owner had invested $1800 in women's shoes and $4200 in men's shoes.

PROGRESS CHECK 3
An automobile dealer has $55,000 invested in compacts and midsize cars. The profit on sales of the compacts is 10%, while the profit on sales of midsize cars is 16%. How much did the dealer invest in compact cars if the overall profit on the total investment was 12%?

Answer
$36,666.67

EXERCISE SET 4.3
1. A part of $8000 is invested at 7% annual interest and the remainder at 8%. If the total annual interest is $590, how much was invested at each rate?
2. A $20,000 scholarship endowment fund is to be invested in two ways: part in a stock paying 5.5% annual interest in dividends and the remainder in a bond paying 7.5%. How much should be invested in each to obtain a net yield of 6.8%?
3. To help pay for his child's college education, a father invests $10,000 in two separate investments: part is in a certificate of deposit paying 8.5% annual interest, the rest is in a mutual fund paying 7%. The annual income on the certificate of deposit is $200 more than the annual income on the mutual fund. How much was invested in each type of investment?
4. A bicycle store selling 3-speed and 10-speed models has $16,000 in inventory. The profit on a 3-speed bicycle is 11%, while the profit on a 10-speed model is 22%. If the profit on the entire stock is 19%, how much was invested in each type of bicycle?
5. A film shop carrying black-and-white film and color film has $4000 in inventory. The profit on black-and-white film is 12%, while the profit on color film is 21%. If the annual profit on color film is $150 less than the annual profit on black-and-white film, how much was invested in each type of film?
6. A widow invested one third of her assets in a certificate of deposit paying 6% annual interest, one sixth of her assets in a mutual fund paying 8%, and the remainder in a stock paying 8.5%. If her total annual income from these investments is $910, what was the total amount invested by the widow?
7. A trust fund has invested $8000 at 6% annual interest. How much additional money should be invested at 8.5% to obtain a return of 8% on the total amount invested?

8. A businessman invested a total of $12,000 in two ventures. In the first one he made a profit of 8% and in the other he lost 4%. If his net profit for the year was $120, how much did he invest in each venture?

9. A retiree invested a certain amount of money at 6% annual interest; a second amount, which is $300 more than the first amount, at 8%; and a third amount, which is 4 times as much as the first amount, at 10%. If the total annual income from these investments is $1860, how much was invested at each rate?

10. A finance company lends a certain amount of money to Firm A at 7% annual interest, an amount $100 less than that lent to Firm A is lent to Firm B at 8%, and an amount $200 more than that lent to Firm A is lent to Firm C at 8.5%. If the total annual income is $126.50, how much was lent to each firm?

4.4
DISTANCE (UNIFORM MOTION) PROBLEMS

Here is the basic formula for solving distance problems.

$$\text{Distance} = \text{rate} \times \text{time}$$
$$\text{or}$$
$$d = r \cdot t$$

For instance, an automobile traveling at an average speed of 50 miles per hour for 3 hours will travel a distance of

$$d = r \cdot t$$
$$= 50 \cdot 3 = 150 \text{ miles}$$

The relationships that permit you to write an equation are sometimes obscured by the words. Here are some questions to ask as you set up a distance problem.

(a) Are there two distances which are equal? Will two objects have traveled the same distance? Is the distance on a return trip the same as the distance going?

(b) Is the sum (or difference) of two distances equal to a constant? When two objects are traveling toward each other, they meet when the sum of the distances traveled by each equals the original distance between them.

EXAMPLE 1
Two trains leave New York for Chicago. The first train travels at an average speed of 60 miles per hour, while the second train, which departs an hour later, travels at an average speed of 80 miles per hour. How long will it take the second train to overtake the first train?

Solution
Since we are interested in the time the second train travels, we choose to

let t = number of hours second train travels
then
$t + 1$ = number of hours first train travels

since the first train departs one hour earlier.

	Rate	×	Time	=	Distance
First train	60		$t + 1$		$60(t + 1)$
Second train	80		t		$80t$

At the moment the second train overtakes the first, they must both have traveled the *same* distance.

$$60(t + 1) = 80t$$

$$60t + 60 = 80t$$

$$60 = 20t$$

$$3 = t$$

It will take the second train 3 hours to catch up with the first train.

PROGRESS CHECK 1

A light plane leaves the airport at 9 A.M. traveling at an average speed of 200 miles per hour. At 11 A.M. a jet plane departs and follows the same route. If the jet travels at an average speed of 600 miles per hour, at what time will the jet overtake the light plane?

Answer
12 noon

WARNING

The units of measurement of rate, time, and distance must be consistent. If a car travels at an average speed of 40 miles per hour for 15 minutes, then the distance covered is

$$d = r \cdot t$$

$$d = 40 \cdot \frac{1}{4} = 10 \text{ miles}$$

since 15 minutes $= \dfrac{1}{4}$ hour.

EXAMPLE 2

A jogger running at the rate of 4 miles per hour takes 45 minutes more than a car traveling at 40 miles per hour to cover a certain course. How long does it take the jogger to complete the course and what is the length of the course?

Solution

Notice that time is expressed in both minutes and hours. Let's choose hours as the unit of time and

then

let $t =$ time for the jogger to complete the course

$$t - \frac{3}{4} = \text{time for the car to complete the course}$$

since the car takes 45 minutes ($= \frac{3}{4}$ hour) less time.

	Rate	×	Time	=	Distance
Jogger	4		t		$4t$
Car	40		$t - \dfrac{3}{4}$		$40\left(t - \dfrac{3}{4}\right)$

Since the jogger and car travel the same distance,

$$4t = 40\left(t - \frac{3}{4}\right) = 40t - 30$$

$$30 = 36t$$

$$\frac{5}{6} = t$$

The jogger takes $\frac{5}{6}$ hour or 50 minutes. The distance traveled is

$$4t = 4 \cdot \frac{5}{6} = \frac{20}{6} = 3\frac{1}{3} \text{ miles}$$

PROGRESS CHECK 2

The winning horse finished the race in 3 minutes, while a losing horse took 4 minutes. If the average rate of the winning horse was 5 feet per second more than the average rate of the slower horse, find the average rates of both horses.

Answer
Winner: 20 feet per second; loser: 15 feet per second

EXAMPLE 3

At 2 P.M. a plane leaves Boston for San Francisco, traveling at an average speed of 500 miles per hour. Two hours later a plane departs from San Francisco to Boston traveling at an average speed of 600 miles per hour. If the cities are 3200 miles apart, at what time do the planes pass each other?

Solution

Let t = the number of hours after 2 P.M. at which the planes meet.

Let's piece together the information that we have.

	Rate	×	Time	=	Distance
From Boston	500		t		$500t$
From San Francisco	600		$t - 2$		$600(t - 2)$

At the moment that the planes pass each other, the sum of the distances traveled by both planes must be 3200 miles.

Thus

$$3200 = 500t + 600(t - 2)$$

$$3200 = 500t + 600t - 1200$$
$$4400 = 1100t$$
$$4 = t$$

The planes meet 4 hours after the departure of the plane from Boston.

PROGRESS CHECK 3
Two cyclists start at the same time from the same place and travel in the same direction. If one cyclist averages 16 miles per hour and the second averages 20 miles per hour, how long will it take for them to be 12 miles apart?

Answer
3 hours

EXERCISE SET 4.4

1. Two trucks leave Philadelphia for Miami. The first truck travels at an average speed of 50 kilometers per hour. The second truck, which leaves 2 hours later, travels at an average speed of 55 kilometers per hour. How long will it take the second truck to overtake the first truck?

2. Jackie either drives or bicycles from home to school. Her average speed when driving is 36 miles per hour and her average speed when bicycling is 12 miles per hour. If it takes her $\frac{1}{2}$ hour less to drive than to bicycle to school, how long does it take to go to school and how far is the school from her home?

3. Professors Roberts and Jones, who live 676 miles apart, are exchanging houses and jobs for four months. They start out at exactly the same time for their new locations and meet after 6.5 hours of driving. If their average speeds differ by 4 miles per hour, what is each professor's average speed?

4. Steve leaves school by moped for spring vacation. Forty minutes later, his roommate, Frank, notices that Steve forgot to take his camera, so he decides to try to catch up with him by car. If Steve's average speed is 25 miles per hour and Frank averages 45 miles per hour, how long does it take Frank to overtake Steve?

5. A tour boat goes to a fishing village and back to the mainland in 6 hours. If the average speed going is 15 miles per hour and the average speed returning is 12 miles per hour, how far from the mainland is the island?

6. Two cars start out from the same point at the same time and travel in opposite directions. If their average speeds are 36 and 44 miles per hour respectively, after how many hours will they be 360 miles apart?

7. An express train and a local train start out from the same point at the same time and travel in opposite directions. The express train travels twice as fast as the local train. If after 4 hours they are 480 kilometers apart, what was the average speed of each train?

8. Two planes start out from the same place at the same time and travel in the same direction. The first plane has an average speed of 400 miles per hour and the second plane has an average speed of 480 miles per hour. After how many hours will they be 340 miles apart?

9. Two cyclists start out at the same time from points that are 395 kilometers apart and travel toward each other. The first cyclist travels at an average speed of 40 kilometers per hour, the second travels at an average speed of 50 kilometers per hour. After how many hours will they be 35 kilometers apart?

10. It takes a student 8 hours to drive from her home back to college, a distance of 580 kilometers. Before lunch her average speed is 80 kilometers per hour and after lunch it is 60 kilometers per hour. How many hours did she travel at each rate?

4.5
MIXTURE PROBLEMS

One type of mixture problem involves mixing commodities, say two or more types of nuts, to obtain a mix that has a desired value. If the commodities are measured in pounds, the relationships we need are

Number of pounds	×	Price per pound	=	Value of the commodity
Pounds in mix		=	Sum of the number of pounds in each commodity	
Value of mix		=	Sum of the values of each commodity	

EXAMPLE 1

How many pounds of Brazilian coffee worth $5 per pound must be mixed with 20 pounds of Colombian coffee worth $4 per pound to produce a mixture worth $4.20 per pound?

Solution

Let n = number of pounds of Brazilian coffee. We display all the information, using cents in place of dollars.

Type of coffee	Number of pounds	×	Price per pound	=	Value in cents
Brazilian	n		500		$500n$
Colombian	20		400		8000
Mixture	$n + 20$		420		$420(n + 20)$

(Note that the weight of the mixture equals the sum of the weights of the Brazilian and Colombian coffees going into the mixture.) Since the value of the mixture is the sum of the values of the two types of coffee,

$$420(n + 20) = 500n + 8000$$
$$420n + 8400 = 500n + 8000$$
$$400 = 80n$$
$$5 = n$$

We must add 5 pounds of Brazilian coffee.

PROGRESS CHECK 1

How many pounds of macadamia nuts worth $4 per pound must be mixed with 4 pounds of cashews worth $2.50 per pound and 6 pounds of pecans worth $3 per pound to produce a mixture which is worth $3.20 per pound?

Answer
5 pounds

EXAMPLE 2

Caramels worth $1.75 per pound are to be mixed with cream chocolates worth $2 per pound to make a 5-pound mixture that will be sold at $1.90 per pound. How many pounds of each are needed?

Solution

Let n = number of pounds of caramels. Displaying all of the information, we have:

Type of candy	Number of pounds	$\times$	Price per pound	$=$	Value in cents
Caramels	n		175		$175n$
Cream chocolates	$5 - n$		200		$200(5 - n)$
Mixture	5		190		950

(Note that the number of pounds of cream chocolates is the weight of the mixture less the weight of the caramels.) Since the value of the mixture is the sum of the values of the two components,

$$950 = 175n + 200(5 - n)$$
$$950 = 175n + 1000 - 200n$$
$$25n = 50$$
$$n = 2$$

We must have 2 pounds of caramels and 3 pounds of cream chocolates.

PROGRESS CHECK 2

How many gallons of oil worth 55¢ per gallon and how many gallons of oil worth 75¢ per gallon must be mixed to obtain 40 gallons of oil worth 60¢ per gallon?

Answer
30 gallons of the 55¢ oil and 10 gallons of the 75¢ oil

A second type of mixture problem involves solutions containing different concentrations of materials. For instance, a 40-gallon drum of a solution which is 75% acid contains $(40)(0.75) = 30$ gallons of acid. If the solutions are measured in gallons, the relationship we need is

Number of gallons of solution	$\times$	% of component A	$=$	Number of gallons of component A

The other relationships we need are really the same as in our first type of mixture problem.

Number of gallons in mixture	$=$	Sum of the number of gallons in each solution
Number of gallons of component A in mixture	$=$	Sum of the number of gallons of component A in each solution

EXAMPLE 3

A 40% acid solution is to be mixed with a 75% acid solution to produce 140 gallons of a solution which is 50% acid. How many gallons of each solution must be used?

Solution

Let n = number of gallons of the 40% acid solution. Then $140 - n$ = number of gallons of the 75% acid solution since the number of gallons in the mixture is the sum of the number of gallons in each solution.

Displaying all the information, we have:

	Number of gallons	×	% acid	=	Number of gallons of acid
40% solution	n		40		$0.40n$
75% solution	$140 - n$		75		$0.75(140 - n)$
Mixture	140		50		70

Since the number of gallons of acid in the mixture is the sum of the number of gallons of acid in each solution, we have

$$70 = 0.40n + 0.75(140 - n)$$
$$70 = 0.40n + 105 - 0.75n$$
$$-35 = -0.35n$$
$$n = 100 \text{ gallons}$$
$$140 - n = 40 \text{ gallons}$$

Thus, we mix 100 gallons of the 40% solution with 40 gallons of the 75% solution to produce 140 gallons of the 50% solution.

PROGRESS CHECK 3

How many gallons of milk that is 22% butter fat must be mixed with how many gallons of cream that is 60% butter fat to produce 19 gallons of a mixture that is 40% butterfat?

Answer
10 gallons of milk and 9 gallons of cream.

EXAMPLE 4

How many ounces of a tin alloy that is 30% tin must be mixed with 15 ounces of a tin alloy that is 12% tin to produce an alloy that is 24% tin?

Solution

Let n = number of ounces of the 30% tin alloy. The alloy is essentially a solution and we can display the information as before.

	Number of ounces	×	% tin	=	Number of ounces of tin
30% alloy	n		30		$0.30n$
12% alloy	15		12		1.8
Mixture	$n + 15$		24		$0.24(n + 15)$

(Note that the number of ounces in the mixture is the sum of the number of ounces in the alloys going into the mixture.) Since the number of ounces of *tin* in the mixture is the sum of the number of ounces of tin in each alloy, we have

$$0.24(n + 15) = 0.30n + 1.8$$
$$0.24n + 3.6 = 0.30n + 1.8$$
$$1.8 = 0.06n$$
$$n = 30$$

Thus, we need to add 30 ounces of the 30% alloy to 15 ounces of the 12% alloy.

PROGRESS CHECK 4

How many pounds of a 25% copper alloy must be added to 50 pounds of a 55% copper alloy to produce an alloy which is 45% copper?

Answer
Add 25 pounds of 25% copper alloy

EXAMPLE 5

A tank contains 40 gallons of water and 10 gallons of alcohol. How many gallons of water must be removed if the remaining solution is to be 30% alcohol?

Solution

Let n = number of gallons of water to be removed. This problem is different since we are removing water. Here is how to display the information.

	Number of gallons	$\times$	% alcohol	$=$	Gallons of alcohol
Original solution	50		20		10
Water removed	n		0		0
New solution	$50 - n$		30		$0.3(50 - n)$

(Note that the water removed has 0% alcohol!) The number of gallons of alcohol in the new solution is the same amount as in the original solution since only water has been removed.

$$0.3(50 - n) = 10$$
$$15 - 0.3n = 10$$
$$5 = 0.3n$$
$$n = 16\frac{2}{3}$$

Thus, we must remove $16\frac{2}{3}$ gallons of water.

PROGRESS CHECK 5

A tank contains 90 quarts of an antifreeze solution that is 50% antifreeze. How much water should be removed to raise the antifreeze level to 60% in the new solution?

Answer
15 quarts of water should be removed.

EXERCISE SET 4.5

1. How many pounds of raisins worth $1.50 per pound must be mixed with 10 pounds of peanuts worth $1.20 per pound to produce a mixture worth $1.40 per pound?

2. How many ounces of Ceylon tea worth $1.50 per ounce and how many ounces of Formosa tea worth $2.00 per ounce must be mixed to obtain a mixture of 8 ounces that is worth $1.85 per ounce?

3. A copper alloy that is 40% copper is to be combined with a copper alloy that is 80% copper to produce 120 kilograms of an alloy that is 70% copper. How many kilograms of each alloy must be used?

4. How many liters of an ammonia solution that is 20% ammonia must be mixed with 20 liters of an ammonia solution that is 48% ammonia to produce a solution that is 36% ammonia?

5. A vat contains 60 gallons of a 15% saline solution. How many gallons of water must be evaporated so that the resulting solution will be 20% saline?

6. How many grams of pure silver must be added to 30 grams of an alloy that is 50% silver to obtain an alloy that is 60% silver?

7. How much water must be added to dilute 10 quarts of a solution containing 18% iodine so that the resulting solution will be 12% iodine?

8. A vat contains 27 gallons of water and 9 gallons of acetic acid. How many gallons of water must be evaporated if the remaining solution is to be 40% acetic acid?

9. How many pounds of a fertilizer worth $3 per pound must be combined with 12 pounds of a weed killer worth $6 per pound and 18 pounds of phosphate worth $6 per pound to produce a mixture worth $4.80 per pound?

10. A producer of packaged frozen vegetables wants to market the product at $1.20 per kilogram. How many kilograms of green beans worth $1 per kilogram must be mixed with 100 kilograms of corn worth $1.30 per kilogram and 90 kilograms of peas worth $1.40 per kilogram to produce the required mix?

KEY IDEAS FOR REVIEW

☐ Many word problems can be analyzed with the help of a table which exhibits the information and the basic equation.

☐ The basic equations for some of the most common types of word problems are

Coin problems

$$\text{Number of coins} \times \text{Number of cents in each coin} = \text{Value in cents}$$

Investment problems

Amount invested × Rate = Interest

Uniform motion problems

Rate × Time = Distance

Mixture problems

Number of pounds × Price per pound = Value

$$\text{Amount of solution} \times \text{\% of component A} = \text{Amount of component A}$$

PROGRESS TEST 4A

1. Translate into algebra: "The number of chairs is 3 less than 4 times the number of tables."
2. Steve is presently 6 years younger than Lisa. If the sum of their ages is 40, how old is each?
3. The width of a rectangle is 4 cm less than twice its length. If the perimeter is 12 cm, find the dimensions of each side.
4. A donation box contains 30 coins consisting of nickels, dimes, and quarters. The number of dimes is 4 more than twice the number of quarters. If the total value of the coins is $2.60, how many coins of each type are there?
5. A fruit grower ships crates of oranges that weigh 30, 50, and 60 pounds each. A certain shipment weighs 1140 pounds. If the number of 30-pound crates is 3 more than one half the number of 50-pound crates and the number of 60-pound crates is 1 less than twice the number of 50-pound crates, how many crates of each type are there?
6. A college fund has invested $12,000 at 7% annual interest. How much additional money must be invested at 9% to obtain a return of 7.8% on the total amount invested?
7. A businessperson invested a certain amount of money at 6.5% annual interest; a second amount, which is $200 more than the first amount, at 7.5%; and a third amount, which is $300 more than twice the first amount, at 9%. If the total annual income from these investments is $1962, how much was invested at each rate?
8. A moped and a car leave from the same point at the same time and travel in opposite directions. The car travels 3 times as fast as the moped. If after 5 hours they are 300 miles apart, what was the average speed of each vehicle?
9. A bush pilot in Australia picks up mail at a remote village and returns to home base in 4 hours. If the average speed going is 150 miles per hour and the average speed returning is 100 miles per hour, how far from the home base is the village?
10. An alloy that is 60% silver is to be combined with an alloy that is 80% silver to produce 120 ounces of an alloy that is 75% silver. How many ounces of each alloy must be used?
11. A beaker contains 150 cubic centimeters of a solution that is 30% acid. How much water must be evaporated so that the resulting solution will be 40% acid?

PROGRESS TEST 4B

1. Translate into algebra: "The number of Democrats is 4 more than one third the number of Republicans."
2. Separate 48 into two parts so that 3 times the smaller plus the larger is 80.
3. One side of a triangle is 2 cm shorter than the third side, while the second side is 3 cm longer than one half the third side. If the perimeter is 15 cm, find the length of each side.
4. An envelope contains 20 discount coupons in $1, $5, and $10 denominations. The number of $5 coupons is twice the number of $10 coupons. If the total value of the coupons is $54, how many coupons of each type are there?
5. A cheese sampler with a total weight of 25 ounces of cheese contains 1-ounce, 2-ounce, and 3-ounce samples. If the number of 1-ounce samples is 3 more than the number of 3-ounce samples, and the number of 2-ounce samples is 1 less than twice the number of 3-ounce samples, how many samples of each weight are there?

6. Part of an $18,000 trust fund is to be invested in a stock paying 6% in dividends and the remainder in a bond paying 7.2% annual interest. How much should be invested in each to obtain a net yield of 7%?

7. A woman invested a certain amount of money at 8% annual interest, and a second amount of money, $2000 greater than the first amount, at 6%. If the annual incomes on the two investments are equal, how much was invested at each rate?

8. Two trains start out at 10 A.M. from stations that are 1120 kilometers apart, and travel toward each other at average speeds of 80 and 60 kilometers per hour. At what time will they pass each other?

9. Two charter buses leave New York for Los Angeles. The first one travels at an average speed of 40 miles per hour. The second one leaves 3 hours later and travels at an average speed of 50 miles per hour. How long will it take the second bus to overtake the first one?

10. How many pounds of lawn seed worth $4.00 per pound must be mixed with 15 pounds of fertilizer worth $3.00 per pound to produce a mixture worth $3.20 per pound?

11. A vat contains 12 gallons of acid and 48 gallons of water. How much acid must be added to make a solution that is 40% acid?

CHAPTER FIVE
ALGEBRAIC FRACTIONS

In algebra we are often faced with complicated fractions such as

$$\frac{1 - \dfrac{1}{x}}{\dfrac{1}{x^2} + \dfrac{1}{x}}$$

which we wish to simplify. At times we also need to find the sum of fractions such as

$$\frac{x}{x - 2} + \frac{2x^2}{x - 3}$$

Our objective in this chapter is to learn to handle typical problems involving algebraic fractions. We will study the rules for basic operations with fractions (addition, subtraction, multiplication, and division). We will see that a simple idea forms the cornerstone for much of our work—*multiplying a number or expression by a fraction equivalent to 1 does not change its value.*

We shall also solve equations involving algebraic fractions, and word problems leading to such equations.

5.1
MULTIPLICATION AND DIVISION OF FRACTIONS

The rule for multiplication of fractions is already familiar to us from arithmetic.

Multiplication of Fractions

$$\frac{a}{b} \cdot \frac{c}{d} = \frac{ac}{bd}$$

That is, when multiplying two given fractions we obtain a new fraction whose numerator is the product of the numerators of the given fractions and whose denominator is the product of the denominators of the given fractions. The same rule holds whether a, b, c, and d are numbers or algebraic expressions.

EXAMPLE 1
Multiply.

(a) $\dfrac{5}{2} \cdot \dfrac{3}{4} = \dfrac{5 \cdot 3}{2 \cdot 4} = \dfrac{15}{8}$

(b) $\dfrac{3}{4} \cdot \dfrac{x-1}{2} = \dfrac{3(x-1)}{4 \cdot 2} = \dfrac{3(x-1)}{8}$

(c) $\dfrac{x-1}{2} \cdot \dfrac{x+1}{x} = \dfrac{(x-1)(x+1)}{2 \cdot x} = \dfrac{x^2-1}{2x}$

(d) $\dfrac{3}{4x} \cdot \dfrac{x-3}{2y} \cdot \dfrac{x+3}{y-1} = \dfrac{3(x-3)(x+3)}{4x \cdot 2y(y-1)} = \dfrac{3(x^2-9)}{8xy(y-1)}$

PROGRESS CHECK 1
Multiply.

(a) $\dfrac{2x}{y} \cdot \dfrac{y+1}{y^2+1}$ (b) $\dfrac{3}{a} \cdot \dfrac{b^2}{2a} \cdot \dfrac{c}{d-3}$

Answers

(a) $\dfrac{2x(y+1)}{y(y^2+1)}$ (b) $\dfrac{3b^2c}{2a^2(d-3)}$

To divide two fractions, we multiply by the reciprocal of the divisor.

Division of Fractions

$$\frac{\dfrac{a}{b}}{\dfrac{c}{d}} = \frac{a}{b} \cdot \frac{d}{c} = \frac{ad}{bc}$$

This rule "works" because we are really multiplying by a cleverly disguised form of 1, namely

$$\frac{\dfrac{d}{c}}{\dfrac{d}{c}} = 1$$

Observe that

$$\frac{\dfrac{a}{b} \cdot \dfrac{d}{c}}{\dfrac{c}{d} \cdot \dfrac{d}{c}} = \frac{\dfrac{ad}{bc}}{\dfrac{cd}{cd}} = \frac{\dfrac{ad}{bc}}{1} = \frac{ad}{bc}$$

EXAMPLE 2
Divide.

(a) $\dfrac{\dfrac{2}{5}}{\dfrac{3}{4}} = \dfrac{2}{5} \cdot \dfrac{4}{3} = \dfrac{2 \cdot 4}{5 \cdot 3} = \dfrac{8}{15}$

(b) $\dfrac{\dfrac{2x}{x-1}}{\dfrac{x-2}{x}} = \dfrac{2x}{x-1} \cdot \dfrac{x}{x-2} = \dfrac{2x^2}{(x-1)(x-2)}$

(c) $\dfrac{\dfrac{3a^3b^2}{2cd}}{\dfrac{c-1}{2a^2b}} = \dfrac{3a^3b^2}{2cd} \cdot \dfrac{2a^2b}{c-1} = \dfrac{6a^5b^3}{2c(c-1)d}$

(d) $\dfrac{2x}{y} \div \dfrac{3y^3}{x-3}$

We will write this problem as a fraction and solve.

$$\dfrac{\dfrac{2x}{y}}{\dfrac{3y^3}{x-3}} = \dfrac{2x}{y} \cdot \dfrac{x-3}{3y^3} = \dfrac{2x(x-3)}{3y^4}$$

(e) $\dfrac{\dfrac{x}{2x+1}}{\dfrac{x-2}{1}} = \dfrac{\dfrac{x}{2x+1}}{\dfrac{x-2}{1}} = \dfrac{x}{2x+1} \cdot \dfrac{1}{x-2} = \dfrac{x}{(2x+1)(x-2)}$

PROGRESS CHECK 2
Divide.

(a) $\dfrac{\dfrac{5(y-1)^2}{x}}{\dfrac{2(x-1)^2}{y}}$

(b) $\dfrac{3a^2b}{a+1} \div \dfrac{2(b-1)}{a-1}$

Answers

(a) $\dfrac{5y(y-1)^2}{2x(x-1)^2}$ (b) $\dfrac{3a^2(a-1)b}{2(a+1)(b-1)}$

Now that we can handle multiplication and division, we can look at the basic rule that allows us to simplify fractions.

Cancellation Principle

$$\frac{ab}{ac} = \frac{b}{c} \quad \text{if} \quad a \neq 0$$

This is not anything new; we have used cancellation in arithmetic before.

$$\frac{2}{\cancel{3}} \cdot \frac{\cancel{3}}{5} = \frac{2}{5}$$

Why does this work? Why are we allowed to "cancel" the number 3? Because $\frac{3}{3} = 1$. In fact, for any nonzero number a, $a/a = 1$, so that if $a \neq 0$ is a factor of both the numerator *and* denominator of two fractions that are being multiplied, we can cancel the number a.

The same principle applies to algebraic expressions. Cancellation of a common factor does not change an algebraic expression. Thus,

$$\frac{x-1}{2\cancel{x^2}} \cdot \frac{\cancel{x^2}}{x-2} = \frac{x-1}{2(x-2)}$$

We summarize a systematic procedure for the simplification of fractions.

Simplifying Fractions

Step 1. Factor the numerator completely.

Step 2. Factor the denominator completely.

Step 3. Cancel factors that are common to both the numerator and denominator.

EXAMPLE 3

Simplify.

(a) $\dfrac{x^2-4}{x^2+5x+6} = \dfrac{\cancel{(x+2)}(x-2)}{(x+3)\cancel{(x+2)}} = \dfrac{x-2}{x+3}$

(b) $\dfrac{\dfrac{3a^2(b-1)}{6c^2d^3}}{\dfrac{(b-1)^2}{2cd^2}} = \dfrac{\cancel{3}a^2\cancel{(b-1)}}{\cancel{6}c^{\cancel{2}}d^{\cancel{3}}} \cdot \dfrac{\cancel{2}c\cancel{d^2}}{\cancel{(b-1)^2}} = \dfrac{a^2}{(b-1)cd}$

PROGRESS CHECK 3

Simplify.

(a) $\dfrac{4 - x^2}{x^2 - x - 6}$ (b) $\dfrac{m^4}{3n^2} \div \left(\dfrac{m^2}{9n} \cdot \dfrac{n}{2m^3}\right)$

Answers

(a) $\dfrac{2 - x}{x - 3}$ (b) $\dfrac{6m^5}{n^2}$

WARNING

(a) Only multiplicative factors of the entire numerator and denominator can be canceled. *Don't* write

$$\frac{2\not{x} - 4}{\not{x}} = 2 - 4 = -2$$

Since x is *not* a *multiplicative factor* of the whole numerator, we may *not* cancel it.

(b) *Don't* write

$$\frac{y^2 - x^2}{y - x} = y - x$$

To simplify correctly, write

$$\frac{y^2 - x^2}{y - x} = \frac{(y + x)\cancel{(y - x)}}{\cancel{y - x}} = y + x$$

We finish our discussion of cancellation with a somewhat subtle technique. Can you do anything with this fraction?

$$\frac{x - 5}{5 - x}$$

At first glance you might say "No, there are no common factors." But if you recognize that $5 - x = -(x - 5)$, then you can see that

$$\frac{x - 5}{5 - x} = \frac{x - 5}{-(x - 5)} = \frac{1\cancel{(x - 5)}}{-1\cancel{(x - 5)}} = \frac{1}{-1} = -1$$

EXAMPLE 4

Simplify.

(a) $\dfrac{x^2 - x - 6}{3x - x^2} = \dfrac{(x - 3)(x + 2)}{x(3 - x)} = \dfrac{\cancel{(x - 3)}(x + 2)}{-x\cancel{(x - 3)}}$

$$= \frac{x + 2}{-x} = -\frac{x + 2}{x}$$

(b) $\dfrac{16 - x^2}{x^2 - 3x - 4} \cdot \dfrac{x + 1}{x + 4} = \dfrac{(4 + x)(4 - x)}{(x + 1)(x - 4)} \cdot \dfrac{\cancel{x + 1}}{\cancel{x + 4}}$

$$= \frac{-\cancel{(x - 4)}}{\cancel{x - 4}} = -1$$

(c) $\dfrac{x^2 - 9}{y} \div \dfrac{x - 3}{2}$

This is rewritten as

$$\dfrac{x^2 - 9}{y} \cdot \dfrac{2}{x - 3} = \dfrac{(x + 3)(x - 3)}{y} \cdot \dfrac{2}{x - 3} = \dfrac{2(x + 3)}{y}$$

PROGRESS CHECK 4
Simplify.

(a) $\dfrac{x^2 + 7x - 8}{x - x^2}$ (b) $\dfrac{9y^2 - x^2}{x^2 + 7x + 6} \cdot \dfrac{x + 1}{x - 3y}$

(c) $\dfrac{8 - 2x}{y} \div \dfrac{x^2 - 16}{y}$

Answers

(a) $-\dfrac{x + 8}{x}$ (b) $-\dfrac{3y + x}{x + 6}$ (c) $-\dfrac{2}{x + 4}$

EXERCISE SET 5.1
Determine whether each of the following is true (T) or false (F).

1. $\dfrac{3}{8} \cdot \dfrac{x + 5}{4} = \dfrac{3(x + 5)}{32}$

2. $\dfrac{2}{3} \cdot \dfrac{x + 4}{7} = \dfrac{2x + 4}{21}$

3. $\dfrac{5}{2} \cdot \dfrac{4x + 7}{15} = \dfrac{2x + 7}{3}$

4. $\dfrac{2x}{5y} \cdot \dfrac{5x + 15}{4} = \dfrac{x(x + 5)}{2y}$

5. $\dfrac{\dfrac{1}{3}}{\dfrac{2x}{x - 1}} = \dfrac{x - 1}{6x}$

6. $\dfrac{\dfrac{x}{2x + 1}}{\dfrac{2}{x}} = \dfrac{2}{2x + 1}$

7. $\dfrac{\dfrac{2}{3}}{\dfrac{4x}{3x + 6}} = \dfrac{x + 6}{2x}$

8. $\dfrac{\dfrac{x^2 - y^2}{xy}}{\dfrac{x + y}{x^2}} = \dfrac{x(x - y)}{y}$

Simplify, if possible, in the following exercises.

9. $\dfrac{6x + 3}{3}$ 10. $\dfrac{8x - 4}{2}$ 11. $\dfrac{3x + 2}{3}$

12. $\dfrac{6x - 3}{5}$ 13. $\dfrac{5}{10x^2 - 15}$ 14. $\dfrac{3}{6x - 12y^2}$

15. $\dfrac{5a^4}{25a^2}$ 16. $\dfrac{18}{27} \cdot \dfrac{a^2 b^4}{a^3 b^2}$ 17. $\dfrac{x + 4}{x^2 - 16}$

18. $\dfrac{y^2 - 25}{y + 5}$ 19. $\dfrac{x^2 - 8x + 16}{x - 4}$ 20. $\dfrac{5x^2 - 45}{2x - 6}$

21. $\dfrac{6x^2 - x - 1}{2x^2 + 3x - 2}$ 22. $\dfrac{2x^3 + x^2 - 3x}{3x^2 + x + 2}$

Compute and simplify your answer in the following.

23. $\dfrac{x+1}{3} \cdot \dfrac{2x+3}{4}$

24. $\dfrac{3x-1}{2y} \cdot \dfrac{2x+3}{3x+1}$

25. $\dfrac{(a-4)}{3} \cdot \dfrac{9(a+4)}{b}$

26. $\dfrac{x+1}{3x+6} \cdot \dfrac{6}{x+1}$

27. $\dfrac{a^2}{4} \div \dfrac{a}{2}$

28. $\dfrac{2a^2b^4}{3c^3} \div \dfrac{a^3b^2}{6c^5}$

29. $\dfrac{2}{3x-6} \div \dfrac{3}{2x-4}$

30. $\dfrac{5x+15}{8} \div \dfrac{3x+9}{4}$

31. $\dfrac{3x^2+x}{2x+4} \cdot \dfrac{4}{x^2+2x}$

32. $\dfrac{a^2-a}{b+1} \cdot \dfrac{2b}{a^3-a^2}$

33. $\dfrac{25-a^2}{b+3} \cdot \dfrac{2b^2+6b}{a-5}$

34. $\dfrac{2xy^2}{x+y} \cdot \dfrac{x+y}{4xy}$

35. $\dfrac{2x}{x-3} \div \dfrac{6x^2}{x+3}$

36. $\dfrac{a-a^2}{b-1} \div \dfrac{a^2-a}{b}$

37. $\dfrac{x^2-4}{x+1} \div \dfrac{2x+3}{2x-4}$

38. $\dfrac{9}{a^2-16} \div \dfrac{3}{a+4}$

39. $\dfrac{x+2}{3y} \div \dfrac{x^2-2x-8}{15y^2}$

40. $\dfrac{3x}{x+2} \div \dfrac{6x^2}{x^2-x-6}$

41. $\dfrac{6x^2-x-2}{2x^2-5x+3} \cdot \dfrac{2x^2-7x+6}{3x^2+x-2}$

42. $\dfrac{6x^2+7x-2}{4x^2-3x-1} \cdot \dfrac{5x^2-3x-2}{3x^2+7x+2}$

43. $\dfrac{x^2+3x-10}{x^2+4x+3} \div \dfrac{x^2+2x-15}{x^2-x-2}$

44. $\dfrac{25-15x}{x^2-4} \div \dfrac{3x^2-8x+5}{x+2}$

45. $(x^2-4) \cdot \dfrac{2x+3}{x^2+2x-8}$

46. $(a^2-2a) \cdot \dfrac{a+1}{6-a-a^2}$

47. $(x^2-2x-15) \div \dfrac{x^2-7x+10}{x^2+1}$

48. $\dfrac{2y^2-5y-3}{y-4} \div (y^2+y-12)$

49. $\dfrac{x^2-4}{x^2+2x-3} \cdot \dfrac{x^2+3x-4}{x^2-7x+10} \cdot \dfrac{x+3}{x^2+3x+2}$

50. $\dfrac{x^2-9}{6x^2+x-1} \cdot \dfrac{2x^2+5x+2}{x^2+4x+3} \cdot \dfrac{x^2-x-2}{x^2-3x}$

5.2
ADDITION AND SUBTRACTION OF FRACTIONS

Here is a basic principle of addition and subtraction that must be remembered.

Addition and Subtraction Principle

We can add or subtract fractions directly only if they have the same denominator.

The process is already familiar to you.

Addition and Subtraction Rule

$$\frac{a}{c} + \frac{b}{c} = \frac{a + b}{c}$$

$$\frac{a}{c} - \frac{b}{c} = \frac{a - b}{c}$$

Here c is the common denominator. We add the numerators $(a + b)$ to find the numerator of the sum and retain the common denominator. Similarly, we subtract b from a to find the numerator of the difference and retain the common denominator. For example,

$$\frac{2}{x} - \frac{4}{x} + \frac{5}{x} = \frac{2 - 4 + 5}{x} = \frac{3}{x}$$

How do we handle the addition of fractions if the denominators are not the same? We must rewrite each fraction as an equivalent fraction so that they all have the same denominator. Although any common denominator will do, we will concentrate on finding the **least common denominator,** or **L.C.D.** Here is the procedure.

TABLE 5.1

Least Common Denominator	Example
	$\dfrac{1}{x^3 - x^2}, \quad \dfrac{-2}{x^3 - x}, \quad \dfrac{3x}{x^2 + 2x + 1}$
Step 1. Factor the denominator of each fraction completely.	*Step 1.* $\dfrac{1}{x^2(x - 1)}, \quad \dfrac{-2}{x(x - 1)(x + 1)},$ $\dfrac{3x}{(x + 1)^2}$
Step 2. Determine the different factors in the denominators of the fractions, and the highest power to which each factor occurs in any denominator.	*Step 2.* <table><tr><td>Factor</td><td>Highest power</td><td>Final factor</td></tr><tr><td>x</td><td>2</td><td>x^2</td></tr><tr><td>$x - 1$</td><td>1</td><td>$x - 1$</td></tr><tr><td>$x + 1$</td><td>2</td><td>$(x + 1)^2$</td></tr></table>
Step 3. The product of the factors determined in Step 2 is the L.C.D.	*Step 3.* The L.C.D. is $x^2(x - 1)(x + 1)^2$

EXAMPLE 1
Find the L.C.D. of the fractions.

$$\frac{y + 2}{2x^2 - 18}, \quad \frac{4y}{3x^3 + 9x^2}, \quad \frac{y + 1}{(x - 3)^2(y - 1)^2}$$

Solution
Factoring each denominator completely we have

$$\frac{y + 2}{2(x + 3)(x - 3)}, \quad \frac{4y}{3x^2(x + 3)}, \quad \frac{y + 1}{(x - 3)^2(y - 1)^2}$$

The different factors and the highest power of each factor in any denominator are

Factor	Highest power	Final factor
2	1	2
3	1	3
x	2	x^2
$x + 3$	1	$x + 3$
$x - 3$	2	$(x - 3)^2$
$y - 1$	2	$(y - 1)^2$

The L.C.D. is then the product

$$6x^2(x + 3)(x - 3)^2(y - 1)^2$$

PROGRESS CHECK 1
Find the L.C.D. of the fractions.

$$\frac{2a}{(3a^2 + 12a + 12)b}, \quad \frac{-7b}{a(4b^2 - 8b + 4)}, \quad \frac{3}{ab^3 + 2b^3}$$

Answer
$12ab^3(a + 2)^2(b - 1)^2$

Having determined the L.C.D., we must convert each fraction to an equivalent fraction with the L.C.D. as its denominator. We have already seen that multiplying a fraction by 1 yields an equivalent fraction. By writing 1 in a properly chosen way, we can *force* each fraction to have the L.C.D. as its denominator. Here is the process.

TABLE 5.2

Addition of Fractions	Example
	$$\frac{4}{3x(x+3)} + \frac{x-1}{x^2(x-2)}$$
Step 1. Find the L.C.D.	L.C.D. is $3x^2(x+3)(x-2)$.
Step 2. Examine the first fraction. Multiply it by a fraction whose numerator and denominator are the same and consist of all factors of the L.C.D. which are missing in the denominator of the first fraction.	first fraction $$\left[\frac{4}{3x(x+3)}\right] \cdot \frac{x(x-2)}{\overparen{x(x-2)}}$$ factors of L.C.D. missing in denominator $3x(x+3)$ $$= \frac{4x(x-2)}{3x^2(x+3)(x-2)}$$ $$= \frac{4x^2-8x}{3x^2(x+3)(x-2)}$$
Step 3. Repeat Step 2 for each fraction.	second fraction $$\left[\frac{x-1}{x^2(x-2)}\right] \cdot \frac{3(x+3)}{\overparen{3(x+3)}}$$ factors of L.C.D. missing in denominator $x^2(x-2)$ $$= \frac{3(x-1)(x+3)}{3x^2(x-2)(x+3)}$$ $$= \frac{3(x^2+2x-3)}{3x^2(x+3)(x-2)}$$ $$= \frac{3x^2+6x-9}{3x^2(x+3)(x-2)}$$
Step 4. The fractions now all have the same denominator. Apply the addition principle. (Do not multiply out the denominators. It may be possible to perform cancellation.)	$$\frac{4}{3x(x+3)} + \frac{x-1}{x^2(x-2)} \quad \text{Original example}$$ $$= \frac{4x^2-8x}{3x^2(x+3)(x-2)} + \frac{3x^2+6x-9}{3x^2(x+3)(x-2)}$$ $$= \frac{4x^2-8x+3x^2+6x-9}{3x^2(x+3)(x-2)}$$ $$= \frac{7x^2-2x-9}{3x^2(x+3)(x-2)}$$

EXAMPLE 2
Find the sum.

(a) $\dfrac{3}{2xy} - \dfrac{4}{x^2} + \dfrac{2}{y^2}$

The L.C.D. is $2x^2y^2$ (verify). Then

$$\frac{3}{2xy} \cdot \frac{xy}{xy} - \frac{4}{x^2} \cdot \frac{2y^2}{2y^2} + \frac{2}{y^2} \cdot \frac{2x^2}{2x^2}$$

$$= \frac{3xy}{2x^2y^2} - \frac{8y^2}{2x^2y^2} + \frac{4x^2}{2x^2y^2}$$

$$= \frac{3xy - 8y^2 + 4x^2}{2x^2y^2}$$

(b) $\dfrac{2x}{x^2 - 4} + \dfrac{1}{x(x + 2)} - \dfrac{1}{x - 2}$

Since $x^2 - 4 = (x + 2)(x - 2)$, the L.C.D. is $x(x + 2)(x - 2)$. Then

$$\frac{2x}{(x + 2)(x - 2)} \cdot \frac{x}{x} + \frac{1}{x(x + 2)} \cdot \frac{x - 2}{x - 2} - \frac{1}{x - 2} \cdot \frac{x(x + 2)}{x(x + 2)}$$

$$= \frac{2x^2}{x(x + 2)(x - 2)} + \frac{x - 2}{x(x + 2)(x - 2)} - \frac{x(x + 2)}{x(x + 2)(x - 2)}$$

$$= \frac{2x^2 + x - 2 - (x^2 + 2x)}{x(x + 2)(x - 2)} = \frac{2x^2 + x - 2 - x^2 - 2x}{x(x + 2)(x - 2)}$$

$$= \frac{x^2 - x - 2}{x(x + 2)(x - 2)} = \frac{(x - 2)(x + 1)}{x(x + 2)(x - 2)} = \frac{x + 1}{x(x + 2)}$$

PROGRESS CHECK 2
Find the sum.

(a) $\dfrac{4r - 3}{9r^3} - \dfrac{2r + 1}{4r^2} + \dfrac{2}{3r}$ (b) $\dfrac{2}{n} + \dfrac{3}{n + 1} - \dfrac{5}{n + 2}$

Answers

(a) $\dfrac{6r^2 + 7r - 12}{36r^3}$ (b) $\dfrac{7n + 4}{n(n + 1)(n + 2)}$

EXERCISE SET 5.2
Find the L.C.D. in the following exercises.

1. $\dfrac{4}{x}, \dfrac{x - 2}{y}$ 2. $\dfrac{x}{x - 1}, \dfrac{x + 4}{x + 2}$

3. $\dfrac{5 - a}{a}, \dfrac{7}{2a}$ 4. $\dfrac{x + 2}{x}, \dfrac{x - 2}{x^2}$

5. $\dfrac{2b}{b - 1}, \dfrac{3}{(b - 1)^2}$ 6. $\dfrac{2 + x}{x^2 - 4}, \dfrac{3}{x - 2}$

7. $\dfrac{4x}{x - 2}, \dfrac{5}{x^2 + x - 6}$ 8. $\dfrac{3}{y^2 - 3y - 4}, \dfrac{2y}{y + 1}$

9. $\dfrac{3}{(x + 1)}, \dfrac{2}{x}, \dfrac{x}{(x - 1)}$ 10. $\dfrac{4}{x}, \dfrac{3}{x - 1}, \dfrac{x}{x^2 - 2x + 1}$

Perform the indicated operations and simplify in the following.

11. $\dfrac{2}{x} + \dfrac{5}{x}$ 12. $\dfrac{12}{b + 1} + \dfrac{3}{b + 1}$

13. $\dfrac{x}{y} + \dfrac{2x}{y}$ 14. $\dfrac{x^2 + 5}{x + 1} - \dfrac{x^2 + 3}{x + 1}$

15. $\dfrac{2a - 3}{a - 2} + \dfrac{1 - a}{a - 2}$

16. $\dfrac{x + 2}{x - 1} - \dfrac{2x - 3}{x - 1}$

17. $\dfrac{x^2 + 4x}{x + 3} - \dfrac{9 + 4x}{x + 3}$

18. $\dfrac{x^2 + x}{x + 2} + \dfrac{2x + 2}{x + 2}$

19. $\dfrac{2y - 16}{y^2 - 16} + \dfrac{2 - y}{y^2 - 16}$

20. $\dfrac{3x - 5}{x - 4} - \dfrac{x + 3}{x - 4}$

21. $\dfrac{8}{a - 2} + \dfrac{4}{2 - a}$

22. $\dfrac{x}{x^2 - 4} + \dfrac{2}{4 - x^2}$

23. $\dfrac{3}{x - 1} - \dfrac{x}{x - 1} + \dfrac{3x - 5}{x - 1}$

24. $\dfrac{y^2}{y^2 - 9} + \dfrac{9}{y^2 - 9} - \dfrac{6y}{y^2 - 9}$

25. $\dfrac{2}{x} + \dfrac{1}{5}$

26. $\dfrac{x}{4} - \dfrac{y}{3}$

27. $5 - \dfrac{2}{x}.$

28. $\dfrac{x - 1}{3} + 2$

29. $\dfrac{1}{x - 1} + \dfrac{2}{x - 2}$

30. $\dfrac{1}{a + 2} + \dfrac{3}{a - 2}$

31. $\dfrac{a}{8b} - \dfrac{b}{12a}$

32. $\dfrac{4}{3x} - \dfrac{5}{xy}$

33. $\dfrac{4x - 1}{6x^3} + \dfrac{2}{3x^2}$

34. $\dfrac{5}{2x + 6} - \dfrac{x}{x + 3}$

35. $\dfrac{2x}{x - 3} + \dfrac{4x - 2}{9 - 3x}$

36. $\dfrac{3}{a^2 - 16} - \dfrac{2}{a - 4}$

37. $\dfrac{x}{x - y} - \dfrac{y}{x + y}$

38. $\dfrac{5x}{2x^2 - 18} + \dfrac{4}{3x^2 - 9}$

39. $\dfrac{x}{x - y} + \dfrac{y}{x + y}$

40. $\dfrac{2x}{x^2 - 9} + \dfrac{5}{3x + 9}$

41. $\dfrac{4}{r} - \dfrac{3}{r + 2}$

42. $2 + \dfrac{4}{a^2 - 4}$

43. $\dfrac{1}{(x - 1)} + \dfrac{2x - 1}{(x - 2)(x + 1)}$

44. $\dfrac{2x}{2x + 1} - \dfrac{x - 1}{(2x + 1)(x - 2)}$

45. $\dfrac{a + 2}{a^2 - a} - \dfrac{2a}{a + 1}$

46. $\dfrac{2x}{x^2 + x - 2} + \dfrac{3}{x + 2}$

47. $\dfrac{2}{x - 2} + \dfrac{x}{x^2 - x - 6}$

48. $\dfrac{2x - 1}{x^2 + 5x + 6} - \dfrac{x - 2}{x^2 + 4x + 3}$

49. $\dfrac{2x - 1}{x^3 - 4x} - \dfrac{x}{x^2 + x - 2}$

50. $\dfrac{2x}{x^2 - 1} + \dfrac{x + 1}{x^2 + 3x - 4}$

51. $\dfrac{2}{x + 2} + \dfrac{3}{x - 2} - \dfrac{5}{x + 3}$

52. $\dfrac{2x}{x + 2} + \dfrac{x}{x - 2} - \dfrac{1}{x^2 - 4}$

53. $\dfrac{2}{y^2 - y} - \dfrac{y}{y + 1} + \dfrac{y + 1}{y}$

54. $\dfrac{3}{x^2 + 5x + 6} - \dfrac{2}{x^2 + 4x + 3} + \dfrac{4}{x^2 + x - 2}$

COMPLEX FRACTIONS

At the beginning of this chapter, we said that we would like to simplify fractions such as

$$\frac{1 - \dfrac{1}{x}}{\dfrac{1}{x^2} + \dfrac{1}{x}}$$

This is an example of a **complex fraction,** which is an algebraic expression with fractions in the numerator or denominator, or in both.

There are two methods commonly used to simplify complex fractions. Fortunately, we already have all the tools needed and will apply both methods to the problem.

EXAMPLE 1
Simplify.

$$\frac{1 - \dfrac{1}{x}}{\dfrac{1}{x^2} + \dfrac{1}{x}}$$

Solution

Method 1	Example
Step 1. Find the L.C.D. of all fractions appearing in the numerator and denominator.	*Step 1.* The L.C.D. of $\dfrac{1}{1}$, $\dfrac{1}{x}$, and $\dfrac{1}{x^2}$ is x^2
Step 2. Multiply the numerator and denominator by the L.C.D. Since this is multiplication by 1, the result is an equivalent fraction. Then simplify.	*Step 2.* $\dfrac{x^2\left(1 - \dfrac{1}{x}\right)}{x^2\left(\dfrac{1}{x^2} + \dfrac{1}{x}\right)} = \dfrac{x^2 - x}{1 + x} = \dfrac{x(x - 1)}{x + 1}$

Method 2	Example
Step 1. Combine the terms in the numerator into a single fraction.	*Step 1.* $1 - \dfrac{1}{x} = \dfrac{x}{x} - \dfrac{1}{x} = \dfrac{x - 1}{x}$
Step 2. Combine the terms in the denominator into a single fraction.	*Step 2.* $\dfrac{1}{x^2} + \dfrac{1}{x} = \dfrac{1}{x^2} + \dfrac{x}{x^2} = \dfrac{1 + x}{x^2}$
Step 3. Apply the rule for division of fractions, that is, multiply the numerator by the reciprocal of the denominator.	*Step 3.* $\dfrac{\dfrac{x - 1}{x}}{\dfrac{1 + x}{x^2}} = \dfrac{x - 1}{\cancel{x}} \cdot \dfrac{\overset{x}{\cancel{x^2}}}{1 + x} = \dfrac{x(x - 1)}{x + 1}$

PROGRESS CHECK 1
Simplify.

(a) $\dfrac{2 + \dfrac{1}{x}}{1 - \dfrac{2}{x}}$ (b) $\dfrac{a - 1}{1 - \dfrac{1}{a}}$

Answers

(a) $\dfrac{2x + 1}{x - 2}$ (b) a

EXAMPLE 2
Simplify.

$$\frac{\dfrac{a}{b} + \dfrac{b}{a}}{\dfrac{1}{a} - \dfrac{1}{b}}$$

Solution
The L.C.D. of all the fractions is ab. Thus,

$$\frac{\dfrac{a}{b} + \dfrac{b}{a}}{\dfrac{1}{a} - \dfrac{1}{b}} = \frac{ab\left(\dfrac{a}{b} + \dfrac{b}{a}\right)}{ab\left(\dfrac{1}{a} - \dfrac{1}{b}\right)} = \frac{\dfrac{a^2 b}{b} + \dfrac{ab^2}{a}}{\dfrac{ab}{a} - \dfrac{ab}{b}}$$

$$= \frac{a^2 + b^2}{b - a} = -\frac{a^2 + b^2}{a - b}$$

PROGRESS CHECK 2
Simplify.

(a) $\dfrac{\dfrac{y}{x} + \dfrac{1}{y}}{\dfrac{x}{y} - \dfrac{1}{x}}$ (b) $\dfrac{\dfrac{1}{x} - y}{\dfrac{1}{y} - x}$

Answers

(a) $\dfrac{x + y^2}{x^2 - y}$ (b) $\dfrac{y}{x}$

EXAMPLE 3
Write as a simple fraction.

$$1 + \frac{1 + \dfrac{1}{x}}{\dfrac{2}{x} - \dfrac{1}{x - 1}}$$

Solution

We first work on the complex fraction to simplify it. The L.C.D. of all the fractions is $x(x - 1)$. We multiply numerator and denominator by the L.C.D.

$$\frac{x(x - 1)\left(1 + \dfrac{1}{x}\right)}{x(x - 1)\left(\dfrac{2}{x} - \dfrac{1}{x - 1}\right)} = \frac{(x^2 - x) + (x - 1)}{2(x - 1) - x} = \frac{x^2 - 1}{x - 2}$$

We now substitute this equivalent fraction in the original problem, and carry out the addition.

$$1 + \frac{1 + \dfrac{1}{x}}{\dfrac{2}{x} - \dfrac{1}{x - 1}} = 1 + \frac{x^2 - 1}{x - 2} = \frac{x - 2}{x - 2} + \frac{x^2 - 1}{x - 2} = \frac{x^2 + x - 3}{x - 2}$$

PROGRESS CHECK 3

Write as a simple fraction.

(a) $\dfrac{\dfrac{1}{\dfrac{1}{x} + 1} - 1}{}$ (b) $2 + \dfrac{x}{1 - \dfrac{1}{x}}$

Answers

(a) $-\dfrac{1}{x + 1}$ (b) $\dfrac{x^2 + 2x - 2}{x - 1}$

EXERCISE SET 5.3

Simplify in the following exercises.

1. $\dfrac{1 + \dfrac{2}{x}}{1 - \dfrac{3}{x}}$

2. $\dfrac{x - \dfrac{1}{x}}{2 + \dfrac{1}{x}}$

3. $\dfrac{3 - \dfrac{4}{x}}{5x}$

4. $\dfrac{2 - \dfrac{1}{x + 1}}{x - 1}$

5. $\dfrac{x + 1}{1 - \dfrac{1}{x}}$

6. $\dfrac{1 - \dfrac{r^2}{s^2}}{1 + \dfrac{r}{s}}$

7. $\dfrac{x^2 - 16}{\dfrac{1}{4} - \dfrac{1}{x}}$

8. $\dfrac{\dfrac{a}{a - b} - \dfrac{b}{a + b}}{a^2 - b^2}$

9. $2 - \dfrac{1}{1 + \dfrac{1}{a}}$

10. $\dfrac{\dfrac{4}{x^2 - 4} + 1}{\dfrac{x}{x^2 + x - 6}}$

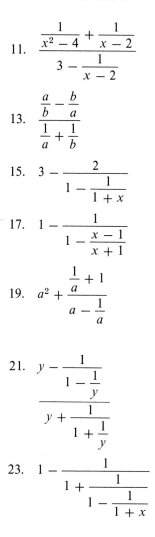

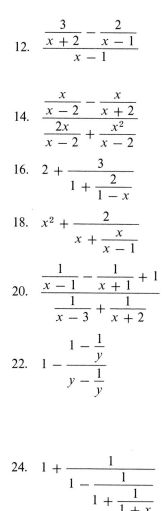

11. $\dfrac{\dfrac{1}{x^2-4}+\dfrac{1}{x-2}}{3-\dfrac{1}{x-2}}$

12. $\dfrac{\dfrac{3}{x+2}-\dfrac{2}{x-1}}{x-1}$

13. $\dfrac{\dfrac{a}{b}-\dfrac{b}{a}}{\dfrac{1}{a}+\dfrac{1}{b}}$

14. $\dfrac{\dfrac{x}{x-2}-\dfrac{x}{x+2}}{\dfrac{2x}{x-2}+\dfrac{x^2}{x-2}}$

15. $3-\dfrac{2}{1-\dfrac{1}{1+x}}$

16. $2+\dfrac{3}{1+\dfrac{2}{1-x}}$

17. $1-\dfrac{1}{1-\dfrac{x-1}{x+1}}$

18. $x^2+\dfrac{2}{x+\dfrac{x}{x-1}}$

19. $a^2+\dfrac{\dfrac{1}{a}+1}{a-\dfrac{1}{a}}$

20. $\dfrac{\dfrac{1}{x-1}-\dfrac{1}{x+1}+1}{\dfrac{1}{x-3}+\dfrac{1}{x+2}}$

21. $\dfrac{y-\dfrac{1}{1-\dfrac{1}{y}}}{y+\dfrac{1}{1+\dfrac{1}{y}}}$

22. $1-\dfrac{1-\dfrac{1}{y}}{y-\dfrac{1}{y}}$

23. $1-\dfrac{1}{1+\dfrac{1}{1-\dfrac{1}{1+x}}}$

24. $1+\dfrac{1}{1-\dfrac{1}{1+\dfrac{1}{1+x}}}$

5.4
RATIO AND PROPORTION

Suppose that in looking about your mathematics classroom you find there are 10 male and 14 female students. If we understand **ratio** to be the quotient of two quantities, we see that the ratio of male students to female students is $\frac{10}{14}$ or $\frac{5}{7}$, which we can also write as $10:14$ (read "10 to 14").

EXAMPLE 1
The entire stock of a small photography store consists of 75 Kodak and 60 Polaroid cameras.
(a)   What is the ratio of Kodak to Polaroid cameras?
(b)   What is the ratio of Polaroid cameras to the entire stock?

Solution

(a) $\dfrac{\text{Kodak}}{\text{Polaroid}}=\dfrac{75}{60}=\dfrac{5}{4}$ (or Kodak : Polaroid $= 5:4$)

(b) $\dfrac{\text{Polaroid}}{\text{Entire stock}} = \dfrac{60}{135} = \dfrac{4}{9}$ (or Polaroid : Entire stock = 4:9)

PROGRESS CHECK 1
The American alligator is an endangered species. On an American alligator reserve, there are 600 females and 900 males. Find the ratio of
(a) males to females
(b) females to males
(c) females to total number of alligators.

Answers

(a) $9:6$ or $\dfrac{3}{2}$ (b) $6:9$ or $\dfrac{2}{3}$ (c) $6:15$ or $\dfrac{2}{5}$

Ratios can often be used to solve word problems. They enable us to set up equations that can then be solved.

EXAMPLE 2
The length and width of a rectangular room are in the ratio 3:4. If the perimeter of the room is 70 feet, what are the dimensions of the room?

Solution
Let $3x$ denote the length of the room. Then the width must be $4x$, since

$$\frac{\text{length}}{\text{width}} = \frac{3}{4} = \frac{3x}{4x}$$

The perimeter P of the rectangle is given by

$$P = \text{length} + \text{length} + \text{width} + \text{width}$$
$$70 = 2(\text{length} + \text{width})$$
$$70 = 2(3x + 4x)$$
$$70 = 14x$$
$$x = 5 \text{ feet}$$

Thus,
$$\text{length} = 3x = 3(5) = 15 \text{ feet}$$
$$\text{width} = 4x = 4(5) = 20 \text{ feet}$$

PROGRESS CHECK 2
The angles of a triangle are in the ratio 2:3:4. Find the measure of each angle.

Answer
40°, 60°, 80°

When two ratios are set equal to each other we have what is called a **proportion.** We shall soon see that proportions will be useful in solving many word problems. First we practice the mechanics of operating with proportions.

EXAMPLE 3
Solve for x.

$$\frac{x}{12} = \frac{4}{9}$$

Solution

Multiplying both sides of the equation by 12, we have

$$\frac{x}{\cancel{12}} \cdot \frac{\cancel{12}}{1} = \frac{4}{9} \cdot \frac{12}{1} = \frac{48}{9} = \frac{16}{3}$$

Thus,

$$x = \frac{16}{3}$$

PROGRESS CHECK 3
Solve for r.

(a) $\dfrac{3}{r-1} = \dfrac{1}{4}$ (b) $\dfrac{4}{2} = \dfrac{2r}{5}$

Answers
(a) $r = 13$ (b) $r = 5$

We now turn to the use of proportions in the solution of word problems.

EXAMPLE 4
ABC University has determined that a student–teacher ratio of $19:2$ is ideal. If there will be 855 students next fall, how large a teaching staff will be required, assuming that the ideal student–faculty ratio will be maintained?

Solution
Let x denote the number of teachers. Then

$$\frac{\text{students}}{\text{teachers}} = \frac{19}{2} = \frac{855}{x}$$

Then multiplying by $2x$ to clear fractions,

$$19x = 2(855) = 1710$$

$$x = \frac{1710}{19} = 90$$

Thus, 90 teachers are required.

PROGRESS CHECK 4
Four out of every five homes in suburban Philadelphia have two telephones. If 25,000 new homes are to be built, how many of these will have two telephones?

Answer
20,000

EXAMPLE 5
A landscaping service advertises that the cost of sodding a 1500-square-foot area is $120. What would be the proportional cost of sodding an area of 3500 square feet?

Solution
Let x be the cost of sodding the larger area. Then

$$\frac{\text{area}}{\text{cost}} = \frac{1500}{120} = \frac{3500}{x}$$

or, after clearing fractions,

$$1500x = 420,000$$

$$x = 280$$

Thus, the proportional cost is $280.

PROGRESS CHECK 5

The cost of an airplane flight of 600 miles is $75. What is the proportional cost of a flight of 1600 miles?

Answer
$200

EXERCISE SET 5.4

Write as a ratio.

1. 5.08 centimeters to 2 inches
2. 2 inches to 5.08 centimeters
3. 8 feet to 2.5 feet
4. 1.3 inches to 12 inches
5. 12 cubic feet to 16 cubic feet
6. 2.12 quarts to 2 liters

7. The sides of a triangle are in the ratio $5:4:2$ and the perimeter of the triangle is 44 centimeters. Find the dimensions of the triangle.
8. An artist wants to make a rectangular frame whose length and width are in the ratio $2:3$. If the amount of framing material that is available is 25 inches, find the dimensions of the frame.

Solve for the unknown in each proportion.

9. $\dfrac{x}{3} = \dfrac{6}{5}$

10. $\dfrac{y}{4} = \dfrac{2}{3}$

11. $\dfrac{5}{2} = \dfrac{3}{4r}$

12. $\dfrac{6}{5r} = \dfrac{2}{3}$

13. $\dfrac{2}{r+1} = \dfrac{1}{3}$

14. $\dfrac{5}{s-1} = \dfrac{3}{2}$

15. Two numbers whose sum is 30 are in the ratio of $3:7$. Find the numbers.
16. A stockbroker charges a commission of $42 on the purchase or sale of 500 shares of stock. What would be the proportional commission on the purchase or sale of 800 shares of stock?
17. A taxpayer pays a state tax of $400 on an income of $12,000. What is the proportional tax on an income of $15,000?
18. In a certain county of Ohio, the ratio of Republicans to Democrats is $3:2$. (a) If there are 1800 Republicans, how many Democrats are there? (b) If there are 1800 Democrats, how many Republicans are there?
19. A toy train manufacturer makes a $\frac{1}{2}$-foot-long locomotive model of a 40-foot actual locomotive. If the same scale is maintained, how long is a sleeping car whose scale model is 0.75 feet long?
20. On a certain map, 2 centimeters represent 25 kilometers. How many kilometers do 8 centimeters represent?
21. If 5 out of 6 people in Philadelphia read *The Bulletin*, how many readers of this paper are there in an area with a population of 66,000 people?
22. A car uses 12 gallons of gasoline to travel 216 miles. Assuming the same type of driving, how many gallons of gasoline will be used on a 441-mile trip?

23. If a copying machine can make 740 copies in 40 minutes, how many copies can it make in 52 minutes?
24. A chemical plant that makes 120 tons daily of a certain product discharges 500 gallons of waste products into a nearby stream. If production is increased to 200 tons daily, how many gallons of waste products will be dumped into the stream?
25. A 150-pound person is given 18 cubic centimeters of a certain drug for a metabolic disorder. How much of the drug will be required by a 200-pound person, assuming the same dose-to-weight ratio?
26. A marketing research firm has determined that two out of five suburban car owners have a station wagon. How many station wagon owners are there among 40,000 suburban car owners?

5.5
EQUATIONS AND
INEQUALITIES WITH FRACTIONS

Suppose we are interested in solving the equation

$$\frac{2}{x-1} + \frac{1}{3} = \frac{1}{x-1}$$

for x. We must first try to clear the equation of all fractions. Once again, the least common denominator (L.C.D.) is exactly what we need. We shall explain the method and illustrate it with this equation.

Solving Equations with Fractions	Example
	$\dfrac{2}{x-1} + \dfrac{1}{3} = \dfrac{1}{x-1}$
Step 1. Find the L.C.D.	L.C.D. is $3(x-1)$.
Step 2. Multiply both sides of the equation by the L.C.D. Then carry out all possible cancellations. This will leave an equation without fractions.	$3(x-1)\left[\dfrac{2}{x-1} + \dfrac{1}{3}\right] = 3(x-1)\left[\dfrac{1}{x-1}\right]$ $3(x-1)\dfrac{2}{x-1} + 3(x-1)\dfrac{1}{3} = 3(x-1)\cdot\dfrac{1}{x-1}$ $6 + (x-1) = 3$
Step 3. Solve the resulting equation.	$5 + x = 3$ $x = -2$
Step 4. Check the answer(s) obtained in Step 3 by substituting in the original equation. Reject any answers that do not satisfy the equation.	Check: $\dfrac{2}{-2-1} + \dfrac{1}{3} \overset{?}{=} \dfrac{1}{-2-1}$ $-\dfrac{2}{3} + \dfrac{1}{3} \overset{?}{=} -\dfrac{1}{3}$ $-\dfrac{1}{3} \overset{\checkmark}{=} -\dfrac{1}{3}$

WARNING

When we multiply by the L.C.D., we must multiply *each term* by the L.C.D. If you are multiplying a side of an equation by the L.C.D. of $2x$, *don't* write

$$2x\left(\frac{1}{2x} + 5x\right) = 2 + 5x$$

The correct procedure is

$$2x\left(\frac{1}{2x} + 5x\right) = 2 + 10x^2$$

Note that it is not enough to clear the given equation of all fractions and proceed to find an answer: *The answer may not be a solution of the original equation.* Substituting the answer in the original equation may produce a denominator of 0, and division by 0 is not permitted. Thus, when dealing with equations that involve fractions, *you must always check that the answer is a solution* by substituting in the original equation. The following example illustrates this point.

EXAMPLE 1
Solve and check.

$$\frac{8x + 1}{x - 2} + 4 = \frac{7x + 3}{x - 2}$$

Solution

Step 1. The L.C.D. is $x - 2$.

Step 2.

$$(x - 2)\left[\frac{8x + 1}{x - 2} + 4\right] = (x - 2)\left[\frac{7x + 3}{x - 2}\right]$$

$$\cancel{(x - 2)}\frac{8x + 1}{\cancel{x - 2}} + (x - 2)\cdot 4 = \cancel{(x - 2)}\frac{7x + 3}{\cancel{x - 2}}$$

$$8x + 1 + 4x - 8 = 7x + 3$$

$$5x = 10$$

Step 3. $x = 2$.

Step 4. Check: 2 is not a solution since substituting 2 in the original equation yields a denominator of 0, and we cannot divide by zero. Thus, the given equation has no solution.

PROGRESS CHECK 1
Solve and check.

(a) $\dfrac{3}{x} - 1 = \dfrac{1}{2} - \dfrac{6}{x}$ (b) $-\dfrac{2x}{x + 1} = 1 + \dfrac{2}{x + 1}$

Answers

(a) $x = 6$ (b) *No solution*

SOLVING INEQUALITIES WITH FRACTIONS

In this chapter we will limit ourselves to inequalities with fractions that only have constants in the denominators. When the L.C.D. is a positive number we may proceed to clear fractions in exactly the same manner as with equations.

EXAMPLE 2
Solve for x.

$$\frac{x}{2} - 9 < \frac{1 - 2x}{3}$$

Solution
Step 1. The L.C.D. is $2 \cdot 3 = 6$.

Step 2. Multiply both sides by the L.C.D.

$$6\left[\frac{x}{2} - 9\right] < 6\left[\frac{1 - 2x}{3}\right]$$

$$\frac{6x}{2} - 54 < \frac{\overset{2}{\cancel{6}}(1 - 2x)}{\cancel{3}}$$

$$3x - 54 < 2(1 - 2x) = 2 - 4x$$

Step 3.

$$3x - 54 < 2 - 4x$$
$$3x + 4x < 2 + 54$$
$$7x < 56$$
$$x < 8$$

PROGRESS CHECK 2
Solve.

(a) $\dfrac{3x - 1}{4} + 1 > 2 + \dfrac{x}{3}$ (b) $\dfrac{2 + 3x}{5} - 1 \le \dfrac{x + 1}{3}$

Answers

(a) $x > 3$ (b) $x \le \dfrac{7}{2}$

Not every inequality has a solution. Here is an example.

EXAMPLE 3
Solve.

$$\frac{2(x + 1)}{3} < \frac{2x}{3} - \frac{1}{6}$$

Solution

The L.C.D. is 6. Then

$$6\left[\frac{2(x+1)}{3}\right] < 6\left[\frac{2x}{3} - \frac{1}{6}\right] \quad \text{Multiplying both sides by 6}$$

$$4(x+1) < 6\left(\frac{2x}{3}\right) - 6 \cdot \frac{1}{6}$$

$$4x + 4 < 4x - 1$$

$$4 < -1$$

Our procedure has led to a contradiction, indicating that there is no solution to the inequality.

PROGRESS CHECK 3

Solve.

$$\frac{2x-3}{2} \geq x + \frac{2}{5}$$

Answer

No solution.

EXERCISE SET 5.5

Solve and check.

1. $\dfrac{x}{2} = \dfrac{5}{3}$

2. $\dfrac{3x}{4} - 5 = \dfrac{1}{4}$

3. $\dfrac{2}{x} + 1 = \dfrac{3}{x}$

4. $\dfrac{5}{a} - \dfrac{3}{2} = \dfrac{1}{4}$

5. $\dfrac{x}{x+3} = \dfrac{3}{5}$

6. $\dfrac{a}{a-2} = \dfrac{3}{5}$

7. $\dfrac{2y-3}{y+3} = \dfrac{5}{7}$

8. $\dfrac{1-4x}{1-2x} = \dfrac{9}{8}$

9. $\dfrac{1}{x-2} + \dfrac{1}{2} = \dfrac{2}{x-2}$

10. $\dfrac{4}{x-4} - 2 = \dfrac{1}{x-4}$

11. $\dfrac{3r+1}{r+3} + 2 = \dfrac{5r-2}{r+3}$

12. $\dfrac{2x-1}{x-5} + 3 = \dfrac{3x-2}{5-x}$

13. $\dfrac{2x-3}{2x+1} + 3 = \dfrac{2}{2x+1}$

14. $\dfrac{4t+3}{2t-1} - 2 = \dfrac{5}{2t-1}$

15. $\dfrac{2}{x-2} + \dfrac{2}{x^2-4} = \dfrac{3}{x+2}$

16. $\dfrac{3}{x-1} + \dfrac{2}{x+1} = \dfrac{5}{x^2-1}$

17. $\dfrac{4-3x}{3x-1} = \dfrac{3}{2} - \dfrac{2x+3}{1-3x}$

18. $\dfrac{3a+2}{a-3} + 1 = \dfrac{a+8}{a-3}$

19. $\dfrac{x}{x-1} - 1 = \dfrac{3}{x+1}$

20. $\dfrac{2}{x-2} + 1 = \dfrac{x+2}{x-2}$

21. $\dfrac{4}{b} - \dfrac{1}{b+3} = \dfrac{3b+2}{b^2+2b-3}$

22. $\dfrac{3}{x^2-2x} + \dfrac{2x-1}{x^2+2x-8} = \dfrac{2}{x+4}$

23. $\dfrac{x}{2} - 3 < \dfrac{1}{2}$

24. $\dfrac{a}{3} - 2 > \dfrac{2}{3}$

25. $\dfrac{x}{2} - 2 \geq x - \dfrac{3}{2}$

26. $\dfrac{t}{3} - 1 \leq t + \dfrac{2}{3}$

27. $\dfrac{4 - x}{3} \geq 2 - x$

28. $\dfrac{x}{3} - 5 < \dfrac{2 - 3x}{4}$

5.6
APPLICATIONS; WORK PROBLEMS

Many applications involve equations or inequalities with algebraic fractions. Now that we have learned how to handle algebraic fractions we can tackle these applications.

EXAMPLE 1
A certain number is 3 times another number. If the sum of their reciprocals is $\frac{20}{3}$, find the numbers.

Solution

then

Let $x =$ the smaller number

$3x =$ the larger number

The reciprocals of these numbers are $\dfrac{1}{x}$ and $\dfrac{1}{3x}$, so that

$$\frac{1}{x} + \frac{1}{3x} = \frac{20}{3}$$

We multiply both sides by the L.C.D., which is $3x$.

$$3x\left[\frac{1}{x} + \frac{1}{3x}\right] = 3x\left(\frac{20}{3}\right)$$

$$3 + 1 = 20x$$

$$4 = 20x$$

$$x = \frac{4}{20} = \frac{1}{5}$$

The numbers are then $\frac{1}{5}$ and $\frac{3}{5}$.

Check: We add the reciprocals,

$$\frac{5}{1} + \frac{5}{3} = \frac{15}{3} + \frac{5}{3} = \frac{20}{3}$$

and verify that $\frac{1}{5}$ and $\frac{3}{5}$ constitute a solution.

PROGRESS CHECK 1
One number is twice another. If the difference of their reciprocals is $\frac{8}{3}$, find the numbers.

Answer

$\dfrac{3}{8}$ *and* $\dfrac{3}{16}$; $-\dfrac{3}{8}$ *and* $-\dfrac{3}{16}$

EXAMPLE 2

The denominator of a fraction is 2 more than its numerator. If $\frac{5}{2}$ is added to the fraction, the result is $\frac{17}{6}$. Find the fraction.

Solution

then

$$\text{Let} \quad x = \text{the numerator of the fraction}$$
$$x + 2 = \text{the denominator of the fraction}$$

The fraction is then $\dfrac{x}{x + 2}$ and we have

$$\frac{x}{x + 2} + \frac{5}{2} = \frac{17}{6}$$

We multiply both sides by the L.C.D., which is $6(x + 2)$.

$$6(x + 2)\left[\frac{x}{x + 2} + \frac{5}{2}\right] = 6(x + 2)\left(\frac{17}{6}\right)$$

$$6x + 15(x + 2) = 17(x + 2)$$

$$21x + 30 = 17x + 34$$

$$4x = 4$$

$$x = 1$$

Then

$$\frac{x}{x + 2} = \frac{1}{3}$$

is the fraction. (Verify that the fraction $\frac{1}{3}$ does satisfy the given conditions.)

PROGRESS CHECK 2

The numerator of a fraction is 3 more than its denominator. If the fraction is subtracted from $\frac{11}{4}$ the result is 1. Find the fraction.

Answer

$\dfrac{7}{4}$

EXAMPLE 3

An airplane flying against the wind travels 150 miles in the same time that it can travel 180 miles with the wind. If the wind speed is 10 miles per hour, what is the speed of the airplane in still air?

Solution

Let $r = $ the rate (or speed) of the airplane in still air. Let's display the information that we have.

	Rate	$\times$	Time	$=$	Distance
With wind	$r + 10$		t		180
Against wind	$r - 10$		t		150

From the equation rate $\times$ time $=$ distance, we know

$$\text{time} = \frac{\text{distance}}{\text{rate}}$$

Since we are told that the time of travel with the wind is the same as that against the wind

$$\text{time with wind} = \text{time against wind}$$

$$\frac{180}{r + 10} = \frac{150}{r - 10}$$

Multiplying both sides of the equation by the L.C.D., which is $(r + 10)(r - 10)$, we have

$$(r + 10)(r - 10)\left[\frac{180}{r + 10}\right] = (r + 10)(r - 10)\left[\frac{150}{r - 10}\right]$$

$$180(r - 10) = 150(r + 10)$$

$$180r - 1800 = 150r + 1500$$

$$30r = 3300$$

$$r = 110$$

The rate of the plane in still air is 110 miles per hour.

PROGRESS CHECK 3
An express train travels 200 miles in the same time that a local train travels 150 miles. If the express train travels 20 miles per hour faster than the local train, find the rate of each train.

Answer
60 and 80 miles per hour

There is a class of problems called **work problems** which lead to equations with fractions. Work problems typically involve two or more people or machines working on the same task. The key to these problems is to express the rate of work per unit of time, whether it be hours, days, weeks, and so on. For example, if a machine can do a job in 5 days, then

$$\text{rate of machine} = \frac{1}{5} \text{ job per day}$$

If this machine were used for two days, it would perform $\frac{2}{5}$ of the job. In summary,

---

If a machine (or person) can complete a job in n days, then

$$\text{Rate of machine (or person)} = \frac{1}{n} \text{ job per day}$$

$$\text{Work done} = \text{Rate} \times \text{Time}$$

EXAMPLE 4
A firm with two factories receives an order for the manufacture of circuit boards. Factory #1 can complete the order in 20 days and factory #2 can complete the order in 30 days. How long will it take to complete the order if both factories are assigned to the task?

Solution

Let x = number of days for completing the job when both factories are assigned. We can display the information in a table.

	Rate	×	Time	=	Work done
Factory #1	$\dfrac{1}{20}$		x		$\dfrac{x}{20}$
Factory #2	$\dfrac{1}{30}$		x		$\dfrac{x}{30}$

Since

$$\begin{array}{c}\text{work done by} \\ \text{factory \#1}\end{array} + \begin{array}{c}\text{work done by} \\ \text{factory \#2}\end{array} = 1 \text{ whole job}$$

we have

$$\frac{x}{20} + \frac{x}{30} = 1$$

Multiplying both sides by the L.C.D., which is 60, we have

$$60\left[\frac{x}{20} + \frac{x}{30}\right] = 60 \cdot 1$$

$$3x + 2x = 60$$

$$5x = 60$$

$$x = 12$$

Thus, it takes 12 days to complete the job if both factories are assigned to the task.

PROGRESS CHECK 4

An electrician can complete a job in 6 hours, while his assistant requires 12 hours to do the same job. How long would it take the electrician and assistant, working together, to do the job?

Answer
4 hours

EXAMPLE 5

Using a small mower, a student begins to mow a lawn at 12 noon, a job which would take him 9 hours. At 1 P.M. another student, using a tractor, joins him and they complete the job together at 3 P.M. How many hours would it take to do the job by tractor only?

Solution

Let x = number of hours to do the job by tractor alone. The small mower worked from 12 noon to 3 P.M., or 3 hours; the tractor was used from 1 P.M. to 3 P.M., or 2 hours. All of the information can be displayed in a table.

	Rate	×	Time	=	Work done
Small mower	$\dfrac{1}{9}$		3		$\dfrac{3}{9}$
Tractor	$\dfrac{1}{x}$		2		$\dfrac{2}{x}$

Since

$$\begin{matrix} \text{work done by} \\ \text{small mower} \end{matrix} + \begin{matrix} \text{work done by} \\ \text{tractor} \end{matrix} = 1 \text{ whole job}$$

$$\frac{3}{9} + \frac{2}{x} = 1$$

To solve, multiply both sides by the L.C.D., which is $9x$.

$$9x\left[\frac{3}{9} + \frac{2}{x}\right] = 9x \cdot 1$$

$$3x + 18 = 9x$$

$$18 = 6x$$

$$x = 3$$

Thus, by tractor alone, the job can be done in 3 hours.

PROGRESS CHECK 5

A printing press can print the morning newspaper in 6 hours. After the press has been in operation for 2 hours, a second press joins in printing the paper and both presses finish the job in 2 more hours. How long would it take the second press to print the morning newspaper if it had to do the entire job alone?

Answer
6 hours

EXERCISE SET 5.6

1. If two-thirds of a certain number is added to one-half of the number, the result is 21. Find the number.
2. If one-fourth of a certain number is added to one-third of the number, the result is 14. Find the number.
3. A certain number is twice another. If the sum of their reciprocals is 4, find the numbers.
4. A certain number is 3 times another. If the difference of their reciprocals is 8, find the numbers.
5. The denominator of a fraction is one more than its numerator. If $\frac{1}{2}$ is added to the fraction, the result is $\frac{5}{4}$. Find the fraction.
6. The numerator of a certain fraction is 3 less than its denominator. If $\frac{1}{10}$ is added to the fraction, the result is $\frac{1}{2}$. Find the fraction.
7. If $\frac{3}{2}$ is added to twice the reciprocal of a certain number, the result is 2. Find the number.
8. If $\frac{1}{3}$ is subtracted from 3 times the reciprocal of a certain number, the result is $\frac{25}{6}$. Find the number.
9. John can mow a lawn in 2 hours and Peter can mow the same lawn in 3 hours. How long would it take to mow the lawn if they worked together?
10. Computer A can carry out an engineering analysis in 4 hours, while computer B can do the same job in 6 hours. How long would it take to complete the job if both computers work together?
11. Jackie, Lisa, and Susan can paint a certain room in 3, 4, and 2 hours, respectively. How long would it take to paint the room if they all work together?
12. A mechanic and assistant, working together, can repair an engine in 3 hours. Working alone, the mechanic can complete the job in 5 hours. How long would it take the assistant to do the job alone?

13. Copying machines A and B, working together, can prepare enough copies of the annual report for the board of directors meeting in 2 hours. Machine A, working alone, requires 3 hours to do the job. How long would it take machine B to do the job by itself?

14. A 5-horsepower snowblower together with a 8-horsepower snowblower can clear a parking lot in 1 hour. The 5-horsepower blower, working alone, can do the job in 3 hours. How long would it take the 8-horsepower blower to do the job by itself?

15. Hoses A and B together can fill a swimming pool in 5 hours. If hose A alone takes 12 hours to fill the pool, how long would it take hose B to fill the pool?

16. A senior copyeditor together with a junior copyeditor can edit a book in 3 days. The junior editor, working alone, takes twice as long to complete the job as the senior editor would require if working alone. How long would it take each editor to complete the job by herself?

17. Hose A can fill a certain vat in 3 hours. After 2 hours of pumping, hose A is turned off. Hose B is then turned on and completes filling the vat in 3 hours. How long would it take hose B alone to fill the vat?

18. A shovel dozer together with a large backhoe can complete a certain excavation project in 4 hours. The shovel dozer is half as fast as the large backhoe. How long would it take each piece of equipment to complete the job by itself?

19. A printing shop starts a job at 10 A.M. using press A. Using this press alone, it would take 8 hours to complete the job. At 2 P.M. press B is also turned on and both presses together finish the job at 4 P.M. How long would it take press B to do the job by itself?

20. A moped covers 40 miles in the same time that a bicycle covers 24 miles. If the rate of the moped is 8 miles per hour faster than the bicycle, find the rate of each vehicle.

21. A boat travels 20 kilometers upstream in the same time that it takes the same boat to travel 30 kilometers downstream. If the rate of the stream is 5 kilometers per hour, find the speed of the boat in still water.

22. An airplane flying against the wind travels 300 miles in the same time that it takes the same plane to travel 400 miles with the wind. If the wind speed is 20 miles per hour, find the speed of the airplane in still air.

23. Car A can travel 20 kilometers per hour faster than car B. If car A covers 240 kilometers in the same time that car B covers 200 kilometers, what is the speed of each car?

24. A sedan is 20 miles per hour slower than a sports car. If the sedan can travel 160 miles in the same time that the sports car travels 240 miles, find the speed of each car.

TERMS AND SYMBOLS

cancellation principle   complex fraction (p. 125) $a:b$ (p. 128)
  (p. 116)   ratio (p. 128)   proportion (p. 129)
L.C.D. (p. 120)

KEY IDEAS FOR REVIEW

☐ Multiplication and division of algebraic fractions follow the same rules as arithmetic fractions.

$$\frac{a}{b} \cdot \frac{c}{d} = \frac{ac}{bd} \qquad \frac{\dfrac{a}{b}}{\dfrac{c}{d}} = \frac{a}{b} \cdot \frac{d}{c} = \frac{ad}{bc}$$

☐ Nonzero *factors* appearing in the numerator and denominator of a fraction may be canceled.

☐ The L.C.D. of two or more algebraic fractions is found by forming the product of all the different factors in the denominators, each factor raised to the highest power that it has in any denominator.

☐ Complex fractions can be simplified by
    Method 1. Multiplying the numerator and denominator by the L.C.D. of all the fractions appearing in the complex fraction.
    Method 2. Reducing the numerator and denominator, independently, into simple fractions and then performing the division.

☐ To solve an equation involving fractions, find the L.C.D. and multiply both sides of the equation by the L.C.D. This will produce an equation without fractions. If the L.C.D. involved the unknown, then the answer *must be checked.*

☐ Inequalities with only constants in the denominator can be handled in the same manner as equations if we choose the L.C.D. to be a positive number.

COMMON ERRORS

1. It's essential to factor completely before attempting cancellation. This will allow you to see which factors are common to the entire numerator and the entire denominator.

2. *Don't* write

$$\frac{xy - 2y}{x} = y - 2y$$

Since x is not a multiplicative factor of the entire numerator, we may not cancel.

3. *Don't* write

$$\frac{x^2 - y^2}{x - y} = x - y$$

After factoring completely, we see that

$$\frac{x^2 - y^2}{x - y} = \frac{(x + y)(x - y)}{x - y} = x + y$$

4. If an equation is multiplied by an L.C.D. involving a variable, you must check that the answer is a solution of the original equation. The answer may result in a zero in a denominator, in which case it is not a solution.

PROGRESS TEST 5A

1. Multiply $\dfrac{-8x^2(4 - x^2)}{2y^2} \cdot \dfrac{3y}{x - 2}$.

2. Divide $\dfrac{2x^2 - x - 6}{x^2 + x - 12}$ by $\dfrac{-x^2 - 2x + 8}{3x^2 - 10x + 3}$.

3. Compute $\dfrac{2x^2 - 5x + 2}{5 - x} \div \dfrac{3x}{x^2 - 6x + 5}$.

4. Find the L.C.D. of $\dfrac{2}{(x - 1)y}, \dfrac{-4}{y^2}, \dfrac{x + 2}{5(x - 1)^2}$.

5. Find $\dfrac{2}{x-5} - \dfrac{11}{5-x}.$

6. Find $\dfrac{4}{y(y+1)} - \dfrac{3y}{y+1} + \dfrac{y-1}{2y}.$

7. Simplify $\dfrac{6x-3}{2-\dfrac{1}{x}}.$

8. Simplify $\dfrac{1-\dfrac{7}{x^2-9}}{\dfrac{x-4}{x^2+x-6}}.$

9. Solve $\dfrac{3}{n+2} = \dfrac{1}{2}.$

10. The interest on a $3000 loan is $125. What is the interest on a $7000 loan at the same interest rate for the same length of time?

11. Solve $\dfrac{-5x-2}{2x+6} - 2 = \dfrac{x+4}{x+3}.$

12. Solve $\dfrac{2x+3}{2} \leq \dfrac{x}{6} - 4.$

13. An apprentice plumber can complete a job in 6 hours. After he has been working on the assignment for 2 hours, he is joined by a master plumber and the two complete the job in 1 more hour. How long would it take the master plumber working alone to do the entire job?

PROGRESS TEST 5B

1. Multiply $\dfrac{14(y-1)}{3(x^2-y^2)} \cdot \dfrac{9(x+y)}{-7y^2}.$

2. Divide $\dfrac{5-x}{3x^2+5x-2}$ by $\dfrac{x^2-4x-5}{2x^2+3x-2}.$

3. Compute $\dfrac{x^4-x^2}{x^2-4} \div \dfrac{x^2-x}{-2x+4}.$

4. Find the L.C.D. of $\dfrac{y-1}{x^2(y+1)}, \quad \dfrac{x-2}{2x(y-1)}, \quad \dfrac{3x}{4(y+1)^2}.$

5. Find $\dfrac{x+1}{3-x} + \dfrac{x-1}{x-3}.$

6. Find $\dfrac{3}{4(v-1)} + \dfrac{v+2}{v^2(v-1)} - \dfrac{v+1}{2v}.$

7. Simplify $\dfrac{2x-4}{\dfrac{2}{x}-1}.$

8. Simplify $\dfrac{\dfrac{x^2-x-6}{2x+2}}{\dfrac{3}{x^2-1}-1}.$

9. Solve $\dfrac{4}{t-1} = \dfrac{2}{3}$.

10. A golfer cards a 36 for the first 8 holes. What is her proportional score for 18 holes?

11. Solve $\dfrac{1-4x}{2x-2} + 6 = \dfrac{2x}{1-x}$.

12. Solve $\dfrac{3-x}{3} + 3 \geq \dfrac{x}{2}$.

13. A fast collator can do a job in 4 hours while a slower-model collator requires 6 hours. They are both assigned to a job, but after one hour the slower collator is reassigned. How long will it take the fast collator to finish the job?

CHAPTER SIX
FUNCTIONS

What is the effect of increased fertilization on the growth of an azalea? If the minimum wage is increased, what will be the impact upon the number of unemployed workers? When a submarine dives, can we calculate the water pressure on the hull at a given depth?

Each of the questions posed seeks a relationship between phenomena. The search for relationships or correspondence is a central concept in man's attempt to study his universe and is used in mathematics, engineering, the natural and biological sciences, the social sciences, and in business and economics.

The concept of a function has been developed as a means of organizing and assisting in the study of relationships. Since graphs are powerful means of exhibiting relationships, we begin with a study of the Cartesian or rectangular coordinate system. We will then formally define a function and will provide a number of ways of viewing the function concept. Function notation will be introduced to provide a convenient means of writing functions.

Finally, we will explore some special types of functional relationships (increasing and decreasing functions) and will see that variation can be viewed as a functional relationship.

6.1
RECTANGULAR COORDINATE SYSTEMS

In Chapter One we associated the system of real numbers with points on the real number line. That is, we saw that there is a one-to-one correspondence between the system of real numbers and points on the real number line.

We will now develop an analogous way to handle points in a plane. We begin by drawing a pair of perpendicular lines intersecting at a point O called the **origin.** One of the lines, called the **x-axis,** is usually drawn in a horizontal position. The other line, called the **y-axis,** is usually drawn vertically.

If we think of the x-axis as a real number line, we may mark off some convenient unit of length with positive numbers to the right of the origin and negative numbers to the left of the origin. Similarly, we may think of the y-axis as a real number line. Again, we may mark off a convenient unit of length (usually the same as the unit of length on the x-axis) with the upward direction representing positive numbers and the downward direction negative numbers. The x and y axes are called **coordinate axes,** and together they constitute a **rectangular coordinate system** or a **Cartesian coordinate system** (Figure 6.1).

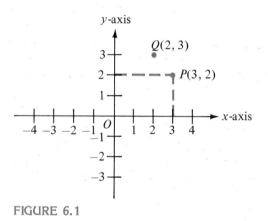

FIGURE 6.1

By using coordinate axes, we can outline a procedure for labeling a point P in the plane (see Figure 6.1). From P, draw a perpendicular to the x-axis and note that it meets the x-axis at $x = 3$. Now draw a perpendicular from P to the y-axis and note that it meets the y-axis at $y = 2$. We say that the **coordinates** of P are given by the **ordered pair** $(3, 2)$. The term "ordered pair" means that the order is significant, that is, the ordered pair $(3, 2)$ is different from the ordered pair $(2, 3)$. In fact, the ordered pair $(2, 3)$ gives the coordinates of the point Q shown in Figure 6.1.

The first number of the ordered pair (a, b) is sometimes called the **abscissa** of P and the second number is called the **ordinate** of P. We will use a simpler terminology. We call a the **x-coordinate** (since we measure it along the x-axis) and b the **y-coordinate** (since we measure it along the y-axis).

Let's recap what we have done. We now have a procedure by which each point P in the plane determines a unique ordered pair of real numbers (a, b). It is customary to write the point P as $P(a, b)$. It is also true that every ordered pair of real numbers (a, b) determines a unique point P in the plane.

We can note a few additional facts using Figure 6.2.

The coordinate axes divide the plane into four **quadrants** which we label I, II, III, and IV. The point $(-3, 2)$ is in Quadrant II; the point $(2, -4)$ is in Quadrant IV.

All points on the x-axis have a y-coordinate of 0. For example, point A has coordinates $(-1, 0)$; point E has coordinates $(3, 0)$.

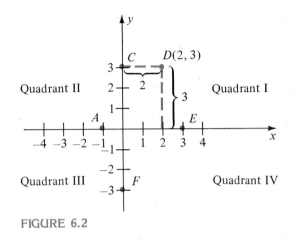

FIGURE 6.2

All points on the y-axis have an x-coordinate of 0. For example, point C has coordinates (0, 3) and point F has coordinates (0, −3).

The x-coordinate of a point is the distance of the point from the y-axis; the y-coordinate is the distance from the x-axis. Point D(2, 3) is 2 units from the y-axis and 3 units from the x-axis.

EXAMPLE 1
Find the coordinates of the points A, B, C, D, and E in Figure 6.3.

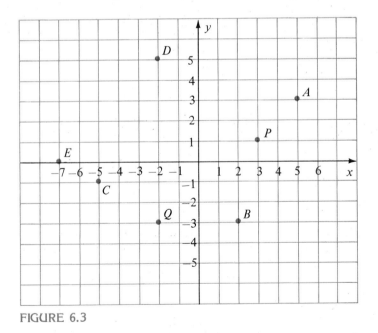

FIGURE 6.3

Solution
$A(5, 3)$, $B(2, -3)$, $C(-5, -1)$, $D(-2, 5)$, $E(-7, 0)$

PROGRESS CHECK 1
Using Figure 6.3 in Example 1, find the coordinates of
(a)   P and Q   (b)   the point one unit to the right and two units below A
(c)   the point three units to the left and four units up from C.

Answers

(a)   P(3, 1),   Q(−2, −3)   (b)   (6, 1)   (c)   (−8, 3)

EXAMPLE 2

Plot the following points and state the quadrant in which each point lies.

A(1, 6)   B(−3, −3)   C(−3, 2)

D(2, −5)   E(4, 0)   F(0, −1)

Solution

See Figure 6.4.

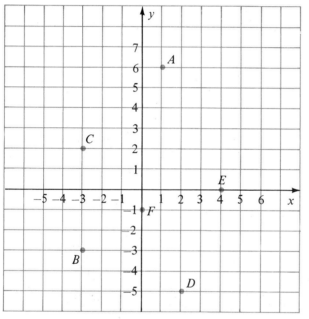

A: Quadrant I
B: Quadrant III
C: Quadrant II
D: Quadrant IV
E: Not in a quadrant
F: Not in a quadrant

FIGURE 6.4

PROGRESS CHECK 2

Plot each of the following points and state the quadrant in which it lies.

L(−4, −2)   R(2, 5)   M(0, 3)

S(−1, 0)   T(−2, 3)   U(3, −1.5)

Answers

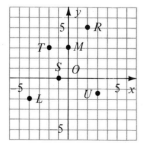

L: Quadrant III
R: Quadrant I
M: Not in a quadrant
S: Not in a quadrant
T: Quadrant II
U: Quadrant IV

GRAPHS OF EQUATIONS

The Cartesian coordinate system provides a means of drawing a "picture" or **graph of an equation in two variables.** In general, by the graph of an equation in two variables x and y we shall mean the set in the plane of all points $P(x, y)$ whose coordinates (x, y) satisfy the given equation.

Let's try to graph $y = x^2 - 4$, an equation in the variables x and y. A solution to this equation is any pair of values that when substituted in the equation in place of x and y yields a true statement. If we choose a value for x, say $x = 3$, and substitute this value of x in the equation, we obtain the corresponding value of y.

$$y = x^2 - 4$$
$$y = (3)^2 - 4 = 5$$

Thus, $x = 3$, $y = 5$ is a solution. Table 6.1 shows a number of solutions. (Verify that these are solutions.)

TABLE 6.1

x	-3	-2	-1	0	1	2	$\dfrac{5}{2}$
$y = x^2 - 4$	5	0	-3	-4	-3	0	$\dfrac{9}{4}$

We can treat the numbers in Table 6.1 as ordered pairs (x, y) and plot the points that they represent. Figure 6.5a shows the points; in Figure 6.5b we have joined the points to form a smooth curve, which is the graph of the equation.

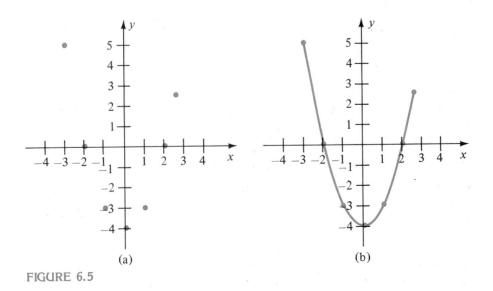

(a)   (b)

FIGURE 6.5

EXAMPLE 3

Graph the equation $y = |x + 1|$.

Solution

Form a table by assigning values to x and calculating the corresponding values of y.

x	-3	-2	-1	0	1	2	3	4
$y = |x + 1|$	2	1	0	1	2	3	4	5

(Verify that the table entries are correct. Remember—we are dealing with absolute value.) Now we plot the points and join the points in a "smooth" curve (Figure 6.6). The curve of $y = |x + 1|$ appears to be two rays intersecting at the point $(-1, 0)$. (A ray is a straight line with one endpoint.)

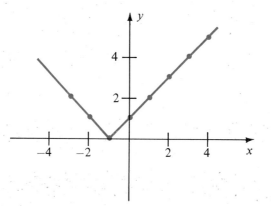

FIGURE 6.6

PROGRESS CHECK 3

Graph the equations.

(a) $y = 4 - x^2$ (b) $y = |x - 2|$

Answers

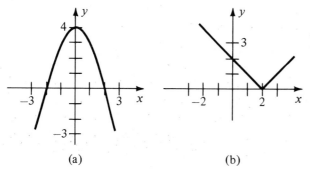

(a)   (b)

EXAMPLE 4
Graph the equation $x = y^2 + 1$.

Solution
We need to find ordered pairs (x, y) which will satisfy the given equation. In this case it is easiest to pick a value of y and find the corresponding value of x from the given equation. Thus, if $y = 0$, then $x = 1$, and if $y = 1$, then $x = 2$. In this manner we obtain the following table.

y	-3	-2	-1	0	1	2	3
x	10	5	2	1	2	5	10
(x, y)	$(10, -3)$	$(5, -2)$	$(2, -1)$	$(1, 0)$	$(2, 1)$	$(5, 2)$	$(10, 3)$

The graph is shown in Figure 6.7.

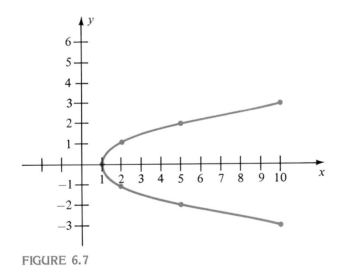

FIGURE 6.7

PROGRESS CHECK 4
Graph the equation $x = 4 - y^2$.

Answer

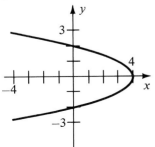

EXERCISE SET 6.1

Find the coordinates of the points *A, B, C, D, E, F, G,* and *H* from the following graphs.

1.

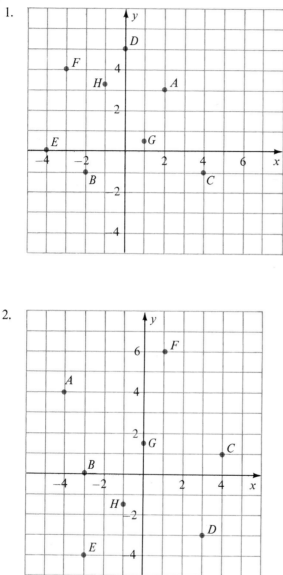

2.

Plot the given points in the following exercises.

3. (2, 3), (−3, −2), (−1, 0), (4, −4), (−1, 1), (0, −2)

4. (−3, 4), (3, 0), (3, 2), (3, −3), (0, 4), (−1, −2)

5. (4, −3), (1, 4), (−5, 2), (0, 0), (−4, 0), (−4, −5)

6. (−4, −2), (2, 1), (−3, 1), (−2, 0), (0, 2), (1, −3)

7. $\left(-\frac{1}{2}, \frac{1}{2}\right)$, (2.5, 1.5), (−7.5, −2.5), $\left(1, -\frac{1}{2}\right)$, $\left(-\frac{1}{2}, 0\right)$, $\left(0, \frac{1}{4}\right)$

8. $\left(-\frac{5}{2}, -\frac{3}{2}\right)$, $\left(0, -\frac{1}{2}\right)$, $\left(\frac{1}{4}, 1\right)$, $\left(-1, \frac{3}{2}\right)$, $\left(0, \frac{3}{2}\right)$, $\left(\frac{1}{2}, 0\right)$

9. Using the figure in Exercise 1, find the coordinates of
   (a) the point 2 units to the left and 1 unit above B
   (b) the point 4 units to the right and 5 units below A.

10. Using the figure in Exercise 2, find the coordinates of
    (a) the point 3 units to the right and 2 units above A
    (b) the point 2 units to the left and 1 unit below H.

Without plotting, name the quadrant in which the given point is located in the following exercises.

11. (2, 4)   12. (−3, 80)   13. (200, −80)

14. (−5, 20)   15. (−8, −26)   16. (40, −20.1)

17. (π, 8)   18. (−2, 0.3)   19. (−84.7, −12.8)

20. (2.84, −80)   21. $\left(\frac{17}{4}, \frac{4}{5}\right)$ 22. (−0.5, 0.3)

23. Which of the following are solutions to $2x - 3y = 12$?
    (a) (0, −4)   (b) (1, 3)   (c) (3, 1)   (d) (3, −2)

24. Which of the following are solutions to $3x + 2y = 18$?
    (a) (−4, 15)   (b) $\left(0, -\frac{3}{2}\right)$ (c) (−9, 0)   (d) (4, 3)

25. Which of the following are solutions to $2x + 3y^2 = 18$?
    (a) (3, −2)   (b) (2, 1)   (c) (9, 0)   (d) (15, 4)

26. Which of the following are solutions to $3x^2 - 2y = 12$?
    (a) (0, −6)   (b) (4, 30)   (c) (2, 0)   (d) (−2, 12)

27. Consider the equation $4x + 3y = 12$. Complete the following table so that each ordered pair (x, y) is a solution of the given equation.

x	1		0		−3	
y		−2		0		2

28. Consider the equation $2x - 3y = 6$. Complete the following table so that each ordered pair (x, y) is a solution of the given equation.

x	6		0		−3	
y		−6		0		2

Graph the given equation in Exercises 29–59.

29. $y = 2x$
30. $y = 3x$
31. $y = 2x + 4$

32. $y = -3x + 5$
33. $3x - 2y = 6$
34. $x = 2y + 3$

35. $3x + 5y = 15$
36. $x = 2$
37. $y = -3$

38. $y = x^2 + 3$
39. $y = 3 - x^2$
40. $y = 3x - x^2$

41. $x = y^2 - 1$
42. $x = 2 - y^2$
43. $y = x^3 + 1$

44. $y = x^3 - 2$
45. $x = y^3 - 1$
46. $x = 2 - y^3$

47. $y = |x - 2|$
48. $y = |x + 3|$
49. $y = |x| + 1$

50. $y = \dfrac{1}{2x + 1}$
51. $xy = 2$
52. $2x^2 + y = 4$

53. $x^2 - y + 8 = 0$

54. $2.23y - 6.47x + 3.41 = 0.$

55. $7.37y + 2.75x = 9.46$

56. $3.17x^2 - 2.02y - 3.73 = 0$

57. $6.59x^2 + 3.72y = -9.82$

58. $4.81y^2 - 3.07x + 4.21 = 0$

59. $8.07y^2 + 0.11x = 3.46$

60. Graph the set of all points whose y-coordinate is 3.
61. What is the equation whose graph is shown below?

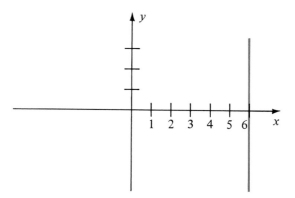

62. The points $A(2, 7)$, $B(4, 3)$, and $C(x, 3)$ determine a right triangle whose hypotenuse is AB. Find x.
63. The points $A(2, 6)$, $B(4, 6)$, $C(4, 8)$, and $D(x, y)$ form a rectangle. Find x and y.

6.2
FUNCTIONS AND FUNCTION
NOTATION

The equation

$$y = 2x + 3$$

can be thought of as a rule that assigns a value to y for every value of x. If we let X denote the set of values which we can assign to x and let Y denote the set of values which the equation assigns to y, we can show the correspondence schematically (Figure 6.8).

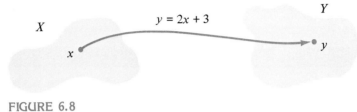

FIGURE 6.8

We are particularly interested in those rules that assign exactly one y in Y for a given x in X. This type of correspondence plays a fundamental role in many mathematical applications and is given a special name.

A **function** is a rule which, for each x in a set X, assigns exactly one y in a set Y. The element y is called the **image** of x. The set X is called the **domain** of the function, and the set of all images is called the **range** of the function.

We can think of the rule defined by the equation $y = 2x + 3$ as a function machine (see Figure 6.9). Each time we drop a value of x into the input hopper, exactly one value of y falls out of the output hopper. If we drop in $x = 5$, the function machine follows the rule and produces $y = 13$. If the rule in the machine drops out more than one value of y for a given x, then it is not a function. Since we are free to choose those values of x that we drop into the machine, we call x the **independent variable;** the value of y that drops out depends upon the choice of x, and y is called the **dependent variable.** We say that the dependent variable is a function of the independent variable; that is, *the output is a function of the input.*

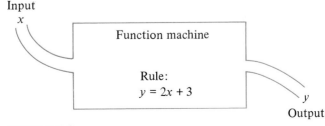

FIGURE 6.9

Let's look at a few schematic presentations. The correspondence in Figure 6.10a is a function; for each x in X there is exactly one corresponding value of y in Y. True, y_1 is the image of both x_1 and x_2, but this does not violate the definition of a function! However, the correspondence in Figure 6.10b is not a function since x_1 is assigned to both y_1 and y_2, which does violate the definition of a function.

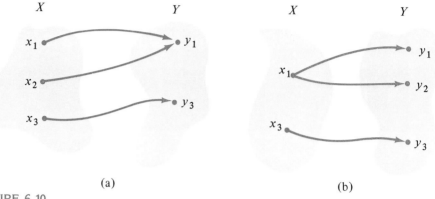

(a)   (b)

FIGURE 6.10

There is a graphic way to test if an equation determines a function. Let's graph the equations $y = x^2$ and $y^2 = x$ in which x is the independent variable. Now draw vertical lines on both graphs in Figure 6.11. No vertical line intersects the graph of $y = x^2$ in more than one point; however, some vertical lines intersect the graph of $y^2 = x$ in two points. This is another way of saying that the equation $y = x^2$ assigns exactly one y for each x and therefore determines y as a function of x. On the other hand, the equation $y^2 = x$ assigns *two* values of y to some values of x and so the correspondence does not determine y as a function of x. Thus, *not every equation in two variables determines one variable as a function of the other.*

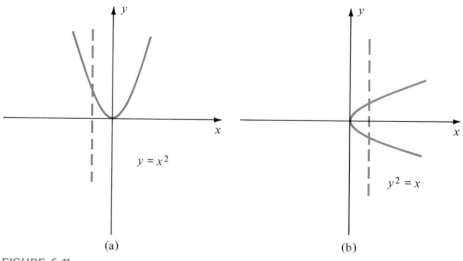

(a)   (b)

FIGURE 6.11

Vertical Line Test

If any vertical line meets the graph of an equation in more than one point, then the equation does not determine a function.

In general, we will consider the domain of a function to be the set of all real numbers for which the function is defined, that is, for which the dependent variable assumes a real value. For example, the domain of the function determined by the equation

$$y = \frac{2}{x - 1}$$

is the set of real numbers other than $x = 1$ since division by 0 is not defined.

The range of a function is, in general, not as easily determined as is the domain. The range is the set of all y values that occur in the correspondence; that is, it is the set of all outputs of the function. For our purposes it will be adequate to determine the range by examining the graph of the function.

EXAMPLE 1

Graph the equation. If the correspondence determines a function, find the domain and range.

$$y = 4 + x, \quad 0 \leq x \leq 5$$

Solution

See Figure 6.12. The graph is a line segment and it is clear that no vertical line meets the graph in more than one point. The equation therefore determines a function.

We are given that $0 \leq x \leq 5$, and since the function is defined for all such x, the domain is $\{x \mid 0 \leq x \leq 5\}$. We see from the graph that the range is $\{y \mid 4 \leq y \leq 9\}$.

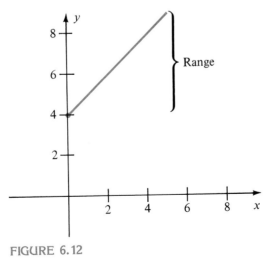

FIGURE 6.12

PROGRESS CHECK 1

Graph the equation $y = x^2 - 4$, $-3 \leq x \leq 3$. If the correspondence determines a function, find the domain and range.

Answer

The desired graph is the portion of the curve shown in Figure 6.5b of Section 6.1 for the values of x between −3 and 3. The domain is $\{x \mid -3 \leq x \leq 3\}$; the range is $\{y \mid -4 \leq y \leq 5\}$.

EXAMPLE 2

Which of the equations whose graphs are shown in Figure 6.13 determine functions?

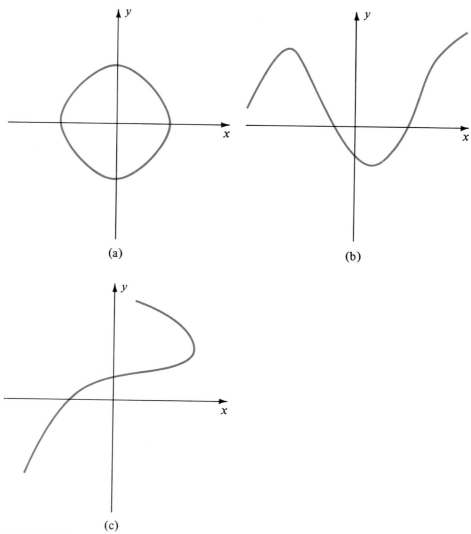

(a)

(b)

(c)

FIGURE 6.13

Solution

(a) Not a function. Some vertical lines meet the graph in more than one point.

(b) A function. Passes the vertical line test.

(c) Not a function. Fails the vertical line test.

PROGRESS CHECK 2
Which of the equations whose graphs are shown in Figure 6.14 determine functions?

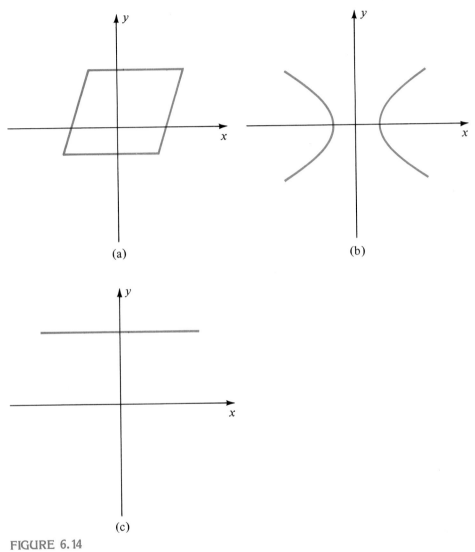

(a)

(b)

(c)

FIGURE 6.14

Answer
(c)

There is a special notation used for functions. We express the functional relation

$$y = 2x + 3$$

by writing

$$f(x) = 2x + 3$$

The symbol "$f(x)$" is read "f of x" and denotes the *output* corresponding to the *input* x. The statement "find the value of y corresponding to $x = 5$"

becomes "find $f(5)$"; that is, $f(5)$ designates the value of $y = f(x)$ when $x = 5$. To find this output, we merely substitute 5 into the expression in place of x. We then obtain

$$y = f(5) = 2(5) + 3 = 13$$

Thus, the output is 13 when the input is 5.

To **evaluate** a function f for a given value of the independent variable x is to find the output $f(x)$ corresponding to the input x.

A function may be denoted by a letter other than f. Thus, F, g, h, and C may all denote a function.

EXAMPLE 3

(a) If $f(x) = 2x^2 - 2x + 1$, find $f(-1)$. We substitute -1 in place of x.

$$f(-1) = 2(-1)^2 - 2(-1) + 1 = 5$$

(b) If $f(t) = 3t - 1$, find $f(a^2)$. We substitute a^2 in place of t.

$$f(a^2) = 3(a^2) - 1 = 3a^2 - 1$$

PROGRESS CHECK 3

(a) If $f(u) = u^3 + 3u - 4$, find $f(-2)$.

(b) If $f(t) = t^2 + 1$, find $f(t - 1)$.

Answers

(a) -18 (b) $t^2 - 2t + 2$

EXAMPLE 4

Let the function f be defined by $f(x) = x^2 - 1$. Find

(a) $f(-2)$ (b) $f(a)$ (c) $f(a + h)$ (d) $f(a + h) - f(a)$

Solution

(a) $f(-2) = (-2)^2 - 1 = 4 - 1 = 3$

(b) $f(a) = a^2 - 1$

(c) $f(a + h) = (a + h)^2 - 1 = a^2 + 2ah + h^2 - 1$

(d) $f(a + h) - f(a) = (a + h)^2 - 1 - (a^2 - 1)$

$$= a^2 + 2ah + h^2 - 1 - a^2 + 1$$

$$= 2ah + h^2$$

PROGRESS CHECK 4

Let the function f be defined by $f(t) = t^2 + t - 1$. Find (a) $f(0)$ (b) $f(a + h)$
(c) $f(a + h) - f(a)$

Answers

(a) -1 (b) $a^2 + 2ah + h^2 + a + h - 1$ (c) $2ah + h^2 + h$

WARNING

(a) *Don't* write

$$f(a + 3) = f(a) + f(3)$$

Function notation is not to be confused with the distributive law.

(b) *Don't* write

$$f(a + 3) = f(a) + 3$$

To evaluate $f(a + 3)$, substitute $a + 3$ for each occurrence of the independent variable.

(c) *Don't* write

$$f(x^2) = f \cdot x^2$$

The use of parentheses in function notation does not imply multiplication.

(d) *Don't* write

$$f(x^2) = [f(x)]^2$$

Squaring x is not the same as squaring $f(x)$.

(e) *Don't* write

$$f(3x) = 3f(x)$$

EXAMPLE 5

The correspondence between Fahrenheit temperature F and Celsius temperature C is given by

$$F(C) = \frac{9}{5}C + 32$$

which is often simply written as $F = \frac{9}{5}C + 32$.

(a) Find the Fahrenheit temperature corresponding to a Celsius reading of 37° (normal body temperature).

(b) Write C as a function of F.

(c) Find the Celsius temperature corresponding to a Fahrenheit reading of 212° (the boiling point of water).

Solution

(a) We substitute $C = 37$ to find $F(37)$.

$$F = \frac{9}{5}(37) + 32 = 98.6$$

Normal body temperature is 98.6° Fahrenheit.

(b) We solve the equation for C.

$$F = \frac{9}{5}C + 32$$

$$F - 32 = \frac{9}{5}C$$

$$C = \frac{5}{9}(F - 32)$$

or

$$C(F) = \frac{5}{9}(F - 32)$$

(c)   We substitute $F = 212$ to find $C(212)$.

$$C = \frac{5}{9}(F - 32)$$

$$C = \frac{5}{9}(212 - 32) = \frac{5}{9}(180) = 100$$

Water boils at $100°$ Celsius.

PROGRESS CHECK 5

Suppose that an object is dropped from a fixed height. If we neglect air resistance, the distance s (in feet) which the object has fallen after t seconds is a function of t given by

$$s(t) = 16t^2$$

(Note that the function does *not* depend upon the mass of the object.) Find the distance traveled by an object when t is (a) 2 seconds (b) 4 seconds. (c) How long does it take an object to fall 400 feet?

Answers
(*a*) *64 feet*   (*b*) *256 feet*   (*c*) *5 seconds*

EXERCISE SET 6.2

In Exercises 1–10, graph the given equation. If the correspondence is a function, determine the domain and range.

1. $y = 3 + x, \quad 0 \le x \le 4$

2. $y = 2 - x, \quad -2 \le x \le 5$

3. $y = x^2 + 1$

4. $y = 9 - x^2$

5. $y = x^2 - 4$

6. $y = -4 - x^2$

7. $y = |x|, \quad -2 \le x \le 3$

8. $y = |x + 1|, \quad -3 \le x \le 1$

9. $y = |2x - 1|, \quad -1 \le x \le 2$

10. $y = |x| - 1, \quad -2 \le x \le 2$

In Exercises 11–18, determine the domain of the given function.

11. $f(x) = 2x^2 + x - 3$

12. $g(t) = \dfrac{1}{t - 2}$

13. $f(v) = \dfrac{1}{(v - 3)(v + 1)}$

14. $g(x) = \dfrac{x - 3}{(x + 2)(x - 4)}$

15. $f(x) = \dfrac{x - 2}{x + 1}$

16. $h(t) = \dfrac{t}{(t - 3)(t + 5)}$

17. $f(x) = \dfrac{5}{x}$

18. $g(s) = \dfrac{5s}{s - 2}$

In Exercises 19–30, determine whether or not the given curve is the graph of a function.

19.

20.

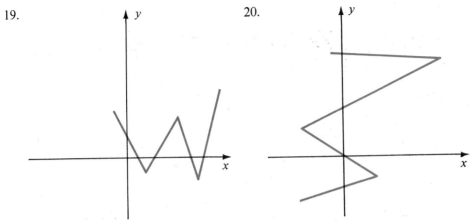

21.

22.

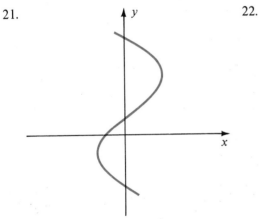

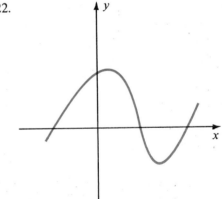

23.

24.

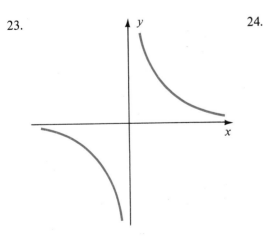

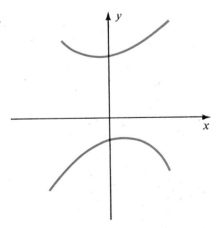

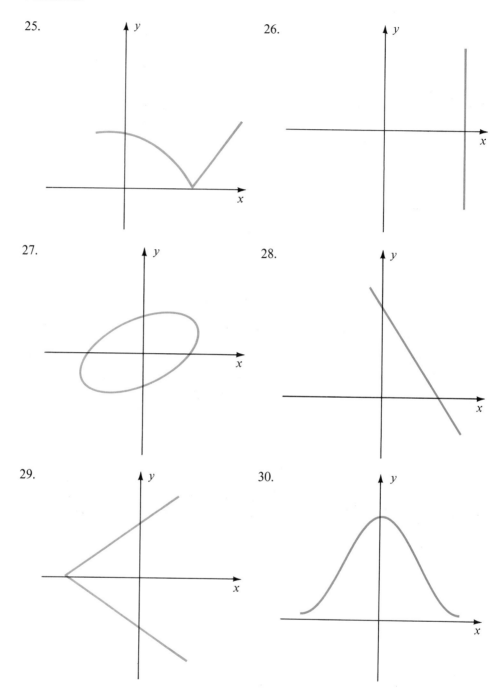

25.

26.

27.

28.

29.

30.

Consider the function f defined by $f(x) = 2x^2 + 5$. Compute the following.

31. $f(0)$ 32. $f(-2)$ 33. $f(a)$

34. $f(3x)$ 35. $3f(x)$ 36. $-f(x)$

Consider the function g defined by $g(x) = x^2 + 2x$. Compute the following.

37. $g(-3)$

38. $g\left(\dfrac{1}{x}\right)$

39. $\dfrac{1}{g(x)}$

40. $g(-x)$

41. $g(a + h)$

42. $\dfrac{g(a + h) - g(a)}{h}$

Consider the function F defined by $F(x) = \dfrac{x^2 + 1}{3x - 1}$. Compute the following.

43. $F(-2.73)$

44. $F(16.11)$

45. $\dfrac{1}{F(x)}$

46. $F(-x)$

47. $2F(2x)$

48. $F(x^2)$

Consider the function r defined by $r(t) = \dfrac{t - 2}{t^2 + 2t - 3}$. Compute the following.

49. $r(-8.27)$

50. $r(2.04)$

51. $r(2a)$

52. $2r(a)$

53. $r(a + 1)$

54. $r(1 + h)$

55. A tour operator who runs charter flights to Rome has established the following pricing schedule. For a group of no more than 100 people, the round trip fare per person is $300. For a group which has more than 100 but less than 150 people the fare will be $250 for each person in excess of 100 people. Write the tour operator's total revenue R as a function of the number of people x in the group.

56. A firm packages and ships 1-pound jars of instant coffee. The cost C of shipping is 40 cents for the first pound and 25 cents for each additional pound.
   (a) Write C as a function of the weight w (in pounds) for $0 < w \le 30$.
   (b) Sketch the graph of the function found in part (a).
   (c) What is the cost of shipping a package containing 24 jars of instant coffee?

57. Suppose that x dollars are invested at 7% interest per year compounded annually. Express the amount A in the account at the end of one year as a function of x.

58. The rate of a car rental firm is $14 daily plus 8¢ per mile that the rented car is driven.
   (a) Express the cost c of renting a car as a function of the number of miles m traveled.
   (b) What is the domain of the function?
   (c) How much would it cost to rent a car for a 100-mile trip?

59. In a wildlife preserve, the population P of eagles depends upon the population x of rodents, its basic food supply. Suppose that P is given by

$$P(x) = 0.002x + 0.004x^2$$

What is the eagle population when the rodent population is (a) 500? (b) 2000?

60. A record club offers the following sale. If 3 records are bought at the regular price of $7.98 each, you may purchase up to 7 more records at half price.
   (a) Express the total cost c to a customer as a function of the number of half-price records r bought.
   (b) What is the domain of this function?
   (c) How much will it cost to buy a total of 8 records?

6.3
GRAPHS OF FUNCTIONS

We have used the graph of an equation to help us find out whether or not the equation determines a function. It is therefore natural that when we speak of the **graph of a function** such as

$$f(x) = -2x + 4$$

we mean the graph of the equation

$$y = -2x + 4$$

We can therefore use the method of plotting points developed in Section 6.1 to plot the graph of a function determined by an equation.

At times, functions are defined other than by equations. In many important applications, a function may be defined by a table, or by several formulas. We illustrate this by several examples.

EXAMPLE 1

The commission earned by a door-to-door cosmetics salesperson is determined as shown in the table.

Weekly sales	Commission
Less than $300	20% of sales
$300 or more but less than $400	$60 + 35% of sales over $300
$400 or more	$95 + 60% of sales over $400

(a) Express the commission C as a function of sales s.
(b) Find the commission if the weekly sales are $425.
(c) Sketch the graph of the function.

Solution

(a) The function C can be described by three equations.

$$C(s) = \begin{cases} 0.2s & \text{if } s < 300 \\ 60 + 0.35(s - 300) & \text{if } 300 \le s < 400 \\ 95 + 0.60(s - 400) & \text{if } s \ge 400 \end{cases}$$

(b) When $s = 425$, we must use the third equation and substitute to determine $C(425)$.

$$C(425) = 95 + 0.6(425 - 400)$$
$$= 95 + 0.6(25)$$
$$= 110$$

The commission on sales of $425 is $110.

(c)   The graph of the function C consists of three line segments (Figure 6.15).

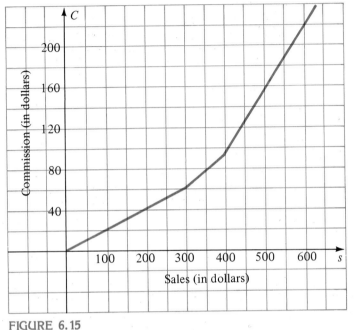

FIGURE 6.15

PROGRESS CHECK 1

The state tax due is given in the following table.

Annual income	Tax due
Under $5000	1% of income
$5000 or more, but less than $15,000	$50 + 2% of income over $5000
$15,000 or more	$250 + 4% of income over $15,000

(a)   Express the tax T as a function of income d.
(b)   Find the tax due if the income is $7000.
(c)   Find the tax due if the income is $18,000.
(d)   Sketch the graph of the function T.

Answers
(a)
$$T(d) = \begin{cases} 0.01d & \text{if } d < 5000 \\ 50 + 0.02(d - 5000) & \text{if } 5000 \le d < 15{,}000 \\ 250 + 0.04(d - 15{,}000) & \text{if } d \ge 15{,}000 \end{cases}$$

(b) $T(7000) = 90$ (c) $T(18{,}000) = 370$

(d)

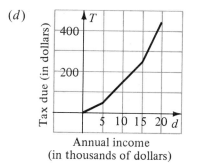

Annual income
(in thousands of dollars)

EXAMPLE 2

Sketch the graph of the function defined by

$$f(x) = \begin{cases} x^2 & \text{if} \quad -2 \leq x \leq 2 \\ 2x + 1 & \text{if} \quad 2 < x \leq 5 \end{cases}$$

Solution

We form a table of points to be plotted.

x	-2	-1	0	1	2	3	4	5
$f(x)$	4	1	0	1	4	7	9	11

See Figure 6.16. Note that the graph has a gap. Also note that the point $(2, 5)$ has been marked with an open circle to indicate that it is not on the graph of the function. Had the point $(2, 5)$ been included, we would have two values of y correspond to $x = 2$ and we would not have a function.

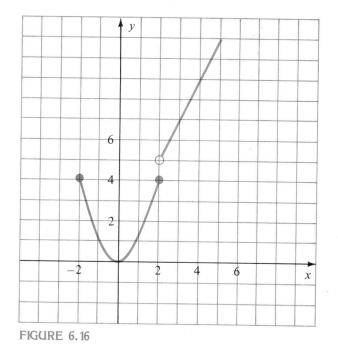

FIGURE 6.16

PROGRESS CHECK 2

Sketch the graph of the function defined by

$$f(x) = \begin{cases} 2x, & -3 \le x < 4 \\ 5, & x \ge 4 \end{cases}$$

Answer

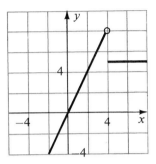

The function

$$f(x) = ax + b$$

is called a **linear function.** We will study this function in detail in the next chapter and will show that its graph is a straight line. For now, we sketch the graphs of a few linear functions to "convince" ourselves that the graphs appear to be straight lines.

EXAMPLE 3

Sketch the graphs of $f(x) = x$ and $g(x) = -x + 2$ on the same coordinate axes.

Solution

We need to graph $y = x$ and $y = -x + 2$. We form a table of values, plot the corresponding points, and connect these by "smooth" curves. See Figure 6.17.

x	$y = x$	$y = -x + 2$
-4	-4	6
-2	-2	4
0	0	2
1	1	1
3	3	-1

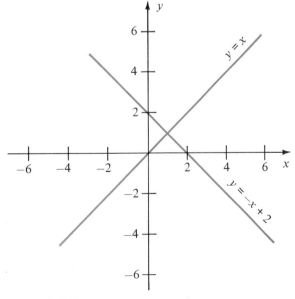

FIGURE 6.17

PROGRESS CHECK 3
Sketch the graphs of $f(x) = 2x + 1$ and $g(x) = -3x + 1$ on the same coordinate axes.

Answer

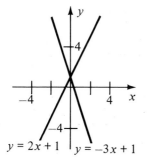

$y = 2x + 1$ $y = -3x + 1$

The function

$$f(x) = ax^2 + bx + c, \quad a \neq 0$$

is called a **quadratic function.** The graph of this function is called a **parabola** and will be studied in detail in a later chapter.

EXAMPLE 4
Sketch the graph of $f(x) = 2x^2 - 4x + 3$.

Solution
We need to graph $y = 2x^2 - 4x + 3$. We form a table of values, plot the corresponding points, and connect these by a "smooth" curve. See Figure 6.18.

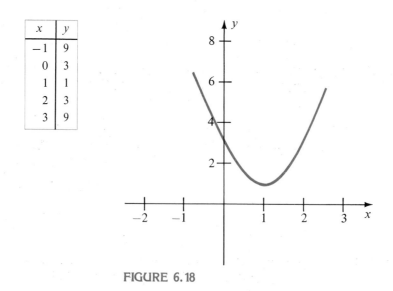

x	y
-1	9
0	3
1	1
2	3
3	9

FIGURE 6.18

PROGRESS CHECK 4
Sketch the graph of $f(x) = -x^2 + 4x - 5$.

Answer

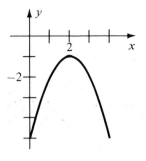

In Chapter Two we were introduced to polynomials and were shown how to determine the degree of a polynomial. Polynomials in one variable are of particular interest because they always determine a function. Here are some examples.

$$P(x) = x^2 - 3x + 1 \qquad P(x) = -2x^5 + \frac{1}{2}x^3 + 2x - 4$$

Linear and quadratic functions are **polynomial functions** of the first and second degree, respectively. The graph of a linear function is always a straight line; the graph of a quadratic function is always a parabola. The graphs of higher-degree polynomials are more complex, but are always "smooth curves" and are often used in mathematics to illustrate a point or to test an idea. The exercises are intended to help you gain experience with the graphs of polynomial functions.

EXERCISE SET 6.3

Sketch the graph of each given function.

1. $f(x) = 3x + 4$
2. $f(x) = 2x - 3$
3. $f(x) = 3 - x$
4. $g(x) = -4 - 3x$
5. $f(x) = 2x^2 + 3$
6. $h(x) = 2x^2 - 3$
7. $g(x) = 5 - 2x^2$
8. $f(x) = -4 - 3x^2$
9. $h(x) = x^2 - 4x + 4$
10. $g(x) = 3x - 2x^2$
11. $f(x) = \dfrac{2}{x - 3}$
12. $f(x) = \dfrac{3}{2x + 1}$
13. $g(x) = \dfrac{3x - 4}{2}$
14. $h(x) = \dfrac{4 - x}{3}$
15. $h(x) = |x - 1|$
16. $f(x) = 2|1 - x|$
17. $f(x) = |x| + 1$
18. $g(x) = 2|x| - 1$
19. $f(x) = 3$
20. $f(x) = -5$
21. $f(x) = \begin{cases} -3x, & -4 \le x \le 2 \\ -4, & x > 2 \end{cases}$
22. $f(x) = \begin{cases} 3, & x < -2 \\ -2x, & x \ge -2 \end{cases}$

23. $g(x) = \begin{cases} \frac{1}{2}x + 1, & x \le 2 \\ -2x + 6, & x > 2 \end{cases}$

24. $f(x) = \begin{cases} |x + 1|, & x < -1 \\ x, & x \ge -1 \end{cases}$

25. $f(x) = \begin{cases} 2, & x < 3 \\ 1, & x > 3 \end{cases}$

26. $g(x) = \begin{cases} 2, & x < 3 \\ 1, & x = 3 \\ -3, & x > 3 \end{cases}$

27. $f(x) = \begin{cases} -4x - 1, & x \le -1 \\ -x + 2, & x > -1 \end{cases}$

28. $f(x) = \begin{cases} x + 1, & x < -1 \\ -x^2 + 1, & x > -1 \end{cases}$

29. $h(x) = \begin{cases} x^2, & x < 1 \\ 2, & x \ge 1 \end{cases}$

30. $g(x) = \begin{cases} -x^2 + 2, & x \le 2 \\ -3x + 1, & x > 2 \end{cases}$

Sketch the graphs of the given functions on the same coordinate axes.

31. $f(x) = x^2$, $g(x) = 2x^2$, $h(x) = \frac{1}{2}x^2$

32. $f(x) = \frac{1}{2}x^2$, $g(x) = \frac{1}{3}x^2$, $h(x) = \frac{1}{4}x^2$

33. $f(x) = 2x^2$, $g(x) = -2x^2$

34. $f(x) = x^2 - 2$, $g(x) = 2 - x^2$

35. $f(x) = x^3$, $g(x) = 2x^3$

36. $f(x) = \frac{1}{2}x^3$, $g(x) = \frac{1}{4}x^3$

37. $f(x) = x^3$, $g(x) = -x^3$

38. $f(x) = -2x^3$, $g(x) = -4x^3$

In Exercises 39–42, sketch the graph of each given function.

39. $f(x) = 0.65x^2 - 0.44$

40. $f(x) = 0.84x^2 + 0.17x - 0.55$

41. $f(x) = 0.15x^3 - 2.1x^2 + 4.6$

42. $f(x) = -3.4x^2 - 1.8x + 6.3$

43. Graph the shipping function of Exercise 56, Section 6.2.

44. Graph the temperature function of Example 5, Section 6.2.

45. The telephone company charges a fee of $6.50 per month for the first 100 message units and an additional fee of 6 cents for each of the next 100 message units. A reduced rate of 5 cents is charged for each message unit after the first 200 units. Express the monthly charge C as a function of the number of message units u. Graph this function.

46. The annual dues of a union are as shown in the table.

Employee's annual salary	Annual dues
Less than $8000	$60
$8000 or more, but less than $15,000	$60 + 1% of the salary in excess of $8000
$15,000 or more	$130 + 2% of the salary in excess of $15,000

Express the annual dues d as a function of the salary. Graph this function.

INCREASING AND DECREASING FUNCTIONS

We say that the straight line in Figure 6.19a is increasing or rising, since the values of y increase as we move from left to right. Since the graph of a function f is obtained by sketching $y = f(x)$, we can give a precise definition of **increasing** and **decreasing functions.**

Increasing and Decreasing Functions

A function f is increasing if $f(x_2) > f(x_1)$ whenever $x_1 < x_2$.

A function f is decreasing if $f(x_2) < f(x_1)$ whenever $x_1 < x_2$.

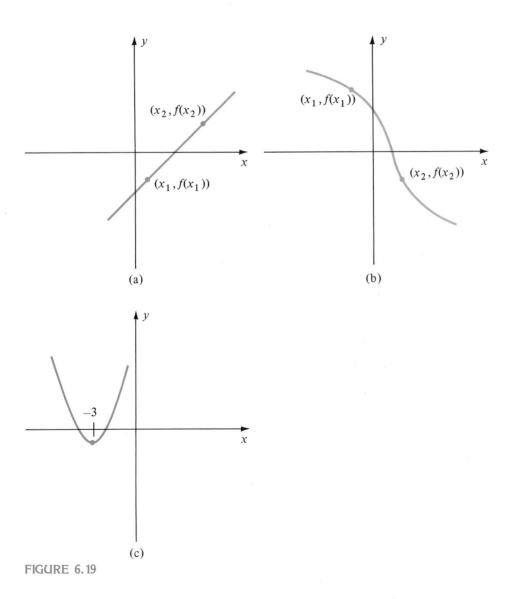

FIGURE 6.19

In other words, if a function is increasing, the dependent variable y assumes larger values as we move from left to right (Figure 6.19a); for a decreasing function (Figure 6.19b), y takes on smaller values as we move from left to right. The function pictured in Figure 6.19c is neither increasing nor decreasing according to this definition. In fact, one portion of the graph is decreasing and another is increasing. We can modify our definition of increasing and decreasing functions so as to apply to *intervals* in the domain. A function may then be increasing in some intervals and decreasing in others.

Returning to Figure 6.19c, we see that the function whose graph is shown is decreasing when $x \le -3$ and increasing when $x \ge -3$. This is the "usual" situation for a function—there are intervals in which the function is increasing and intervals in which it is decreasing. Of course, there is another possibility. The function may have the same value over an interval, in which case we call it a **constant function** over that interval.

EXAMPLE 1
Given the function $f(x) = 1 - x^2$, determine where the function is increasing and where it is decreasing.

Solution
We obtain the graph of $y = 1 - x^2$ by plotting several points. See Figure 6.20. From the graph we see that

$$f \text{ is increasing when } x \le 0$$
$$f \text{ is decreasing when } x \ge 0$$

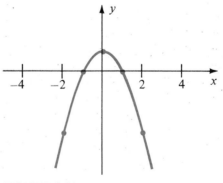

FIGURE 6.20

PROGRESS CHECK 1
Given $f(x) = x^2 + 2x$, determine where the function is increasing and where the function is decreasing.

Answer
Increasing when $x \ge -1$. Decreasing when $x \le -1$.

EXAMPLE 2
The function f is defined by

$$f(x) = \begin{cases} |x| & \text{if } x \le 2 \\ -3 & \text{if } x > 2 \end{cases}$$

Find the values of x for which the function is increasing, decreasing, and constant.

Solution
We sketch the graph of f by plotting a number of points. See Figure 6.21. From the graph we determine that

f is increasing if $0 \le x \le 2$
f is decreasing if $x \le 0$
f is constant and has value -3 if $x > 2$

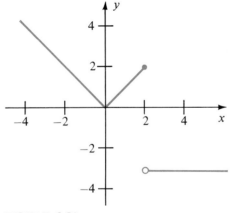

FIGURE 6.21

PROGRESS CHECK 2
The function f is defined by

$$f(x) = \begin{cases} 2x + 1 & \text{if } x < -1 \\ 0 & \text{if } -1 \le x \le 3 \\ -2x + 1 & \text{if } x > 3 \end{cases}$$

Find the values of x for which the function is increasing, decreasing, and constant.

Answer
Increasing if $x < -1$. Decreasing if $x > 3$. Constant if $-1 \le x \le 3$.

EXERCISE SET 6.4
For each given function determine the values of x where the function is increasing, decreasing, and constant.

1. $f(x) = \dfrac{1}{3}x + 2$

2. $f(x) = 3 - \dfrac{1}{2}x$

3. $f(x) = x^2 + 1$

4. $f(x) = x^2 - 4$

5. $f(x) = 9 - x^2$

6. $f(x) = x^2 - 3x$

7. $f(x) = 4x - x^2$

8. $f(x) = x^2 - 2x + 4$

9. $f(x) = \frac{1}{2}x^2 + 3$

10. $f(x) = 4 - \frac{1}{2}x^2$

11. $f(x) = (x - 2)^2$

12. $f(x) = |x - 1|$

13. $f(x) = |2x + 1|$

14. $f(x) = \begin{cases} x, & x < 2 \\ 2, & x \geq 2 \end{cases}$

15. $f(x) = \begin{cases} 2x, & x > -1 \\ -x - 1, & x \leq -1 \end{cases}$

16. $f(x) = \begin{cases} x + 1, & x > 2 \\ 1, & -1 \leq x \leq 2 \\ -x + 1, & x < -1 \end{cases}$

17. $f(x) = (x + 1)^3$

18. $f(x) = \frac{1}{2}(x - 1)^3$

19. $f(x) = (x - 2)^3$

20. $f(x) = x^3 + 1$

21. $f(x) = x + \frac{1}{x}$

22. $f(x) = \frac{1}{x - 1}$

23. $f(x) = x^4 + 1$

24. $f(x) = (x - 1)^4$

25. A manufacturer of skis finds that the profit made from selling x pairs of skis per week is given by $P(x) = 80x - x^2 - 60$. For what values of x is $P(x)$ increasing? decreasing?

26. A psychologist who is training chimpanzees to understand human speech finds that the number $N(t)$ of words learned after t weeks of training is given by $N(t) = 80t - t^2$, $0 \leq t \leq 80$. For what values of t does $N(t)$ increase? decrease?

27. It has been found that x hours after a dosage of a standard drug has been given to a person, the change in blood pressure is given by

$$P(x) = \frac{x^3}{3} - \frac{7}{2}x^2 + 10x, \quad 0 \leq x \leq 6$$

During the six-hour period of observation, when will the blood pressure be increasing? decreasing?

28. Suppose that the profit made by a moped manufacturer from selling x mopeds per week is given by $P(x) = x^2 - 1000x + 500$. For what values of x is $P(x)$ increasing? decreasing?

6.5
DIRECT AND INVERSE VARIATION

Two functional relationships occur so frequently that they are given distinct names. They are direct variation and inverse variation. Two quantities are said to **vary directly** if an increase in one causes a proportional increase in the other. In the table

x	1	2	3	4
y	3	6	9	12

we see that an increase in x causes a proportional increase in y. If we look at the ratios y/x we have

$$\frac{y}{x} = \frac{3}{1} = \frac{6}{2} = \frac{9}{3} = \frac{12}{4} = 3$$

or $y = 3x$. The ratio y/x remains constant for all values of y and $x \neq 0$. This is an example of the

Principle of Direct Variation

y varies directly as x means $y = kx$ for some constant k.

As another example, y varies directly as the square of x means $y = kx^2$ for some constant k. Direct variation, then, involves a constant k which is called the **constant of variation.**

EXAMPLE 1
Write the appropriate equation, solve for the constant of variation k, and use this k to relate the variables.
(a) d varies directly as t, and $d = 15$ when $t = 2$.

(b) y varies directly as the cube of x, and $y = 24$ when $x = -2$.

Solution
(a)   Using the principle of direct variation, the functional relationship is

$$d = kt \quad \text{for some constant } k$$

Substituting the values $d = 15$ and $t = 2$, we have

$$15 = k \cdot 2$$

$$k = \frac{15}{2}$$

Therefore,

$$d = \frac{15}{2}t$$

(b)   Using the principle of direct variation, the functional relationship is

$$y = kx^3 \quad \text{for some constant } k$$

Substituting the values $y = 24$, $x = -2$, we have

$$24 = k \cdot (-2)^3 = -8k$$

$$k = -3$$

Thus,

$$y = -3x^3$$

PROGRESS CHECK 1
(a)   If P varies directly as the square of V, and $P = 64$ when $V = 16$, find the constant of variation.
(b)   The circumference C of a circle varies directly as the radius r. If $C = 25.13$ when $r = 4$, express C as a function of r.

Answers

(a) $\dfrac{1}{4}$ (b) $C = 6.2825r$

Two quantities are said to **vary inversely** if an increase in one causes a proportional decrease in the other. In the table

x	1	2	3	4
y	24	12	8	6

we see that an increase in x causes a proportional decrease in y. If we look at the product xy we have

$$xy = 1 \cdot 24 = 2 \cdot 12 = 3 \cdot 8 = 4 \cdot 6 = 24$$

or $y = 24/x$. In general, we have the

Principle of Inverse Variation

y varies inversely as x means $y = \dfrac{k}{x}$ for some constant k.

Once again, k is called the constant of variation.

EXAMPLE 2

Write the appropriate equation, solve for the constant of variation k, and use k to relate the variables.

(a) m varies inversely as d, and $m = 9$ when $d = -3$.

(b) y varies inversely as the square of x, and $y = 10$ when $x = 10$.

Solution

(a)   The principle of inverse variation tells us that

$$m = \frac{k}{d}$$

for some constant k. Substituting $m = 9$ and $d = -3$ yields $k = -27$. Thus,

$$m = \frac{-27}{d}$$

(b)   The functional relationship is

$$y = \frac{k}{x^2} \quad \text{for some constant } k$$

Substituting $y = 10$ and $x = 10$,

$$10 = \frac{k}{(10)^2} = \frac{k}{100}$$

$$k = 1000$$

Thus,

$$y = \frac{1000}{x^2}$$

PROGRESS CHECK 2
If v varies inversely as the cube of w, and $v = 2$ when $w = -2$, find the constant of variation.

Answer
-16

An equation of variation can involve more than two variables. We say that a quantity **varies jointly** as two or more other quantities if it varies directly as their product.

EXAMPLE 3
Express as an equation: P varies jointly as R, S, and the square of T.

Solution
Since P must vary directly as $R \cdot S \cdot T^2$, we have $P = k \cdot R \cdot S \cdot T^2$ for some constant k.

PROGRESS CHECK 3
Express as an equation: m varies jointly as p and q, and inversely as d.

Answer
$$m = k \cdot \frac{pq}{d}$$

EXAMPLE 4
Find the constant of variation if x varies jointly as y and z, and $x = 30$ when $y = 2$ and $z = 3$.

Solution
We have

$$x = k \cdot y \cdot z \quad \text{for some constant } k$$

and substitute for x, y, and z.

$$30 = k \cdot 2 \cdot 3$$
$$30 = 6k$$
$$k = 5$$

Thus,

$$x = 5yz$$

PROGRESS CHECK 4
Find the constant of variation if x varies jointly as y and the cube of z, and inversely as t, and $x = -\frac{1}{4}$ when $y = -1$, $z = -2$, and $t = 4$.

Answer
$-\dfrac{1}{8}$

EXERCISE SET 6.5

1. In the following table, y varies directly as x.

x	2	3	4	6	8	12		
y	8	12	16	24			80	120

    (a) Find the constant of variation.
    (b) Write an equation showing that y varies directly as x.
    (c) Complete the blanks in the table.
2. In the following table, y varies inversely as x.

x	1	2	3	6	9	12	15	18		
y	6	3	2	1	$\frac{2}{3}$	$\frac{1}{2}$			$\frac{1}{4}$	$\frac{1}{10}$

    (a) Find the constant of variation.
    (b) Write an equation showing that y varies inversely as x.
    (c) Complete the blanks in the table.
3. If y varies directly as x, and $y = -\dfrac{1}{4}$ when $x = 8$,

    (a) find the constant of variation.
    (b) find y when $x = 12$.
4. If C varies directly as the square of s, and $C = 12$ when $s = 6$,
    (a) find the constant of variation.
    (b) find C when $s = 9$.
5. If s varies directly as the square of t, and $s = 10$ when $t = 10$,
    (a) find the constant of variation.
    (b) find s when $t = 5$.
6. If V varies directly as the cube of T, and $V = 16$ when $T = 4$,
    (a) find the constant of variation.
    (b) find V when $T = 6$.
7. If y varies inversely as x, and $y = -\dfrac{1}{2}$ when $x = 6$,

    (a) find the constant of variation.
    (b) find y when $x = 12$.
8. If V varies inversely as the square of p, and $V = \dfrac{2}{3}$ when $p = 6$,

    (a) find the constant of variation.
    (b) find V when $p = 8$.
9. If K varies inversely as the cube of r, and $K = 8$ when $r = 4$,
    (a) find the constant of variation.
    (b) find K when $r = 5$.
10. If T varies inversely as the cube of u, and $T = 2$ when $u = 2$,
    (a) find the constant of variation.
    (b) find T when $u = 5$.

11. If M varies directly as the square of r, and inversely as the square of s, and $M = 4$ when $r = 4$ and $s = 2$,
    (a) write the appropriate equation relating M, r, and s.
    (b) find M when $r = 6$ and $s = 5$.

12. If f varies jointly as u and v, and $f = 36$ when $u = 3$ and $v = 4$,
    (a) write the appropriate equation connecting f, u, and v.
    (b) find f when $u = 5$ and $v = 2$.

13. If T varies jointly as p and the cube of v, and inversely as the square of u, and $T = 24$ when $p = 3$, $v = 2$, and $u = 4$,
    (a) write the appropriate equation connecting T, p, v, and u.
    (b) find T when $p = 2$, $v = 3$, and $u = 36$.

14. If A varies jointly as the square of b and the square of c, and inversely as the cube of d, and $A = 18$ when $b = 4$, $c = 3$, and $d = 2$,
    (a) write the appropriate equation relating A, b, c, and d.
    (b) find A when $b = 9$, $c = 4$, and $d = 3$.

15. The distance s an object falls from rest in t seconds varies directly as the square of t. If an object falls 144 feet in 3 seconds,
    (a) how far does it fall in 4 seconds?
    (b) how long does it take to fall 400 feet?

16. In a certain state the income tax paid by a person varies directly as the income. If the tax is $20 per month when the monthly income is $1600, find the tax due when the monthly income is $900.

17. The resistance R of a conductor varies inversely as the area A of its cross section. If $R = 20$ ohms when $A = 8$ square centimeters, find R when $A = 12$ square centimeters.

18. The pressure P of a certain enclosed gas varies directly as the temperature T and inversely as the volume V. Suppose that 300 cubic feet of gas exert a pressure of 20 pounds per square foot when the temperature is 500°K (absolute temperature measured in the Kelvin scale). What is the pressure of this gas when the temperature is lowered to 400°K and the volume is increased to 500 cubic feet?

19. The intensity of illumination I from a source of light varies inversely as the square of the distance d from the source. If the intensity is 200 candlepower when the source is 4 feet away,
    (a) what is the intensity when the source is 6 feet away?
    (b) how close should the source be to provide an intensity of 50 candlepower?

20. The weight of a body in space varies inversely as the square of its distance from the center of the earth. If a body weighs 400 pounds on the surface of the earth, how much does it weigh 1000 miles above the surface of the earth? (Assume that the radius of the earth is 4000 miles.)

21. The equipment cost of a printing job varies jointly as the number of presses and the number of hours that the presses are run. When 4 presses are run for 6 hours, the equipment cost is $1200. If the equipment cost for 12 hours of running is $3600, how many presses are being used?

22. The current I in a wire varies directly as the electromotive force E and inversely as the resistance R. If a current of 36 amperes is obtained with a wire that has resistance of 10 ohms, and the electromotive force is 120 volts, find the current produced when $E = 220$ volts and $R = 30$ ohms.

23. The illumination from a light source varies directly as the intensity of the source and inversely as the square of the distance from the source. If the illumination is 50 candlepower per square foot when 2 feet away from a light source whose intensity is 400 candlepower, what is the illumination when 4 feet away from a source whose intensity is 3840 candlepower?

24. If f varies directly as u and inversely as the square of v, what happens to f if both u and v are doubled?

TERMS AND SYMBOLS

origin (p. 146)
x-axis (p. 146)
y-axis (p. 146)
coordinate axes (p. 146)
rectangular coordinate
system (p. 146)
Cartesian coordinate
system (p. 146)
coordinates of a point
(p. 146)
ordered pair (p. 146)
abscissa (p. 146)
ordinate (p. 146)
x-coordinate (p. 146)
y-coordinate (p. 146)
quadrant (p. 146)
graph of an equation in
two variables (p. 149)

solution to an equation in
two variables (p. 149)
function (p. 155)
image (p. 155)
domain (p. 155)
range (p. 155)
independent variable
(p. 155)
dependent variable
(p. 155)
vertical line test (p. 157)
$f(x)$ (p. 159)
"evaluate" a function
(p. 160)
graph of a function
(p. 166)

linear function (p. 169)
quadratic function
(p. 170)
parabola (p. 170)
polynomial function
(p. 171)
increasing function
(p. 173)
decreasing function
(p. 173)
constant function (p. 174)
direct variation (p. 177)
constant of variation
(p. 177)
inverse variation (p. 178)
joint variation (p. 179)

KEY IDEAS FOR REVIEW

☐ In a rectangular coordinate system, every ordered pair of real numbers (a, b) corresponds to a point in the plane and every point in the plane corresponds to an ordered pair of real numbers.

☐ An equation in two variables can be graphed by plotting points that satisfy the equation and then joining the points to form a smooth curve.

☐ A function is a rule that assigns exactly one element y of a set Y to each element x of a set X. The domain is the set of inputs and the range is the set of outputs.

☐ A graph represents a function $y = f(x)$ if no vertical line meets the graph in more than one point.

☐ Function notation gives both the definition of the function and the value or expression at which to evaluate the function. Thus, if the function f is defined by $f(x) = x^2 + 2x$, then the notation $f(3)$ denotes the result of replacing the independent variable x by 3 wherever it appears: $f(3) = 3^2 + 2(3) = 15$.

☐ To graph $f(x)$, simply graph the equation $y = f(x)$.

☐ An equation is not the only way to define a function. Sometimes a function is defined by a table or chart, or by several equations. Moreover, not every equation determines a function.

☐ The graph of a function can have holes or gaps, and can be defined in "pieces."

☐ Polynomials in one variable determine functions and have "smooth" curves as their graphs.

☐ As we move from left to right, the graph of an increasing function rises and the graph of a decreasing function falls.

☐ The graph of a constant function neither rises nor falls; it is horizontal.

☐ Direct and inverse variation are functional relationships.

☐ We say that y varies directly as x if $y = kx$ for some constant k. We say that y varies inversely as x if $y = \dfrac{k}{x}$ for some constant k.

☐ We say that y varies jointly as two or more other quantities if it varies directly as their product.

COMMON ERRORS

1. Function notation is not distributive. *Don't* write

$$f(a + 3) = f(a) + f(3)$$

or

$$f(a + 3) = f(a) + 3$$

Instead, substitute $a + 3$ for the independent variable.

2. *Don't* write

$$f(x^2) = [f(x)]^2$$

or

$$f(x^2) = f \cdot x^2$$

Again, the notation $f(x^2)$ denotes the output when we replace the independent variable by x^2.

3. It is legitimate for the graph of a function to have holes or gaps. Don't force the graph of every function to be "continuous."

PROGRESS TEST 6A

1. Sketch the graph of $y = -2x^3 + 1$.

2. Find the domain of the function $f(x) = \dfrac{2x}{x + 3}$.

3. Find the domain of the function $g(y) = \dfrac{2}{y^2 - 4}$.

4. Use the vertical line test to determine if the equations

$$y = \begin{cases} |x|, & -3 \le x \le 3 \\ x^2, & x > 0 \end{cases}$$

define a function.

5. Is the following curve the graph of a function?

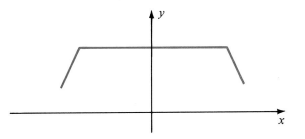

6. Evaluate $f\left(-\dfrac{1}{2}\right)$ for $f(x) = x^2 - 3x + 1$.

7. Evaluate $f(2t)$ for $f(x) = \dfrac{1 - x^2}{1 + x}$.

8. Evaluate $\dfrac{f(a + h) - f(a)}{h}$ for $f(x) = 2x^2 + 3$.

9. Sketch the graph of the function

$$f(x) = \begin{cases} x - 1, & -5 \le x \le -1 \\ x^2, & -1 < x \le 2 \\ -2, & 2 < x \le 5 \end{cases}$$

10. Sketch the graph of $f(x) = x^2 - 4x + 2$.
11. If R varies directly as q, and $R = 20$ when $q = 5$, find R when $q = 40$.
12. If S varies inversely as the cube of t, and $S = 8$ when $t = -1$, find S when $t = -2$.
13. If P varies jointly as q and r, and inversely as the square of t, and $P = -3$ when $q = 2$, $r = -3$, and $t = 4$, find P when $q = -1$, $r = \frac{1}{2}$, and $t = 4$.
14. Determine the intervals where the function $f(x) = x^2 - 2x + 1$ is increasing, decreasing, and constant.
15. Determine the intervals where the function

$$f(x) = \begin{cases} |x - 1|, & x < 3 \\ -1, & x \geq 3 \end{cases}$$

is increasing, decreasing, and constant.

PROGRESS TEST 6B

1. Sketch the graph of $y = \frac{1}{2}x^3 - 1$.

2. Find the domain of the function $f(t) = \dfrac{t}{2t - 1}$.

3. Find the domain of the function $g(x) = \dfrac{4x}{1 - x^2}$.

4. Use the vertical line test to determine if the equations

$$y = \begin{cases} |x - 1|, & x \leq 5 \\ 4, & x \geq 5 \end{cases}$$

   define a function.
5. Is the following curve the graph of a function?

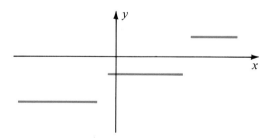

6. Evaluate $f(-1)$ for $f(x) = \dfrac{x^2 + 2}{x}$.

7. Evaluate $f\left(\dfrac{a}{2}\right)$ for $f(x) = (1 + x)^2$.

8. Evaluate $\dfrac{f(a + h) - f(a)}{h}$ for $f(x) = -x^2 + 2$.

9. Sketch the graph of the function

$$f(x) = \begin{cases} |x| + 1, & -4 \leq x \leq 1 \\ 2x, & 1 \leq x \leq 4 \\ 4, & x > 4 \end{cases}$$

10. Sketch the graph of $f(x) = -x^2 + 2x$.
11. If L varies directly as the cube of r, and $L = 2$ when $r = -\frac{1}{2}$, find L when $r = 4$.
12. If A varies inversely as the square of b, and $A = -2$ when $b = 4$, find A when $b = 3$.
13. If T varies jointly as a and the square of b, and inversely as the cube of c, and $T = 64$ when $a = -1$, $b = \frac{1}{2}$, and $c = 2$, find T when $a = 2$, $b = 4$, and $c = -1$.
14. Determine the intervals where the function $f(x) = |x - 2|$ is increasing, decreasing, and constant.
15. Determine the intervals where the function

$$f(x) = \begin{cases} 5, & x < -1 \\ |x|, & -1 \le x \le 3 \\ -1, & x > 3 \end{cases}$$

is increasing, decreasing, and constant.

THE STRAIGHT LINE

In Chapter Six we said that functions of the form $f(x) = ax + b$ are called linear functions and saw that the graphs of such functions appeared to be straight lines. In this chapter we will demonstrate that these conjectures are well founded—the graph of a linear function is indeed a straight line.

The concept of slope is introduced and is used to develop two important forms of the equation of the straight line, the point–slope form and the slope–intercept form. Horizontal, vertical, parallel, and perpendicular lines are also explored.

Finally, the graphs of linear inequalities are discussed and a simple technique for sketching such graphs is developed.

7.1
SLOPE OF THE STRAIGHT LINE

Consider a straight line L that is not parallel to the y-axis (see Figure 7.1) and let $P_1(x_1, y_1)$ and $P_2(x_2, y_2)$ be any two distinct points on L. We have indicated the increments or changes $x_2 - x_1$ and $y_2 - y_1$ in the x and y coordinates from P_1 to P_2. Note that the increment $y_2 - y_1$ can be positive, negative, or zero, while the increment $x_2 - x_1$ can be negative or positive. If we choose any other pair of points, say $P_3(x_3, y_3)$ and $P_4(x_4, y_4)$, we will, in

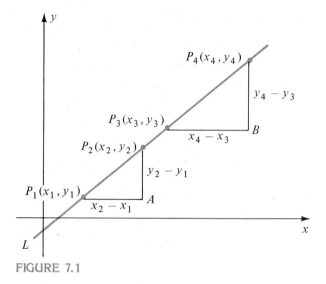

FIGURE 7.1

general, obtain different increments in the x and y coordinates. However, since triangles P_1AP_2 and P_3BP_4 are similar, the ratio

$$\frac{y_2 - y_1}{x_2 - x_1}$$

will be the same as the ratio

$$\frac{y_4 - y_3}{x_4 - x_3}$$

We call this ratio the **slope of the line** L and denote it by m.

Slope of a Line

The slope of a line that is not vertical is given by

$$m = \frac{y_2 - y_1}{x_2 - x_1}$$

where $P_1(x_1, y_1)$ and $P_2(x_2, y_2)$ are any two points on the line.

For a *vertical* line, $x_1 = x_2$, so that in the equation for m, the denominator $x_2 - x_1 = 0$. Since we cannot divide by 0, we say that a vertical line has no slope.

EXAMPLE 1

Find the slope of the line that passes through these points.

(a) (4, 2) and (1, −2)   (b) (−5, 2) and (−2, −1)

Solution

(a)   We may choose either point as (x_1, y_1) and the other as (x_2, y_2). Our choice is

$$(x_1, y_1) = (4, 2)$$
$$(x_2, y_2) = (1, -2)$$

Then

$$m = \frac{y_2 - y_1}{x_2 - x_1} = \frac{-2 - 2}{1 - 4} = \frac{-4}{-3} = \frac{4}{3}$$

If we had reversed the choice we would have

$$(x_1, y_1) = (1, -2)$$
$$(x_2, y_2) = (4, 2)$$

and

$$m = \frac{y_2 - y_1}{x_2 - x_1} = \frac{2 - (-2)}{4 - 1} = \frac{4}{3}$$

which is the same result. Reversing the choice does not affect the value of m. (Why?)

(b)   Let $(x_1, y_1) = (-5, 2)$ and $(x_2, y_2) = (-2, -1)$. Then

$$m = \frac{y_2 - y_1}{x_2 - x_1} = \frac{-1 - 2}{-2 - (-5)} = \frac{-3}{3} = -1$$

PROGRESS CHECK 1

Find the slope of the line that passes through these points.

(a) $(2, -4)$ and $(4, 1)$ (b) $(-1, -3)$ and $(-2, -5)$

Answers

(a) $\frac{5}{2}$ (b)   2

WARNING

Once you have chosen a point as (x_1, y_1), you must be consistent. If you have $(x_1, y_1) = (5, 2)$ and $(x_2, y_2) = (1, 6)$, *don't* write

$$m = \frac{6 - 2}{5 - 1} = \frac{4}{4} = 1$$

This answer has the wrong sign because

$$\frac{y_2 - y_1}{x_1 - x_2} \neq \frac{y_2 - y_1}{x_2 - x_1}$$

Slope is a means of measuring the steepness of a line. In Figure 7.2 we have indicated several lines with positive and negative slopes m. We can summarize our observations.

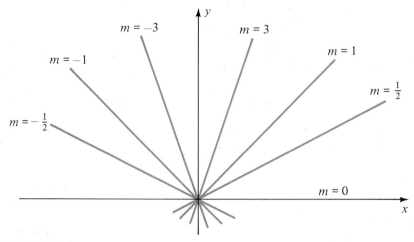

FIGURE 7.2

(a) When $m > 0$, the line is the graph of an increasing function.
(b) When $m < 0$, the line is the graph of a decreasing function.
(c) When $m = 0$, the line is the graph of a constant function.
(d) m does not exist for a vertical line, and a vertical line is not the graph of any function.

EXAMPLE 2

Find the slope of the line through the given points and state whether the line is rising or falling.

(a) $(1, 2)$ and $(-2, -7)$

Let
$$(x_1, y_1) = (1, 2)$$
$$(x_2, y_2) = (-2, -7)$$

Then
$$m = \frac{y_2 - y_1}{x_2 - x_1} = \frac{-7 - 2}{-2 - 1} = \frac{-9}{-3} = 3$$

Since m is positive, the line *rises* from left to right (see Figure 7.3).

(b) $(1, 0)$ and $(3, -2)$

Let
$$(x_1, y_1) = (1, 0)$$
$$(x_2, y_2) = (3, -2)$$

Then
$$m = \frac{y_2 - y_1}{x_2 - x_1} = \frac{-2 - 0}{3 - 1} = \frac{-2}{2} = -1$$

Since m is negative, the line *falls* from left to right (see Figure 7.4).

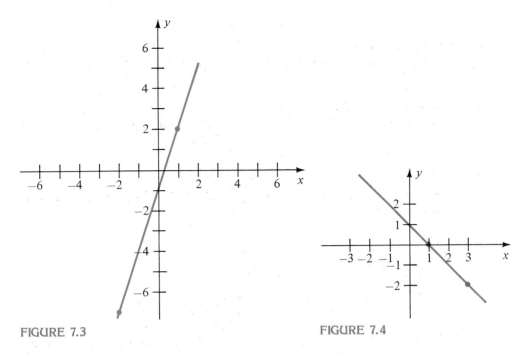

FIGURE 7.3   FIGURE 7.4

PROGRESS CHECK 2

Find the slope of the line through the given points and state whether the line is rising, falling, or constant.

(a) $(-1, -3)$ and $(0, 1)$ (b) $(-1, -4)$ and $(3, -4)$

(c) $(2, 0)$ and $(4, -1)$

Answers

(a) 4; *rising*   (b) 0; *constant*   (c) $-\dfrac{1}{2}$; *falling*

EXERCISE SET 7.1

Find the slope of the line passing through the given points and state whether the line is rising or falling.

1. $(2, 3)$ and $(-1, -3)$ 2. $(1, 2)$ and $(-2, 5)$

3. $(1, -4)$ and $(-1, -2)$ 4. $(2, -3)$ and $(3, 2)$

5. $(-2, 3)$ and $(0, 0)$ 6. $\left(\dfrac{1}{2}, 2\right)$ and $\left(\dfrac{3}{2}, 1\right)$

7. $(2, 4)$ and $(-3, 4)$ 8. $(-2, 2)$ and $(-2, -4)$

Find the slope of the lines in Figure 7.5.

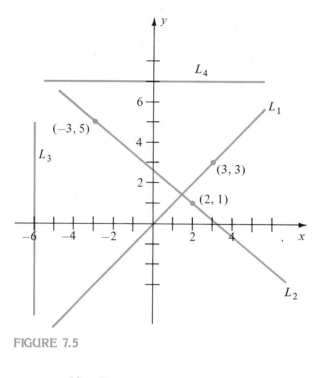

FIGURE 7.5

9. L_1 10. L_2 11. L_3 12. L_4

EQUATIONS OF THE STRAIGHT LINE

We can apply the concept of slope to develop important forms of the equations of a straight line. In Figure 7.6, the point $P_1(x_1, y_1)$ lies on a line L whose slope is m. If $P(x, y)$ is any other point on L, then we may use P and P_1

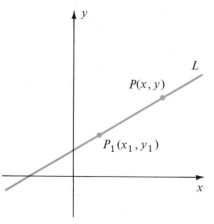

FIGURE 7.6

to compute m, that is,

$$m = \frac{y - y_1}{x - x_1}$$

which can be written in the form

$$y - y_1 = m(x - x_1)$$

Since (x_1, y_1) satisfies this equation, every point on L satisfies this equation. Conversely, any point satisfying this equation must lie on the line L since there is only one line through $P_1(x_1, y_1)$ with slope m. This result is so important that the equation is given a special name.

Point-Slope Form

$$y - y_1 = m(x - x_1)$$

is an equation of the line with slope m that passes through the point (x_1, y_1).

EXAMPLE 1
Find an equation of the line whose slope is -2 and passes through the point $(4, -1)$. Sketch the line.

Solution
We have $m = -2$ and $(x_1, y_1) = (4, -1)$. Using the point–slope form,

$$y - y_1 = m(x - x_1)$$
$$y - (-1) = -2(x - 4)$$
$$y + 1 = -2x + 8$$
$$y = -2x + 7$$

To sketch the line, we need a second point. We can substitute a value of x, such as $x = 0$, in the equation of the line to obtain another point. See Figure 7.7.

x	y
4	-1
0	7

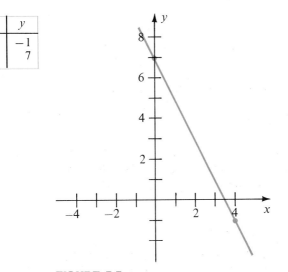

FIGURE 7.7

PROGRESS CHECK 1
Find an equation of the line whose slope is 3 and passes through the point $(-1, -5)$.

Answer
$y = 3x - 2$

It is also possible to use the point–slope form to find an equation of a line when we know two points on the line.

EXAMPLE 2
Find an equation of the line that passes through $(6, -2)$ and $(-4, 3)$.

Solution
First, we find the slope. If we let

$$(x_1, y_1) = (6, -2)$$
$$(x_2, y_2) = (-4, 3)$$

then

$$m = \frac{y_2 - y_1}{x_2 - x_1} = \frac{3 - (-2)}{-4 - 6} = \frac{5}{-10} = -\frac{1}{2}$$

Next, the point–slope form is used with $m = -\dfrac{1}{2}$ and $(x_1, y_1) = (6, -2)$.

$$y - y_1 = m(x - x_1)$$
$$y - (-2) = -\frac{1}{2}(x - 6)$$
$$y + 2 = -\frac{1}{2}x + 3$$
$$y = -\frac{1}{2}x + 1$$

PROGRESS CHECK 2
Find an equation of the line through the points $(3, 0)$ and $(-15, -6)$.

Answer
$y = \dfrac{1}{3}x - 1$

There is another form of the equation of the straight line that is very useful. In Figure 7.8, the line L meets the y-axis at the point $(0, b)$, and is assumed to have slope m. Then we can let $(x_1, y_1) = (0, b)$ and use the point–slope form.

$$y - y_1 = m(x - x_1)$$
$$y - b = m(x - 0)$$
$$y = mx + b$$

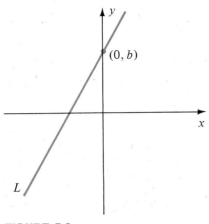

FIGURE 7.8

We call b the **y-intercept** and we now have the result.

Slope–Intercept Form

The graph of the equation
$$y = mx + b$$
is a straight line with slope m and y-intercept b.

Since the form $f(x) = ax + b$ defines a linear function, we have shown that the graph of a linear function is a straight line.

EXAMPLE 3
Find an equation of the line with slope -1 and y-intercept 5.

Solution
We substitute $m = -1$ and $b = 5$ in the equation
$$y = mx + b$$
to obtain
$$y = -x + 5$$

PROGRESS CHECK 3
Find an equation of the line with slope $\frac{1}{2}$ and y-intercept -4.

Answer
$$y = \frac{1}{2}x - 4$$

EXAMPLE 4

Find the slope and y-intercept of the line $y = -2x - 4$.

Solution

This example illustrates the most important use of the slope–intercept form—to find the slope and y-intercept directly from the equation. We have to align corresponding coefficients of x and the constant terms.

$$y = -2x - 4$$
$$y = mx + b$$

Then $m = -2$ is the slope and $b = -4$ is the y-intercept.

PROGRESS CHECK 4

Find the slope and y-intercept of the line $y = -\frac{1}{3}x + 14$.

Answer

Slope $= m = -\frac{1}{3}$; y-intercept $= b = 14$.

EXAMPLE 5

Find the slope and y-intercept of the line $y - 3x + 1 = 0$.

Solution

The equation must be written in the form $y = mx + b$. That is, we must solve for y.

$$y = 3x - 1$$

We see that $m = 3$ is the slope, and $b = -1$ is the y-intercept.

PROGRESS CHECK 5

Find the slope and y-intercept of the line $2y + x - 3 = 0$.

Answer

Slope $= m = -\frac{1}{2}$; y-intercept $= b = \frac{3}{2}$.

Throughout this chapter we have been dealing with first-degree equations in two variables. The **general first-degree equation** in x and y can always be written in the form

$$Ax + By + C = 0$$

where A, B, and C are constants and A and B are not both zero. We can rewrite this equation as

$$By = -Ax - C$$

If $B \neq 0$, the equation becomes

$$y = -\frac{A}{B}x - \frac{C}{B}$$

which we recognize as having a straight line graph with slope $-A/B$ and y-intercept $-C/B$. If $B = 0$, the equation becomes $Ax + C = 0$ whose graph is a vertical line. We have thereby proved the following.

> The graph of the first-degree equation
>
> $Ax + By + C = 0$, A, B not both zero
>
> is a straight line.

WARNING

(1) Given the equation $y + 2x - 3 = 0$, *don't* write $m = 2$, $b = -3$. You must rewrite the equation by solving for y

$$y = -2x + 3$$

and then obtain $m = -2$, $b = 3$.

(2) Given the equation $y = 5x - 6$, *don't* write $m = 5$, $b = 6$. The sign is part of the answer, that is, $m = 5$, $b = -6$ is the correct answer.

EXERCISE SET 7.2

1. Which of the following are linear equations in x and y?

   (a) $3x + 2y = 4$

   (b) $xy = 2$

   (c) $2x^2 - y = 5$

   (d) $2\left(x - \dfrac{3}{2}\right) + 5y = 4$

2. Which of the following are linear equations in x and y?

   (a) $2x^2 + y = 7$

   (b) $3x - 2y = 7$

   (c) $\dfrac{1}{2}(2x^2 - 4) + 4y = 2$

   (d) $x = 2y - 3$

Express each of the following equations in the form $Ax + By = C$ and state the values of A, B, and C.

3. $y = 2x - 3$

4. $y - 2 = \dfrac{3}{2}(x - 4)$

5. $y = 3$

6. $x = \dfrac{3}{4}y - 1$

7. $3\left(x - \dfrac{1}{3}\right) - 2y = 6$

8. $2x + 5y + 7 = 0$

9. $x = \dfrac{1}{2}$

10. $y + 1 = -\dfrac{1}{2}(x - 2)$

Graph each of the following.

11. $y = 2x + 1$

12. $y = 3x - 2$

13. $y = -2x + 3$

14. $y = 3x$

15. $x = 2y + 1$

16. $x = y + 2$

17. $x = -2y + 3$

18. $x = \dfrac{1}{2}y$

19. $y + 2x = 4$

20. $2y - x = 0$

21. $x + 2y + 3 = 0$

22. $x - 3y + 6 = 0$

Find the point–slope form of the line satisfying the given conditions.

23. Its slope is 2 and it passes through the point $(-1, 3)$.

24. Its slope is $-\dfrac{1}{2}$ and it passes through the point $(1, -2)$.

25. Its slope is 3 and it passes through the point $(0, 0)$.
26. Its slope is 0 and it passes through the point $(-1, 3)$.
27. It passes through the points $(2, 4)$ and $(-3, -6)$.
28. It passes through the points $(-3, 5)$ and $(1, 7)$.
29. It passes through the points $(0, 0)$ and $(3, 2)$.
30. It passes through the points $(-2, 4)$ and $(3, 4)$.

Find the slope–intercept form of the line satisfying the following properties.

31. Its slope is 3 and y-intercept is 2.
32. Its slope is -3 and y-intercept is -3.
33. Its slope is 0 and y-intercept is 2.

34. Its slope is $-\dfrac{1}{2}$ and y-intercept is $\dfrac{1}{2}$.

Find the slope and y-intercept of the following lines.

35. $y = 3x + 2$ 36. $y = -\dfrac{2}{3}x - 4$

37. $x = -5$ 38. $x = 3$

39. $y = 3$ 40. $y = -4$

41. $3x + 4y = 5$ 42. $2x + 3y = 6$

43. $2x - 5y + 3 = 0$ 44. $3x + 4y + 2 = 0$

45. $x = \dfrac{2}{3}y + 2$ 46. $x = -\dfrac{1}{2}y + 3$

In each of the following tell whether the given line rises from left to right or falls from left to right.

47. $y = 2x + 3$ 48. $y = -\dfrac{3}{2}x + 5$

49. $y = \dfrac{3}{4}x - 2$ 50. $y = -\dfrac{4}{5}x - 6$

51. $x = 2y - 5$ 52. $x = 3 - 4y$

Find the slope–intercept form of the line determined by the points in each of the following.

53. $(-1, 2)$ and $(3, 5)$ 54. $(-2, -3)$ and $(3, 4)$

55. $(32.65, -17.47)$ and $(-4.76, 19.24)$

56. $(0, 14.38)$ and $(-7.62, 3.04)$

57. $(-6.45, -12.42)$ and $(8.44, 0)$

58. $(0, 0)$ and $(-4.47, 9.31)$

Find equations of the following lines.

59.

60.

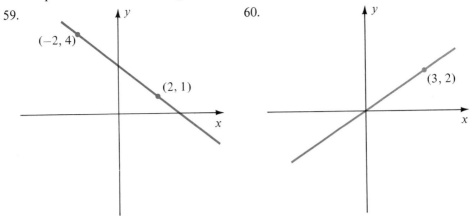

In each of the Exercises 61–64, write an equation relating the variables. Identify the variables.

61. A rental firm charges $8 plus $1.50 per hour for use of a power aerator.
62. A taxi charges 60 cents plus 35 cents per mile.
63. A brokerage firm charges $25 plus 12 cents per share.
64. A theater may be rented for $200 plus $1.25 per person.

65. The Celsius (C) and Fahrenheit (F) temperature scales are related by a linear equation. Water boils at 212°F or 100°C, and freezes at 32°F or 0°C.
   (a) Write a linear equation expressing F in terms of C.
   (b) What is the Fahrenheit temperature when the Celsius temperature is 20°?
66. The college bookstore sells a textbook costing $10 for $13.50 and a textbook costing $12 for $15.90. If the markup policy of the bookstore is linear, write an equation that relates sales price S and cost C. What is the cost of a textbook that sells for $22?
67. An appliance manufacturer finds that it had sales of $200,000 five years ago and sales of $600,000 this year. If the growth in sales is assumed to be linear, what will the sales be five years from now?
68. A product that sold for $250 three years ago sells for $325 this year. If price increases are assumed to be linear, how much will the product sell for six years from now?

7.3

FURTHER PROPERTIES OF THE STRAIGHT LINE

Horizontal and vertical lines are special cases that deserve particular attention. In Figure 7.9a we have a vertical line through the point (3, 2). Choose any other point on the line and answer the question: What is the x-coordinate of the point? You now see that every point on this vertical line has an x-coordinate of 3. The equation of the line is $x = 3$, since x remains constant. If we take a second point on this line, say (3, 4), we see that the slope is

$$m = \frac{y_2 - y_1}{x_2 - x_1} = \frac{4 - 2}{3 - 3} = \frac{2}{0}$$

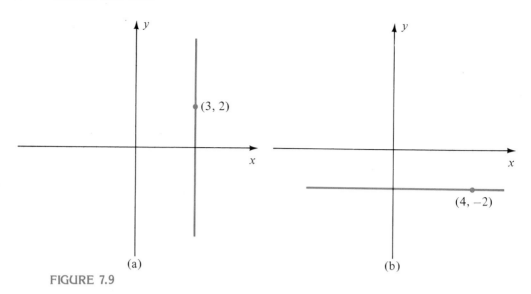

FIGURE 7.9

Since we cannot divide by 0, we say that the slope is undefined.

Vertical Lines

The equation of the vertical line through (a, b) is

$$x = a$$

The slope of a vertical line is undefined.

Returning to Figure 7.9b, we see that the coordinates of all points on the horizontal line through $(4, -2)$ have the form $(x, -2)$. The equation of the line is then $y = -2$ since y remains constant. If we choose a second point on the line, say $(6, -2)$, we find that the slope is

$$m = \frac{y_2 - y_1}{x_2 - x_1} = \frac{-2 - (-2)}{6 - 4} = \frac{0}{2} = 0$$

Horizontal Lines

The equation of the horizontal line through (a, b) is

$$y = b$$

The slope of a horizontal line is 0.

EXAMPLE 1

Find the equations of the horizontal and vertical lines through the point $(-4, 7)$.

Solution

The horizontal line has the equation $y = 7$. The vertical line has the equation $x = -4$.

PROGRESS CHECK 1
Find the equations of the horizontal and vertical lines through the point $(5, -6)$.

Answer
Horizontal line: $y = -6$. *Vertical line:* $x = 5$.

EXAMPLE 2
Find the equation of the line passing through the points $(4, -1)$ and $(-5, -1)$.

Solution
Let $(x_1, y_1) = (4, -1)$, $(x_2, y_2) = (-5, -1)$. The slope is

$$m = \frac{y_2 - y_1}{x_2 - x_1} = \frac{-1 - (-1)}{-5 - 4} = \frac{0}{-9} = 0$$

The equation then is

$$y - y_1 = m(x - x_1)$$
$$y - (-1) = 0(x - 4)$$
$$y + 1 = 0$$
$$y = -1$$

There is another way of solving this problem. Since both points have the same y-coordinate, we are dealing with a horizontal line. The equation of a horizontal line through $(a, -1)$ is $y = -1$.

PROGRESS CHECK 2
Find the equation of the line passing through $(6, -1)$ and $(6, 7)$.

Answer
$x = 6$

PARALLEL AND PERPENDICULAR LINES

In Figure 7.10 we have sketched two lines that are parallel. Clearly, the two lines have the same "steepness" or slope. In general, we can say that

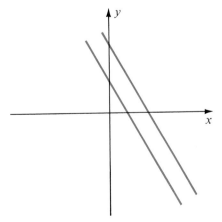

FIGURE 7.10

Parallel lines have the same slope.

Lines that have the same slope are parallel.

EXAMPLE 3

Find the slope of every line parallel to the line $y = -\frac{2}{3}x + 5$.

Solution

Since $y = -\frac{2}{3}x + 5$ is in the form $y = mx + b$, we see that the slope $m = -\frac{2}{3}$. Every line parallel to $y = -\frac{2}{3}x + 5$ has the same slope, so we conclude that all such lines also have slope $m = -\frac{2}{3}$.

PROGRESS CHECK 3

Find the slope of every line parallel to the line $y - 4x + 5 = 0$.

Answer

$m = 4$

EXAMPLE 4

Find an equation of the line passing through the point $(2, -1)$ and parallel to $y = \frac{1}{2}x - 5$.

Solution

The slope of the line $y = \frac{1}{2}x - 5$ and of every line parallel to it is $m = \frac{1}{2}$. Letting $(x_1, y_1) = (2, -1)$ we have

$$y - y_1 = m(x - x_1)$$

$$y - (-1) = \frac{1}{2}(x - 2)$$

$$y + 1 = \frac{1}{2}x - 1$$

$$y = \frac{1}{2}x - 2$$

PROGRESS CHECK 4

Find an equation of the line through the point $(-8, 4)$ and parallel to $2y - 2x + 17 = 0$.

Answer

$y = x + 12$

Slope can also be used to determine if two lines are perpendicular.

If two lines with slopes m_1 and m_2 are perpendicular, then $m_2 = -\dfrac{1}{m_1}$.

If the slopes m_1 and m_2 of two lines satisfy $m_2 = -\dfrac{1}{m_1}$, then the two lines are perpendicular.

This criterion for perpendicularity applies only when neither line is vertical. It can be established by a geometric argument (see Exercise 37). The following example illustrates the use of this criterion.

EXAMPLE 5
Find an equation of the line passing through the point $(-3, 4)$ that is perpendicular to the line $y = 3x - 2$. Sketch both lines.

Solution
The line $y = 3x - 2$ has slope $m_1 = 3$. The line we seek has slope $m_2 = -1/m_1 = -1/3$ and passes through $(x_1, y_1) = (-3, 4)$. See Figure 7.11.

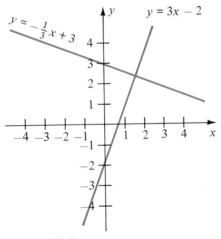

FIGURE 7.11

Thus,

$$y - y_1 = m(x - x_1)$$

$$y - 4 = -\frac{1}{3}[x - (-3)]$$

$$y - 4 = -\frac{1}{3}(x + 3) = -\frac{1}{3}x - 1$$

$$y = -\frac{1}{3}x + 3$$

PROGRESS CHECK 5
Find an equation of the line passing through the point $(2, -3)$ that is perpendicular to the line $2y + 4x - 1 = 0$.

Answer

$$y = \frac{1}{2}x - 4$$

EXAMPLE 6

State whether each pair of lines is parallel, perpendicular, or neither.

(a) $y - 3x + 11 = 0;$ $2y = 6x - 4$

(b) $y = -4x + 1;$ $2y = -4x + 9$

(c) $3y - x - 7 = 0;$ $2y + 6x + 8 = 0$

Solution

In each case, we must determine the slope of each of the lines.

(a) $y - 3x + 11 = 0$ $2y = 6x - 4$

$$y = 3x - 11 \qquad\qquad y = 3x - 2$$

$$m_1 = 3 \qquad\qquad\qquad m_2 = 3$$

The lines are parallel since they have the same slope.

(b) $y = -4x + 1$ $2y = -4x + 9$

$$y = -2x + \frac{9}{2}$$

$$m_1 = -4 \qquad\qquad m_2 = -2$$

Since $m_1 \neq m_2$, the lines cannot be parallel. Also, since

$$-\frac{1}{m_1} = -\frac{1}{-4} = \frac{1}{4} \neq m_2$$

the lines are not perpendicular. Therefore, the lines are neither parallel nor perpendicular.

(c) $3y - x - 7 = 0$ $2y + 6x + 8 = 0$

$$3y = x + 7 \qquad\qquad\qquad 2y = -6x - 8$$

$$y = \frac{1}{3}x + \frac{7}{3} \qquad\qquad\qquad y = -3x - 4$$

$$m_1 = \frac{1}{3} \qquad\qquad\qquad\qquad m_2 = -3$$

Since

$$m_2 = -3 \quad \text{and} \quad -\frac{1}{m_1} = -3$$

we see that

$$m_2 = -\frac{1}{m_1}$$

and the lines are perpendicular.

PROGRESS CHECK 6

State whether each pair of lines is parallel, perpendicular, or neither.

(a) $4y - 6x = 11;$ $3y + 2x - 7 = 0$

(b) $9y - x + 16 = 0;$ $3y = 9x + 4$

(c) $5y = x - 4;$ $25y - 5x + 17 = 0$

Answers

(*a*) *perpendicular*   (*b*) *neither*   (*c*) *parallel*

EXERCISE SET 7.3

Write an equation of the line satisfying the given conditions.

1. It is horizontal and passes through the point $(3, 2)$.
2. It is horizontal and passes through the point $(-2, 4)$.
3. It is vertical and passes through the point $(-2, 3)$.
4. It is vertical and passes through the point $(3, -2)$.

For each of the following points, write an equation of (a) the horizontal line passing through the point, and (b) the vertical line passing through the point.

5. $(-6, 3)$	6. $(-5, -2)$	7. $(4, -5)$
8. $(11, -14)$	9. $(0, 0)$	10. $(0, -4)$
11. $(-7, 0)$	12. $(-1, -1)$	13. $(0, 5)$
14. $(5, 0)$		

Write an equation of the line shown in each graph.

15.   16.

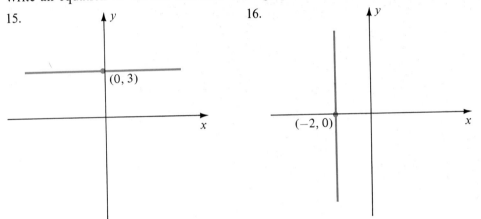

In the following exercises, let L be the line determined by P_1 and P_2, and let L' be the line determined by P_3 and P_4. Determine whether L and L' are parallel and sketch both L and L'.

17. $P_1(1, -1)$, $P_2(3, 4)$ $P_3(2, 3)$, $P_4(-1, 8)$
18. $P_1(2, 1)$, $P_2(4, 4)$ $P_3(0, -2)$, $P_4(-2, -5)$
19. $P_1(1, 3)$, $P_2(0, 5)$ $P_3(-4, 8)$, $P_4(-2, 4)$
20. $P_1(4, 2)$, $P_2(6, -1)$ $P_3(4, 5)$, $P_4(1, 8)$

Find the slope of every line parallel to the given line in the following exercises.

21. $y = -\dfrac{2}{5}x + 4$ 22. $x - 2y + 5 = 0$

23. $3x + 2y = 6$ 24. $x = 3y - 2$

25. Find an equation of the line that is parallel to the line $y = \frac{3}{2}x + 5$ and has y-intercept of -2.
26. Find an equation of the line that is parallel to the line $y = -\frac{1}{2}x - 2$ and has y-intercept of 3.
27. Find an equation of the line passing through the point $(1, 3)$ and parallel to $y = -3x + 2$.

28. Find an equation of the line passing through the point $(-1, 2)$ and parallel to $3y + 2x = 6$.

Find the slope of every line perpendicular to the line whose slope is given in each of the following.

29. 2   30. -3 31. $-\dfrac{1}{2}$ 32. $-\dfrac{3}{4}$

33. Find an equation of the line passing through the point $(-3, 2)$ that is perpendicular to the line $3x + 5y = 2$.

34. Find an equation of the line passing through the point $(-1, -3)$ that is perpendicular to the line $3y + 4x - 5 = 0$.

35. State whether each pair of lines is parallel, perpendicular, or neither.

    (a) $3x + 2y = 7$; $3y - 2x = 4$

    (b) $y - 3x + 1 = 0$; $3y + x = 8$

    (c) $y = \dfrac{2}{3}x + 3$; $6y - 4x + 8 = 0$

36. State whether each pair of lines is parallel, perpendicular, or neither.

    (a) $y - 3x + 1 = 0$; $x = \dfrac{y}{3} + 2$

    (b) $2x + 5y = 1$; $x + y = 2$

    (c) $3x + 2y = 6$; $12y - 8x + 7 = 0$

37. In the accompanying figure, perpendicular lines l_1 and l_2 with slopes m_1 and m_2, respectively, intersect at a point Q. A perpendicular from Q intersects the x-axis at the point C.

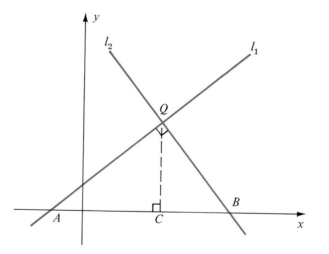

    (a)   Show that $\triangle ACQ$ is similar to $\triangle BCQ$. (*Hint:* $\angle CAQ$ and $\angle BQC$ are both complementary to $\angle AQC$.)

    (b)   Show that

$$\frac{\overline{QC}}{\overline{AC}} = \frac{\overline{CB}}{\overline{QC}}$$

(*Hint:* Corresponding sides of similar triangles are in proportion.)

(c)   Show that

$$m_1 = \frac{\overline{QC}}{\overline{AC}}, \quad m_2 = \frac{\overline{QC}}{\overline{CB}}$$

(*Hint:* Use the definition of slope.)

(d)   Use (b) and (c) to show that

$$m_2 = -\frac{1}{m_1}$$

When we graph a linear equation, say

$$y = 2x - 1$$

we can readily see that the graph of the line divides the plane into two regions called **half planes.**

If, in the equation $y = 2x - 1$, we replace the equal sign by any of the symbols $<$, $>$, $\leq$, or $\geq$, we have a **linear inequality in two variables.** By the **graph of a linear inequality** such as

$$y < 2x - 1$$

we mean the set of all points whose coordinates satisfy the inequality. Since the coordinates of every point on the line L in Figure 7.12 satisfy the *equation* $y = 2x - 1$, we readily see that the coordinates of those points in the half plane below the line must satisfy the *inequality $y < 2x - 1$.* Similarly, the coordinates of those points in the half plane above the line must satisfy the *inequality $y > 2x - 1$.* This suggests that the graph of a linear inequality in two variables is a half plane and leads to a straightforward method for graphing linear inequalities.

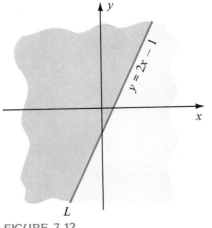

FIGURE 7.12

Graphing linear inequalities	Example: $y \le x - 1$
Step 1. Replace the inequality by an equal sign and plot the line. (a) If the inequality is $\le$ or $\ge$, plot a solid line (points on the line will satisfy the inequality). (b) If the inequality is $<$ or $>$, plot a dashed line (points on the line will not satisfy the inequality).	*Step 1.* $y = x - 1$
Step 2. Choose any point that is not on the line, as a test point. If the origin is not on the line, it is the most convenient choice.	*Step 2.* Choose $(0, 0)$ as a test point.
Step 3. Substitute the coordinates of the test point into the inequality. (a) If the test point satisfies the inequality, then the coordinates of every point in the half plane that contains the test point will satisfy the inequality.	*Step 3.* Substituting $(0, 0)$ in $$y \le x - 1$$ $$0 \le 0 - 1 \quad (?)$$ $$0 \le -1$$
(b) If the test point does not satisfy the inequality, then the half plane on the other side of the line contains all the points satisfying the inequality.	is false. Since $(0, 0)$ is in the half plane above the line and does not satisfy the inequality, all the points below the line will satisfy the inequality.

EXAMPLE 1

Graph $2x - 3y > 6$.

Solution

We first graph the line $2x - 3y = 6$ by plotting two points. We draw a dashed or broken line to indicate that $2x - 3y = 6$ is not part of the graph (see Figure 7.13).

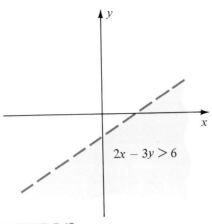

FIGURE 7.13

Since $(0, 0)$ is not on the line, we can use it as a test point.

$$2x - 3y > 6$$
$$2(0) - 3(0) > 6 \quad (?)$$
$$0 - 0 > 6 \quad (?)$$
$$0 > 6$$

is false. Since $(0, 0)$ is in the half plane above the line, the graph consists of the half plane below the line.

PROGRESS CHECK 1

Graph the inequalities.

(a) $y \le 2x + 1$ (b) $y + 3x > -2$ (c) $y \ge -x + 1$

Answers

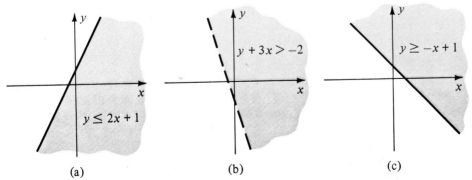

EXAMPLE 2
Graph (a) $y < x$ (b) $2x \geq 5$

Solution
(a) Since the origin lies on the line $y = x$, we choose another test point, say $P(0, 1)$ above the line. Since $(0, 1)$ does not satisfy the inequality, the graph of the inequality is the half plane below the line. See Figure 7.14a.
(b) The graph of $2x = 5$ is a vertical line and the graph of $2x \geq 5$ is the half plane to the right of the line and also the line itself. See Figure 7.14b.

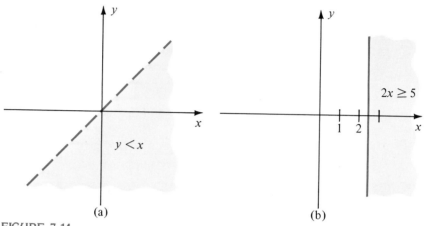

FIGURE 7.14

PROGRESS CHECK 2
Graph the inequalities.
(a) $2y \geq 7$ (b) $x < -2$ (c) $1 \leq y < 3$

Answers

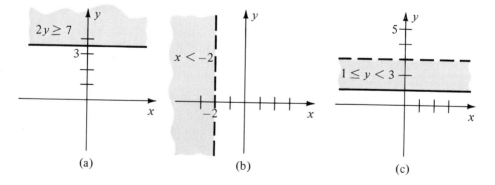

EXERCISE SET 7.4
Graph the given inequality in the following exercises.

1. $y \leq x + 2$
2. $y \geq x + 3$
3. $y > x - 4$
4. $y < x - 5$
5. $y \leq 4 - x$
6. $y \geq 2 - x$

7. $y > x$

8. $y \leq 2x$

9. $y \geq \frac{1}{2}x - 3$

10. $y \leq 3 - \frac{2}{3}x$

11. $y < 1 - \frac{5}{2}x$

12. $y > \frac{1}{3}x + 2$

13. $3x - 5y > 15$

14. $2y - 3x < 12$

15. $3x + 8y + 24 \leq 0$

16. $2x - 5y - 10 > 0$

17. $x \leq 4$

18. $3x > -2$

19. $y > -3$

20. $5y \leq 25$

21. $x < 0$

22. $y \geq 0$

23. $x > 0$

24. $y < 0$

25. $-2 \leq x \leq 3$

26. $-6 < y < -2$

27. $1 < y < 4$

28. $0 \leq x \leq 6$

Give the linear inequality whose graph is shown.

29.

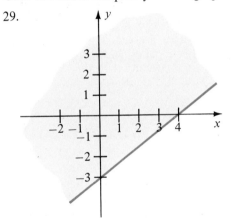

30.

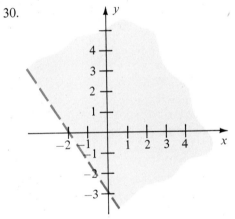

31.

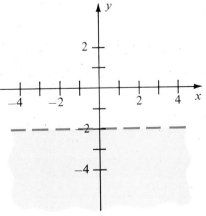

32.

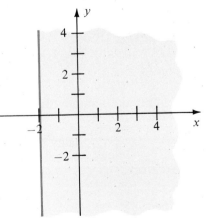

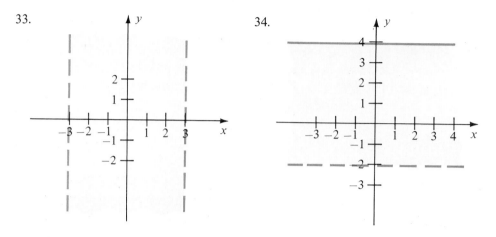

35. A steel producer makes two types of steel, regular and special. A ton of regular steel requires 2 hours in the open-hearth furnace and a ton of special steel requires 5 hours. Let x and y denote the number of tons of regular and special steel, respectively, made per day. If the open-hearth is available at most 15 hours per day, write an inequality that must be satisfied by x and y. Graph this inequality.

36. A patient is placed on a diet that restricts caloric intake to 1500 calories per day. The patient plans to eat x ounces of cheese, y slices of bread, and z apples on the first day of the diet. If cheese contains 100 calories per ounce, bread 110 calories per slice, and apples 80 calories each, write an inequality that must be satisfied by x, y, and z.

TERMS AND SYMBOLS

slope of a line (p. 188)
**increasing function
    (p. 190)**
**decreasing function
    (p. 190)**
constant function (p. 190)
point–slope form (p. 193)
y-intercept (p. 195)

**slope–intercept form
    (p. 195)**
**general first-degree
    equation (p. 196)**
vertical line (p. 200)
horizontal line (p. 200)
parallel lines (p. 202)

perpendicular lines (p. 202)
half plane (p. 207)
**linear inequality in two
    variables (p. 207)**
**graph of a linear
    inequality in two
    variables (p. 207)**

KEY IDEAS FOR REVIEW

☐ The graph of the linear function $f(x) = ax + b$ is a straight line.
☐ Any two points $P_1(x_1, y_1)$ and $P_2(x_2, y_2)$ on a line can be used to find the slope
    $$m = \frac{y_2 - y_1}{x_2 - x_1}.$$
☐ Positive slope indicates that a line is rising; negative slope indicates that a line is falling.
☐ The slope of a horizontal line is 0; the slope of a vertical line is undefined.
☐ The point–slope form of the equation of a line is $y - y_1 = m(x - x_1)$.
☐ The slope–intercept form of the equation of a line is $y = mx + b$.
☐ The graph of the general first-degree equation $Ax + By + C = 0$ is always a straight line.
☐ The equation of a horizontal line through the point (a, b) is $y = b$.

☐ The equation of a vertical line through the point (a, b) is $x = a$.

☐ Parallel lines have the same slope.

☐ The slopes of perpendicular lines are negative reciprocals of each other.

☐ The graph of a linear inequality in two variables is a half plane.

COMMON ERRORS

1. When calculating the slope

$$m = \frac{y_2 - y_1}{x_2 - x_1}$$

by using two points, you must be consistent in labeling the points. For example, if we have $(x_1, y_1) = (2, 3)$ and $(x_2, y_2) = (4, 5)$, then

$$m = \frac{5 - 3}{4 - 2} = \frac{2}{2} = 1$$

Don't write

$$m = \frac{5 - 3}{2 - 4} = \frac{2}{-2} = -1$$

2. The equation of the horizontal line through (a, b) is $y = b$. The equation of the vertical line through (a, b) is $x = a$. *Don't reverse these.*

3. The equation $y - 6x + 2 = 0$ must be rewritten in the form $y = 6x - 2$ before you can determine the slope $m = 6$ and intercept $b = -2$. Notice also that the intercept includes the sign, that is, the intercept is -2, not 2.

4. If a line L has slope $-\frac{1}{3}$, then every line perpendicular to L has slope 3, *not* -3. For instance, the line $2y = x - 1$ has slope $\frac{1}{2}$; every line perpendicular to this line has slope -2, *not* 2.

5. Do not confuse zero slope and "no slope." A horizontal line has zero slope, but a vertical line does not have slope.

PROGRESS TEST 7A

1. Find the slope of the line through the points $(6, -3)$ and $(-6, -3)$.

2. Find the slope of the line through the points $\left(-1, \frac{3}{2}\right)$ and $\left(\frac{1}{2}, \frac{1}{2}\right)$.

3. Find an equation of the line with slope -7 that passes through the point $(-1, 2)$.

4. Find an equation of the line through the points $(6, 8)$ and $(-4, 7)$.

5. Find the slope and y-intercept of the line $2y + 5x - 6 = 0$.

6. Find the slope of the line $3y = 2x - 1$, and determine whether the line is rising or falling.

7. Find the slope of the line $2y + \frac{1}{2}x - 5 = 0$, and determine whether the line is rising or falling.

8. A telephone company charges 25 cents for the first minute and 20 cents for each additional minute. Write an equation relating the total cost C of a phone call and the number of minutes t the parties are connected.

9. Find an equation of the vertical line through the point $\left(-11, \frac{3}{4}\right)$.

10. Find an equation of the horizontal line through the point $\left(4, -\frac{7}{3}\right)$.

11. Find an equation of the line through $(0, 2)$ that is parallel to the line $y - 3x - 2 = 0$.
12. Find the slope of every line perpendicular to the line $2y + 7x - 6 = 0$.
13. Find an equation of the line through $(-1, -2)$ that is perpendicular to the line $3y = 4x - 1$.
14. Graph the inequality $y > x - 2$.
15. Graph the inequality $x \leq -3$.

PROGRESS TEST 7B

1. Find the slope of the line through the points $(-4, -2)$ and $(-4, 6)$.
2. Find the slope of the line through the points $\left(\frac{1}{3}, -\frac{4}{3}\right)$ and $\left(-\frac{1}{2}, \frac{1}{3}\right)$.
3. Find an equation of the line with slope 5 that passes through the point $(-6, -7)$.
4. Find an equation of the line through the points $(-9, 2)$ and $(3, -6)$.
5. Find the slope and y-intercept of the line $3y - 12x + 5 = 0$.
6. Find the slope of the line $2y - 7x + 10 = 0$, and determine whether the line is rising or falling.
7. Find the slope of the line $4y + 5x - 3 = 0$, and determine whether the line is rising or falling.
8. The U.S. Postal Service charges 15 cents for the first ounce and 13 cents for every additional ounce of a first class letter. Write an equation relating the total cost C of a mailing and the number of ounces n when n is a natural number.
9. Find an equation of the horizontal line through the point $\left(-1, \frac{5}{4}\right)$.
10. Find an equation of the vertical line through the point $\left(3, -\frac{2}{3}\right)$.
11. Find an equation of the line through $(-5, -7)$ that is parallel to the line $y - 2x = 6$.
12. Find the slope of every line perpendicular to the line $-3x + 5y - 2 = 0$.
13. Find an equation of the line through $\left(\frac{1}{2}, -\frac{3}{2}\right)$ that is perpendicular to the line $3y - x = 0$.
14. Graph the inequality $2y + 3x \leq 1$.
15. Graph the inequality $y > -1$.

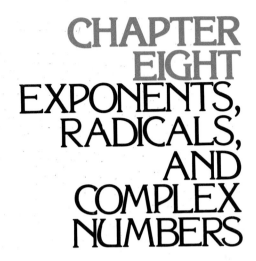

CHAPTER EIGHT
EXPONENTS, RADICALS, AND COMPLEX NUMBERS

We have previously worked with exponential notation such as x^n when n is a positive integer. We now seek to expand our capability to handle any exponent, whether zero, a positive integer, a negative integer, or a rational number. We will see that the same rules apply to all these cases.

Radicals are an alternate way of writing rational exponent forms. Since the solutions of polynomial equations frequently involve radicals, we will learn to manipulate and simplify radical forms as background to our study of polynomial equations of degree greater than 1.

We will also see that the real number system is inadequate to provide a solution to all polynomial equations. It is necessary to create a new type of number, called a complex number, which we explore at the end of the chapter.

8.1
POSITIVE INTEGER EXPONENTS

In Chapter Two we defined a^n by

$$a^n = \underbrace{a \cdot a \cdot \ldots \cdot a}_{n \text{ factors}}$$

where n is a natural number and a is any real number. We then saw that if m and n are natural numbers and a is any real number,

$$a^m \cdot a^n = a^{m+n}$$

We shall now develop additional rules for exponents. We see that

$$(a^2)^3 = \underbrace{a^2 \cdot a^2 \cdot a^2}_{3 \text{ factors}} = a^6$$

and, in general, if m and n are any natural numbers and a is any real number, then

$$(a^m)^n = a^{mn}$$

EXAMPLE 1

(a) $(2^2)^3 = 2^{2 \cdot 3} = 2^6$

(b) $(x^4)^3 = x^{4 \cdot 3} = x^{12}$

(c) $(x^2 \cdot x^3)^4 = (x^5)^4 = x^{5 \cdot 4} = x^{20}$

(d) $(a^2)^n = a^{2n}$

(e) $(r^n)^{2n} = r^{n \cdot 2n} = r^{2n^2}$

(f) $[(x + 2)^4]^2 = (x + 2)^{4 \cdot 2} = (x + 2)^8$

PROGRESS CHECK 1
Simplify.

(a) $(x^3)^4$ (b) $(x^7)^7$ (c) $(x^2)^3 \cdot x^5$

(d) $(y^n)^2$ (e) $[(2a - 2)^3]^5$ (f) $x^4(x^2)^3$

Answers

(a) x^{12} (b) x^{49} (c) x^{11} (d) y^{2n} (e) $(2a - 2)^{15}$ (f) x^{10}

Now we turn to a^m/a^n. We will use our old friend, the cancellation principle, to simplify. We see that

$$\frac{a^4}{a^2} = \frac{\cancel{a} \cdot \cancel{a} \cdot a \cdot a}{\cancel{a} \cdot \cancel{a}} = \frac{a^2}{1} = a^2$$

and

$$\frac{a^2}{a^4} = \frac{\cancel{a} \cdot \cancel{a}}{\cancel{a} \cdot \cancel{a} \cdot a \cdot a} = \frac{1}{a^2}$$

We can conclude that the result depends upon which exponent is larger. If m and n are natural numbers and a is any real number, then

$$\frac{a^m}{a^n} = a^{m-n} \quad \text{if} \quad m > n$$

$$\frac{a^m}{a^n} = \frac{1}{a^{n-m}} \quad \text{if} \quad n > m$$

$$\frac{a^m}{a^n} = 1 \quad \text{if} \quad m = n$$

EXAMPLE 2

(a) $\dfrac{5^7}{5^4} = 5^{7-4} = 5^3 = 125$

(b) $\dfrac{(-3)^2}{(-3)^3} = \dfrac{1}{(-3)^{3-2}} = \dfrac{1}{-3} = -\dfrac{1}{3}$

(c) $\dfrac{-x^{17}}{x^{11}} = -x^{17-11} = -x^6$

(d) $\dfrac{-(2y-1)^6}{(2y-1)^8} = \dfrac{-1}{(2y-1)^{8-6}} = \dfrac{-1}{(2y-1)^2}$

(e) $\dfrac{x^{2n+1}}{x^n} = x^{2n+1-n} = x^{n+1}$

(f) $\dfrac{(x^3)^4}{x^7(x^2)^5} = \dfrac{x^{12}}{x^{17}} = \dfrac{1}{x^5}$

PROGRESS CHECK 2

Simplify, using only positive exponents.

(a) $\dfrac{6^3}{6^9}$ (b) $\dfrac{-2^4}{2^2}$ (c) $\dfrac{a^{14}}{a^8}$

(d) $\dfrac{b^{17}}{b^{25}}$ (e) $\dfrac{y^n}{y^{2n}}$ (f) $\dfrac{-2(x+1)^4}{(x+1)^{11}}$

Answers

(a) $\dfrac{1}{6^6}$ (b) -4 (c) a^6 (d) $\dfrac{1}{b^8}$ (e) $\dfrac{1}{y^n}$ (f) $\dfrac{-2}{(x+1)^7}$

Thus far, we have developed rules for exponents when there is a common base. It is also easy to develop rules when there is a common exponent. For example,

$$(ab)^3 = (ab)(ab)(ab) = (a \cdot a \cdot a) \cdot (b \cdot b \cdot b)$$
$$= a^3 b^3$$

and, in general, if m is any natural number and a and b are any real numbers, then

$$(ab)^m = a^m b^m$$

Similarly,

$$\left(\frac{a}{b}\right)^3 = \left(\frac{a}{b}\right)\left(\frac{a}{b}\right)\left(\frac{a}{b}\right) = \frac{a \cdot a \cdot a}{b \cdot b \cdot b} = \frac{a^3}{b^3}$$

and, in general, if m is any natural number and a and b are any real numbers, then

$$\left(\frac{a}{b}\right)^m = \frac{a^m}{b^m}$$

There is a helpful technique for keeping track of the sign. If we have $(-ab)^m$, we treat this as $[(-1)ab]^m$ and then have

$$(-ab)^m = [(-1)ab]^m = (-1)^m a^m b^m$$

The sign of the result will then depend upon whether m is even or odd. If m is even, $(-1)^m = 1$; and if m is odd, $(-1)^m = -1$. Thus, if m is even, the result is $a^m b^m$; if m is odd, the result is $-a^m b^m$. For example, $(-xy)^3 = -x^3 y^3$, while $(-xy)^6 = x^6 y^6$.

EXAMPLE 3

(a) $\left(\dfrac{3}{2}\right)^2 = \dfrac{3^2}{2^2} = \dfrac{9}{4}$ (b) $(xy)^6 = x^6 y^6$

(c) $\left(\dfrac{r}{s}\right)^{11} = \dfrac{r^{11}}{s^{11}}$ (d) $(2x)^3 = 2^3 \cdot x^3 = 8x^3$

(e) $-\left(\dfrac{2}{x}\right)^3 = -\dfrac{2^3}{x^3} = -\dfrac{8}{x^3}$

(f) $(2x^2 y)^4 = 2^4 (x^2)^4 \cdot y^4 = 16x^8 y^4$

(g) $\left(\dfrac{-ab^2}{c^3}\right)^3 = \dfrac{(-1)^3 a^3 \cdot (b^2)^3}{(c^3)^3} = \dfrac{-a^3 b^6}{c^9}$

(h) $\dfrac{(24y)^3}{(12x)^3} = \left(\dfrac{24y}{12x}\right)^3 = \left(\dfrac{2y}{x}\right)^3 = \dfrac{2^3 y^3}{x^3} = \dfrac{8y^3}{x^3}$

PROGRESS CHECK 3
Simplify.

(a) $\left(\dfrac{2}{3}\right)^3$ (b) $(2xy)^2$ (c) $\left(\dfrac{2x^2}{y^3}\right)^4$

(d) $\left(-\dfrac{2}{5}xy^2\right)^2$ (e) $(-xy)^3$ (f) $\left(\dfrac{x^5 y^2}{y}\right)^3$

Answers

(a) $\dfrac{8}{27}$ (b) $4x^2 y^2$ (c) $\dfrac{16x^8}{y^{12}}$ (d) $\dfrac{4}{25}x^2 y^4$ (e) $-x^3 y^3$ (f) $x^{15} y^3$

EXERCISE SET 8.1

Evaluate. Identify the base and exponent.

1. $\left(\dfrac{1}{3}\right)^4$ 2. $\left(\dfrac{2}{5}\right)^3$ 3. $(-2)^5$ 4. $(-0.2)^4$

Write, using exponents.

5. $(-5)(-5)(-5)(-5)$

6. $\left(\dfrac{1}{2}\right)\left(\dfrac{1}{2}\right)\left(\dfrac{1}{2}\right)\left(\dfrac{1}{3}\right)\left(\dfrac{1}{3}\right)$

7. $4 \cdot 4 \cdot x \cdot x \cdot y \cdot y \cdot y$

8. $3x \cdot x \cdot x + 2 \cdot y \cdot y \cdot y \cdot y$

The right-hand side of each of the following is incorrect. Correct that side.

9. $x^2 \cdot x^4 = x^8$

10. $(y^2)^5 = y^7$

11. $\dfrac{b^6}{b^2} = b^3$

12. $\dfrac{x^2}{x^6} = x^4$

13. $(2x)^4 = 2x^4$

14. $\left(\dfrac{4}{3}\right)^4 = \dfrac{4}{3^4}$

Simplify using the properties of exponents.

15. $\left(-\dfrac{1}{2}\right)^4\left(-\dfrac{1}{2}\right)^3$

16. $x^3 \cdot x^6$

17. $(x^m)^{3m}$

18. $(y^4)^{2n}$

19. $\dfrac{3^6}{3^2}$

20. $\dfrac{(-4)^6}{(-4)^{10}}$

21. $-\left(\dfrac{x}{y}\right)^3$

22. $\left[\dfrac{(2x+1)^2}{(3x-2)^3}\right]^3$

23. $y^2 \cdot y^6$

24. $-3r^3r^3$

25. $(x^3)^5 \cdot x^4$

26. $[(-2)^5]^4(-2)^3$

27. $\dfrac{x^{12}}{x^8}$

28. $\dfrac{x^4}{x^8}$

29. $(-2x^2)^5$

30. $-(2x^2)^5$

31. $x^{3n} \cdot x^n$

32. $(-2)^m(-2)^n$

33. $(3^2)^4$

34. $(x^6)^5$

35. $\dfrac{x^n}{x^{n+2}}$

36. $\dfrac{x^{3+n}}{x^n}$

37. $\left(\dfrac{3x^3}{y^2}\right)^5$

38. $\left(\dfrac{2x^4y^3}{y^2}\right)^3$

39. $(-5x^3)(-6x^5)$

40. $(4a^4)(-2a^3)$

41. $(x^2)^3(y^2)^4(x^3)^7$

42. $(a^3)^4(b^2)^5(b^3)^6$

43. $\dfrac{(r^2)^4}{(r^4)^2}$

44. $\dfrac{[(x+1)^3]^2}{[(x+1)^5]^3}$

45. $[(3b+1)^5]^5$

46. $[(2x+3)^3]^7$

47. $\dfrac{(2a+b)^4}{(2a+b)^6}$

48. $\dfrac{5(3x-y)^8}{(3x-y)^2}$

49. $(3x^3y)^3$

50. $\left(\dfrac{3}{2}x^2y^3\right)^n$

51. $\dfrac{(3x^2)^3}{(-2x)^3}$

52. $\dfrac{(-2a^2b)^4}{(-3ab^2)^3}$

53. $(2x+1)^3(2x+1)^7$

54. $(3x-2)^6(3x-2)^5$

55. $\dfrac{y^3(y^3)^4}{(y^4)^6}$

56. $\dfrac{2[(3x-1)^3]^5[(3x-1)^2]^4}{[(3x-1)^2]^7}$

57. $(-2a^2b^3)^{2n}$

58. $\left(-\dfrac{2}{3}a^2b^3c^2\right)^3$

59. $\left(\dfrac{-2a^2b^3}{c^2}\right)^3$

60. $\left(\dfrac{8x^3y}{6xy^4}\right)^3$

Evaluate.

61. $(1.27)^2(3.65)^2$

62. $(-4.73)^3(-0.22)^3$

63. $\dfrac{-(6.14)^2(2.07)^2}{(7.93)^2}$

64. $\dfrac{(1.77)^2}{(2.85)^2(8.19)^2}$

65. $(2x - 1)^3(x + 1)^3$ when $x = 1.73$

66. $\dfrac{(3 - x)^2}{(x + 3)^2}$ when $x = 2.25$

67. $(5.46^2)^2$

68. $(3.29^2)^2$

8.2
INTEGER EXPONENTS

We would like to expand our rules for exponents to include zero and negative exponents.

We begin with a^0, where $a \neq 0$ since 0^0 has no mathematical meaning. We will assume that the previous rules for exponents apply to a^0 and see if this leads us to a definition of a^0. For example, applying the rule $a^m a^n = a^{m+n}$ yields

$$a^m \cdot a^0 = a^{m+0} = a^m$$

Dividing both sides by a^m, we see that we must have

$$a^0 = 1$$

We now *define* a^0 in this way. The student can easily verify that the other rules for exponents also hold. For example,

$$\left(\frac{a}{b}\right)^0 = 1 \quad \text{and} \quad \left(\frac{a}{b}\right)^0 = \frac{a^0}{b^0} = \frac{1}{1} = 1$$

EXAMPLE 1

(a) $3^0 = 1$ (b) $(-4)^0 = 1$ (c) $\left(\dfrac{2}{5}\right)^0 = 1$

(d) $4(xy)^0 = 4(1) = 4$ (e) $\dfrac{-2}{(t^2 - 1)^0} = \dfrac{-2}{1} = -2$

PROGRESS CHECK 1
Simplify.

(a) $(-4x)^0$ (b) $-3(r^2s)^0$ (c) $9\left(\dfrac{2}{7}\right)^0$ (d) $-4(x^2 - 5)^0$

Answers
(a)   *1*   (b)   *−3*   (c)   *9*   (d)   *−4*

The same approach will lead us to a meaning for negative exponents such as a^{-m}. For consistency, we must have

$$a^m \cdot a^{-m} = a^{m-m} = a^0 = 1$$

Thus,

$$a^m \cdot a^{-m} = 1$$

If we divide both sides by a^m, we obtain

$$a^{-m} = \frac{1}{a^m}$$

Had we divided by a^{-m} we would have obtained

$$a^m = \frac{1}{a^{-m}}$$

Thus, a^{-m} is the reciprocal of a^m, and a^m is the reciprocal of a^{-m}. Again, the student should verify that all the rules for exponents hold with this definition of a^{-m}. For example,

$$(a^{-m})^n = \left(\frac{1}{a^m}\right)^n = \frac{1}{a^{mn}} = a^{-mn}$$

The rules for negative exponents can be expressed in another way.

A factor moves from numerator to denominator (or from denominator to numerator) by changing the sign of the exponent.

EXAMPLE 2

Simplify using positive exponents.

(a) $3^{-2} = \dfrac{1}{3^2} = \dfrac{1}{9}$

(b) $\dfrac{1}{2^{-3}} = 2^3 = 8$

(c) $-x^{-7} = -(x^{-7}) = -\dfrac{1}{x^7}$

(d) $\dfrac{-2}{(a-1)^{-2}} = -2(a-1)^2$

(e) $(2x)^{-3} = \dfrac{1}{(2x)^3} = \dfrac{1}{8x^3}$

PROGRESS CHECK 2

Simplify using positive exponents.

(a) 4^{-2} (b) $\dfrac{1}{3^{-2}}$ (c) $\dfrac{r^{-6}}{s^{-2}}$ (d) $x^{-2}y^{-3}$

Answers

(a) $\dfrac{1}{16}$ (b) 9 (c) $\dfrac{s^2}{r^6}$ (d) $\dfrac{1}{x^2y^3}$

EXAMPLE 3

Simplify, using only positive exponents in the answer.

(a) $x^2x^{-6} = x^{2-6} = x^{-4} = \dfrac{1}{x^4}$

(b) $\dfrac{4a^3}{6a^{-5}} = \dfrac{2}{3}a^3a^5 = \dfrac{2}{3}a^8$

(c) $\dfrac{-2x^{-2}y^4}{8x^3y^{-1}} = -\dfrac{y^4y^1}{4x^3x^2} = -\dfrac{y^5}{4x^5}$

or

$\dfrac{-2x^{-2}y^4}{8x^3y^{-1}} = -\dfrac{y^{4-(-1)}}{4x^{3-(-2)}} = -\dfrac{y^5}{4x^5}$

PROGRESS CHECK 3

Simplify, using only positive exponents in the answer.

(a) y^4y^{-6} (b) $\dfrac{x^7}{x^{-5}}$ (c) $\dfrac{-3x^4y^{-2}}{9x^{-8}y^6}$

Answers

(a) $\dfrac{1}{y^2}$ (b) x^{12} (c) $-\dfrac{x^{12}}{3y^8}$

EXAMPLE 4

Simplify, using only positive exponents in the answer.

(a) $(x^2y^{-3})^{-5} = (x^2)^{-5}(y^{-3})^{-5} = x^{-10}y^{15} = \dfrac{y^{15}}{x^{10}}$

(b) $\left(\dfrac{a^{-5}b^2}{c^{-3}}\right)^{-4} = \dfrac{(a^{-5}b^2)^{-4}}{(c^{-3})^{-4}} = \dfrac{a^{20}b^{-8}}{c^{12}} = \dfrac{a^{20}}{b^8c^{12}}$

(c) $(m^{-1} + n^{-1})^2 = \left(\dfrac{1}{m} + \dfrac{1}{n}\right)^2 = \left(\dfrac{n + m}{mn}\right)^2$

$\qquad\qquad = \dfrac{n^2 + 2mn + m^2}{m^2n^2}$

PROGRESS CHECK 4

Simplify, using only positive exponents in the answer.

(a) $(x^{-4}y^{-3})^4$ (b) $\left(\dfrac{x^{-3}}{x^{-4}}\right)^{-1}$ (c) $(x^{-1} - y^{-1})^2$

Answers

(a) $\dfrac{1}{x^{16}y^{12}}$ (b) $\dfrac{1}{x}$ (c) $\dfrac{x^2 - 2xy + y^2}{x^2y^2}$

WARNING

Don't confuse negative numbers and negative exponents.

(a) $2^{-4} = \dfrac{1}{2^4}$

Don't write

$$2^{-4} = -2^4$$

(b) $(-2)^{-3} = \dfrac{1}{(-2)^3} = \dfrac{1}{-8} = -\dfrac{1}{8}$

Don't write

$$(-2)^{-3} = \dfrac{1}{2^3} = \dfrac{1}{8}$$

We can now summarize the "laws of exponents."

$$a^m \cdot a^n = a^{m+n} \qquad\qquad \left(\dfrac{a}{b}\right)^n = \dfrac{a^n}{b^n}$$

$$\dfrac{a^m}{a^n} = a^{m-n} = \dfrac{1}{a^{n-m}} \qquad a^0 = 1$$

$$(a^m)^n = a^{mn}$$

$$(ab)^n = a^n b^n \qquad a^{-n} = \dfrac{1}{a^n}$$

EXERCISE SET 8.2

Simplify and write the answer using only positive exponents.

1. 2^0

2. $(-1.2)^0$

3. $(xy)^0$

4. $2(a^2 - 1)^0$

5. $\dfrac{3}{(2x^2 + 1)^0}$

6. 2^{-4}

7. $(-3)^{-3}$

8. $\dfrac{1}{3^{-4}}$

9. $\dfrac{2}{4^{-3}}$

10. x^{-5}

11. $(-x)^3$

12. $-x^{-5}$

13. $\dfrac{1}{y^{-6}}$

14. $\dfrac{1}{x^{-7}}$

15. $(2a)^{-6}$

16. $-6(3x + 2y)^{-5}$

17. $\dfrac{-4}{(5a - 3b)^{-2}}$

18. $3^4 3^{-6}$

19. $5^{-3} 5^5$

20. $x^{-4} x^2$

21. $4y^5 y^{-2}$

22. $a^4 a^{-4}$

23. $x^{-4} x^{-5} x^2$

24. $-3a^{-3} a^{-6} a^4$

25. $(3^2)^{-3}$

26. $(4^{-2})^2$

27. $[(-2)^3]^3$

28. $(x^{-2})^4$

29. $(x^{-3})^{-3}$

30. $[(x + y)^{-2}]^2$

31. $\dfrac{2^2}{2^{-3}}$

32. $\dfrac{3^{-5}}{3^2}$

33. $\dfrac{x^8}{x^{-10}}$

34. $\dfrac{3x^{-7}}{x^4}$

35. $\dfrac{4x^{-3}}{x^{-2}}$

36. $\dfrac{2x^4y^{-2}}{x^2y^{-3}}$

37. $2x^3x^{-3}$

38. $(x^4y^{-2})^{-1}$

39. $(2a^2b^{-3})^{-3}$

40. $\left(\dfrac{6}{9}x^{-3}y^2\right)^2$

41. $(3a^{-2}b^{-3})^{-2}$

42. $\dfrac{1}{(2xy)^{-2}}$

43. $\left(\dfrac{a^{-3}b^3}{c^{-2}}\right)^{-3}$

44. $\left(-\dfrac{1}{2}x^3y^{-4}\right)^{-3}$

45. $\dfrac{a^2b^{-4}}{c}$

46. $\dfrac{(x^{-2})^2}{(3y^{-2})^3}$

47. $\dfrac{3a^5b^{-2}}{9a^{-4}b^2}$

48. $\dfrac{8x^{-3}y^{-4}}{2x^{-4}y^{-3}}$

49. $\left(\dfrac{x^2}{x^3}\right)^{-1}$

50. $\left(\dfrac{x^3}{x^{-2}}\right)^2$

51. $\left(\dfrac{2a^2b^{-4}}{a^{-3}c^{-3}}\right)^2$

52. $\left(\dfrac{3xy^{-1}}{2x^{-1}y^2}\right)^{-2}$

53. $\dfrac{2x^{-3}y^2}{x^{-3}y^{-3}}$

54. $(a^{-2}b^2)^{-1}$

55. $\left(\dfrac{y^{-2}}{y^{-3}}\right)^{-1}$

56. $\left(\dfrac{8x^{-3}y}{6xy^{-4}}\right)^{-3}$

57. $\left(\dfrac{-2a^{-2}b^3}{c^{-2}}\right)^{-2}$

58. $\left(-\dfrac{3}{2}a^2b^{-2}c^{-2}\right)^{-3}$

59. $\dfrac{(a+b)^{-1}}{(a-b)^{-2}}$

60. $(a^{-1}+b^{-1})^{-1}$

Evaluate.

61. $(1.20^2)^{-1}$

62. $(-3.67^2)^{-1}$

63. $\left(\dfrac{7.65^{-1}}{7.65^2}\right)^2$

64. $\left(\dfrac{4.46^2}{4.46^{-1}}\right)^{-1}$

65. $\dfrac{(1.7)^{-2}(2.1)^2}{(1.7)(2.1)^{-1}}$

66. $\dfrac{(3.42+2.01)^{-1}}{(3.42-2.01)^{-2}}$

67. $\dfrac{(2x+1)^{-1}}{(x-1)^{-2}}$ when $x = 8.65$

68. $\dfrac{(3x-1)(x+2)^2}{(4-x)^{-2}}$ when $x = 4.03$

8.3
RATIONAL
EXPONENTS AND RADICALS

Suppose a square whose sides are of length a has an area of 25 square inches (Figure 8.1). We can write the equation

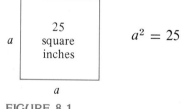

$a^2 = 25$

FIGURE 8.1

and seek a number a whose square is 25. We say that a is a **square root of b** if $a^2 = b$. Similarly, we say that a is a **cube root of b** if $a^3 = b$ and, in general, if n is a natural number, we say that

$$a \text{ is an } \textbf{n\text{th root of } b} \text{ if } a^n = b.$$

Thus, 5 is a square root of 25 since $5^2 = 25$, and -2 is a cube root of -8 since $(-2)^3 = -8$.

Clearly, 5 is not the only square root of 25. Since $(-5)^2 = 25$, -5 is also a square root of 25. This leads us to ask: How many real nth roots of a given number b are there? We can experiment with some numbers and display the results in a table (Table 8.1).

TABLE 8.1

b	Square root	b	Cube root	b	Fourth root
4·	2, -2	8	2	1	1, -1
25	5, -5	-8	-2	16	2, -2
64	8, -8	27	3	81	3, -3
81	9, -9	-27	-3	256	4, -4

But what about the square root of -4? Is there a real number a such that $a^2 = -4$? No, since the square of a real number is always nonnegative. We summarize these results in Table 8.2.

TABLE 8.2

b	n	Number of nth roots of b such that $b = a^n$	Form of nth roots
>0	Even	2	a, $-a$
<0	Even	None	None
>0	Odd	1	$a > 0$
<0	Odd	1	$a < 0$
0	All	1	0

(a)   If $b > 0$ and n is even, there are two real nth roots of b, and one is the negative of the other.

(b)   If $b < 0$ and n is even, there are no real nth roots of b.

(c)   If n is odd, there is one real nth root of b, and its sign is the same as the sign of b.

(d)   If $b = 0$, then the nth root of b is 0.

We can provide a meaning to powers with rational exponents which will enable us to apply the laws of exponents developed earlier in this chapter. If we compute $(b^{1/n})^n$ using the laws of exponents, then

$$(b^{1/n})^n = b^{n/n} = b$$

But a is an nth root of b if $a^n = b$. Then for every natural number n, $b^{1/n}$ is an nth root of b. If n is even and b is positive, we see from Table 8.2 that there are two numbers a such that $a^n = b$. For example,

$$4^2 = 16 \quad \text{and} \quad (-4)^2 = 16$$

Thus, there are two candidates for $16^{1/2}$, namely, 4 and -4. The positive one is called the **principal square root,** and we say that $16^{1/2} = 4$. In summary, for every natural number n,

$$b^{1/n} \text{ is the principal } n\text{th root of } b.$$

EXAMPLE 1
Evaluate.

(a) $144^{1/2} = 12$ (b) $8^{1/3} = 2$ (c) $(-8)^{1/3} = -2$

(d) $-25^{1/2} = -5$ (e) $(-25)^{1/2}$ not a real number

(f) $\left(\dfrac{1}{16}\right)^{1/4} = \dfrac{1}{2}$ (g) $-\left(\dfrac{1}{16}\right)^{1/4} = -\dfrac{1}{2}$

PROGRESS CHECK 1
Evaluate.

(a) $(27)^{1/3}$ (b) $(-27)^{1/3}$ (c) $36^{1/2}$

(d) $(-36)^{1/2}$ (e) $-36^{1/2}$ (f) $\left(\dfrac{1}{8}\right)^{1/3}$

Answers

(a) *3*   (b) -3 (c) *6*   (d) *not a real number*   (e) -6 (f) $\dfrac{1}{2}$

WARNING

(1)   Note that

$$-36^{1/2} = -(36)^{1/2} = -6$$

but $(-36)^{1/2}$ is undefined.

(2)   *Don't* write $16^{1/2} = -4$. The definition of an nth root requires that you write $16^{1/2} = 4$.

Now we are prepared to define $b^{m/n}$ where m is an integer, n is a natural number, and $b \geq 0$ when n is even. If the rules for exponents are to hold for a rational exponent such as $4^{3/2}$, we must have

$$4^{3/2} = 4^{(1/2)(3)} = (4^{1/2})^3 = 2^3 = 8$$

and

$$4^{3/2} = 4^{(3)(1/2)} = (4^3)^{1/2} = (64)^{1/2} = 8$$

To achieve this, we must define

$$b^{m/n} = (b^{1/n})^m = (b^m)^{1/n}$$

where $b \geq 0$ when n is even. All of the laws of exponents continue to hold when the exponents are fractions.

EXAMPLE 2
Simplify.

(a) $(64)^{2/3} = [(64)^{1/3}]^2 = 4^2 = 16$

(b) $(-8)^{4/3} = [(-8)^{1/3}]^4 = (-2)^4 = 16$

(c) $x^{1/2} \cdot x^{3/4} = x^{1/2+3/4} = x^{5/4}$

(d) $(x^{3/4})^2 = x^{(3/4)(2)} = x^{3/2}$

(e) $\left(\dfrac{x^8}{y^{-6}}\right)^{-1/2} = \dfrac{(x^8)^{-1/2}}{(y^{-6})^{-1/2}} = \dfrac{x^{(8)(-1/2)}}{y^{(-6)(-1/2)}} = \dfrac{x^{-4}}{y^3} = \dfrac{1}{x^4 y^3}$

(f) $(3x^{2/3}y^{-5/3})^3 = 3^3 \cdot x^{(2/3)(3)}y^{(-5/3)(3)} = 27x^2 y^{-5} = \dfrac{27x^2}{y^5}$

PROGRESS CHECK 2
Simplify.

(a) $27^{4/3}$

(b) $(-64)^{2/3}$

(c) $\left(\dfrac{1}{4}\right)^{5/2}$

(d) $\dfrac{2x^{1/5}}{x^{4/5}}$

(e) $(a^{1/2}b^{-2})^{-2}$

(f) $\left(\dfrac{x^{1/3}y^{2/3}}{z^{5/6}}\right)^{12}$

Answers

(a) 81   (b) 16   (c) $\dfrac{1}{32}$ (d) $\dfrac{2}{x^{3/5}}$ (e) $\dfrac{b^4}{a}$ (f) $\dfrac{x^4 y^8}{z^{10}}$

The symbol $\sqrt{b}$ is an alternate way of writing $b^{1/2}$; that is, $\sqrt{b}$ denotes the nonnegative square root of b. The symbol $\sqrt{}$ is called a **radical sign,** and $\sqrt{b}$ is called the **principal square root of b.** Thus,

$$\sqrt{25} = 5, \qquad \sqrt{0} = 0, \qquad \sqrt{-25} \text{ is undefined}$$

In general, the symbol $\sqrt[n]{b}$ denotes the **principal nth root of b** and is defined this way.

$$\sqrt[n]{b} = b^{1/n} = a \quad \text{where} \quad a^n = b$$

Since $b^{1/n}$ is undefined if n is even and $b < 0$, $\sqrt[n]{b}$ is also undefined under these conditions. As a matter of convenience, in Sections 8.4 and 8.5 we shall assume that any variable under a radical sign is *nonnegative* when an even root is indicated.

In short, $\sqrt[n]{b}$ is the **radical form** of $b^{1/n}$. We can switch back and forth from rational exponent form to radical form and vice versa. For instance,

$$7^{2/3} = (7^2)^{1/3} = \sqrt[3]{7^2}$$

and

$$7^{2/3} = (7^{1/3})^2 = (\sqrt[3]{7})^2$$

In general,

$$b^{m/n} = (b^m)^{1/n} = \sqrt[n]{b^m}$$
$$b^{m/n} = (b^{1/n})^m = (\sqrt[n]{b})^m$$

EXAMPLE 3
Write in radical form.

(a) $32^{2/5} = \sqrt[5]{3^2} = \sqrt[5]{9}$

(b) $y^{1/3} = \sqrt[3]{y}$

(c) $(2x^3y^2)^{3/7} = \sqrt[7]{(2x^3y^2)^3}$

(d) $x^{-3/2} = \dfrac{1}{x^{3/2}} = \dfrac{1}{\sqrt{x^3}}$

(e) $(-8a)^{3/5} = \sqrt[5]{(-8a)^3}$

(f) $(x^2 + y^2)^{1/3} = \sqrt[3]{x^2 + y^2}$

PROGRESS CHECK 3
Write in radical form.

(a) $5^{2/3}$

(b) $x^{4/5}$

(c) $(3ab^5)^{8/3}$

(d) $(5xy)^{1/2}$

(e) $m^{-5/4}$

(f) $(x + y)^{5/2}$

Answers

(a) $\sqrt[3]{5^2}$

(b) $\sqrt[5]{x^4}$

(c) $\sqrt[3]{(3ab^5)^8}$

(d) $\sqrt{5xy}$

(e) $\dfrac{1}{\sqrt[4]{m^5}}$

(f) $\sqrt{(x + y)^5}$

EXAMPLE 4
Write in rational exponent form.

(a) $\sqrt[3]{10^2} = 10^{2/3}$

(b) $\sqrt{y^3} = y^{3/2}$

(c) $\sqrt[5]{(5m^4n^2)^2} = (5m^4n^2)^{2/5}$

(d) $\dfrac{1}{\sqrt[7]{x^4}} = \dfrac{1}{x^{4/7}} = x^{-4/7}$

(e) $\sqrt[3]{(-2x)^5} = (-2x)^{5/3}$

(f) $\sqrt{2x + 3y} = (2x + 3y)^{1/2}$

PROGRESS CHECK 4
Write in rational exponent form.

(a) $\sqrt[6]{9^5}$

(b) $\sqrt[3]{a^4}$

(c) $\sqrt[4]{(2rs^3)^3}$

(d) $\dfrac{1}{\sqrt[5]{x^4}}$

(e) $\sqrt[7]{\left(-\dfrac{1}{2}y\right)^4}$

(f) $\sqrt[4]{2x^2 + y^3}$

Answers

(a) $9^{5/6}$

(b) $a^{4/3}$

(c) $(2rs^3)^{3/4}$

(d) $x^{-4/5}$

(e) $\left(-\dfrac{1}{2}y\right)^{4/7}$

(f) $(2x^2 + y^3)^{1/4}$

WARNING ——
Don't write $\sqrt{49} = \pm 7$. The symbol $\sqrt[n]{b}$ indicates the principal nth root of b and is nonnegative when n is even. To indicate both real square roots of 49, you must write $\pm\sqrt{49} = \pm 7$.
——

We must define

$$\sqrt{x^2} = |x|$$

It is a common error to write $\sqrt{x^2} = x$. This leads to the conclusion that $\sqrt{(-6)^2} = -6$, which contradicts our earlier definition that states that the

symbol $\sqrt{}$ represents the *nonnegative* square root of a number. It is there-
fore essential to write $\sqrt{x^2} = |x|$, unless we know that $x \geq 0$, in which case
we may write $\sqrt{x^2} = x$.

EXERCISE SET 8.3

Simplify and write answers using only positive exponents.

1. $16^{3/4}$

2. $4^{5/2}$

3. $(-125)^{-1/3}$

4. $(-64)^{-2/3}$

5. $\left(\dfrac{-8}{125}\right)^{2/3}$

6. $\left(\dfrac{-8}{27}\right)^{5/3}$

7. $c^{1/4} \cdot c^{-2/3}$

8. $x^{-1/2} \cdot x^{1/3}$

9. $\dfrac{2x^{1/3}}{x^{-3/4}}$

10. $\dfrac{y^{-2/3}}{y^{1/5}}$

11. $\left(\dfrac{x^{7/2}}{x^{2/3}}\right)^{-6}$

12. $\left(\dfrac{x^{3/2}}{x^{2/3}}\right)^{1/6}$

13. $\dfrac{125^{4/3}}{125^{2/3}}$

14. $\dfrac{32^{-1/5}}{32^{3/5}}$

15. $(x^2y^3)^{1/3}$

16. $(x^{1/3}y^2)^6$

17. $(16a^4b^2)^{3/2}$

18. $(x^6y^4)^{-1/2}$

19. $\left(\dfrac{x^{15}}{y^{10}}\right)^{3/5}$

20. $\left(\dfrac{x^8}{y^{12}}\right)^{3/4}$

21. $\left(\dfrac{x^{18}}{y^{-6}}\right)^{2/3}$

22. $\left(\dfrac{a^{-3/2}}{b^{-1/2}}\right)^{-5/2}$

23. $\left(\dfrac{x^3}{y^{-6}}\right)^{-4/3}$

24. $\left(\dfrac{x^9}{y^{-3/2}}\right)^{-2/3}$

Write in radical form.

25. $\left(\dfrac{1}{4}\right)^{2/5}$

26. $(-6)^{2/3}$

27. $x^{2/3}$

28. $a^{3/4}$

29. $(12y)^{-4/3}$

30. $(-8x^2)^{2/5}$

31. $(12x^3y^{-2})^{2/3}$

32. $\left(\dfrac{9}{4}a^{-4}b^2\right)^{-7/4}$

33. $\left(\dfrac{8}{3}x^{-2}y^{-4}\right)^{-3/2}$

34. $\left(\dfrac{2}{5}x^3y^2z^3\right)^{2/3}$

35. $(2x^2y^2 + z^3)^{3/4}$

36. $(3x^2y - z^3)^{-4/3}$

Write in exponent form.

37. $\sqrt[4]{8^3}$

38. $\sqrt[5]{3^2}$

39. $\sqrt[4]{\left(\dfrac{1}{8}\right)^3}$

40. $\dfrac{1}{\sqrt[5]{(-8)^2}}$

41. $\dfrac{1}{\sqrt[3]{x^7}}$

42. $\dfrac{1}{\sqrt[5]{3x^4}}$

43. $\dfrac{1}{\sqrt[4]{\dfrac{4}{9}a^3}}$

44. $\sqrt[5]{(2a^2b^3)^4}$

45. $\sqrt[6]{(3a^4b^6)^5}$

46. $\sqrt{(2x^{-4}y^3)^7}$

47. $\sqrt[4]{(3x^3 + y^5)^2}$

48. $\dfrac{1}{\sqrt[5]{(2a^2 + 3b^3)^3}}$

Select the correct answer in each of the following.

49. $\sqrt{121} =$ (a)   12   (b)   10   (c)   11   (d) -11
    (e)   not a real number

50. $\sqrt[3]{-27} =$ (a) 3 (b) $\dfrac{1}{3}$ (c) $-\dfrac{1}{3}$ (d) -3

(e) not a real number

51. $-\sqrt[3]{-\dfrac{1}{8}} =$ (a) -2 (b) 2 (c) $\dfrac{1}{2}$ (d) $-\dfrac{1}{2}$

(e) not a real number

52. $\sqrt[4]{-16} =$ (a) 2 (b) -2 (c) $\dfrac{1}{2}$ (d) $-\dfrac{1}{2}$

(e) not a real number

Evaluate.

53. $\sqrt{\dfrac{4}{9}}$ 54. $\sqrt{\dfrac{25}{4}}$ 55. $\sqrt{-36}$

56. $\sqrt[4]{-81}$ 57. $\sqrt[3]{\dfrac{1}{27}}$ 58. $\sqrt[3]{-\dfrac{1}{125}}$

59. $\sqrt{(-5)^2}$ 60. $\sqrt{\left(-\dfrac{1}{3}\right)^2}$ 61. $\sqrt{\left(\dfrac{1}{2}\right)^2}$

62. $\sqrt{\left(\dfrac{5}{4}\right)^2}$ 63. $\sqrt{\left(-\dfrac{7}{2}\right)^2}$ 64. $\sqrt{(-7)^2}$

65. $(14.43)^{3/2}$ 66. $-(2.46)^{3/2}$ 67. $(10.46)^{2/3}$

68. $(8.97)^{4/3}$ 69. $\dfrac{(6.47)^{1/3}}{(6.47)^{4/3}}$ 70. $\dfrac{(3.75)^{3/2}}{(3.75)^{1/2}}$

71. $\sqrt{3x^2 + 4y^3}$ when $x = 1.6$, $y = 5.7$

72. $\sqrt{(a^2 + b^4)^3}$ when $a = 2.5$, $b = 6.7$

8.4
EVALUATING AND SIMPLIFYING RADICALS

There are three properties of radicals that are useful in simplifying radical expressions. By switching from radical to exponent form we see that

$$\sqrt[n]{b^n} = (b^n)^{1/n} = b^{n/n} = b$$

Thus,

$$\sqrt[n]{b^n} = b$$

For example, $\sqrt[6]{3^6} = 3$, and $\sqrt[3]{x^3} = x$. If n is even we define $\sqrt[n]{b^n} = |b|$

The following properties are also easily established by switching from radical to exponent form.

$\sqrt[n]{ab} = (ab)^{1/n}$	$\sqrt[n]{\dfrac{a}{b}} = \left(\dfrac{a}{b}\right)^{1/n}$	Exponent form of radicals
$= a^{1/n}b^{1/n}$	$= \dfrac{a^{1/n}}{b^{1/n}}$	Property of exponents
$= \sqrt[n]{a}\,\sqrt[n]{b}$	$= \dfrac{\sqrt[n]{a}}{\sqrt[n]{b}}$	Radical form of exponents

In summary,

$$\sqrt[n]{b^n} = b \text{ (if } n \text{ is even } \sqrt[n]{b^n} = |b|)$$
$$\sqrt[n]{ab} = \sqrt[n]{a}\sqrt[n]{b}$$
$$\sqrt[n]{\frac{a}{b}} = \frac{\sqrt[n]{a}}{\sqrt[n]{b}}$$

These properties can be used to evaluate and simplify radical forms. The key to the process is to think in terms of perfect squares when dealing with square roots, to think in terms of perfect cubes when dealing with cube roots, and so on. Here are some examples.

EXAMPLE 1
Simplify.

(a) $\sqrt{72} = \sqrt{36 \cdot 2} = \sqrt{36}\sqrt{2} = 6\sqrt{2}$

(b) $\sqrt[3]{-54} = \sqrt[3]{(-27)(2)} = \sqrt[3]{-27}\sqrt[3]{2} = -3\sqrt[3]{2}$

(c) $2\sqrt[3]{8x^3y} = 2\sqrt[3]{8}\sqrt[3]{x^3}\sqrt[3]{y} = 2(2)(x)\sqrt[3]{y} = 4x\sqrt[3]{y}$

(d) $\sqrt{\frac{16x}{9y^2}} = \frac{\sqrt{16x}}{\sqrt{9y^2}} = \frac{\sqrt{16}\sqrt{x}}{\sqrt{9}\sqrt{y^2}} = \frac{4\sqrt{x}}{3y}$

PROGRESS CHECK 1
Simplify.

(a) $\sqrt{45}$ (b) $\sqrt[3]{-81}$ (c) $-3\sqrt[4]{x^3y^4}$

Answers

(a) $3\sqrt{5}$ (b) $-3\sqrt[3]{3}$ (c) $-3y\sqrt[4]{x^3}$

There are other techniques we can use to simplify radical forms. If we have $\sqrt[3]{x^4}$, we can write this as

$$\sqrt[3]{x^4} = \sqrt[3]{x^3 \cdot x} = \sqrt[3]{x^3}\sqrt[3]{x} = x\sqrt[3]{x}$$

In general,

$\sqrt[n]{x^m}$ can always be simplified so that the exponent within the radical is less than n.

Another possibility is illustrated by $\sqrt[4]{x^{10}}$ which can be written as

$$\sqrt[4]{x^{10}} = x^{10/4} = x^{5/2} = \sqrt{x^5}$$

In general,

$\sqrt[n]{x^m}$ can always be simplified so that m and n have no common factors.

EXAMPLE 2
Simplify.

(a) $\sqrt{x^3y^3} = \sqrt{(x^2 \cdot x)(y^2 \cdot y)} = \sqrt{x^2y^2}\sqrt{xy} = xy\sqrt{xy}$

(b) $\sqrt{12x^5} = \sqrt{(4 \cdot 3)(x^4 \cdot x)} = \sqrt{4x^4}\sqrt{3x} = 2x^2\sqrt{3x}$

(c) $\sqrt[3]{x^7y^6} = \sqrt[3]{x^6 \cdot x}\sqrt[3]{y^6} = \sqrt[3]{x^6}\sqrt[3]{y^6}\sqrt[3]{x} = x^2y^2\sqrt[3]{x}$

PROGRESS CHECK 2
Simplify.

(a) $\sqrt{4xy^5}$ (b) $\sqrt[3]{16x^4y^6}$ (c) $\sqrt[4]{16x^8y^5}$

Answers

(a) $2y^2\sqrt{xy}$ (b) $2xy^2\sqrt[3]{2x}$ (c) $2x^2y\sqrt[4]{y}$

It is always possible to rewrite a fraction so that the denominator is free of radicals, a process called **rationalizing the denominator.** The key to this process is to change the fraction so that the denominator is of the form $\sqrt[n]{a^n}$ which reduces to a.

$$\frac{5}{\sqrt{7}} = \frac{5}{\sqrt{7}} \cdot \frac{\sqrt{7}}{\sqrt{7}}$$

$$= \frac{5\sqrt{7}}{\sqrt{7^2}} = \frac{5\sqrt{7}}{7}$$

Multiply numerator and denominator by $\sqrt{7}$ to form a perfect square under the radical sign in the denominator.

If we are given

$$\frac{1}{\sqrt[3]{2x^2}}$$

we seek a factor that will change the expression under the radical sign in the denominator into the perfect cube $\sqrt[3]{2^3x^3}$. The factor is $\sqrt[3]{2^2x}$ since $\sqrt[3]{2x^2} \cdot \sqrt[3]{2^2x} = \sqrt[3]{2^3x^3}$. Thus,

$$\frac{1}{\sqrt[3]{2x^2}} = \frac{1}{\sqrt[3]{2x^2}} \cdot \frac{\sqrt[3]{2^2x}}{\sqrt[3]{2^2x}} = \frac{\sqrt[3]{4x}}{\sqrt[3]{2^3x^3}} = \frac{\sqrt[3]{4x}}{2x}$$

EXAMPLE 3
Rationalize the denominator.

(a) $\dfrac{1}{\sqrt{3}} = \dfrac{1}{\sqrt{3}} \cdot \dfrac{\sqrt{3}}{\sqrt{3}} = \dfrac{\sqrt{3}}{3}$

(b) $\dfrac{6xy^2}{5\sqrt{2x}} = \dfrac{6xy^2}{5\sqrt{2x}} \cdot \dfrac{\sqrt{2x}}{\sqrt{2x}} = \dfrac{6xy^2\sqrt{2x}}{5 \cdot 2x} = \dfrac{3}{5}y^2\sqrt{2x}$

(c) $\dfrac{-3x^5y^6}{\sqrt[4]{x^3y}} = \dfrac{-3x^5y^6}{\sqrt[4]{x^3y}} \cdot \dfrac{\sqrt[4]{xy^3}}{\sqrt[4]{xy^3}}$ The multiplier $\sqrt[4]{xy^3}$ will produce $\sqrt[4]{x^4y^4}$ in the denominator.

$= \dfrac{-3x^5y^6\,\sqrt[4]{xy^3}}{\sqrt[4]{x^4y^4}}$

$= \dfrac{-3x^5y^6\,\sqrt[4]{xy^3}}{xy}$

$= -3x^4y^5\,\sqrt[4]{xy^3}$

PROGRESS CHECK 3
Rationalize the denominator.

(a) $\sqrt{\dfrac{3}{7}}$ (b) $\dfrac{2}{\sqrt{6}}$ (c) $\dfrac{-9xy^3}{\sqrt{3xy}}$ (d) $\dfrac{5xy}{\sqrt[4]{2x^2y}}$

Answers

(a) $\dfrac{1}{7}\sqrt{21}$ (b) $\dfrac{1}{3}\sqrt{6}$ (c) $-3y^2\sqrt{3xy}$ (d) $\dfrac{5}{2}\sqrt[4]{8x^2y^3}$

A radical expression is said to be **simplified** when

(a) $\sqrt[n]{x^m}$ has $m < n$.

(b) $\sqrt[n]{x^m}$ has no common factors between m and n.

(c) The denominator is rationalized.

EXAMPLE 4
Write in simplified form.

(a) $\sqrt[4]{x^2y^5} = \sqrt[4]{x^2}\,\sqrt[4]{y^4 \cdot y} = \sqrt{x}\,\sqrt[4]{y^4}\,\sqrt[4]{y} = y\sqrt{x}\,\sqrt[4]{y}$

(b) $\sqrt{\dfrac{8x^3}{y}} = \dfrac{\sqrt{(4x^2)(2x)}}{\sqrt{y}} = \dfrac{\sqrt{4x^2}\,\sqrt{2x}}{\sqrt{y}} = \dfrac{2x\,\sqrt{2x}}{\sqrt{y}}$

$= \dfrac{2x\,\sqrt{2x}}{\sqrt{y}} \cdot \dfrac{\sqrt{y}}{\sqrt{y}} = \dfrac{2x\,\sqrt{2xy}}{y}$

(c) $\sqrt[6]{\dfrac{x^3}{y^2}} = \dfrac{\sqrt[6]{x^3}}{\sqrt[6]{y^2}} = \dfrac{\sqrt{x}}{\sqrt[3]{y}} = \dfrac{\sqrt{x}}{\sqrt[3]{y}} \cdot \dfrac{\sqrt[3]{y^2}}{\sqrt[3]{y^2}} = \dfrac{\sqrt{x}\,\sqrt[3]{y^2}}{y}$

PROGRESS CHECK 4
Write in simplified form.

(a) $\sqrt{\dfrac{18x^6}{y}}$ (b) $\sqrt[3]{ab^4c^7}$ (c) $\dfrac{-2xy^3}{\sqrt[4]{32x^3y^5}}$

Answers

(a) $\dfrac{3x^3\sqrt{2y}}{y}$ (b) $bc^2\sqrt[3]{abc}$ (c) $-\dfrac{y}{2}\sqrt[4]{8xy^3}$

EXERCISE SET 8.4

Simplify each of the following and write the answer in simplified form. Every variable represents a positive real number.

1. $\sqrt{48}$

2. $\sqrt{100}$

3. $\sqrt[3]{54}$

4. $\sqrt{80}$

5. $\sqrt[3]{40}$

6. $\sqrt{\dfrac{8}{27}}$

7. $\sqrt[3]{\dfrac{8}{27}}$

8. $\sqrt[4]{16}$

9. $\sqrt[3]{\dfrac{8}{125}}$

10. $\sqrt{x^3}$

11. $\sqrt{x^8}$

12. $\sqrt[3]{y^7}$

13. $\sqrt[3]{a^{11}}$

14. $\sqrt[4]{b^{14}}$

15. $\sqrt{x^6}$

16. $\sqrt{48x^9}$

17. $\sqrt{98b^{10}}$

18. $\sqrt[3]{24x^8}$

19. $\sqrt[3]{108y^{16}}$

20. $\sqrt[4]{96x^{10}}$

21. $\sqrt{20x^4}$

22. $\sqrt{x^5y^4}$

23. $\sqrt{a^{10}b^7}$

24. $\sqrt{x^5y^3}$

25. $\sqrt[3]{x^6y^8}$

26. $\sqrt[3]{x^{14}y^{17}}$

27. $\sqrt[4]{a^5b^{10}}$

28. $\sqrt{9x^8y^5}$

29. $\sqrt{72x^7y^{11}}$

30. $\sqrt[3]{48x^6y^9}$

31. $\sqrt[3]{24b^{10}c^{14}}$

32. $\sqrt[4]{16x^8y^5}$

33. $\sqrt[4]{48b^{10}c^{12}}$

34. $\sqrt{20x^5y^7z^4}$

35. $\sqrt[3]{40a^8b^4c^5}$

36. $\sqrt[3]{72x^6y^9z^8}$

37. $\sqrt{\dfrac{1}{5}}$

38. $\dfrac{2}{\sqrt{5}}$

39. $\dfrac{4}{3\sqrt{11}}$

40. $\dfrac{3}{\sqrt{6}}$

41. $\dfrac{2}{\sqrt{12}}$

42. $\dfrac{1}{\sqrt{x}}$

43. $\dfrac{1}{\sqrt{3y}}$

44. $\sqrt{\dfrac{2}{y}}$

45. $\dfrac{4x^2}{\sqrt{2x}}$

46. $\sqrt{\dfrac{2}{x^2}}$

47. $\dfrac{6xy}{\sqrt{2x}}$

48. $\dfrac{8a^2b^2}{2\sqrt{2b}}$

49. $\dfrac{-5x^4y^5}{\sqrt[3]{x^2y^2}}$

50. $\dfrac{-4a^8b^6}{\sqrt[4]{a^3b^2}}$

51. $\dfrac{5ab}{\sqrt[3]{2ab^2}}$

52. $\sqrt{\dfrac{12x^4}{y}}$

53. $\sqrt{\dfrac{20x^5}{y^3}}$

54. $\dfrac{\sqrt{8a^5}}{\sqrt{3b}}$

55. $\sqrt[3]{\dfrac{x^5}{y^2}}$

56. $\sqrt[3]{\dfrac{2a^2}{3b}}$

57. $\sqrt[4]{\dfrac{8x^2}{3y}}$

58. $\sqrt{a^3b^5}$

59. $\sqrt{x^7y^5}$

60. $\sqrt[3]{x^2y^7}$

61. $\sqrt[4]{32x^8y^6}$

62. $\sqrt{9x^7y^5}$

63. $\sqrt[4]{48x^8y^6z^2}$

64. $\dfrac{xy}{\sqrt[3]{54x^2y^4}}$

65. $\dfrac{7xy^2}{\sqrt[4]{48x^7y^3}}$

66. $\dfrac{3a^3b^3}{\sqrt[3]{2a^2y^5}}$

67. $\dfrac{x}{\sqrt[3]{96x^6y^9}}$

68. $\dfrac{xy}{\sqrt[3]{32x^6y^4}}$

69. $\dfrac{4a^4b^2}{\sqrt[3]{125a^9b^4}}$

8.5
OPERATIONS WITH RADICALS

We can add or subtract expressions involving radical forms that are exactly the same. For example,

$$2\sqrt{2} + 3\sqrt{2} = (2 + 3)\sqrt{2} = 5\sqrt{2}$$
$$3\sqrt[3]{x^2y} - 7\sqrt[3]{x^2y} = (3 - 7)\sqrt[3]{x^2y} = -4\sqrt[3]{x^2y}$$

(Note that in each example we displayed an intermediate step to show that the distributive law is really the key to the addition process.)

EXAMPLE 1
Simplify and combine terms.

(a) $\sqrt{27} - \sqrt{12} = \sqrt{9 \cdot 3} - \sqrt{4 \cdot 3} = 3\sqrt{3} - 2\sqrt{3} = \sqrt{3}$

(b) $7\sqrt{5} + 4\sqrt{3} - 9\sqrt{5} = (7 - 9)\sqrt{5} + 4\sqrt{3} = -2\sqrt{5} + 4\sqrt{3}$

(c) $\sqrt[3]{x^2y} - \dfrac{1}{2}\sqrt{xy} - 3\sqrt[3]{x^2y} + 4\sqrt{xy}$

$= (1 - 3)\sqrt[3]{x^2y} + \left(-\dfrac{1}{2} + 4\right)\sqrt{xy} = -2\sqrt[3]{x^2y} + \dfrac{7}{2}\sqrt{xy}$

PROGRESS CHECK 1
Simplify and combine terms.

(a) $\sqrt{125} - \sqrt{80}$ (b) $\sqrt[3]{24} - \sqrt{8} - \sqrt[3]{81} + \sqrt{32}$

(c) $2\sqrt[5]{xy^3} - 4\sqrt[5]{x^3y} - 5\sqrt[5]{xy^3} - 2\sqrt[5]{x^3y}$

Answers

(a) $\sqrt{5}$ (b) $-\sqrt[3]{3} + 2\sqrt{2}$ (c) $-3\sqrt[5]{xy^3} - 6\sqrt[5]{x^3y}$

WARNING
Don't write

$$\sqrt{9} + \sqrt{16} = \sqrt{25}$$

You can perform addition only with identical radical forms. *This is one of the most common mistakes made by students in algebra!*

It is easy to simplify the product of $\sqrt[n]{a}$ and $\sqrt[n]{b}$ when $m = n$. Thus,

$$\sqrt[5]{x^2y} \cdot \sqrt[5]{xy} = \sqrt[5]{x^3y^2}$$

but $\sqrt[3]{x^2y} \cdot \sqrt[5]{xy}$ cannot be simplified in this manner.
Products of the form

$$(\sqrt{2x} - 5)(\sqrt{2x} + 3)$$

can be handled by forming all four products and then simplifying.

$$(\sqrt{2x} - 5)(\sqrt{2x} + 3) = \sqrt{2x} \cdot \sqrt{2x} + 3\sqrt{2x} - 5\sqrt{2x} - 15$$
$$= 2x - 2\sqrt{2x} - 15$$

EXAMPLE 2
Multiply and simplify.

(a) $2\sqrt[3]{xy^2} \cdot \sqrt[3]{x^2y^2} = 2\sqrt[3]{x^3y^4} = 2xy\sqrt[3]{y}$

(b) $(\sqrt{3} + 2)(\sqrt{3} - 2) = \sqrt{3} \cdot \sqrt{3} - 2\sqrt{3} + 2\sqrt{3} - 4$
$$= 3 - 4 = -1$$

(c) $(\sqrt{2} - \sqrt{5})(\sqrt{2} + \sqrt{5}) = \sqrt{2} \cdot \sqrt{2} + \sqrt{2} \cdot \sqrt{5} - \sqrt{5} \cdot \sqrt{2} - \sqrt{5} \cdot \sqrt{5}$
$$= 2 - 5 = -3$$

(d) $(2\sqrt{x} + 3\sqrt{y})^2 = (2\sqrt{x})^2 + 2 \cdot 2\sqrt{x} \cdot 3\sqrt{y} + (3\sqrt{y})^2$
$$= 4x + 12\sqrt{xy} + 9y$$

PROGRESS CHECK 2
Multiply and simplify.

(a) $\sqrt[7]{a^3b^4} \cdot \sqrt[7]{3ab^3}$ (b) $\sqrt{3}(\sqrt{2} - 4)$

(c) $(\sqrt{2} - 5)(\sqrt{2} + 1)$ (d) $(\sqrt[3]{3x} - 4)(\sqrt[3]{3x} + 2)$

(e) $(\sqrt{x} - \sqrt{2y})^2$

Answers

(a) $b\sqrt[7]{3a^4}$ (b) $\sqrt{6} - 4\sqrt{3}$ (c) $-3 - 4\sqrt{2}$

(d) $\sqrt[3]{(3x)^2} - 2\sqrt[3]{3x} - 8$ (e) $x + 2y - 2\sqrt{2xy}$

Note that
$$(\sqrt{m} + \sqrt{n})(\sqrt{m} - \sqrt{n}) = m - n$$

That is, products of this form are free of radicals. This idea is used to rationalize denominators as in this example.

$$\frac{7}{\sqrt{2} + \sqrt{3}} = \frac{7}{\sqrt{2} + \sqrt{3}} \cdot \frac{\sqrt{2} - \sqrt{3}}{\sqrt{2} - \sqrt{3}} = \frac{7(\sqrt{2} - \sqrt{3})}{2 - 3}$$
$$= \frac{7(\sqrt{2} - \sqrt{3})}{-1} = -7(\sqrt{2} - \sqrt{3})$$

EXAMPLE 3
Rationalize the denominator.

(a) $\dfrac{4}{\sqrt{5} - \sqrt{2}} = \dfrac{4}{\sqrt{5} - \sqrt{2}} \cdot \dfrac{\sqrt{5} + \sqrt{2}}{\sqrt{5} + \sqrt{2}}$ Multiply numerator and denominator by $\sqrt{5} + \sqrt{2}$.

$$= \frac{4(\sqrt{5} + \sqrt{2})}{5 - 2} = \frac{4}{3}(\sqrt{5} + \sqrt{2})$$

(b) $\dfrac{5}{\sqrt{x} + 2} = \dfrac{5}{\sqrt{x} + 2} \cdot \dfrac{\sqrt{x} - 2}{\sqrt{x} - 2}$ Multiply numerator and denominator by $\sqrt{x} - 2$.

$$= \frac{5(\sqrt{x} - 2)}{x - 4}$$

(c) $\dfrac{-1}{2 - \sqrt{3y}} = \dfrac{-1}{2 - \sqrt{3y}} \cdot \dfrac{2 + \sqrt{3y}}{2 + \sqrt{3y}} = \dfrac{-2 - \sqrt{3y}}{4 - 3y}$

PROGRESS CHECK 3
Rationalize the denominator.

(a) $\dfrac{-6}{\sqrt{2} + \sqrt{6}}$ (b) $\dfrac{2}{\sqrt{3x} - 1}$ (c) $\dfrac{-4}{\sqrt{x} + \sqrt{5y}}$ (d) $\dfrac{4}{\sqrt{x} - \sqrt{y}}$

Answers

(a) $\dfrac{3}{2}(\sqrt{2} - \sqrt{6})$ (b) $\dfrac{2(\sqrt{3x} + 1)}{3x - 1}$ (c) $\dfrac{-4(\sqrt{x} - \sqrt{5y})}{x - 5y}$

(d) $\dfrac{4(\sqrt{x} + \sqrt{y})}{x - y}$

EXERCISE SET 8.5

Simplify and combine terms in the following. Every variable represents a positive real number.

1. $2\sqrt{3} + 5\sqrt{3}$

2. $3\sqrt{5} - 5\sqrt{5}$

3. $4\sqrt[3]{11} - 6\sqrt[3]{11}$

4. $\dfrac{1}{2}\sqrt[3]{7} - 2\sqrt[3]{7}$

5. $3\sqrt{x} + 4\sqrt{x}$

6. $3\sqrt[3]{a} - \dfrac{2}{5}\sqrt[3]{a}$

7. $3\sqrt{2} + 5\sqrt{2} - 2\sqrt{2}$

8. $\sqrt[3]{4} - 3\sqrt[3]{4} + \sqrt{6} - 2\sqrt{6}$

9. $3\sqrt{y} + 2\sqrt{y} - \dfrac{1}{2}\sqrt{y}$

10. $\dfrac{1}{2}\sqrt[3]{x} - 2\sqrt[4]{x} + \dfrac{1}{3}\sqrt[3]{x} - \dfrac{1}{2}\sqrt[4]{x}$

11. $\sqrt{24} + \sqrt{54}$

12. $\sqrt{75} - \sqrt{150}$

13. $2\sqrt{27} + \sqrt{12} - \sqrt{48}$

14. $\sqrt{20} - 4\sqrt{45} + \sqrt{80}$

15. $\sqrt[3]{40} + \sqrt{45} - \sqrt[3]{135} + 2\sqrt{80}$

16. $3\sqrt[3]{128} + \sqrt{128} - 2\sqrt{72} - \sqrt[3]{54}$

17. $3\sqrt[3]{xy^2} + 2\sqrt[3]{xy^2} - 4\sqrt[3]{x^2y}$

18. $\sqrt[4]{x^2y^2} + 2\sqrt[4]{x^2y^2} - \dfrac{2}{3}\sqrt{x^2y^2}$

19. $2\sqrt[3]{a^4b} + \dfrac{1}{2}\sqrt{ab} - \dfrac{1}{2}\sqrt[3]{a^4b} + 2\sqrt{ab}$

20. $\sqrt{\dfrac{xy}{3}} + \sqrt{6xy}$

21. $\sqrt[5]{2x^3y^2} - 3\sqrt[5]{2x^3y^2} + 4\sqrt[5]{2x^3y^2}$

22. $\sqrt{2abc} - 3\sqrt{8abc} + \sqrt{\dfrac{abc}{2}}$

23. $-\sqrt[3]{xy^4} - 2y\sqrt[3]{xy} - \dfrac{1}{2}\sqrt{xy^4}$

24. $\sqrt{xy} + xy - 3\sqrt{xy} - 2\sqrt{x^2y^2}$

25. $2\sqrt{5} - (3\sqrt{5} + 4\sqrt{5})$

26. $2\sqrt{18} - (3\sqrt{12} - 2\sqrt{75})$

27. $x\sqrt[3]{x^4y} - (\sqrt[3]{x^5y} - 2x\sqrt{x^4y})$

28. $2(\sqrt{x^3y^7} - xy\sqrt[3]{y}) - 3\sqrt[3]{x^3y^7}$

Multiply and simplify in the following.

29. $\sqrt{3}(\sqrt{3} + 4)$

30. $\sqrt{5}(2 - \sqrt{5})$

31. $\sqrt{6}(\sqrt{2} + 2\sqrt{3})$

32. $\sqrt{8}(\sqrt{2} - \sqrt{3})$

33. $3\sqrt[3]{x^2y} \cdot \sqrt[3]{x \cdot y^2}$

34. $2\sqrt[4]{a^2b^3} \cdot \sqrt[4]{a^3b^2}$

35. $-4\sqrt[5]{x^2y^3} \cdot \sqrt[5]{x^4y^2}$

36. $\sqrt[3]{3a^2b} \cdot \sqrt[3]{9a^3b^4}$

37. $(\sqrt{2} + 3)(\sqrt{2} - 2)$

38. $(\sqrt{7} + 5)(\sqrt{7} - 3)$

39. $(2\sqrt{3} - 3)(3\sqrt{2} + 2)$

40. $(\sqrt{5} - 1)(\sqrt{5} + 2)$

41. $(\sqrt{2} - \sqrt{3})^2$

42. $(\sqrt{3} + \sqrt{5})^2$

43. $(\sqrt{8} - 2\sqrt{2})(\sqrt{2} + 2\sqrt{8})$

44. $(\sqrt{x} + 2\sqrt{y})(\sqrt{x} - 3\sqrt{y})$

45. $(\sqrt{3x} + \sqrt{2y})(\sqrt{3x} - 2\sqrt{2y})$

46. $(\sqrt{2a} + \sqrt{b})(\sqrt{3ab} + 2)$

47. $(\sqrt[3]{2x} + 3)(\sqrt[3]{2x} - 3)$

48. $(\sqrt[3]{xy^2} - 3)(\sqrt[3]{x^2y} + 2)$

49. $(\sqrt[3]{a^4} + \sqrt[3]{b^2})(\sqrt[3]{a^2} - 2\sqrt[3]{b^4})$

50. $(\sqrt[4]{x^2y^3} + 2\sqrt[4]{x^5y})(\sqrt[4]{x^2} - \sqrt[4]{y^3})$

Rationalize the denominator.

51. $\dfrac{3}{\sqrt{2} + 3}$

52. $\dfrac{2}{\sqrt{5} - 2}$

53. $\dfrac{-3}{\sqrt{7} - 9}$

54. $\dfrac{2}{\sqrt{3} - 4}$

55. $\dfrac{2}{\sqrt{x} + 3}$

56. $\dfrac{3}{\sqrt{x} - 5}$

57. $\dfrac{-3}{3\sqrt{a} + 1}$

58. $\dfrac{-4}{2\sqrt{x} - 2}$

59. $\dfrac{4}{2 - \sqrt{2y}}$

60. $\dfrac{-3}{5 + \sqrt{5y}}$

61. $\dfrac{\sqrt{8}}{\sqrt{2} + 2}$

62. $\dfrac{\sqrt{3}}{\sqrt{3} - 5}$

63. $\dfrac{\sqrt{2} + 1}{\sqrt{2} - 1}$

64. $\dfrac{\sqrt{5} - 1}{\sqrt{5} + 1}$

65. $\dfrac{\sqrt{5} + \sqrt{3}}{\sqrt{5} - \sqrt{3}}$

66. $\dfrac{\sqrt{6} + \sqrt{2}}{\sqrt{3} - \sqrt{2}}$

67. $\dfrac{\sqrt{x}}{\sqrt{x} + \sqrt{y}}$

68. $\dfrac{2\sqrt{a}}{\sqrt{2x} + \sqrt{y}}$

69. $\dfrac{\sqrt{a} + 1}{2\sqrt{a} - \sqrt{b}}$

70. $\dfrac{4\sqrt{x} - \sqrt{y}}{\sqrt{x} + \sqrt{y}}$

71. $\dfrac{2\sqrt{x} + \sqrt{y}}{\sqrt{2x} - \sqrt{y}}$

72. $\dfrac{3\sqrt{a} - \sqrt{3b}}{\sqrt{2a} + \sqrt{b}}$

8.6
COMPLEX NUMBERS

Algebra is very much concerned with finding solutions to equations and, in particular, to polynomial equations. Yet there is no real number that satisfies a simple polynomial equation such as

$$x^2 = -4$$

since the square of a real number is always nonnegative.

To resolve this problem, mathematicians created a new number system built upon an "imaginary unit" i defined by $i = \sqrt{-1}$. This number i has the property that when we square both sides of the equation we have $i^2 = -1$, a result which we cannot obtain with real numbers. In summary,

$$i = \sqrt{-1}$$
$$i^2 = -1$$

We also assume that i behaves according to all the algebraic laws we have already developed (with the exception of the rules for inequalities). This allows us to simplify higher powers of i. Thus,

$$i^3 = i^2 \cdot i = (-1)i = -i$$
$$i^4 = i^2 \cdot i^2 = (-1)(-1) = 1$$

Now it's easy to simplify i^n when n is any natural number. Since $i^4 = 1$, we simply seek the highest multiple of 4 which is less than or equal to n. For example,

$$i^5 = i^4 \cdot i = (1) \cdot i = i$$
$$i^{27} = i^{24} \cdot i^3 = (i^4)^6 \cdot i^3 = (1)^6 \cdot i^3 = i^3 = -i$$
$$i^{62} = i^{60} \cdot i^2 = (i^4)^{15} \cdot i^2 = (1)^{15} \cdot i^2 = i^2 = -1$$

EXAMPLE 1
Simplify.

(a) $i^{101} = i^{100} \cdot i = (i^4)^{25} \cdot i = (1)^{25} \cdot i = i$

(b) $-i^{74} = -i^{72} \cdot i^2 = -(i^4)^{18} \cdot i^2 = -(1)^{18} \cdot i^2$
$$= (-1)(-1) = 1$$

(c) $i^{36} = (i^4)^9 = (1)^9 = 1$

(d) $i^{51} = i^{48} \cdot i^3 = (i^4)^{12} \cdot i^3 = (1)^{12} \cdot i^3 = i^3 = -i$

PROGRESS CHECK 1
Simplify.

(a) i^{22} (b) i^{15} (c) i^{29} (d) i^{200}

Answers
(a) -1 (b) $-i$ (c) i (d) 1

It's easy to write square roots of negative numbers in terms of i. For example,

$$\sqrt{-25} = i\sqrt{25} = 5i$$

and, in general,

$$\sqrt{-a} = i\sqrt{a} \quad \text{for} \quad a > 0$$

Any number of the form xi, where x is a real number, is called an **imaginary number.**

Having created imaginary numbers, mathematicians next combined real and imaginary numbers. We say that $a + bi$, where a and b are real numbers, is a **complex number.** The number a is called the **real part** of $a + bi$, and b is called the **imaginary part.** Here are some examples of complex numbers.

$3 + 2i$	$a = 3, b = 2$	$2 - i$	$a = 2, b = -1$
5	$a = 5, b = 0$	$-2i$	$a = 0, b = -2$

$$2 - 3i \qquad a = 2, b = -3 \qquad\qquad -5 - \frac{1}{2}i \qquad a = -5, b = -\frac{1}{2}$$

$$-\frac{1}{3} \qquad a = -\frac{1}{3}, b = 0 \qquad\qquad \frac{4}{5} + \frac{1}{5}i \qquad a = \frac{4}{5}, b = \frac{1}{5}$$

Note that every real number a can be written as a complex number by choosing $b = 0$. Thus,

$$a = a + 0i$$

EXAMPLE 2

Write as a complex number, in the form $a + bi$, where a and b are real numbers.

(a) $-\frac{1}{2} = -\frac{1}{2} + 0i$

(b) $\sqrt{-9} = i\sqrt{9} = 3i = 0 + 3i$

(c) $-1 - \sqrt{-4} = -1 - i\sqrt{4} = -1 - 2i$

PROGRESS CHECK 2

Write as a complex number.

(a)   0.2   (b) $-\sqrt{-3}$ (c) $3 - \sqrt{-9}$

Answers

(a) $0.2 + 0i$ (b) $0 - i\sqrt{3}$ (c) $3 - 3i$

WARNING

Don't write

$$\sqrt{-4} \cdot \sqrt{-9} = \sqrt{36} = 6$$

The rule $\sqrt{ab} = \sqrt{a} \cdot \sqrt{b}$ holds only when $a \geq 0$ and $b \geq 0$. Instead, write

$$\sqrt{-4} \cdot \sqrt{-9} = 2i \cdot 3i = 6i^2 = -6$$

We say two complex numbers are **equal** if their real parts are equal and their imaginary parts are equal, that is

$$a + bi = c + di \quad \text{if and only if}$$
$$a = c \quad \text{and} \quad b = d$$

Thus, $x + 3i = 6 - yi$ if and only if $x = 6$ and $y = -3$.

We add and subtract the complex numbers $3 + 4i$ and $2 - i$ by combining like terms; that is, we combine the real parts and we combine the imaginary parts.

$$(3 + 4i) + (2 - i) = (3 + 2) + (4 - 1)i = 5 + 3i$$

$$(3 + 4i) - (2 - i) = (3 - 2) + (4 + 1)i = 1 + 5i$$

In general,

$$(a + bi) + (c + di) = (a + c) + (b + d)i$$
$$(a + bi) - (c + di) = (a - c) + (b - d)i$$

Thus, the sum or difference of two complex numbers is again a complex number.

EXAMPLE 3
Perform the indicated operations.

(a) $(7 - 2i) + (4 - 3i) = (7 + 4) + (-2 - 3)i = 11 - 5i$

(b) $3i + (-7 + 5i) = (0 - 7) + (3 + 5)i = -7 + 8i$

(c) $(-11 + 5i) - (9 + i) = (-11 - 9) + (5 - 1)i = -20 + 4i$

(d) $14 - (3 - 8i) = (14 - 3) + (0 + 8)i = 11 + 8i$

PROGRESS CHECK 3
Perform the indicated operations.

(a) $(-9 + 3i) + (6 - 2i)$ (b) $(-17 + i) + 15$

(c) $(2 - 3i) - (9 - 4i)$ (d) $7i - (-3 + 9i)$

Answers

(a) $-3 + i$ (b) $-2 + i$ (c) $-7 + i$ (d) $3 - 2i$

Multiplication of complex numbers is analogous to multiplication of polynomials. The distributive law is used to form all the products, and the substitution $i^2 = -1$ is used to simplify. For example,

$$5i(2 - 3i) = 5i(2) - (5i)(3i) = 10i - 15i^2$$
$$= 10i - 15(-1) = 15 + 10i$$

$$(2 + 3i)(3 - 5i) = 2(3 - 5i) + 3i(3 - 5i)$$
$$= 6 - 10i + 9i - 15i^2$$
$$= 6 - i - 15(-1)$$
$$= 21 - i$$

In general,
$$(a + bi)(c + di) = a(c + di) + bi(c + di)$$
$$= ac + adi + bci + bdi^2$$
$$= ac + (ad + bc)i + bd(-1)$$
$$= (ac - bd) + (ad + bc)i$$

Thus,

$$(a + bi)(c + di) = (ac - bd) + (ad + bc)i$$

This result is significant since it shows that the product of two complex numbers is again a complex number.

EXAMPLE 4
Find the product of $(2 - 3i)$ and $(7 + 5i)$.

Solution

$$
\begin{aligned}
(2 - 3i)(7 + 5i) &= 2(7 + 5i) - 3i(7 + 5i) \\
&= 14 + 10i - 21i - 15i^2 \\
&= 14 - 11i - 15(-1) \\
&= 29 - 11i
\end{aligned}
$$

PROGRESS CHECK 4
Find the product.

(a) $(-3 - i)(4 - 2i)$ (b) $(-4 - 2i)(2 - 3i)$

Answers

(a) $-14 + 2i$ (b) $-14 + 8i$

The complex number $a - bi$ is called the **conjugate** of the complex number $a + bi$. We see that

$$
\begin{aligned}
(a + bi)(a - bi) &= a(a - bi) + bi(a - bi) \\
&= a^2 - abi + abi - b^2i^2 \\
&= a^2 - b^2(-1) \\
&= a^2 + b^2
\end{aligned}
$$

which is a real number.

> The product of a complex number and its conjugate is a real number.
> $$(a + bi)(a - bi) = a^2 + b^2$$

EXAMPLE 5
Multiply by the conjugate of the given complex number.
(a) $6 - i$. The conjugate is $6 + i$, and we have
$$(6 - i)(6 + i) = 36 - i^2 = 36 + 1 = 37$$
(b) $-3i$. The conjugate of $0 - 3i$ is $0 + 3i$. Thus,
$$(-3i)(3i) = -9i^2 = 9$$

PROGRESS CHECK 5
Multiply by the conjugate of the given complex number.

(a) $3 + 2i$ (b) $-\sqrt{2} - 3i$ (c) $4i$

Answers

(a) *13*   (b) *11*   (c) *16*

The quotient of two complex numbers

$$
\frac{q + ri}{s + ti}
$$

can be written in the form $a + bi$ by multiplying both numerator and denominator by $s - ti$, the conjugate of the denominator.

EXAMPLE 6
Write in the form $a + bi$.

(a) $\dfrac{-2 + 3i}{3 - i} = \dfrac{-2 + 3i}{3 - i} \cdot \dfrac{3 + i}{3 + i}$ Multiplying by the conjugate $3 + i$

$$= \dfrac{-6 - 2i + 9i + 3i^2}{9 + 3i - 3i - i^2} = \dfrac{-6 + 7i + 3(-1)}{9 - (-1)}$$

$$= \dfrac{-9 + 7i}{10} = -\dfrac{9}{10} + \dfrac{7}{10}i$$

(b) $\dfrac{2i}{2 + 5i} = \dfrac{2i}{2 + 5i} \cdot \dfrac{2 - 5i}{2 - 5i}$ Multiplying by the conjugate $2 - 5i$

$$= \dfrac{4i - 10i^2}{4 - 10i + 10i - 25i^2} = \dfrac{4i - 10(-1)}{4 - 25(-1)}$$

$$= \dfrac{10 + 4i}{29} = \dfrac{10}{29} + \dfrac{4}{29}i$$

PROGRESS CHECK 6
Write in the form $a + bi$.

(a) $\dfrac{4 - 2i}{5 + 2i}$ (b) $\dfrac{3}{2 - 3i}$ (c) $\dfrac{-3i}{3 + 5i}$

Answers

(a) $\dfrac{16}{29} - \dfrac{18}{29}i$ (b) $\dfrac{6}{13} + \dfrac{9}{13}i$ (c) $-\dfrac{15}{34} - \dfrac{9}{34}i$

The reciprocal of $q + ri$ is the quotient

$$\dfrac{1}{q + ri}$$

which can also be written in the form $a + bi$. Here is how it is done.

EXAMPLE 7
Write the reciprocal in the form $a + bi$.

(a) $2 - 2i$. The reciprocal is

$$\dfrac{1}{2 - 2i}$$

Multiplying both numerator and denominator by the conjugate $2 + 2i$, we have

$$\dfrac{1}{2 - 2i} \cdot \dfrac{2 + 2i}{2 + 2i} = \dfrac{2 + 2i}{4 - 4i^2} = \dfrac{2 + 2i}{4 + 4}$$

$$= \dfrac{2 + 2i}{8} = \dfrac{1}{4} + \dfrac{1}{4}i$$

Verify that the product of the original complex number and its reciprocal equals 1, that is, $(2 - 2i)(\frac{1}{4} + \frac{1}{4}i) = 1$.

(b) $3i$. The reciprocal is $\dfrac{1}{3i}$ and

$$\frac{1}{3i} = \frac{1}{3i} \cdot \frac{-3i}{-3i} = \frac{-3i}{-9i^2} = \frac{-3i}{9} = -\frac{1}{3}i$$

It is easy to see that $(3i)\left(-\dfrac{1}{3}i\right) = -i^2 = 1$.

PROGRESS CHECK 7

Write the reciprocal in the form $a + bi$.

(a) $3 - i$ (b) $1 + 3i$ (c) $-2i$

Answers

(a) $\dfrac{3}{10} + \dfrac{1}{10}i$ (b) $\dfrac{1}{10} - \dfrac{3}{10}i$ (c) $\dfrac{1}{2}i$

Why have mathematicians created complex numbers? Because the real number system is inadequate. In the next chapter we will show that complex numbers are indispensable in solving second-degree equations. Beyond that, advanced mathematics in science and engineering has many uses for complex numbers. The real number system simply isn't enough.

EXERCISE SET 8.6

Simplify.

1. i^{60} 2. i^{58} 3. i^{27}

4. i^{83} 5. $-i^{48}$ 6. $-i^{54}$

7. $-i^{33}$ 8. $-i^{95}$ 9. i^{-15}

10. i^{-84} 11. $-i^{-26}$ 12. $-i^{39}$

13. $-i^{-25}$ 14. $i^{8/3}\,[= (i^8)^{1/3}]$

Write as a complex number.

15. 2   16. -4 17. $-\dfrac{1}{2}$

18. -0.3 19. $\sqrt{-16}$ 20. $\sqrt{-25}$

21. $-\sqrt{-5}$ 22. $\sqrt{-8}$ 23. $-\sqrt{-36}$

24. $-\sqrt{-18}$ 25. $2 + \sqrt{-16}$ 26. $3 - \sqrt{-49}$

27. $-\dfrac{3}{2} - \sqrt{-72}$ 28. $-2 + \sqrt{-128}$ 29. $0.3 - \sqrt{-98}$

30. $-0.5 + \sqrt{-32}$

Compute and write the answer in the form $a + bi$.

31. $2i + (3 - i)$ 32. $-3i + (2 - 5i)$

33. $2 + (6 - i)$ 34. $3 - (2 - 3i)$

35. $2 + 3i + (3 - 2i)$ 36. $(3 - 2i) - \left(2 + \dfrac{1}{2}i\right)$

37. $-3 - 5i - (2 - i)$

38. $\left(\dfrac{1}{2} - i\right) + \left(1 - \dfrac{2}{3}i\right)$

39. $(2i)(4i)$

40. $(-3i)(6i)$

41. $-2i(3 + i)$

42. $3i(2 - i)$

43. $i\left(-\dfrac{1}{2} + i\right)$

44. $\dfrac{i}{2}\left(\dfrac{4 - i}{2}\right)$

45. $(2 + 3i)(2 + 3i)$

46. $(1 + i)(-3 + 2i)$

47. $(2 - i)(2 + i)$

48. $(5 + i)(2 - 3i)$

49. $(-2 - 2i)(-4 - 3i)$

50. $(2 + 5i)(1 - 3i)$

51. $(3 - 2i)(2 - i)$

52. $(4 - 3i)(2 + 3i)$

Multiply by the conjugate and simplify.

53. $2 - i$

54. $3 + i$

55. $3 + 4i$

56. $2 - 3i$

57. $-4 - 2i$

58. $5 + 2i$

Perform the indicated operations and write the answer in the form $a + bi$.

59. $\dfrac{2 + 5i}{1 - 3i}$

60. $\dfrac{1 + 3i}{2 - 5i}$

61. $\dfrac{3 - 4i}{3 + 4i}$

62. $\dfrac{4 - 3i}{4 + 3i}$

63. $\dfrac{3 - 2i}{2 - i}$

64. $\dfrac{2 - 3i}{3 - i}$

65. $\dfrac{2 + 5i}{3i}$

66. $\dfrac{5 - 2i}{-3i}$

67. $\dfrac{4i}{2 + i}$

68. $\dfrac{-2i}{3 - i}$

Find the reciprocal and write the answer in the form $a + bi$.

69. $3 + 2i$

70. $4 + 3i$

71. $\dfrac{1}{2} - i$

72. $1 - \dfrac{1}{3}i$

73. $-7i$

74. $-5i$

75. $\sqrt{2} - i$

76. $2 - \sqrt{3}\,i$

77. Show that the reciprocal of $a + bi$ is

$$\dfrac{a}{a^2 + b^2} - \dfrac{b}{a^2 + b^2}i$$

TERMS AND SYMBOLS

square root (p. 225)
cube root (p. 225)
nth root (p. 225)
$\sqrt{a}$, $\sqrt[n]{a}$ (p. 227)
radical sign (p. 227)
principal square root
  (p. 227)
principal nth root (p. 227)

radical form (p. 227)
rationalizing the
  denominator (p. 232)
simplified radical (p. 233)
imaginary unit (p. 238)
i (p. 238)
imaginary number
  (p. 239)

complex number (p. 239)
$a + bi$ (p. 239)
real part (p. 239)
imaginary part (p. 239)
conjugate of a complex
  number (p. 242)
reciprocal of a complex
  number (p. 243)

KEY IDEAS FOR REVIEW

☐ The laws of exponents hold for all rational exponents.

☐ Zero as an exponent produces a result of 1. Thus, $a^0 = 1$.

☐ Since

$$a^{-m} = \frac{1}{a^m}, \quad \text{and} \quad a^m = \frac{1}{a^{-m}}$$

we can change the sign of the exponent by moving the factor from numerator to denominator or denominator to numerator.

☐ Radicals are an alternate means of writing rational exponent forms. Thus,

$$a^{m/n} = \sqrt[n]{a^m} = (\sqrt[n]{a})^m \quad \text{if} \quad a \geq 0$$

☐ The identities

$$(\sqrt[n]{a})^n = \sqrt[n]{a^n} = a$$
$$\sqrt[n]{ab} = \sqrt[n]{a}\sqrt[n]{b}$$
$$\sqrt[n]{\frac{a}{b}} = \frac{\sqrt[n]{a}}{\sqrt[n]{b}}$$

are useful in simplifying expressions with radicals when $a \geq 0$ and $b \geq 0$.

☐ A radical expression is in simplified form if

(a) $\sqrt[n]{x^m}$ has $m < n$.

(b) $\sqrt[n]{x^m}$ has no common factors between m and n.

(c) the denominator does not contain a radical.

☐ Addition and subtraction of radical expressions can be performed only if exactly the same radical form is involved.

☐ The product $(\sqrt[n]{a})(\sqrt[m]{b})$ can be simplified if $m = n$.

☐ $i = \sqrt{-1}$ and $i^2 = -1$ can be used to simplify an expression of the form i^n and to rewrite $\sqrt{-a}$ as $i\sqrt{a}$.

☐ Every complex number can be written in the form $a + bi$, where a and b are real numbers.

☐ The sum, difference, product, and quotient of two complex numbers can always be expressed in the form $c + di$. (Of course, we cannot divide by zero.)

☐ The real number system is a subset of the complex number system since any real number a can be written as $a + 0i$.

COMMON ERRORS

1. *Don't* perform intermediate steps on any polynomial, no matter how complicated, that is raised to the zero power, since 1 is the answer.

$$(y^2 + x^2 - 2y + 4)^0 = 1$$

2. Don't confuse negative *numbers* and negative *exponents*. *Don't* write

$$(-2)^{-3} = \frac{1}{2^3} = \frac{1}{8}$$

When the factor is moved to the denominator, only the sign of the exponent changes.

$$(-2)^{-3} = \frac{1}{(-2)^3} = \frac{1}{-8} = -\frac{1}{8}$$

3. *Don't* write

$$\sqrt{4} + \sqrt{60} = \sqrt{64}$$

You can add only if the radical forms are identical.

$$\sqrt{2x} + 5\sqrt{2x} = 6\sqrt{2x}$$

4. *Don't* write

$$\sqrt{-2} \cdot \sqrt{-8} = \sqrt{16} = 4$$

When dealing with negative numbers under the radical sign, first write the expressions in terms of the imaginary unit i.

$$\sqrt{-2} \cdot \sqrt{-8} = (i\sqrt{2})(i\sqrt{8}) = i^2\sqrt{16} = (-1)(4) = -4$$

5. *Don't* write

$$\sqrt{25} = \pm 5$$

The number indicated by $\sqrt{25}$ is the *positive* square root of 25. Thus, $\sqrt{25} = 5$.

6. Remember that

$$\sqrt{x^2} = |x|$$

Don't write

$$\sqrt{x^2} = x$$

unless you know that $x \geq 0$.

PROGRESS TEST 8A

In Problems 1–7, simplify and write answers using only positive exponents.

1. $\dfrac{(x + 1)^{2n-1}}{(x + 1)^{n+1}}$

2. $\left(\dfrac{1}{2}x^3y^2\right)^4$

3. $\left(\dfrac{-2xy^3}{5x^2y}\right)^3$

4. $\dfrac{5(a^2 - 2a + 1)^0}{(x^{-2})^3}$

5. $(2x^{2/5} \cdot 4x^{2/5})^{-2}$

6. $\dfrac{x^{-2/3}y^{1/2}}{x^{-2}y^{-3/2}}$

7. $\left(\dfrac{27x^{-6}z^0}{125y^{-3/2}}\right)^{-2/3}$

8. Write $(2y - 1)^{5/2}$ in radical form.

9. Write $\sqrt[3]{6y^5}$ in rational exponent form.

10. Simplify $-\sqrt{32x^4y^9}$.

11. Simplify $\sqrt{\dfrac{x}{5}}$.

12. Simplify $\dfrac{-4a^2b\sqrt[3]{a^7b^6}}{\sqrt[3]{a}}$.

13. Simplify $\sqrt[3]{24} - 3\sqrt[3]{81}$.

14. Simplify $2\sqrt{xy^2} - 6\sqrt{x^2y} + 3\sqrt{xy^2} - 2\sqrt{x^2y}$.
15. Multiply and simplify $3(2\sqrt{x} - 3\sqrt{y})(2\sqrt{x} + 3\sqrt{y})$.
16. Write in the form $a + bi$: $4 - \sqrt{-4}$.
17. Compute $2 + \sqrt{-27} - 3\sqrt{-3}$.
18. Simplify $-i^{47}$.
19. Write in the form $a + bi$: $(-4 - 2i) \div (3 - 4i)$.
20. Write the reciprocal of $4 + 3i$ in the form $a + bi$.

PROGRESS TEST 8B

In Problems 1–7, simplify and write answers using only positive exponents.

1. $(x^{3n-1})^2$

2. $\left(-\dfrac{1}{3}a^2b\right)^3$

3. $\left(\dfrac{-3y^{-2}}{x^{-3}}\right)^{-3}$

4. $\dfrac{4(x+1)^5(2x+1)^0}{(x+1)^6}$

5. $(-2y^{1/3} \cdot 3y^{1/6})^{-3}$

6. $\dfrac{x^{1/2}y^{-1/5}}{x^{-2}y^{3/5}}$

7. $\left(\dfrac{27x^{-3/2}y^{4/5}}{x^{5/2}y^{1/5}}\right)^{-2/5}$

8. Write $(3x+1)^{2/3}$ in radical form.
9. Write $\sqrt[5]{4x^3}$ in rational exponent form.
10. Simplify $-3\sqrt[4]{16x^6z^2}$.

11. Simplify $\sqrt{\dfrac{2x^3}{y}}$.

12. Simplify $\dfrac{-2x^3y^2\sqrt{x^5y^6}}{\sqrt{xy}}$.

13. Simplify $2\sqrt[3]{54} - 4\sqrt[3]{16}$.
14. Simplify $-\sqrt{(a-1)b^2} + 2\sqrt{(a-1)^3b} - 2\sqrt{(a-1)b^2} - 3\sqrt{(a-1)^3b}$.
15. Multiply and simplify $-2(3\sqrt{a} - \sqrt{b})(2\sqrt{a} - 3\sqrt{b})$.
16. Write in the form $a + bi$: $2 + \sqrt{-8}$.
17. Compute $-3 - 3\sqrt{-8} + 2\sqrt{-18}$.
18. Simplify $-3i^{25}$.
19. Write in the form $a + bi$: $(2 - i) \div (2 - 3i)$.
20. Write the reciprocal of $3 - 5i$ in the form $a + bi$.

CHAPTER NINE
SECOND-DEGREE EQUATIONS AND INEQUALITIES

The function

$$f(x) = ax^2 + bx + c, \quad a \neq 0$$

is called a **quadratic function.** We are interested in finding the zeros of the function, that is, the values of x for which $f(x) = 0$. This is equivalent to finding the roots of the equation

$$ax^2 + bx + c = 0, \quad a \neq 0$$

which we call a **quadratic equation** or **second-degree equation in one variable.** We will look at techniques for solving quadratic equations and at applications which lead to this algebraic form. We will also study methods of attacking second-degree inequalities in one variable.

9.1
SOLVING QUADRATIC EQUATIONS
THE FORM $ax^2 + c = 0$

When the quadratic equation $ax^2 + bx + c = 0$ has the coefficient $b = 0$, we have an equation of the form

$$2x^2 - 10 = 0$$

which is very simple to solve. We begin by isolating x^2.

$$2x^2 - 10 = 0$$
$$2x^2 = 10$$
$$x^2 = 5$$

At this point we ask: Is there a number whose square is 5? There are actually two such numbers: $\sqrt{5}$ and $-\sqrt{5}$.

$$x^2 = 5$$
$$x = \pm\sqrt{5}$$

(Note that we have used the shorthand notation, $\pm\sqrt{5}$, as a way of indicating $\sqrt{5}$ and $-\sqrt{5}$.)

EXAMPLE 1
Solve the equation.

(a) $3x^2 - 8 = 0$

We isolate and solve for x.

$$3x^2 - 8 = 0$$
$$3x^2 = 8$$
$$x^2 = \frac{8}{3}$$
$$x = \pm\sqrt{\frac{8}{3}} = \pm\frac{2\sqrt{2}}{\sqrt{3}}$$

To attain simplest radical form, we must rationalize the denominator.

$$x = \pm\frac{2\sqrt{2}}{\sqrt{3}} \cdot \frac{\sqrt{3}}{\sqrt{3}} = \pm\frac{2\sqrt{6}}{3} \quad \text{or} \quad \pm\frac{2}{3}\sqrt{6}$$

(b) $4x^2 + 11 = 0$

We isolate and solve for x.

$$4x^2 + 11 = 0$$
$$4x^2 = -11$$
$$x^2 = -\frac{11}{4}$$
$$x = \pm\sqrt{\frac{-11}{4}} = \pm\frac{i\sqrt{11}}{2}$$

We see that the solutions to a quadratic equation may be complex numbers.

(c) $(x - 5)^2 = -9$

Although this equation is not strictly of the form $ax^2 + c = 0$, we can use the same approach. We see that

$$(x - 5)^2 = -9$$

which implies that

$$x - 5 = \pm\sqrt{-9}$$

$$x - 5 = \pm 3i \qquad \text{Since } (3i)^2 = (-3i)^2 = -9$$

$$x = 5 \pm 3i \qquad \text{Add 5 to both sides.}$$

Thus, the solutions of the given equation are the complex numbers

$$x = 5 + 3i \quad \text{and} \quad x = 5 - 3i$$

PROGRESS CHECK 1

Solve each equation.

(a) $4x^2 - 9 = 0$ (b) $2x^2 - 15 = 0$ (c) $5x^2 + 13 = 0$ (d) $(2x - 7)^2 + 5 = 0$

Answers

(a) $\pm\dfrac{3}{2}$ (b) $\pm\dfrac{\sqrt{30}}{2}$ (c) $\pm\dfrac{i\sqrt{65}}{5}$ (d) $\dfrac{7 \pm i\sqrt{5}}{2}$

SOLVING BY FACTORING

Under what circumstances can the product of two real numbers be 0? Stated more formally, if a and b are real numbers and $ab = 0$, what must be true of a and b? A little thought will convince you that at least one of the numbers must be zero! In fact, this result is very easy to prove. If we assume that $a \neq 0$, then we can divide both sides by a.

$$ab = 0$$

$$\frac{\cancel{a}b}{\cancel{a}} = \frac{0}{a}$$

$$b = 0$$

Similarly, if we assume $b \neq 0$ we would conclude that $a = 0$.

> If a and b are real numbers and $ab = 0$, then $a = 0$ or $b = 0$ (or both $a = 0$ and $b = 0$).

This simple theorem provides us with a means for solving quadratic equations whenever we can factor the quadratic into linear factors.

EXAMPLE 2

Solve by factoring.

(a) $x^2 - 2x - 3 = 0$

Factoring, we have

$$x^2 - 2x - 3 = 0$$

$$(x - 3)(x + 1) = 0$$

But the product of two real numbers can equal 0 only if at least one of them is 0. Thus, either

$$x - 3 = 0 \qquad \text{or} \qquad x + 1 = 0$$

$$x = 3 \qquad \text{or} \qquad x = -1$$

You can verify that 3 and -1 are roots of the equation by substituting these values in the equation.

(b) $2x^2 - 3x - 2 = 0$

Factoring, we have

$$2x^2 - 3x - 2 = 0$$

$$(2x + 1)(x - 2) = 0$$

Either

$$2x + 1 = 0 \qquad \text{or} \qquad x - 2 = 0$$

$$x = -\frac{1}{2} \qquad \text{or} \qquad x = 2$$

(c) $3x^2 - 4x = 0$

Factoring, we have

$$3x^2 - 4x = 0$$

$$x(3x - 4) = 0$$

Either

$$x = 0 \qquad \text{or} \qquad 3x - 4 = 0$$

$$x = 0 \qquad \text{or} \qquad x = \frac{4}{3}$$

(d) $x^2 + x + 1 = 0$

If we attempt to factor $x^2 + x + 1$, we see that the only possible factors with integer coefficients are $(x + 1)$ and $(x - 1)$, but if we try all the possibilities, $(x + 1)(x + 1)$ or $(x - 1)(x - 1)$ or $(x + 1)(x - 1)$, we find that none of them works. We will look at methods for handling this situation later in this section.

PROGRESS CHECK 2
Solve by factoring.
(a) $x^2 + 3x - 10 = 0$ (b) $3x^2 - 11x - 4 = 0$
(c) $4x^2 - x = 0$ (d) $2x^2 + 4x + 1 = 0$

Answers

(a) $-5, 2$ (b) $-\frac{1}{3}, 4$ (c) $0, \frac{1}{4}$ (d) *cannot be factored*

COMPLETING THE SQUARE

We saw in Example 2d that the method of factoring doesn't always work. However, by completing the square, we can find solutions to *any* quadratic equation. Given an equation such as

$$x^2 + 10x = 2$$

we seek a constant k such that the addition of k^2 will "complete" a perfect square in the left-hand side.

$$x^2 + 10x + k^2 = (x + k)^2 = x^2 + 2kx + k^2$$

Then we must have

$$10x = 2kx$$

from which we see that the constant k that we seek is exactly $\frac{1}{2}$ of the coefficient of x. Thus, in our example, $k = \frac{10}{2} = 5$ and $k^2 = 5^2$ is added to each side of the equation.

$$x^2 + 10x + 5^2 = 2 + 5^2$$
$$x^2 + 10x + 25 = 2 + 25$$
$$(x + 5)^2 = 27$$

In Example 1 we saw that this form is easily solved.

$$(x + 5)^2 = 27$$
$$x + 5 = \pm\sqrt{27}$$
$$x = -5 \pm \sqrt{27}$$

EXAMPLE 3

Solve by completing the square.

$$2x^2 - 5x + 4 = 0$$

Solution

We now outline and explain each step of the process.

Completing the square	Example
Step 1. Divide the equation by the coefficient of x^2.	*Step 1.* $x^2 - \dfrac{5}{2}x + 2 = 0$
Step 2. Rewrite the equation with the constant on the right-hand side.	*Step 2.* $x^2 - \dfrac{5}{2}x = -2$
Step 3. Complete the square, $(x + k)^2$, where k is half the coefficient of x. Balance the equation by adding k^2 to the right-hand side. Simplify.	*Step 3.* $x^2 - \dfrac{5}{2}x + \left(-\dfrac{5}{4}\right)^2$ $\qquad = -2 + \left(\dfrac{-5}{4}\right)^2$ $\left(x - \dfrac{5}{4}\right)^2 = \dfrac{-32}{16} + \dfrac{25}{16}$ $\left(x - \dfrac{5}{4}\right)^2 = -\dfrac{7}{16}$
Step 4. Take the square root of both sides of the equation and solve for x.	*Step 4.* $x - \dfrac{5}{4} = \pm\sqrt{\dfrac{-7}{16}} = \dfrac{\pm i\sqrt{7}}{4}$ $x = \dfrac{5}{4} \pm \dfrac{i\sqrt{7}}{4}$ or $x = \dfrac{5 \pm i\sqrt{7}}{4}$

PROGRESS CHECK 3
Solve by completing the square.

(a) $x^2 - 3x + 2 = 0$ (b) $3x^2 - 4x + 2 = 0$

Answers

(a) 1, 2   (b) $\dfrac{2 \pm i\sqrt{2}}{3}$

WARNING
In the equation

$$x^2 + 8x = -2$$

completing the square on the left-hand side produces $(x + 4)^2$. But this is $4^2 = 16$ more than the original left-hand side. *Don't* forget to balance the equation by adding 16 to the right-hand side.

$$(x + 4)^2 = -2 + 16$$

EXERCISE SET 9.1
Solve the following equations.

1. $3x^2 - 27 = 0$

2. $4x^2 - 64 = 0$

3. $4x^2 - 25 = 0$

4. $49y^2 - 9 = 0$

5. $5y^2 - 25 = 0$

6. $6x^2 - 12 = 0$

7. $(x - 3)^2 = -2$

8. $(s + 3)^2 = 4$

9. $(2r + 5)^2 = 8$

10. $(3x - 4)^2 = -6$

11. $(2y + 4)^2 + 3 = 0$

12. $(3p - 2)^2 + 6 = 0$

13. $(3x - 5)^2 - 8 = 0$

14. $(4t + 1)^2 - 3 = 0$

15. $2x^2 + 8 = 0$

16. $6y^2 + 96 = 0$

17. $9x^2 + 64 = 0$

18. $81x^2 + 25 = 0$

19. $2y^2 + 12 = 0$

20. $9x^2 + 45 = 0$

Solve by factoring.

21. $x^2 - 3x + 2 = 0$

22. $x^2 - 6x + 8 = 0$

23. $x^2 + x - 2 = 0$

24. $3r^2 - 4r + 1 = 0$

25. $x^2 + 6x = -8$

26. $x^2 + 6x + 5 = 0$

27. $y^2 - 4y = 0$

28. $2x^2 - x = 0$

29. $2x^2 - 5x = -2$

30. $2s^2 - 5s - 3 = 0$

31. $t^2 - 4 = 0$

32. $4x^2 - 9 = 0$

33. $6x^2 - 5x + 1 = 0$

34. $6x^2 - x = 2$

Solve by completing the square.

35. $x^2 - 2x = 8$

36. $t^2 - 2t = 15$

37. $2r^2 - 7r = 4$

38. $9x^2 + 3x = 2$

39. $3x^2 + 8x = 3$ 40. $2y^2 + 4y = 5$

41. $2y^2 + 2y = -1$ 42. $3x^2 - 4x = -3$

43. $4x^2 - x = 3$ 44. $2x^2 + x = 2$

45. $3x^2 + 2x = -1$ 46. $3u^2 - 3u = -1$

Solve by any method.

47. $x^2 + x - 12 = 0$ 48. $x^2 - 2x - 8 = 0$

49. $3y^2 + y = 0$ 50. $4x^2 - 4x - 3 = 0$

51. $2x^2 + 2x - 5 = 0$ 52. $2t^2 + 2t + 3 = 0$

53. $3x^2 + 4x - 4 = 0$ 54. $x^2 + 2x = 0$

55. $2x^2 + 5x + 4 = 0$ 56. $2r^2 - 3r + 2 = 0$

57. $4u^2 - 1 = 0$ 58. $x^2 + 2 = 0$

59. $4x^2 + 2x + 3 = 0$ 60. $4s^2 + 4s - 15 = 0$

9.2
THE QUADRATIC FORMULA

Let's apply the method of completing the square to the general quadratic equation

$$ax^2 + bx + c = 0, \; a > 0$$

Following the steps we illustrated in the last section, we have

$ax^2 + bx + c = 0$

$x^2 + \dfrac{b}{a} x + \dfrac{c}{a} = 0$ Divide the equation by the coefficient of x^2.

$x^2 + \dfrac{b}{a}x = -\dfrac{c}{a}$ Rewrite with the constant on the right-hand side.

$x^2 + \dfrac{b}{a}x + \left(\dfrac{b}{2a}\right)^2 = \left(\dfrac{b}{2a}\right)^2 - \dfrac{c}{a}$ Complete the square and balance the equation.

$\left(x + \dfrac{b}{2a}\right)^2 = \dfrac{b^2}{4a^2} - \dfrac{c}{a} = \dfrac{b^2 - 4ac}{4a^2}$ Simplify.

$x + \dfrac{b}{2a} = \pm\sqrt{\dfrac{b^2 - 4ac}{4a^2}}$ Solve for x.

$x = -\dfrac{b}{2a} \pm \dfrac{\sqrt{b^2 - 4ac}}{2a}$

$x = \dfrac{-b \pm \sqrt{b^2 - 4ac}}{2a}$

By applying the method of completing the square to the standard form of the quadratic equation, we have derived a *formula* that gives us the roots or solutions of *any* quadratic equation in one variable.

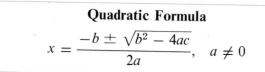

Quadratic Formula

$$x = \frac{-b \pm \sqrt{b^2 - 4ac}}{2a}, \quad a \neq 0$$

That is, the roots of the quadratic equation $ax^2 + bx + c = 0$ are

$$\frac{-b + \sqrt{b^2 - 4ac}}{2a} \quad \text{and} \quad \frac{-b - \sqrt{b^2 - 4ac}}{2a}$$

EXAMPLE 1
Solve by the quadratic formula.

(a) $2x^2 - 3x - 3 = 0$

We begin by identifying a, b, and c.

$$a = 2$$
$$b = -3$$
$$c = -3$$

We can now write the quadratic formula and substitute.

$$x = \frac{-b \pm \sqrt{b^2 - 4ac}}{2a}$$
$$= \frac{-(-3) \pm \sqrt{(-3)^2 - 4(2)(-3)}}{2(2)}$$
$$= \frac{3 \pm \sqrt{33}}{4}$$

(b) $5x^2 - 3x = -2$

To use the quadratic formula, you must write the equation in the standard form $5x^2 - 3x + 2 = 0$. Now we can identify a, b, and c.

$$a = 5$$
$$b = -3$$
$$c = 2$$

We write the quadratic formula and substitute.

$$x = \frac{-b \pm \sqrt{b^2 - 4ac}}{2a}$$
$$= \frac{-(-3) \pm \sqrt{(-3)^2 - 4(5)(2)}}{2(5)}$$
$$= \frac{3 \pm \sqrt{-31}}{10} = \frac{3 \pm i\sqrt{31}}{10}$$

PROGRESS CHECK 1

Solve by the quadratic formula.

(a) $x^2 - 8x = -10$ (b) $4x^2 - 2x + 1 = 0$

(c) $2x^2 - x - 1 = 0$

Answers

(a) $4 \pm \sqrt{6}$ (b) $\dfrac{1 \pm i\sqrt{3}}{4}$ (c) $1, -\dfrac{1}{2}$

WARNING

There are a number of errors that students make in using the quadratic formula.

(a) To solve $x^2 - 3x = -4$, you must write the equation in the form

$$x^2 - 3x + 4 = 0$$

to properly identify a, b, and c.

(b) Since a, b, and c are coefficients, you must include the sign. If

$$x^2 - 3x + 4 = 0$$

then $b = -3$. *Don't* write $b = 3$.

(c) The quadratic formula is

$$x = \frac{-b \pm \sqrt{b^2 - 4ac}}{2a}$$

Don't write

$$x = -b \pm \frac{\sqrt{b^2 - 4ac}}{2a}$$

The term $-b$ must also be divided by $2a$.

Now that you have a formula that works for any quadratic equation, you may be tempted to use it all the time. However, if you see an equation of the form

$$x^2 = 15$$

it is certainly easier to immediately supply the answer: $x = \pm\sqrt{15}$. Similarly, if you are faced with

$$x^2 + 3x + 2 = 0$$

it is faster to solve if you see that

$$x^2 + 3x + 2 = (x + 1)(x + 2)$$

The method of completing the square is generally not used for solving quadratic equations once you have learned the quadratic formula. However, we will need to use this technique when we graph second-degree equations in Chapter Twelve. The technique of completing the square is occasionally helpful in a variety of areas of applications.

EXERCISE SET 9.2

In each of the following exercises, a, b, and c denote the coefficients of the general quadratic equation $ax^2 + bx + c = 0$. Select the correct answer.

1. $2x^2 + 3x - 4 = 0$

    (a) $a = 2$, $b = 3$, $c = 4$

    (b) $a = 3$, $b = 2$, $c = 4$

    (c) $a = -2$, $b = 4$, $c = 3$

    (d) $a = 2$, $b = 3$, $c = -4$

2. $3x^2 - 2x = 5$

    (a) $a = 3$, $b = 2$, $c = 5$

    (b) $a = 2$, $b = 3$, $c = 5$

    (c) $a = 3$, $b = -2$, $c = -5$

    (d) $a = -3$, $b = 2$, $c = -5$

3. $r = 5r^2 - \dfrac{2}{3}$

    (a) $a = 1$, $b = 5$, $c = -\dfrac{2}{3}$

    (b) $a = -1$, $b = 5$, $c = \dfrac{2}{3}$

    (c) $a = 5$, $b = -1$, $c = -\dfrac{2}{3}$

    (d) $a = 5$, $b = 1$, $c = \dfrac{2}{3}$

4. $2r^2 + r = 0$

    (a) $a = -2$, $b = 1$, $c = 0$

    (b) $a = 0$, $b = 2$, $c = 1$

    (c) $a = 2$, $b = -1$, $c = 0$

    (d) $a = 2$, $b = 1$, $c = 0$

In each of the following, identify a, b, and c in the general quadratic equation $ax^2 + bx + c = 0$.

5. $3x^2 - 2x + 5 = 0$

6. $2x^2 + x = 4$

7. $s = 2s^2 + 5$

8. $2x^2 + 5 = 0$

9. $3x^2 - \dfrac{1}{3}x = 0$

10. $4y^2 = 2y + 1$

Solve by the quadratic formula.

11. $x^2 + 5x + 6 = 0$

12. $x^2 + 2x - 8 = 0$

13. $x^2 - 8x = -15$

14. $6r^2 - r = 1$

15. $2x^2 + 3x = 0$

16. $2x^2 + 3x + 3 = 0$

17. $5x^2 - 4x + 3 = 0$

18. $2x^2 - 3x - 2 = 0$

19. $5y^2 - 4y + 5 = 0$

20. $x^2 - 5x = 0$

21. $x^2 + x = 12$

22. $2x^2 + 5x - 6 = 0$

23. $3x^2 + x - 2 = 0$

24. $2x^2 + 4x - 3 = 0$

25. $x^2 - 9 = 0$

26. $5x^2 + 2 = 0$

27. $3y^2 - 4 = 0$

28. $2x^2 + 2x + 5 = 0$

29. $4u^2 + 3u = 0$

30. $4x^2 - 1 = 0$

ROOTS OF A QUADRATIC EQUATION: THE DISCRIMINANT

By analyzing the quadratic formula

$$x = \frac{-b \pm \sqrt{b^2 - 4ac}}{2a}$$

we can learn a great deal about the roots of the quadratic equation $ax^2 + bx + c = 0$. The key to the analysis is the **discriminant** $b^2 - 4ac$ found under the radical.

(a) If $b^2 - 4ac$ is negative, then we have the square root of a negative number and both values of x will be complex numbers; in fact, they will be complex conjugates of each other.

(b) If $b^2 - 4ac$ is positive, then we have the square root of a positive number and both values of x will be real.

(c) If $b^2 - 4ac$ is 0, then $x = -b/2a$, which we call a **double root** or **repeated root** of the quadratic equation. For example, if $x^2 - 10x + 25 = 0$, then $b^2 - 4ac = 0$ and $x = -b/2a = 10/2 = 5$. Moreover

$$x^2 - 10x + 25 = (x - 5)(x - 5) = 0$$

We call $x = 5$ a double root because the factor $(x - 5)$ is a double factor of $x^2 - 10x + 25 = 0$.

If the roots of the quadratic equation are real and a, b, and c are rational numbers, the discriminant enables us to determine whether the roots are rational or irrational. Since $\sqrt{k}$ is a rational number only if k is a perfect square, we see that the quadratic formula produces a rational result only if $b^2 - 4ac$ is a perfect square. We summarize these results.

The quadratic equation $ax^2 + bx + c = 0$ has exactly two roots, the nature of which are determined by the discriminant $b^2 - 4ac$.

Discriminant $b^2 - 4ac$	Roots
Positive	Two real roots
Negative	Two complex roots (conjugates)
0	A double root
Rational $\{$ A perfect square	Rational
a, b, c $\{$ Not a perfect square	Irrational

EXAMPLE 1

Without solving, determine the nature of the roots of the quadratic equation.

(a) $3x^2 - 4x + 6 = 0$

We evaluate $b^2 - 4ac$ using $a = 3$, $b = -4$, $c = 6$.

$$b^2 - 4ac = (-4)^2 - 4(3)(6) = 16 - 72 = -56$$

The discriminant is negative and the equation has two complex roots.

(b) $2x^2 - 7x = -1$

Rewrite the equation in the standard form

$$2x^2 - 7x + 1 = 0$$

and then substitute $a = 2$, $b = -7$, $c = 1$ in the discriminant. Thus,

$$b^2 - 4ac = (-7)^2 - 4(2)(1) = 49 - 8 = 41$$

The discriminant is positive and is not a perfect square; thus, the roots are real, unequal, and irrational.

(c) $4x^2 + 12x + 9 = 0$

Setting $a = 4$, $b = 12$, $c = 9$, we evaluate the discriminant

$$b^2 - 4ac = (12)^2 - 4(4)(9) = 144 - 144 = 0$$

The discriminant is 0 and there is a double real root.

PROGRESS CHECK 1

Without solving, determine the nature of the roots of the quadratic equation by using the discriminant.

(a) $4x^2 - 20x + 25 = 0$ (b) $5x^2 - 6x = -2$

(c) $10x^2 = x + 2$ (d) $x^2 + x - 1 = 0$

Answers
(a)   double real root   (b)   two complex roots
(c)   two real, rational roots   (d)   two real, irrational roots

If we seek the zeros of the quadratic function

$$f(x) = ax^2 + bx + c, \quad a \neq 0 \tag{1}$$

we need only set $f(x) = 0$ and solve the resulting quadratic equation

$$ax^2 + bx + c = 0 \tag{2}$$

It is important to recognize that the solutions to Equation (2) are the same as the x-intercepts of the graph of the function in Equation (1) since these points have coordinates of the form $(x_1, 0)$.

In Section 6.3 we saw that the graph of a second-degree function is a parabola. The discriminant of the quadratic equation (2) therefore tells us the

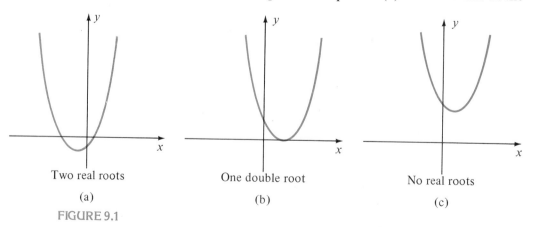

Two real roots   One double root   No real roots
(a)   (b)   (c)

FIGURE 9.1

number of x-intercepts of the parabola. The possibilities are two real roots (Figure 9.1a), a double root (Figure 9.1b), and two complex roots (Figure 9.1c). We can now summarize.

The x-intercepts of the graph of a quadratic function are the real roots of the function. The number of x-intercepts is related to the discriminant as follows.

Discriminant	*Graph of the function*
> 0	Meets the x-axis at two points.
$= 0$	Meets the x-axis at one point.
< 0	Does not meet the x-axis.

EXERCISE SET 9.3

Select the correct answer.

1. In $3x^2 - 2x + 4 = 0$, the discriminant is (a) 52   (b) 44   (c) $\sqrt{44}$
   (d) -52 (e) -44.
2. In $x^2 - 6x + 9 = 0$, the discriminant is (a) 72   (b) -72 (c) -36
   (d) 0   (e) $\sqrt{72}$.
3. In $3y^2 = 2y - 1$, the discriminant is (a) $\sqrt{8}$ (b) -8 (c) 4
   (d) $\sqrt{-8}$ (e) $-2\sqrt{3}$.
4. In $3x^2 + 4x = 0$, the discriminant is (a) 4   (b) -4 (c) 0
   (d) 16   (e) 7.
5. In $r = 5r^2 + 3$, the discriminant is (a) -61 (b) 59   (c) 61
   (d) -59 (e) $\sqrt{61}$.
6. In $2t = -4t^2 - 3$, the discriminant is (a) 44   (b) $2\sqrt{11}$ (c) -44
   (d) -23 (e) 41.
7. In $4x^2 - 8 = 0$, the discriminant is (a) -64 (b) 64   (c) -128
   (d) 128   (e) $\sqrt{128}$.
8. In $4y^2 + 9 = 0$, the discriminant is (a) 65   (b) -144 (c) -65
   (d) 144   (e) 12.
9. $4x^2 - 3x + 5 = 0$ has (a) two real roots   (b) a double root   (c) two complex roots.
10. $y^2 - 10y + 25 = 0$ has (a) two real roots   (b) a double root   (c) two complex roots.
11. $5x^2 = x + 1$ has (a) two real roots   (b) a double root   (c) two complex roots.
12. $4r^2 - 2r = 0$ has (a) two real roots   (b) a double root   (c) two complex roots.
13. $x = 2x^2 - 2$ has (a) two real roots   (b) a double root   (c) two complex roots.
14. $8x^2 + 24 = 0$ has (a) two real roots   (b) a double root   (c) two complex roots.

Without solving, determine the nature of the roots of each quadratic equation.

15. $x^2 - 2x + 3 = 0$ 16. $3x^2 + 2x - 5 = 0$ 17. $4x^2 - 12x + 9 = 0$

18. $2x^2 + x + 5 = 0$ 19. $-3x^2 + 2x + 5 = 0$ 20. $-3y^2 + 2y - 5 = 0$

21. $3x^2 + 2x = 0$ 22. $4x^2 + 20x + 25 = 0$ 23. $2r^2 = r - 4$

24. $3x^2 = 5 - x$ 25. $3x^2 + 6 = 0$ 26. $4x^2 - 25 = 0$

27. $6r = 3r^2 + 1$ 28. $4x = 2x^2 + 3$

29. $12x = 9x^2 + 4$ 30. $4s^2 = -4s - 1$

Determine the number of x-intercepts of the graph of each of the following functions.

31. $f(x) = 3x^2 + 4x - 1$ 32. $f(x) = 2x^2 - 3x + 2$

33. $f(x) = x^2 + x + 3$ 34. $f(x) = 2x^2 - 6x + 1$

35. $f(x) = 4x^2 - 12x + 9$ 36. $f(x) = 9x^2 + 12x + 4$

37. $f(x) = 4x^2 + 3$ 38. $f(x) = -x^2 + 1$

9.4
APPLICATIONS

In earlier chapters we carefully avoided those work problems, number problems, business problems, geometric problems, and other applications that resulted in second-degree equations. Now that you can solve quadratic equations, we can look at these applications.

EXAMPLE 1
Working together, two cranes can unload a ship in 4 hours. The slower crane, working alone, requires 6 hours more than the faster crane to do the job. How long does it take each crane to do the job by itself?

Solution
Let x = number of hours for faster crane to do the job. Then $x + 6$ = number of hours for slower crane to do the job.

Displaying the information in a table, we have

	Rate $\times$ Time = Work		
Faster crane	$\dfrac{1}{x}$	4	$\dfrac{4}{x}$
Slower crane	$\dfrac{1}{x + 6}$	4	$\dfrac{4}{x + 6}$

Since the job is completed in 4 hours when the two cranes work together, we must have

$$\begin{array}{ccccc} \text{work done by} & + & \text{work done by} & = & \text{1 whole job} \\ \text{faster crane} & & \text{slower crane} & & \end{array}$$

or

$$\frac{4}{x} + \frac{4}{x + 6} = 1$$

To solve, we multiply by the L.C.D., which is $x(x + 6)$.

$$x(x + 6)\left[\frac{4}{x} + \frac{4}{x + 6}\right] = x(x + 6)$$

$$4(x + 6) + 4x = x^2 + 6x$$

$$0 = x^2 - 2x - 24$$

$$0 = (x + 4)(x - 6)$$

$$x = -4 \quad \text{or} \quad x = 6$$

The solution $x = -4$ is rejected since it makes no sense to speak of negative hours of work. Thus,

$x = 6$ is the number of hours in which the
faster crane can do the job alone
$x + 6 = 12$ is the number of hours in which the
slower crane can do the job alone

PROGRESS CHECK 1
A storage tank can be filled in 6 hours when two pipes are used. The larger-diameter pipe, used alone, requires 5 hours less than the smaller-diameter pipe to fill the tank. How many hours does each pipe require to fill the tank when working alone?

Answer
The larger pipe requires 10 hours. The smaller pipe requires 15 hours.

EXAMPLE 2
The length of a pool is 3 times its width, and the pool is surrounded by a grass walk 4 feet wide. If the total area enclosed by the walk is 684 square feet, find the dimensions of the pool.

Solution
A diagram is useful in solving geometric problems (see Figure 9.2). If we let $x = $ width of pool, then $3x = $ length of pool, and the region enclosed by the walk has length $3x + 8$ and width $x + 8$. The area is then

$$\text{length} \times \text{width} = \text{area}$$
$$(3x + 8)(x + 8) = 684$$
$$3x^2 + 32x + 64 = 684$$
$$3x^2 + 32x - 620 = 0$$
$$(3x + 62)(x - 10) = 0$$
$$x = 10 \qquad \text{Reject } x = -\frac{62}{3}.$$
$$3x = 30$$

The dimensions of the pool are 10 feet by 30 feet.

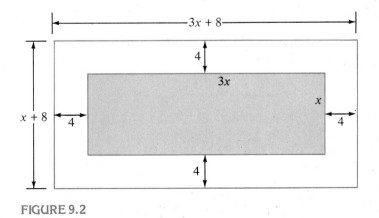

FIGURE 9.2

PROGRESS CHECK 2

The altitude of a triangle is 2 centimeters less than the base. If the area of the triangle is 24 square centimeters, find the base and altitude of the triangle.

Answer
Base = 8 cm. Altitude = 6 cm.

EXAMPLE 3

The larger of two positive numbers exceeds the smaller by 2. If the sum of the squares of the two numbers is 74, find the two numbers.

Solution

If we

then

$$\text{let} \quad x = \text{the larger number}$$
$$x - 2 = \text{the smaller number}$$

The sum of the squares is then

$$x^2 + (x - 2)^2 = 74$$
$$x^2 + x^2 - 4x + 4 = 74$$
$$2x^2 - 4x - 70 = 0$$
$$x^2 - 2x - 35 = 0$$
$$(x + 5)(x - 7) = 0$$
$$x = 7 \qquad \text{Reject } x = -5.$$

The numbers are then 7 and 5. Verify that the sum of the squares is indeed 74.

PROGRESS CHECK 3

The sum of a number and its reciprocal is $\dfrac{17}{4}$. Find the number.

Answer

4 or $\dfrac{1}{4}$

EXAMPLE 4

An investor purchased a number of shares of stock for a total of $600. If the investor had paid $2 less per share, the number of shares would have increased by 10. How many shares were bought?

Solution

and

$$\text{Let} \quad n = \text{the number of shares purchased}$$
$$p = \text{the price paid per share}$$

Since

$$\begin{array}{ccc} \text{number of} \\ \text{shares} \end{array} \quad \times \quad \begin{array}{c} \text{price per} \\ \text{share} \end{array} \quad = \quad \begin{array}{c} \text{total dollars} \\ \text{invested} \end{array}$$

we must have

and

$$n \cdot p = 600$$
$$(n + 10)(p - 2) = 600$$

Substituting $p = 600/n$ we obtain

$$(n + 10)\left(\frac{600}{n} - 2\right) = 600$$

$$600 + \frac{6000}{n} - 2n - 20 = 600$$

$$
\begin{array}{ll}
6000 - 2n^2 - 20n = 0 & \text{Multiplying by } n \\
n^2 + 10n - 3000 = 0 & \text{Dividing by } -2 \\
(n - 50)(n + 60) = 0 & \\
n = 50 & \text{Reject } n = -60.
\end{array}
$$

The investor purchased 50 shares of stock.

PROGRESS CHECK 4

A business machine dealer purchased a number of used printing calculators at an auction for a total expenditure of $240. After giving one of the calculators to his daughter, he sold the remaining calculators at a profit of $15 each, for a profit of $35 on the entire transaction. How many printing calculators did he buy?

Answer
6 calculators

EXERCISE SET 9.4

1. Working together, computers A and B can complete a data processing job in 2 hours. Computer A working alone can do the job in 3 hours less than computer B working alone. How long does it take each computer to do the job by itself?
2. A graphics designer and her assistant working together complete an advertising layout in 6 days. The assistant working alone can complete the job in 16 more days than the designer working alone. How long does it take each person to do the job by herself?
3. A roofer and his assistant working together can finish a roofing job in 4 hours. The roofer working alone can finish the job in 6 hours less than the assistant working alone. How long does it take each person to do the job alone?
4. A mounting board 16 inches by 20 inches is used to mount a photograph. How wide is the uniform border if the photograph occupies $\frac{3}{5}$ of the area of the mounting board?
5. The length of a rectangle exceeds twice its width by 4 feet. If the area of the rectangle is 48 square feet, find the dimensions.
6. The length of a rectangle is 4 centimeters less than twice its width. Find the dimensions if the area of the rectangle is 96 square centimeters.
7. The area of a rectangle is 48 square centimeters. If the length and width are each increased by 4 centimeters, the area of the larger rectangle is 120 square centimeters. Find the dimensions of the original rectangle.
8. The base of a triangle is 2 feet more than twice its altitude. If the area is 12 square feet, find the base and altitude.
9. Find the width of a strip that has been mowed around a rectangular lawn which is 60 feet by 80 feet if one half of the lawn has not yet been mowed.
10. The sum of the reciprocals of two consecutive numbers is $\frac{7}{12}$. Find the numbers.
11. The sum of a number and its reciprocal is $\frac{26}{5}$. Find the number.
12. The difference of a number and its reciprocal is $\frac{35}{6}$. Find the number.
13. The smaller of two numbers is 4 less than the larger. If the sum of their squares is 58, find the numbers.

14. The sum of the reciprocals of two consecutive odd numbers is $\frac{8}{15}$. Find the numbers.

15. The sum of the reciprocals of two consecutive even numbers is $\frac{7}{24}$. Find the numbers.

16. A number of students rented a car for a one-week camping trip for $160. If another student had joined the original group, each person's share of expenses would have been reduced by $8. How many students were there in the original group?

17. An investor placed an order totaling $1200 for a certain number of shares of a stock. If the price of each share of stock were $2 more, the investor would get 30 shares less for the same amount of money. How many shares did the investor buy?

18. A fraternity charters a bus for a ski trip at a cost of $360. If 6 more students join the trip, each person's cost decreases by $2. How many students were in the original group of travelers?

19. A salesman worked a certain number of days to earn $192. If he had been paid $8 more per day he would have earned the same amount of money in two fewer days. How many days did he work?

20. A freelance photographer works a certain number of days for a newspaper to earn $480. If she had been paid $8 less per day, she would have earned the same amount in two more days. What was her daily rate of pay?

9.5
FORMS LEADING TO QUADRATICS

Certain types of equations which do not appear to be quadratic can be transformed into quadratic equations that can be solved by the methods discussed in this chapter. One such form that leads to a quadratic equation is the **radical equation.** To solve an equation such as

$$x = \sqrt{x + 12}$$

it seems natural to square both sides of the equation.

$$x^2 = x + 12$$

We now have a quadratic equation which is easily solved.

$$x^2 - x - 12 = 0$$

$$(x + 3)(x - 4) = 0$$

$$x = -3 \quad \text{or} \quad x = 4$$

Checking these solutions by substituting in the original equation, we have

$$-3 \overset{?}{=} \sqrt{-3 + 12} \qquad 4 \overset{?}{=} \sqrt{4 + 12}$$

$$-3 \overset{?}{=} \sqrt{9} \qquad 4 \overset{?}{=} \sqrt{16}$$

$$-3 \neq 3 \qquad 4 \overset{\checkmark}{=} 4$$

Thus, 4 is a solution and -3 is not a solution of the original equation. We say that -3 is an **extraneous solution,** which was introduced when we raised each side of the original equation to the second power. This is an illustration of the following general theorem.

The solution set of the equation

$$f(x) = g(x)$$

is a subset of the solution set of the equation

$$[f(x)]^n = [g(x)]^n$$

where n is a natural number.

This suggests that we can solve radical equations if we observe a precaution.

If both sides of an equation are raised to the same power, the solutions of the resulting equation must be checked to see that they satisfy the original equation.

Those answers which do not satisfy the original equation are called "extraneous."

EXAMPLE 1
Solve $x - \sqrt{x - 2} = 4$.

Solution
Isolate the radical on one side of the equation before solving.

$$x - 4 = \sqrt{x - 2}$$
$$x^2 - 8x + 16 = x - 2 \qquad \text{Square both sides.}$$
$$x^2 - 9x + 18 = 0$$
$$(x - 3)(x - 6) = 0$$
$$x = 3 \quad \text{or} \quad x = 6$$

Checking by substituting in the original equation we have

$$3 - \sqrt{3 - 2} \overset{?}{=} 4 \qquad 6 - \sqrt{6 - 2} \overset{?}{=} 4$$
$$3 - 1 \overset{?}{=} 4 \qquad 6 - \sqrt{4} \overset{?}{=} 4$$
$$2 \neq 4 \qquad 4 \overset{\checkmark}{=} 4$$

We conclude that 6 is a solution of the original equation, and 3 is rejected as an extraneous solution.

PROGRESS CHECK 1
Solve $x - \sqrt{1 - x} = -5$.

Answer
-3

EXAMPLE 2
Solve $\sqrt{2x - 4} - \sqrt{3x + 4} = -2$.

Solution

Before squaring, rewrite the equation so that a radical is on each side of the equation.

$$\sqrt{2x - 4} = \sqrt{3x + 4} - 2$$

$$2x - 4 = (3x + 4) - 4\sqrt{3x + 4} + 4 \qquad \text{Square both sides.}$$

$$-x - 12 = -4\sqrt{3x + 4} \qquad \text{Isolate the radical.}$$

$$x^2 + 24x + 144 = 16(3x + 4) \qquad \text{Square both sides.}$$

$$x^2 - 24x + 80 = 0$$

$$(x - 20)(x - 4) = 0$$

$$x = 20, \quad x = 4$$

Verify that both 20 and 4 are solutions of the original equation.

PROGRESS CHECK 2

Solve $\sqrt{5x - 1} - \sqrt{x + 2} = 1$.

Answer

2

Although the equation

$$x^4 - x^2 - 2 = 0$$

is not a quadratic in the variable x, it is a quadratic in the variable x^2.

$$(x^2)^2 - (x^2) - 2 = 0$$

This may be seen more clearly by replacing x^2 by a new variable $u = x^2$, which gives us

$$u^2 - u - 2 = 0$$

a quadratic equation in the variable u. Solving,

$$(u + 1)(u - 2) = 0$$

$$u = -1 \quad \text{or} \quad u = 2$$

Since $x^2 = u$, we must next solve the equations

$$x^2 = -1 \qquad x^2 = 2$$

$$x = \pm i \qquad x = \pm\sqrt{2}$$

The original equation has four solutions: i, $-i$, $\sqrt{2}$, and $-\sqrt{2}$.

The technique we have used is called a **substitution of variable.** Although simple in concept, this is a powerful method which is commonly used in calculus. We will apply this same technique to a variety of examples.

EXAMPLE 3

Indicate an appropriate substitution of variable that will lead to a quadratic equation.

(a) $2x^6 + 7x^3 - 8 = 0$

The substitution $u = x^3$ results in the quadratic equation $2u^2 + 7u - 8 = 0$.

(b) $y^{2/3} - 3y^{1/3} - 10 = 0$

The substitution $u = y^{1/3}$ results in the equation $u^2 - 3u - 10 = 0$.

(c) $\dfrac{6}{z^2} - \dfrac{11}{z} + 3 = 0$

Substituting $u = \dfrac{1}{z}$ we obtain $6u^2 - 11u + 3 = 0$.

(d) $\left(\dfrac{1}{x} - 1\right)^2 + 6\left(\dfrac{1}{x} - 1\right) - 7 = 0$

Substituting $u = \dfrac{1}{x} - 1$, we have $u^2 + 6u - 7 = 0$.

PROGRESS CHECK 3

Indicate an appropriate substitution of variable, and solve each of the following equations.

(a) $3x^4 - 10x^2 - 8 = 0$ (b) $4x^{2/3} + 7x^{1/3} - 2 = 0$

(c) $\dfrac{2}{x^2} + \dfrac{1}{x} - 10 = 0$ (d) $\left(1 + \dfrac{2}{x}\right)^2 - 8\left(1 + \dfrac{2}{x}\right) + 15 = 0$

Answers

(a) $u = x^2;$ $\pm 2, \pm \dfrac{i\sqrt{6}}{3}$ (b) $u = x^{1/3};$ $\dfrac{1}{64},$ -8

(c) $u = \dfrac{1}{x};$ $-\dfrac{2}{5}, \dfrac{1}{2}$ (d) $u = 1 + \dfrac{2}{x};$ $1, \dfrac{1}{2}$

EXERCISE SET 9.5

Solve for x.

1. $x + \sqrt{x + 5} = 7$ 2. $x - \sqrt{13 - x} = 1$
3. $x + \sqrt{2x - 3} = 3$ 4. $x - \sqrt{4 - 3x} = -8$
5. $2x + \sqrt{x + 1} = 8$ 6. $3x - \sqrt{1 + 3x} = 1$
7. $\sqrt{3x + 4} - \sqrt{2x + 1} = 1$ 8. $\sqrt{4x + 1} - \sqrt{x + 4} = 3$
9. $\sqrt{2x - 1} + \sqrt{x - 4} = 4$ 10. $\sqrt{5x + 1} + \sqrt{4x - 3} = 7$
11. $\sqrt{x + 3} + \sqrt{2x - 3} = 6$ 12. $\sqrt{x - 1} - \sqrt{3x - 2} = -1$
13. $\sqrt{8x + 20} - \sqrt{7x + 11} = 1$ 14. $\sqrt{6x + 12} - \sqrt{5x + 6} = 1$

In Exercises 15–24 indicate an appropriate substitution that will lead to a quadratic equation.

15. $3x^4 + 5x^2 - 5 = 0$ 16. $-8y^8 + 5y^6 + 4 = 0$

17. $3x^{4/3} + 5x^{2/3} + 3 = 0$ 18. $-3y^{6/5} + y^{3/5} - 8 = 0$

19. $\dfrac{5}{y^4} + \dfrac{2}{y^2} - 3 = 0$ 20. $\dfrac{2}{z^6} + \dfrac{5}{z^3} + 6 = 0$

21. $\dfrac{2}{x^{4/3}} + \dfrac{1}{x^{2/3}} + 4 = 0$ 22. $\dfrac{4}{x^{8/5}} - \dfrac{3}{x^{4/5}} + 2 = 0$

23. $\left(2 + \dfrac{3}{x}\right)^2 - 5\left(2 + \dfrac{3}{x}\right) - 5 = 0$ 24. $\left(1 - \dfrac{2}{x}\right)^4 + 3\left(1 - \dfrac{2}{x}\right)^2 - 8 = 0$

In Exercises 25–32 indicate an appropriate substitution of variable and solve each of the equations.

25. $3x^4 + 5x^2 - 2 = 0$

26. $2x^6 + 15x^3 - 8 = 0$

27. $\dfrac{6}{x^2} + \dfrac{1}{x} - 2 = 0$

28. $\dfrac{2}{x^4} - \dfrac{3}{x^2} - 9 = 0$

29. $2x^{2/5} + 5x^{1/5} + 2 = 0$

30. $3x^{4/3} - 4x^{2/3} - 4 = 0$

31. $2\left(\dfrac{1}{x} + 1\right)^2 - 3\left(\dfrac{1}{x} + 1\right) - 20 = 0$

32. $3\left(\dfrac{1}{x} - 2\right)^2 + 2\left(\dfrac{1}{x} - 2\right) - 1 = 0$

9.6
SECOND-DEGREE INEQUALITIES

Unfortunately, the quadratic formula can't provide solutions to second-degree inequalities such as

$$x^2 - 2x > 15$$

However, if we rewrite the inequality in the form

$$x^2 - 2x - 15 > 0$$

and factor

$$(x + 3)(x - 5) > 0$$

we are in a better position to analyze the possibilities. Under what circumstances will the product of the factor $(x + 3)$ and the factor $(x - 5)$ be positive? By the rules of algebra, a product of real numbers is positive only if both factors have the same sign.

Let's form a table of the values of the linear factor $(x + 3)$.

x	-50	-10	-5	-4	-3	-2	0	5	10	50
$(x + 3)$	-47	-7	-2	-1	0	1	3	8	13	53

Something interesting has happened at $x = -3$: The factor $(x + 3)$ is negative when $x < -3$ and positive when $x > -3$. Similarly, the factor $(x - 5)$ is negative when $x < 5$ and positive when $x > 5$. In general,

The linear factor $ax + b$ equals 0 at the **critical value** $x = -\dfrac{b}{a}$ and has opposite signs to the left and right of the critical value on a number line.

Graphing these results for $(x + 3)$ and $(x - 5)$ we have

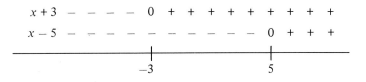

Recall that we want the values of x for which

$$(x + 3)(x - 5) > 0$$

that is, the values of x for which both factors have the same sign. From the graph we see that $(x + 3)$ and $(x - 5)$ have the same sign when $x < -3$ or $x > 5$. The solution of $x^2 - 2x > 15$ is the set

$$\{x \mid x < -3 \quad \text{or} \quad x > 5\}$$

EXAMPLE 1
Solve the inequality $x^2 \leq -3x + 4$ and then graph the solution set on a real number line.

Solution
We rewrite the inequality and factor.

$$x^2 \leq -3x + 4$$
$$x^2 + 3x - 4 \leq 0$$
$$(x - 1)(x + 4) \leq 0$$

The critical values are found by setting each factor equal to 0.

$$x - 1 = 0 \qquad x + 4 = 0$$
$$x = 1 \qquad x = -4$$

We mark the critical values on a real number line and analyze the *signs* of each factor to the left and to the right of each critical value.

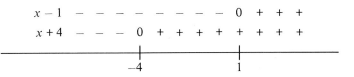

Since we are interested in values of x for which $(x - 1)(x + 4) \leq 0$, we seek values of x for which the factors $(x - 1)$ and $(x + 4)$ have opposite signs or are 0. From the graph, we see that when $-4 \leq x \leq 1$, the conditions are satisfied. Graphing the result we have

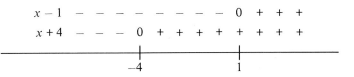

PROGRESS CHECK 1

Solve the inequality $2x^2 \geq 5x + 3$ and graph the solution set on a real number line.

Answer

$$\left\{ x \mid x \leq -\frac{1}{2} \text{ or } x \geq 3 \right\}$$

Although

$$\frac{ax + b}{cx + d} < 0$$

is not a second-degree inequality, the solution to this inequality is the same as the solution to the inequality $(ax + b)(cx + d) < 0$ since both expressions are negative (< 0) when $(ax + b)$ and $(cx + d)$ have opposite signs.

EXAMPLE 2

Solve the inequality $\dfrac{y + 1}{2 - y} \leq 0$.

Solution

The factors equal 0 at the critical value $y = -1$ and $y = 2$. Analyzing signs of $y + 1$ and $2 - y$ we have

$$
\begin{array}{lcccccccccccccccc}
y + 1 & - & - & - & - & 0 & + & + & + & + & + & + & + & + & + & + & + \\
2 - y & + & + & + & + & + & + & + & + & + & + & + & + & 0 & - & - & - \\
\end{array}
$$

$$\underset{-1 \qquad\qquad\qquad\qquad 2}{\rule{8cm}{0.4pt}}$$

Since

$$\frac{y + 1}{2 - y}$$

can be negative (< 0) only if the factors $(y + 1)$ and $(2 - y)$ have opposite signs, the solution set is $\{y \mid y \leq -1 \text{ or } y > 2\}$. Note that $y = 2$ would result in division by 0.

PROGRESS CHECK 2

Solve the inequality $\dfrac{2x - 3}{1 - 2x} \geq 0$.

Answer

$$\left\{ x \mid \frac{1}{2} < x \leq \frac{3}{2} \right\}$$

EXAMPLE 3

Solve the inequality $(x - 2)(2x + 5)(3 - x) < 0$.

Solution

Although this is a third-degree inequality, the same approach will work. The critical values are $x = 2$, $x = -\frac{5}{2}$, and $x = 3$. Graphing, we have

```
x - 2   – – – – – – – – – – – 0 + + + + + + + +
2x + 5  – – – – 0 + + + + + + + + + + + + +
3 - x   + + + + + + + + + + + + + 0 – – –
```

$$-\frac{5}{2} \qquad 2 \qquad 3$$

The product of three factors is negative when an odd number of factors are negative. The solution set is then $\{x \mid -\frac{5}{2} < x < 2 \text{ or } x > 3\}$.

PROGRESS CHECK 3
Solve the inequality $(2y - 9)(6 - y)(y + 5) \geq 0$.

Answer
$$\left\{ y \mid y \leq -5 \text{ or } \frac{9}{2} \leq y \leq 6 \right\}$$

EXERCISE SET 9.6
In the following inequalities select the values that are solutions.

1. $x^2 - 3x - 4 > 0$ (a) $x = 3$ (b) $x = 5$ (c) $x = 0$
   (d) $x = -2$ (e) $x = 6$
2. $x^2 - 7x + 12 \leq 0$ (a) $x = 3$ (b) $x = 3.5$ (c) $x = 5$
   (d) $x = 3.8$ (e) $x = 2$
3. $x^2 + 7x + 10 \geq 0$ (a) $x = -2$ (b) $x = -3$ (c) $x = 0$
   (d) $x = -6$ (e) $x = 4$
4. $2x^2 - 3x - 2 < 0$ (a) $x = 0$ (b) $x = -3$ (c) $x = 2$
   (d) $x = -1$ (e) $x = 3$
5. $2x^2 - x > 0$ (a) $x = 0$ (b) $x = \frac{1}{2}$ (c) $x = -1$ (d) $x = \frac{1}{4}$
   (e) $x = 2$
6. $3x^2 + x \leq 0$ (a) $x = 0$ (b) $x = 2$ (c) $x = -2$
   (d) $x = -4$ (e) $x = 1$

Find the critical values in each inequality.

7. $x^2 + 5x + 6 > 0$ 8. $x^2 + 3x + 4 \leq 0$

9. $2x^2 - x - 1 < 0$ 10. $3x^2 - 4x - 4 \geq 0$

11. $4x - 2x^2 < 0$ 12. $r^2 + 4r \geq 0$

13. $\dfrac{x + 5}{x + 3} < 0$ 14. $\dfrac{x - 6}{x + 4} \geq 0$

15. $\dfrac{2r + 1}{r - 3} \leq 0$ 16. $\dfrac{x - 1}{2x - 3} > 0$

17. $\dfrac{3s + 2}{2s - 1} > 0$ 18. $\dfrac{4x + 5}{x^2} < 0$

Solve and graph the solution set of each inequality in Exercises 19–46.

19. $x^2 + x - 6 > 0$ 20. $x^2 - 3x - 10 \geq 0$

21. $2x^2 - 3x - 5 < 0$ 22. $3x^2 - 4x - 4 \leq 0$

23. $2x^2 + 7x + 6 > 0$ 24. $2y^2 + 3y + 1 < 0$

25. $\dfrac{2r + 3}{2r - 1} < 0$ 26. $\dfrac{3x + 2}{2x - 3} \geq 0$

27. $\dfrac{x - 1}{x + 1} > 0$ 28. $\dfrac{2x - 1}{x + 2} \leq 0$

29. $6x^2 + 8x + 2 \geq 0$ 30. $2x^2 + 5x + 2 \leq 0$

31. $\dfrac{2x + 2}{x + 1} \geq 0$ 32. $\dfrac{3s + 1}{2s + 4} < 0$

33. $3x^2 - 5x + 2 \leq 0$ 34. $2x^2 - 9x + 10 > 0$

35. $\dfrac{5y - 2}{3y - 2} \leq 0$ 36. $\dfrac{4x - 3}{x - 3} > 0$

37. $x^2 - 2x + 1 > 0$ 38. $4r^2 - 4r + 1 < 0$

39. $\dfrac{2x + 1}{2x - 3} \geq 0$ 40. $\dfrac{4x - 2}{2x} < 0$

41. $(x + 2)(3x - 2)(x - 1) > 0$

42. $(x - 4)(2x + 5)(2 - x) \leq 0$

43. $(y - 3)(2 - y)(2y + 4) \geq 0$

44. $(2x + 5)(3x - 2)(x + 1) < 0$

45. $(x - 3)(1 + 2x)(3x + 5) > 0$

46. $(1 - 2x)(2x + 1)(x - 3) \leq 0$

47. A manufacturer of solar heaters finds that when x units are made and sold per week, its profit (in thousands of dollars) is given by $x^2 - 50x - 5000$.
    (a) At least how many units must be manufactured and sold each week to make a profit?
    (b) For what values of x is the firm losing money?

48. Repeat Exercise 47 if the profit is given by $x^2 - 180x - 4000$.

TERMS AND SYMBOLS

quadratic function
  (p. 249)
zeros of a function
  (p. 249)
second-degree equation in
  one variable (p. 249)
quadratic equation
  (p. 249)

solving by factoring
  (p. 251)
completing the square
  (p. 252)
quadratic formula (p. 256)
discriminant (p. 259)
double root (p. 259)
repeated root (p. 259)

radical equation (p. 266)
extraneous solution
  (p. 266)
substitution of variable
  (p. 268)
second-degree inequality
  (p. 270)
critical value (p. 270)

KEY IDEAS FOR REVIEW

☐ A quadratic equation of the form $ax^2 + c = 0$ has solutions $x = \pm\sqrt{-\dfrac{c}{a}}$.

☐ If the product of two real numbers is 0, at least one of the numbers must be 0. Thus, $ab = 0$ if and only if $a = 0$ or $b = 0$, or both $a = 0$ and $b = 0$.

☐ If a quadratic equation can be written as a product of linear factors

$$(rx + s)(ux + v) = 0$$

then

$$x = -\frac{s}{r} \quad \text{and} \quad x = -\frac{v}{u}$$

are the roots of the quadratic.

☐ The quadratic formula

$$x = \frac{-b \pm \sqrt{b^2 - 4ac}}{2a}$$

provides us with a pair of solutions to the quadratic equation $ax^2 + bx + c = 0$.

☐ The expression $b^2 - 4ac$ found under the radical in the quadratic formula is called the discriminant and determines the nature of the roots of the quadratic equation.

☐ If both sides of an equation are raised to the same power, then the resulting equation may have extraneous roots which are not solutions of the original equation.

☐ Certain forms, such as radical equations, can be solved by raising both sides of the equation to a power. The solutions of the resulting equation must be checked to see that they satisfy the original equation.

☐ The method of substitution of variable can be used to convert certain forms to quadratic equations.

☐ A product of linear factors is negative only if an odd number of factors is negative. A product of linear factors is positive if either none or an even number of factors is negative.

☐ The linear factor $ax + b$ equals 0 at the critical value

$$x = -\frac{b}{a}$$

and has opposite signs to the left and right of the critical value on a number line.

☐ To solve an inequality involving products and quotients of linear factors for which the right-hand side is 0, analyze the signs to the left and right of each critical value. The product and quotient of linear factors can be negative (< 0) only if an odd number of factors is negative; they can be positive (> 0) only if an even number of factors is negative.

☐ To solve a second-degree inequality, write it as a product of linear factors with the right-hand side equal to 0.

COMMON ERRORS

1. The equation $2x^2 = 10$ has as its solutions $x = \pm\sqrt{5}$. Remember to write $\pm$ before the radical when solving a quadratic.

2. The quadratic equation $3x^2 + x = 0$ can be factored as

$$x(3x + 1) = 0$$

The solutions are then $x = 0$ and $x = -\frac{1}{3}$. Remember that each linear factor yields a root; in particular, the factor x yields the root $x = 0$.

3. When solving by factoring make sure that one side of the equation is zero. Note that

$$(x + 1)(x - 2) = 2$$

does *not* imply that $x + 1 = 2$ or $x - 2 = 2$.

4. To complete the square in

$$4(x^2 + 6x \quad) = 7$$

we must add $(\frac{6}{2})^2 = 3^2 = 9$ within the parentheses. To balance the equation, we must add $4 \cdot 9 = 36$ to the right-hand side.

$$4(x^2 + 6x + 9) = 7 + 36$$

Don't write

$$4(x^2 + 6x + 9) = 7 + 9$$

5. The quadratic formula is

$$x = \frac{-b \pm \sqrt{b^2 - 4ac}}{2a}$$

Don't use the formula as

$$x = -b \pm \frac{\sqrt{b^2 - 4ac}}{2a}$$

6. Proper use of the quadratic formula requires that the equation be written in the form $ax^2 + bx + c = 0$. Remember that a, b, and c are coefficients and therefore include the sign.

7. To solve the inequality

$$x^2 + 2x \geq 3$$

you must make one side of the equation zero

$$x^2 + 2x - 3 \geq 0$$

and then factor. *Don't* write

$$x(x + 2) \geq 3$$

and then attempt to analyze the signs. The inequality does *not* imply that $x \geq 3$ or $x + 2 \geq 3$.

8. When solving

$$(x + 3)(2x - 1) < 0$$

don't write

$$(x + 3) < 0 \quad \text{or} \quad 2x - 1 < 0$$

You must analyze the linear factors $(x + 3)$ and $(2x - 1)$ and find the values of x for which the factors have opposite signs.

9. If both sides of an equation are raised to a power, some of the solutions of the resulting equation may be extraneous, that is, they may not satisfy the original equation. Always substitute all answers in the original equation to see if the answers are or are not solutions.

PROGRESS TEST 9A

1. Solve $3x^2 + 7 = 0$.
2. Solve $x - \sqrt{12 - 2x} = 2$.
3. Solve $(2x - 3)^2 = 16$.
4. Solve $\left(\frac{x}{2} - 1\right)^2 = -15$.

5. Solve $2x^2 - 7x = 4$ by factoring.
6. Solve $3x^2 - 5x - 2 = 0$ by factoring.
7. Solve $3x^2 - 6x = 8$ by completing the square.
8. Use the discriminant to determine the nature of the roots of the equation $3x^2 - 2x - 5 = 0$.
9. Use the discriminant to determine the nature of the roots of the equation $2x^2 - 4x + 7 = 0$.

10. Solve $2x^2 + 3x - \dfrac{1}{2} = 0$ by the quadratic formula.

11. Solve $3x^2 + 2x = 0$ by the quadratic formula.
12. Solve $2x^2 + 5x \leq 3$.

13. Solve $\dfrac{x + 1}{x - 1} \leq 0$.

14. The length of a rectangle is 5 meters greater than its width. If the area of the rectangle is 546 square meters, find the dimensions of the rectangle.
15. A faster assembly line can fill an order in 6 fewer hours than it takes a slower assembly line to fill the same order. Working together, they can fill the order in 3 hours. How long would it take each assembly line to fill the order by itself?

PROGRESS TEST 9B

1. Solve $4x^2 - 9 = 0$.
2. Solve $x - \sqrt{11 - 2x} = 4$.

3. Solve $\left(\dfrac{x}{3} - 2\right)^2 = 25$.

4. Solve $\left(3x + \dfrac{1}{2}\right)^2 = 10$.

5. Solve $x^2 - x - 12 = 0$ by factoring.
6. Solve $2x^2 + 3x + 1 = 0$ by factoring.
7. Solve $2x^2 - 6x + 5 = 0$ by completing the square.
8. Use the discriminant to determine the nature of the roots of the equation $4x^2 - 12x + 9 = 0$.
9. Use the discriminant to determine the nature of the roots of the equation $3x^2 - 4x + 3 = 0$.
10. Solve $2x^2 - 5x + 4 = 0$ by the quadratic formula.
11. Solve $3x^2 + 5 = 0$ by the quadratic formula.
12. Solve $2x^2 + x \geq 10$.

13. Solve $\dfrac{2x - 1}{2x + 1} \geq 0$.

14. One leg of a right triangle exceeds the other by 3 meters. If the hypotenuse is 15 meters long, find the legs of the triangle.
15. The formula

$$s = \frac{n(n + 1)}{2}$$

gives the sum of the first n natural numbers $1, 2, 3, \ldots$. How many consecutive numbers must be added to obtain a sum of 325?

CHAPTER TEN
ROOTS OF POLYNOMIALS

In Section 6.3 we observed that the polynomial function

$$f(x) = ax + b \tag{1}$$

is called a linear function, and the polynomial function

$$g(x) = ax^2 + bx + c, \quad a \neq 0 \tag{2}$$

is called a quadratic function. To facilitate the study of polynomial functions in general, we now introduce the notation

$$P(x) = a_n x^n + a_{n-1} x^{n-1} + \cdots + a_1 x + a_0, \quad a_n \neq 0 \tag{3}$$

to represent a **polynomial function of degree _n_.** Note that the coefficients a_k may be real or complex numbers and that the subscript k of the coefficient a_k is the same as the exponent of x in x^k.

If $a \neq 0$ in Equation (1), we set the polynomial function equal to zero and obtain the linear equation

$$ax + b = 0$$

which has precisely one solution, $-b/a$. If we set the polynomial function in Equation (2) equal to zero, we have the quadratic equation

$$ax^2 + bx + c = 0$$

that has the two solutions given by the quadratic formula. If we set the polynomial function in Equation (3) equal to zero we have the **polynomial equation of degree n**

$$a_n x^n + a_{n-1} x^{n-1} + \cdots + a_1 x + a_0 = 0 \qquad (4)$$

Our attention in this chapter will turn to finding the roots or solutions of Equation (4). These solutions are also known as the **roots** or **zeros of the polynomial.** We will attempt to answer the following questions for a polynomial equation of degree n.

How many roots does a polynomial have in the field of complex numbers?

How many of the roots of a polynomial are real numbers?

If the coefficients of a polynomial are integers, how many of the roots are rational numbers?

Is there a relationship between the roots and factors of a polynomial?

These questions have attracted the attention of mathematicians since the sixteenth century. A method for finding the roots of polynomial equations of degree 3 was published around 1535 and is known as Cardan's formula despite the fact that Girolamo Cardano stole the result from his friend, Nicolo Tartaglia. Shortly afterward a method that is attributed to Ferrari was published for solving polynomial equations of degree 4.

The search for formulas giving the roots of a polynomial of degree 5 or higher, in terms of its coefficients, continued into the nineteenth century. At that time the Norwegian mathematician N. H. Abel and the French mathematician Evariste Galois proved that no such formulas are possible. Galois' work on this problem was completed a year before his death in a duel at age 21. His proof made use of the new concepts of group theory. The work was so advanced that his teachers wrote it off as being unintelligible gibberish.

10.1
POLYNOMIAL DIVISION AND
SYNTHETIC DIVISION

To find the roots of a polynomial, it will be necessary to divide the polynomial by a second polynomial. There is a procedure for polynomial division that parallels the long division process of arithmetic. In arithmetic, if we divide a real number p by the real number $d \neq 0$, we obtain a quotient q and a remainder r so that we can write

$$\frac{p}{d} = q + \frac{r}{d} \qquad (1)$$

where

$$r < d \qquad (2)$$

This result can also be written in the form

$$p = qd + r, \quad r < d$$

For example,

$$\frac{7284}{13} = 560 + \frac{4}{13}$$

or

$$7284 = (560)(13) + 4$$

In the long division process for polynomials, we divide the dividend $P(x)$ by the Divisor $D(x) \neq 0$ to obtain a quotient $Q(x)$ and a remainder $R(x)$. We then have

$$\frac{P(x)}{D(x)} = Q(x) + \frac{R(x)}{D(x)} \tag{3}$$

where

$$\text{degree of } R(x) < \text{degree of } D(x) \tag{4}$$

Note that Equations (1) and (3) have the same form. Equation (2) requires that the remainder be less than the divisor, and the parallel requirement for polynomials in Equation (4) is that the *degree* of the remainder be less than that of the divisor.

We illustrate the long division process for polynomials by an example.

EXAMPLE 1
Divide $3x^3 - 7x^2 + 1$ by $x - 2$.

Solution

Polynomial Division	
Step 1. Arrange the terms of both polynomials by descending powers of x. If a power is missing, write the term with a zero coefficient.	$x - 2\overline{)3x^3 - 7x^2 + 0x + 1}$
Step 2. Divide the first term of the dividend by the first term of the divisor. The answer is written above the first term of the dividend.	$\dfrac{3x^2}{x - 2\overline{)3x^3 - 7x^2 + 0x + 1}}$
Step 3. Multiply the divisor by the quotient obtained in Step 2 and then subtract the product.	$\begin{array}{r} 3x^2 \\ \hline x - 2\overline{)3x^3 - 7x^2 + 0x + 1} \\ 3x^3 - 6x^2 \\ \hline -x^2 + 0x + 1 \end{array}$
Step 4. Repeat Steps 2 and 3 until the degree of the remainder is less than the degree of the divisor.	$\begin{array}{r} 3x^2 - x - 2 \\ \hline x - 2\overline{)3x^3 - 7x^2 + 0x + 1} \\ 3x^3 - 6x^2 \\ \hline -x^2 + 0x + 1 \\ -x^2 + 2x \\ \hline -2x + 1 \\ -2x + 4 \\ \hline -3 \end{array}$
Step 5. Write the answer in the form of Equation (3).	$\dfrac{3x^3 - 7x^2 + 0x + 1}{x - 2}$ $= 3x^2 - x - 2 - \dfrac{3}{x - 2}$

PROGRESS CHECK 1

Divide $4x^2 - 3x + 6$ by $x + 2$.

Answer

$4x - 11 + \dfrac{28}{x + 2}$

Our work in this chapter will frequently require division of a polynomial by a first-degree polynomial $x - r$ where r is a constant. Fortunately, there is a shortcut called **synthetic division** that simplifies this task. To demonstrate synthetic division we will do Example 1 again, writing only the coefficients.

$$
\begin{array}{r}
\mathbf{3} \quad \mathbf{-1} \quad \mathbf{-2} \\
-2 \; \overline{)3 \quad -7 \quad\; 0 \quad\; 1} \\
\mathbf{3} \quad -6 \\
\hline
\mathbf{-1} \quad\; 0 \quad\; 1 \\
-1 \quad\; 2 \\
\hline
\mathbf{-2} \quad\; 1 \\
-2 \quad\; 4 \\
\hline
-3
\end{array}
$$

Note that the boldface numerals are duplicated. We can use this to our advantage and simplify the process as follows.

$$
\begin{array}{r}
\underline{-2}\; \big|\; 3 \quad -7 \quad\; 0 \quad\; 1 \\
-6 \quad\; 2 \quad\; 4 \\
\hline
3 \quad -1 \quad -2 \quad -3
\end{array}
$$

$\underbrace{}$ $\underset{\uparrow}{}$

coefficients   remainder
of the
quotient

We copied the leading coefficient (3) of the dividend in the third row, multiplied it by the divisor (-2), and wrote the result (-6) in the second row under the next coefficient. The numbers in the second column were subtracted to obtain $-7 - (-6) = -1$. The procedure is repeated until the third row is of the same length as the first row.

Since subtraction is more apt to produce errors than is addition, we can modify this process slightly. If the divisor is $x - r$, we will write r instead of $-r$ in the box and use addition in each step instead of subtraction. Repeating our example, we have

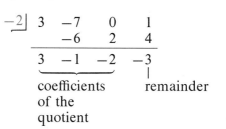

$$
\begin{array}{r}
\underline{2}\; \big|\; 3 \quad -7 \quad\; 0 \quad\; 1 \\
6 \quad -2 \quad -4 \\
\hline
3 \quad -1 \quad -2 \quad -3
\end{array}
$$

EXAMPLE 2

Divide $4x^3 - 2x + 5$ by $x + 2$ using synthetic division.

Solution

Synthetic Division

Step 1. If the divisor is $x - r$, write r in the box. Arrange the coefficients of the dividend by descending power of x, supplying a zero coefficient for every missing power.	*Step 1.* $\underline{-2}$ 4 0 -2 5
Step 2. Copy the leading coefficient in the third row.	*Step 2.* $\underline{-2}$ 4 0 -2 5 4
Step 3. Multiply the last entry in the third row by the number in the box and write the result in the second row under the next coefficient. Add the numbers in that column.	*Step 3.* $\underline{-2}$ 4 0 -2 5 -8 4 -8
Step 4. Repeat Step 3 until there is an entry in the third row for each entry in the first row. The last number in the third row is the remainder; the other numbers are the coefficients of the quotient.	*Step 4.* $\underline{-2}$ 4 0 -2 5 -8 16 -28 4 -8 14 -23 $$\frac{4x^3 - 2x + 5}{x + 2}$$ $$= 4x^2 - 8x + 14 - \frac{23}{x + 2}$$

PROGRESS CHECK 2

Use synthetic division to obtain the quotient $Q(x)$ and the constant remainder R when $2x^4 - 10x^2 - 23x + 6$ is divided by $x - 3$.

Answer

$Q(x) = 2x^3 + 6x^2 + 8x + 1;\ R = 9$

WARNING

In synthetic division, when dividing by $x - r$ we place r in the box. Thus, when the divisor is $x + 3$, we place -3 in the box since $x + 3 = x - (-3)$. Similarly, when the divisor is $x - 3$, we place $+3$ in the box since $x - 3 = x - (+3)$.

EXERCISE SET 10.1

In Exercises 1–10 use polynomial division to find the quotient $Q(x)$ and the remainder $R(x)$ when the first polynomial is divided by the second polynomial.

1. $x^2 - 7x + 12,\quad x - 5$
2. $x^2 + 3x + 3,\quad x + 2$
3. $2x^3 - 2x,\quad x^2 + 2x - 1$
4. $3x^3 - 2x^2 + 4,\quad x^2 - 2$
5. $3x^4 - 2x^2 + 1,\quad x + 3$
6. $x^5 - 1,\quad x^2 - 1$
7. $2x^3 - 3x^2,\quad x^2 + 2$
8. $3x^3 - 2x - 1,\quad x^2 - x$
9. $x^4 - x^3 + 2x^2 - x + 1,\quad x^2 + 1$
10. $2x^4 - 3x^3 - x^2 - x - 2,\quad x - 2$

In Exercises 11–20 use synthetic division to find the quotient $Q(x)$ and the constant remainder R when the first polynomial is divided by the second polynomial.

11. $x^3 - x^2 - 6x + 5,\quad x + 2$
12. $2x^3 - 3x^2 - 4,\quad x - 2$

13. $x^4 - 81$, $x - 3$ 14. $x^4 - 81$, $x + 3$

15. $3x^3 - x^2 + 8$, $x + 1$ 16. $2x^4 - 3x^3 - 4x - 2$, $x - 1$

17. $x^5 + 32$, $x + 2$ 18. $x^5 + 32$, $x - 2$

19. $6x^4 - x^2 + 4$, $x - 3$ 20. $8x^3 + 4x^2 - x - 5$, $x + 3$

10.2
THE REMAINDER AND FACTOR THEOREMS

From our work with the division process in the previous section, we may surmise that division of a polynomial $P(x)$ by $x - r$ results in a quotient $Q(x)$ and a constant remainder R such that

$$P(x) = (x - r) \cdot Q(x) + R$$

Since this identity holds for all real values of x, it must hold when $x = r$. Consequently,

$$P(r) = (r - r) \cdot Q(r) + R$$
$$P(r) = 0 \cdot Q(r) + R$$

or

$$P(r) = R$$

We have proved the Remainder Theorem.

Remainder Theorem

If a polynomial $P(x)$ is divided by $x - r$, then the remainder is $P(r)$.

EXAMPLE 1

Determine the remainder when $P(x) = 2x^3 - 3x^2 - 2x + 1$ is divided by $x - 3$.

Solution

By the Remainder Theorem, the remainder is $R = P(3)$. We then have

$$R = P(3) = 2(3)^3 - 3(3)^2 - 2(3) + 1 = 22$$

We may verify this result by using synthetic division.

$$
\begin{array}{r|rrrr}
3 & 2 & -3 & -2 & 1 \\
  &   & 6 & 9 & 21 \\
\hline
  & 2 & 3 & 7 & \mathbf{22}
\end{array}
$$

The numeral in boldface is the remainder, so we have verified that $R = 22$.

PROGRESS CHECK 1

Determine the remainder when $3x^2 - 2x - 6$ is divided by $x + 2$.

Answer
10

The Remainder Theorem can be used to tabulate values from which we can sketch the graph of a function. The most efficient scheme for performing the calculations is a streamlined form of synthetic division in which the addition is performed without writing the middle row. Repeating Example 1 in this condensed form we have

$$\begin{array}{r|rrrr} & 2 & -3 & -2 & 1 \\ 3 & 2 & 3 & 7 & 22 \end{array}$$

Then the point $(3, 22)$ lies on the graph of $y = 2x^3 - 3x^2 - 2x + 1$. In general, we may choose a value a of the independent variable and use synthetic division to find the remainder $P(a)$. Then $(a, P(a))$ is a point on the graph of $P(x)$.

EXAMPLE 2

Sketch the graph of $P(x) = 2x^3 - 3x^2 - 2x + 1$ for $-3 \le x \le 3$.

Solution

To sketch the graph of $y = P(x)$, we will allow x to assume integer values from -3 to $+3$. The remainder is found by using the condensed form of synthetic division and is the y-coordinate corresponding to the chosen value of x.

	2	-3	-2	1	(x, y)
-3	2	-9	25	-74	$(-3, -74)$
-2	2	-7	12	-23	$(-2, -23)$
-1	2	-5	3	-2	$(-1, -2)$
0	2	-3	-2	1	$(0, 1)$
1	2	-1	-3	-2	$(1, -2)$
2	2	1	0	1	$(2, 1)$
3	2	3	7	22	$(3, 22)$

The ordered pairs shown at the right of each row are the coordinates of points on the graph shown in Figure 1a.

PROGRESS CHECK 2

Sketch the graph of $P(x) = x^4 + 2x^3 - 6x - 9$ for $-3 \le x \le 3$.

Answer
See Figure 1b.

Let's assume that a polynomial $P(x)$ can be written as a product of polynomials, that is,

$$P(x) = D_1(x)D_2(x) \ldots D_n(x)$$

where $D_i(x)$ is a polynomial of degree greater than zero. Then $D_i(x)$ is called a **factor** of $P(x)$. If we focus on $D_1(x)$ and let

$$Q(x) = D_2(x)D_3(x) \ldots D_n(x)$$

then

$$P(x) = D_1(x)Q(x)$$

which demonstrates the following rule.

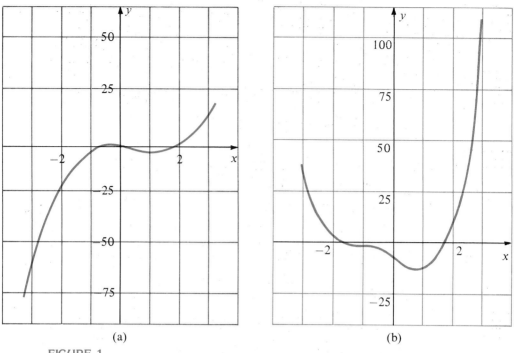

(a)   (b)

FIGURE 1

The polynomial $D(x)$ is a factor of a polynomial $P(x)$ if and only if the division of $P(x)$ by $D(x)$ results in a remainder of zero.

We can now combine this rule and the Remainder Theorem to prove the Factor Theorem.

Factor Theorem

A polynomial $P(x)$ has a factor $x - r$ if and only if

$$P(r) = 0.$$

If $x - r$ is a factor of $P(x)$ then division of $P(x)$ by $x - r$ must result in a remainder of 0. By the Remainder Theorem, the remainder is $P(r)$, and hence $P(r) = 0$. Conversely, if $P(r) = 0$, then the remainder is 0 and $P(x) = (x - r)Q(x)$ for some polynomial $Q(x)$ of degree one less than that of $P(x)$. By definition, $x - r$ is then a factor of $P(x)$.

EXAMPLE 3
Show that $x + 2$ is a factor of

$$P(x) = x^3 - x^2 - 2x + 8$$

Solution
By the Factor theorem, $x + 2$ is a factor if $P(-2) = 0$. Using synthetic division to evaluate $P(-2)$,

$$
\begin{array}{r|rrrr}
-2 & 1 & -1 & -2 & 8 \\
   &   & -2 & 6 & -8 \\
\hline
   & 1 & -3 & 4 & 0
\end{array}
$$

we see that $P(-2) = 0$. Thus, $x + 2$ is a factor of $P(x)$.

PROGRESS CHECK 3
Show that $x - 1$ is a factor of $P(x) = 3x^6 - 3x^5 - 4x^4 + 6x^3 - 2x^2 - x + 1$.

EXERCISE SET 10.2
In Exercises 1–6 use the Remainder Theorem and synthetic division to find $P(r)$.

1. $P(x) = x^3 - 4x^2 + 1$, $r = 2$
2. $P(x) = x^4 - 3x^2 - 5x$, $r = -1$
3. $P(x) = x^5 - 2$, $r = -2$
4. $P(x) = 2x^4 - 3x^3 + 6$, $r = 2$
5. $P(x) = x^6 - 3x^4 + 2x^3 + 4$, $r = -1$
6. $P(x) = x^6 - 2$, $r = 1$

In Exercises 7–12 use the Remainder Theorem to determine the remainder when $P(x)$ is divided by $x - r$.

7. $P(x) = x^3 - 2x^2 + x - 3$, $x - 2$
8. $P(x) = 2x^3 + x^2 - 5$, $x + 2$
9. $P(x) = -4x^3 + 6x - 2$, $x - 1$
10. $P(x) = 6x^5 - 3x^4 + 2x^2 + 7$, $x + 1$
11. $P(x) = x^5 - 30$, $x + 2$
12. $P(x) = x^4 - 16$, $x - 2$

In Exercises 13–18 use the Remainder Theorem and synthetic division to sketch the graph of the given polynomial for $-3 \le x \le 3$.

13. $P(x) = x^3 + x^2 + x + 1$
14. $P(x) = 3x^4 + 5x^3 + x^2 + 5x - 2$
15. $P(x) = 2x^3 + 3x^2 - 5x - 6$
16. $P(x) = x^3 + 3x^2 - 4x - 12$
17. $P(x) = x^4 - 10x^3 + 1$
18. $P(x) = 4x^4 + 4x^3 - 9x^2 - x + 2$

In Exercises 19–26 use the Factor Theorem to decide whether or not the first polynomial is a factor of the second polynomial.

19. $x - 2$, $x^3 - x^2 - 5x + 6$
20. $x - 1$, $x^3 + 4x^2 - 3x + 1$
21. $x + 2$, $x^4 - 3x - 5$
22. $x + 1$, $2x^3 - 3x^2 + x + 6$
23. $x + 3$, $x^3 + 27$
24. $x + 2$, $x^4 + 16$
25. $x + 2$, $x^4 - 16$
26. $x - 3$, $x^3 + 27$

In Exercises 27–30, use synthetic division to determine the value of k or r as requested.

27. Determine the values of r for which division of $x^2 - 2x - 1$ by $x - r$ has a remainder of 2.

28. Determine the values of r so that

$$
\frac{x^2 - 6x - 1}{x - r}
$$

has a remainder of -9.

29. Determine the values of k for which $x - 2$ is a factor of $x^3 - 3x^2 + kx - 1$.

30. Determine the values of k for which $2k^2x^3 + 3kx^2 - 2$ is divisible by $x - 1$.

31.   Use the Factor Theorem to show that $x - 2$ is a factor of $P(x) = x^8 - 256$.

32.   Use the Factor Theorem to show that $P(x) = 2x^4 + 3x^2 + 2$ has no factor of the form $x - r$, where r is a real number.

33.   Use the Factor Theorem to show that $x - y$ is a factor of $x^n - y^n$, where n is a natural number.

10.3
FACTORS AND ROOTS

In Chapter Nine we saw that the roots of quadratic equations may be complex numbers. Since complex numbers play a key role in providing solutions of polynomial equations, we will now review the material of Section 8.6 and explore further properties of this number system.

Recall that the complex number $a - bi$ is called the **complex conjugate** (or simply the **conjugate**) of the complex number $a + bi$. It is easy to verify that the product of a complex number and its conjugate is a real number, that is,

$$(a + bi)(a - bi) = a^2 + b^2$$

We have also seen that the quotient of two complex numbers

$$\frac{q + ri}{s + ti}$$

can be written in the form $a + bi$ by multiplying both numerator and denominator by $s - ti$, the conjugate of the denominator. Of course, the reciprocal of the complex number $s + ti$ is the quotient $1/(s + ti)$, which can also be written as a complex number by using the same technique.

EXAMPLE 1
Write the reciprocal of $2 - 5i$ in the form $a + bi$.

Solution
The reciprocal is $1/(2 - 5i)$. Multiplying both numerator and denominator by the conjugate $2 + 5i$, we have

$$\frac{1}{2 - 5i} \cdot \frac{2 + 5i}{2 + 5i} = \frac{2 + 5i}{2^2 + 5^2} = \frac{2 + 5i}{29} = \frac{2}{29} + \frac{5}{29}i$$

Verify that $(2 - 5i)\left(\frac{2}{29} + \frac{5}{29}i\right) = 1$.

PROGRESS CHECK 1
Write the following in the form $a + bi$.

(a) $\dfrac{4 - 2i}{5 + 2i}$ (b) $\dfrac{1}{2 - 3i}$ (c) $\dfrac{-3i}{3 + 5i}$

Answers

(a) $\dfrac{16}{29} - \dfrac{18}{29}i$ (b) $\dfrac{2}{13} + \dfrac{3}{13}i$ (c) $-\dfrac{15}{34} - \dfrac{9}{34}i$

If we let $z = a + bi$, it is customary to write the conjugate $a - bi$ as $\bar{z}$. We will have need to use the following properties of complex numbers and their conjugates.

If z and w are complex numbers, then

(1) $\bar{z} = \bar{w}$ if and only if $z = w$.
(2) $\bar{z} = z$ if and only if z is a real number.
(3) $\overline{z + w} = \bar{z} + \bar{w}$
(4) $\overline{z \cdot w} = \bar{z} \cdot \bar{w}$
(5) $\overline{z^n} = \bar{z}^n$, n a positive integer

To prove properties (1)–(5), let $z = a + bi$ and $w = c + di$. Properties (1) and (2) follow directly from the definition of equality of complex numbers. To prove property (3), we note that $z + w = (a + c) + (b + d)i$. Then, by the definition of a complex conjugate,

$$\overline{z + w} = (a + c) - (b + d)i$$
$$= (a - bi) + (c - di)$$
$$= \bar{z} + \bar{w}$$

Properties (4) and (5) can be proved in a similar manner, although a rigorous proof of property (5) requires the use of mathematical induction.

EXAMPLE 2

If $z = 1 + 2i$ and $w = 3 - i$, verify that

(a) $\overline{z + w} = \bar{z} + \bar{w}$ (b) $\overline{z \cdot w} = \bar{z} \cdot \bar{w}$ (c) $\overline{z^2} = \bar{z}^2$

Solution

(a) Adding, we get $z + w = 4 + i$. Therefore $\overline{z + w} = 4 - i$. Also,

$$\bar{z} + \bar{w} = (1 - 2i) + (3 + i) = 4 - i$$

Thus, $\overline{z + w} = \bar{z} + \bar{w}$.

(b) Multiplying, we get $z \cdot w = (1 + 2i)(3 - i) = 5 + 5i$. Therefore $\overline{z \cdot w} = 5 - 5i$. Also,

$$\bar{z} \cdot \bar{w} = (1 - 2i)(3 + i) = 5 - 5i$$

Thus, $\overline{z \cdot w} = \bar{z} \cdot \bar{w}$.

(c) Squaring, we get

$$z^2 = (1 + 2i)(1 + 2i) = -3 + 4i$$

Therefore $\overline{z^2} = -3 - 4i$.

Also,

$$\bar{z}^2 = (1 - 2i)(1 - 2i) = -3 - 4i$$

Thus, $\overline{z^2} = \bar{z}^2$.

PROGRESS CHECK 2

If $z = 2 + 3i$ and $w = \frac{1}{2} - 2i$, verify that

(a) $\overline{z + w} = \overline{z} + \overline{w}$ (b) $\overline{z \cdot w} = \overline{z} \cdot \overline{w}$ (c) $\overline{z^2} = \overline{z}^2$ (d) $\overline{w^3} = \overline{w}^3$

We are now in a position to answer some of the questions posed in the introduction to this chapter. By using the Factor Theorem, we can show that there is a close relationship between the factors and the roots of the polynomial $P(x)$. By definition, r is a root of $P(x)$ if and only if $P(r) = 0$. But the Factor Theorem tells us that $P(r) = 0$ if and only if $x - r$ is a factor of $P(x)$. This leads to the following alternate statement of the Factor Theorem.

Factor Theorem

A polynomial $P(x)$ has a root r if and only if $x - r$ is a factor of $P(x)$.

EXAMPLE 3

Find a polynomial $P(x)$ of degree 3 whose roots are -1, 1, and -2.

Solution

By the Factor Theorem, $x + 1$, $x - 1$, and $x + 2$ are factors of $P(x)$. The product

$$P(x) = (x + 1)(x - 1)(x + 2) = x^3 + 2x^2 - x - 2$$

is a polynomial of degree 3 with the desired roots. Note that multiplying $P(x)$ by any nonzero real number results in another polynomial that has the same roots. For example, the polynomial

$$R(x) = 5x^3 + 10x^2 - 5x - 10$$

also has -1, 1, and -2 as its roots. Thus, the answer is not unique.

PROGRESS CHECK 3

Find a polynomial $P(x)$ of degree 3 whose roots are 2, 4, and -3.

Answer
$x^3 - 3x^2 - 10x + 24$

Does a polynomial always have a root? The answer was supplied by Carl Friedrich Gauss in 1799. The proof of his theorem, however, is beyond the scope of this book.

The Fundamental Theorem of Algebra

Every polynomial $P(x)$ of degree $n \geq 1$ has at least one complex root.

Gauss, who is considered by many to have been the greatest mathematician of all time, supplied the proof at age 22. The importance of the theorem is reflected in its title. We now see why it was necessary to create the complex numbers and that we need not create any other number system beyond the complex numbers in order to solve polynomial equations.

How many roots does a polynomial of degree n have? The next theorem will bring us closer to an answer.

Linear Factor Theorem

A polynomial $P(x)$ of degree $n \geq 1$ can be written as the product of n linear factors.

$$P(x) = a(x - r_1)(x - r_2) \cdots (x - r_n)$$

Note that a is the leading coefficient of $P(x)$ and that $r_1, r_2, \ldots, r_n$ are, in general, complex numbers.

To prove this theorem, we first note that the Fundamental Theorem of Algebra guarantees us the existence of a root r_1. By the Factor Theorem, $x - r_1$ is a factor and consequently

$$P(x) = (x - r_1)Q_1(x) \tag{1}$$

where $Q_1(x)$ is a polynomial of degree $n - 1$. If $n - 1 \geq 1$, then $Q_1(x)$ must have a root r_2. Thus

$$Q_1(x) = (x - r_2)Q_2(x) \tag{2}$$

where $Q_2(x)$ is of degree $n - 2$. Substituting in Equation (1) for $Q_1(x)$ we have

$$P(x) = (x - r_1)(r - r_2)Q_2(x) \tag{3}$$

This process is repeated n times until $Q_n(x) = a$ is of degree 0. Hence,

$$P(x) = a(x - r_1)(x - r_2) \cdots (x - r_n) \tag{4}$$

Since a is the leading coefficient of the polynomial on the right side of Equation (4), it must also be the leading coefficient of $P(x)$.

It is now easy to establish the following which may be thought of as an alternate form of the Fundamental Theorem of Algebra.

Number of Roots Theorem

If $P(x)$ is a polynomial of degree $n \geq 1$, then $P(x)$ has precisely n roots among the complex numbers.

We may prove this theorem as follows. If we write $P(x)$ in the form of Equation (4), we see that $r_1, r_2, \ldots, r_n$ are roots of the equation $P(x) = 0$ and hence there exist n roots. If there is an additional root r that is distinct from the roots $r_1, r_2, \ldots, r_n$, then $r - r_1, r - r_2, \ldots, r - r_n$ are all different from 0. Substituting r for x in Equation (4) yields

$$P(r) = a(r - r_1)(r - r_2) \cdots (r - r_n) \tag{5}$$

which cannot equal 0 since the product of nonzero numbers cannot equal 0. Thus, $r_1, r_2, \ldots, r_n$ are roots of $P(x)$ and there are no other roots. Hence, $P(x)$ has precisely n roots.

It is important to recognize that the roots of a polynomial need not be distinct from each other. The polynomial

$$P(x) = x^2 - 2x + 1$$

can be written in the factored form

$$P(x) = (x - 1)(x - 1)$$

which shows that the roots of $P(x)$ are 1 and 1. Since a root is associated with a factor and a factor may be repeated, we may have repeated roots. If the factor $x - r$ appears k times, we say that r is a **root of multiplicity k.**

EXAMPLE 4
Find all roots of the polynomial

$$P(x) = \left(x - \frac{1}{2}\right)^3 (x + i)(x - 5)^4$$

Solution
The distinct roots are $\frac{1}{2}$, $-i$, and 5. Further, $\frac{1}{2}$ is a root of multiplicity 3; $-i$ is a root of multiplicity 1; 5 is a root of multiplicity 4.

PROGRESS CHECK 4
Find all roots of the polynomial $P(x) = (x + 3)^2(x - 1 + 2i)$.

Answer
−3 is a root of multiplicity 2; 1 − 2i is a root of multiplicity 1

If we know that r is a root of $P(x)$, then we may write

$$P(x) = (x - r)Q(x)$$

and note that the roots of $Q(x)$ are also roots of $P(x)$. We call $Q(x) = 0$ the **depressed equation** since $Q(x)$ is of lower degree than $P(x)$. In the next example we illustrate the use of the depressed equation in finding the roots of a polynomial.

EXAMPLE 5
If 4 is a root of the polynomial $P(x) = x^3 - 8x^2 + 21x - 20$, find the other roots.

Solution
Since 4 is a root of $P(x)$, $x - 4$ is a factor of $P(x)$. Therefore,

$$P(x) = (x - 4)Q(x)$$

To find the depressed equation, we compute $Q(x) = P(x)/(x - 4)$ by synthetic division.

$$
\begin{array}{r|rrrr}
4 & 1 & -8 & 21 & -20 \\
  &   & 4 & -16 & 20 \\
\hline
  & 1 & -4 & 5 & 0 \\
\end{array}
$$

$\underbrace{}_{\text{Coefficients of } Q(x)}$ $\underset{\text{Remainder}}{|}$

The depressed equation is

$$x^2 - 4x + 5 = 0$$

Using the quadratic formula, the roots of the depressed equation are found to be $2 + i$ and $2 - i$. The roots of $P(x)$ are then seen to be 4, $2 + i$, and $2 - i$.

PROGRESS CHECK 5

If -2 is a root of the polynomial $P(x) = x^3 - 7x - 6$, find the remaining roots.

Answer

$-1, 3$

EXAMPLE 6

If -1 is a root of multiplicity 2 of $P(x) = x^4 + 4x^3 + 2x^2 - 4x - 3$, find the remaining roots and write $P(x)$ as a product of linear factors.

Solution

Since -1 is a double root of $P(x)$, then $(x + 1)^2$ is a factor of $P(x)$. Therefore,

$$P(x) = (x + 1)^2 Q(x)$$

or

$$P(x) = (x^2 + 2x + 1)Q(x)$$

Using polynomial division, we can divide both sides of the last equation by $x^2 + 2x + 1$ to obtain

$$Q(x) = \frac{x^4 + 4x^3 + 2x^2 - 4x - 3}{x^2 + 2x + 1}$$

$$= x^2 + 2x - 3$$

$$= (x - 1)(x + 3)$$

The roots of the depressed equation $Q(x) = 0$ are 1 and -3 and these are the remaining roots of $P(x)$. By the Linear Factor Theorem,

$$P(x) = (x + 1)^2(x - 1)(x + 3)$$

PROGRESS CHECK 6

If -2 is a root of multiplicity 2 of $P(x) = x^4 + 4x^3 + 5x^2 + 4x + 4$, write $P(x)$ as a product of linear factors.

Answer

$P(x) = (x + 2)(x + 2)(x + i)(x - i)$

We know from the quadratic formula that if a quadratic equation with real coefficients has a complex root $a + bi$, then the conjugate $a - bi$ is the other root. The following theorem extends this result to a polynomial of degree n with real coefficients.

Conjugate Roots Theorem

If $P(x)$ is a polynomial of degree $n \geq 1$ with real coefficients, and if $a + bi, b \neq 0$, is a root of $P(x)$, then the complex conjugate $a - bi$ is also a root of $P(x)$.

PROOF OF CONJUGATE ROOTS THEOREM (Optional)

To prove the Conjugate Roots Theorem, we let $z = a + bi$ and make use of the properties of complex conjugates developed earlier in this section. We may write

$$P(x) = a_n x^n + a_{n-1} x^{n-1} + \cdots + a_1 x + a_0 \qquad (6)$$

and, since z is a root of $P(x)$,

$$a_n z^n + a_{n-1} z^{n-1} + \cdots + a_n z + a_0 = 0 \qquad (7)$$

But if $z = w$, then $\bar{z} = \bar{w}$. Applying this property of complex numbers to both sides of Equation (7), we have

$$\overline{a_n z^n + a_{n-1} z^{n-1} + \cdots + a_1 z + a_0} = \bar{0} = 0 \qquad (8)$$

We also know that $\overline{z + w} = \bar{z} + \bar{w}$. Applying this property to the left side of Equation (8) we see that

$$\overline{a_n z^n} + \overline{a_{n-1} z^{n-1}} + \cdots + \overline{a_1 z} + \overline{a_0} = 0 \qquad (9)$$

Further, $\overline{z \cdot w} = \bar{z} \cdot \bar{w}$ so that we may rewrite Equation (9) as

$$\overline{a_n}\, \overline{z^n} + \overline{a_{n-1}}\, \overline{z^{n-1}} + \cdots + \overline{a_1}\, \bar{z} + \overline{a_0} = 0 \qquad (10)$$

Since a_i are all real numbers, we know that $\overline{a_i} = a_i$. Finally, we use the property $\overline{z^n} = \bar{z}^n$ to rewrite Equation (10) as

$$a_n \bar{z}^n + a_{n-1} \bar{z}^{n-1} + \cdots + a_1 \bar{z} + a_0 = 0$$

which establishes that $\bar{z}$ is a root of $P(x)$.

EXAMPLE 7
Find a polynomial $P(x)$ with real coefficients that is of degree 3 and whose roots include -2 and $1 - i$.

Solution
Since $1 - i$ is a root, it follows from the Conjugate Roots Theorem that $1 + i$ is also a root of $P(x)$. By the Factor Theorem, $(x + 2)$, $[x - (1 - i)]$ and $[x - (1 + i)]$ are factors of $P(x)$. Therefore,

$$
\begin{aligned}
P(x) &= (x + 2)[x - (1 - i)][x - (1 + i)] \\
&= (x + 2)(x^2 - 2x + 2) \\
&= x^3 - 2x + 4
\end{aligned}
$$

PROGRESS CHECK 7
Find a polynomial $P(x)$ with real coefficients that is of degree 4 and whose roots include i and $-3 + i$.

Answer
$P(x) = x^4 + 6x^3 + 11x^2 + 6x + 10$

The following is a corollary of the Conjugate Roots Theorem.

A polynomial $P(x)$ of degree $n \geq 1$ with real coefficients can be written as a product of linear and quadratic factors with real coefficients so that the quadratic factors have no real roots.

By the Linear Factor Theorem, we may write

$$P(x) = a(x - r_1)(x - r_2) \cdots (x - r_n)$$

where $r_1, r_2, \ldots, r_n$ are the n roots of $P(x)$. Of course, some of these roots may be complex numbers. A complex root $a + bi$, $b \neq 0$, may be paired with its conjugate $a - bi$ to provide the quadratic factor

$$[x - (a + bi)][x - (a - bi)] = x^2 - 2ax + a^2 + b^2$$

that has real coefficients. Thus, a quadratic factor with real coefficients results from each pair of complex conjugate roots; a linear factor with real coefficients results from each real root. Further, the discriminant of the quadratic factor $x^2 - 2ax + a^2 + b^2$ is $-4b^2$ and is therefore always negative, which shows that the quadratic factor has no real roots.

EXERCISE SET 10.3

In Exercises 1–6 multiply by the conjugate and simplify.

1. $2 - i$
2. $3 + i$
3. $3 + 4i$

4. $2 - 3i$
5. $-4 - 2i$
6. $5 + 2i$

In Exercises 7–15 perform the indicated operations and write the answer in the form $a + bi$.

7. $\dfrac{2 + 5i}{1 - 3i}$
8. $\dfrac{1 + 3i}{2 - 5i}$
9. $\dfrac{3 - 4i}{3 + 4i}$

10. $\dfrac{4 - 3i}{4 + 3i}$
11. $\dfrac{3 - 2i}{2 - i}$
12. $\dfrac{2 - 3i}{3 - i}$

13. $\dfrac{2 + 5i}{3i}$
14. $\dfrac{5 - 2i}{-3i}$
15. $\dfrac{4i}{2 + i}$

In Exercises 16–21 find the reciprocal and write the answer in the form $a + bi$.

16. $3 + 2i$
17. $4 + 3i$
18. $\frac{1}{2} - i$

19. $1 - \frac{1}{3}i$
20. $-7i$
21. $-5i$

22. Prove that the multiplicative inverse of the complex number $a + bi$ (a and b not both 0) is

$$\frac{a}{a^2 + b^2} - \frac{b}{a^2 + b^2}i$$

23. If z and w are complex numbers, prove that $\overline{z \cdot w} = \overline{z} \cdot \overline{w}$.

24. If z is a complex number, verify that $\overline{z^2} = \overline{z}^2$ and $\overline{z^3} = \overline{z}^3$.

In Exercises 25–30 find a polynomial $P(x)$ of lowest degree that has the indicated roots.

25. $2, -4, 4$
26. $5, -5, 1, -1$
27. $-1, -2, -3$

28. $-3, \sqrt{2}, -\sqrt{2}$
29. $4, 1 \pm \sqrt{3}$
30. $1, 2, 2 \pm \sqrt{2}$

In Exercises 31–34 find the polynomial $P(x)$ of lowest degree that has the indicated roots and satisfies the given condition. (*Hint:* Write $P(x)$ in the form

$$P(x) = a(x - r_1)(x - r_2) \cdots (x - r_n)$$

where $r_1, r_2, \ldots, r_n$, are the indicated roots and a is a real number to be determined.)

31. $\frac{1}{2}, \frac{1}{2}, -2;$ $P(2) = 3$
32. $3, 3, -2, 2;$ $P(4) = 12$

33. $\sqrt{2}, -\sqrt{2}, 4;$ $P(-1) = 5$
34. $\frac{1}{2}, -2, 5;$ $P(0) = 5$

In Exercises 35–42 find the roots of the given equation.

35. $(x - 3)(x + 1)(x - 2) = 0$ 36. $(x - 3)(x^2 - 3x - 4) = 0$

37. $(x + 2)(x^2 - 16) = 0$ 38. $(x^2 - x)(x^2 - 2x + 5) = 0$

39. $(x^2 + 3x + 2)(2x^2 + x) = 0$ 40. $(x^2 + x + 4)(x - 3)^2 = 0$

41. $(x - 5)^3(x + 5)^2 = 0$ 42. $(x + 1)^2(x + 3)^4(x - 2) = 0$

In Exercises 43–46 find a polynomial that has the indicated roots and no others.

43. -2 of multiplicity 3

44.   1 of multiplicity 2, -4 of multiplicity 1

45. $\frac{1}{2}$ of multiplicity 2, -1 of multiplicity 2

46. -1 of multiplicity 2, 0 and 2 each of multiplicity 1

In Exercises 47–52 use the given root(s) to help in finding the remaining roots of the equation.

47. $x^3 - 3x - 2 = 0$; -1 48. $x^3 - 7x^2 + 4x + 24 = 0$;   3

49. $x^3 - 8x^2 + 18x - 15 = 0$;   5   50. $x^3 - 2x^2 - 7x - 4 = 0$; -1

51. $x^4 + x^3 - 12x^2 - 28x - 16 = 0$; -2

52. $x^4 - 2x^2 + 1$;   1 is a double root

In Exercises 53–58 find a polynomial that has the indicated roots and no others.

53. $1 + 3i, -2$ 54. $1, -1, 2 - i$

55. $1 + i, 2 - i$ 56. $-2, 3, 1 + 2i$

57. -2 is a root of multiplicity 2, $3 - 2i$

58.   3 is a triple root, $-i$

In Exercises 59–64 use the given root(s) to help in writing the given equation as a product of linear and quadratic factors with real coefficients.

59. $x^3 - 7x^2 + 16x - 10 = 0$; $3 - i$

60. $x^3 + x^2 - 7x + 65 = 0$; $2 + 3i$

61. $x^4 + 4x^3 + 13x^2 + 18x + 20 = 0$; $-1 - 2i$

62. $x^4 + 3x^3 - 5x^2 - 29x - 30 = 0$; $-2 + i$

63. $x^5 + 3x^4 - 12x^3 - 42x^2 + 32x + 120 = 0$; $-3 - i, -2$

64. $x^5 - 8x^4 + 29x^3 - 54x^2 + 48x - 16 = 0$; $2 + 2i, 2$

65.   Write a polynomial $P(x)$ with complex coefficients that has the root $a + bi$, $b \neq 0$, and does not have $a - bi$ as a root.

66.   Prove that a polynomial equation of degree 4 with real coefficients has 4 real roots, 2 real roots, or no real roots.

67.   Prove that a polynomial equation of odd degree with real coefficients has at least one real root.

10.4
REAL AND RATIONAL ROOTS

In this section we will restrict our investigation to polynomials with real coefficients. Our first objective is to obtain some information concerning the number of positive real roots and the number of negative real roots of such polynomials.

If the terms of a polynomial with real coefficients are written in descending order, then a **variation in sign** occurs whenever two successive terms have opposite signs. In determining the number of variations in sign, we ignore terms with zero coefficients. The polynomial

$$4x^5 - 3x^4 - 2x^2 + 1$$

has two variations in sign. The French mathematician René Descartes (1596–1650), who provided us with the foundations of analytic geometry, also gave us a theorem that relates the nature of the real roots of polynomials to the variations in sign. The proof of Descartes' theorem is outlined in Exercises 39–44.

Descartes' Rule of Signs

If $P(x)$ is a polynomial with real coefficients, then
(I)   the number of positive roots is either equal to the number of variations in sign of $P(x)$, or is less than the number of variations in sign by an even number, and
(II) the number of negative roots is either equal to the number of variations in sign of $P(-x)$ or is less than the number of variations in sign by an even number.

If it is determined that a polynomial of degree n has r real roots, then the remaining $n - r$ roots must be complex numbers.

To apply Descartes' Rule of Signs to the polynomial

$$P(x) = 3x^5 + 2x^4 - x^3 + 2x - 3$$

we first note that there are 3 variations in sign as indicated. Thus, there are either 3 positive roots or there is 1 positive root. Next, we form $P(-x)$,

$$P(-x) = 3(-x)^5 + 2(-x)^4 - (-x)^3 + 2(-x) - 3$$
$$= -3x^5 + 2x^4 + x^3 - 2x - 3$$

which can be obtained by negating the coefficients of the odd power terms. We see that $P(-x)$ has two variations in sign and conclude that $P(x)$ has either 2 negative roots or no negative roots.

EXAMPLE 1
Use Descartes' Rule of Signs to analyze the roots of the equation

$$2x^5 + 7x^4 + 3x^2 - 2 = 0$$

Solution
Since

$$P(x) = 2x^5 + 7x^4 + 3x^2 - 2$$

has 1 variation in sign, there is precisely 1 positive root. The polynomial $P(-x)$ is formed

$$P(-x) = -2x^5 + 7x^4 + 3x^2 - 2$$

and is seen to have 2 variations in sign, so that $P(-x)$ has either 2 negative roots or no negative roots. Since $P(x)$ has 5 roots, the possibilities are

$$\text{1 positive root, 2 negative roots, 2 complex roots}$$

$$\text{1 positive root, 0 negative roots, 4 complex roots}$$

PROGRESS CHECK 1

Use Descartes' Rule of Signs to analyze the nature of the roots of the equation

$$x^6 + 5x^4 - 4x^2 - 3 = 0$$

Answer
1 positive root, 1 negative root, 4 complex roots

The following theorem provides the basis for a systematic search for the rational roots of polynomials with integer coefficients.

Rational Root Theorem

If the coefficients of the polynomial

$$P(x) = a_n x^n + a_{n-1}x^{n-1} + \cdots + a_1 x + a_0 \quad (a_n \neq 0)$$

are all integers and p/q is a rational root, in lowest terms, then
(I) p is a factor of the constant term a_0, and
(II) q is a factor of the leading coefficient a_n.

PROOF OF RATIONAL ROOT THEOREM (Optional)

Since p/q is a root of $P(x)$, then $P(p/q) = 0$. Thus,

$$a_n\left(\frac{p}{q}\right)^n + a_{n-1}\left(\frac{p}{q}\right)^{n-1} + \cdots + a_1\left(\frac{p}{q}\right) + a_0 = 0 \tag{1}$$

Multiplying Equation (1) by q^n, we have

$$a_n p^n + a_{n-1}p^{n-1}q + \cdots + a_1 pq^{n-1} + a_0 q^n = 0 \tag{2}$$

or

$$a_n p^n + a_{n-1}p^{n-1}q + \cdots + a_1 pq^{n-1} = -a_0 q^n \tag{3}$$

Taking the common factor p out of the left-hand side of Equation (3) yields

$$p(a_n p^{n-1} + a_{n-1}p^{n-2}q + \cdots + a_1 q^{n-1}) = -a_0 q^n \tag{4}$$

Since $a_1, a_2, \ldots, a_n, p$, and q are all integers, the quantity in parentheses in the left-hand side of Equation (4) is an integer. Division of the left-hand side by p results in an integer and we conclude that p must also be a factor of the right-hand side, $-a_0 q^n$. But p and q have no common factors since, by hypothesis, p/q is in lowest terms. Hence, p must be a factor of a_0 which proves part (1) of the Rational Root Theorem.

We may also rewrite Equation (2) in the form

$$q(a_{n-1}p^{n-1} + a_{n-2}p^{n-2}q + \cdots + a_1 pq^{n-2} + a_0 q^{n-1}) = -a_n p^n \tag{5}$$

An argument similar to the preceding one now establishes part (II) of the theorem.

EXAMPLE 2

Find the rational roots of the equation

$$8x^4 - 2x^3 + 7x^2 - 2x - 1 = 0$$

Solution

If p/q is a rational root in lowest terms, then p is a factor of 1 and q is a factor of 8. We can now list the possibilities:

$$\text{possible numerators: } \pm 1 \quad \text{(the factors of 1)}$$

$$\text{possible denominators: } \pm 1, \pm 2, \pm 4, \pm 8 \quad \text{(the factors of 8)}$$

$$\text{possible rational roots: } \pm 1, \pm \frac{1}{2}, \pm \frac{1}{4}, \pm \frac{1}{8}$$

Synthetic division can be used to test if these numbers are roots. Trying $x = 1$ and $x = -1$, we find that they are not roots. Trying $\frac{1}{2}$ we have

$$
\begin{array}{r|rrrrr}
\frac{1}{2} & 8 & -2 & 7 & -2 & -1 \\
  &   & 4 & 1 & 4 & 1 \\
\hline
  & 8 & 2 & 8 & 2 & 0 \\
\end{array}
$$

which demonstrates that $\frac{1}{2}$ is a root. Similarly,

$$
\begin{array}{r|rrrrr}
-\frac{1}{4} & 8 & -2 & 7 & -2 & -1 \\
  &   & -2 & 1 & -2 & 1 \\
\hline
  & 8 & -4 & 8 & -4 & 0 \\
\end{array}
$$

which shows that $-\frac{1}{4}$ is also a root. The student may verify that these roots are not repeated and that none of the other possible rational roots will result in a zero remainder when synthetic division is employed. We can conclude that the other two roots are a pair of complex conjugates.

PROGRESS CHECK 2

Find the rational roots of the equation

$$9x^4 - 12x^3 + 13x^2 - 12x + 4 = 0$$

Answer

$$\frac{2}{3}, \frac{2}{3}$$

EXAMPLE 3

Find all roots of the equation

$$8x^5 + 12x^4 + 14x^3 + 13x^2 + 6x + 1 = 0$$

Solution

We first list the possible rational roots.

$$\text{possible numerators: } \pm 1 \quad \text{(factors of 1)}$$

$$\text{possible denominators: } \pm 1, \pm 2, \pm 4, \pm 8 \quad \text{(factors of 8)}$$

$$\text{possible rational roots: } \pm 1, \pm \frac{1}{2}, \pm \frac{1}{4}, \pm \frac{1}{8}$$

We next employ Descartes' Rule of Signs. Since $P(x)$ has no variations in sign, there are no positive roots. $P(-x)$ has 5 variations in sign, indicating that there are either 5 negative roots, 3 negative roots, or 1 negative root. Using synthetic division to test the possible negative rational roots, we find that $-\frac{1}{2}$ is a root.

$$
\begin{array}{r|rrrrrr}
-\frac{1}{2} & 8 & 12 & 14 & 13 & 6 & 1 \\
& & -4 & -4 & -5 & -4 & -1 \\
\hline
& 8 & 8 & 10 & 8 & 2 & 0
\end{array}
$$

coefficients of depressed equation

We can now use the depressed equation and continue testing, with the same list of possible negative roots. Once again, $-\frac{1}{2}$ is seen to be a root.

$$
\begin{array}{r|rrrrr}
-\frac{1}{2} & 8 & 8 & 10 & 8 & 2 \\
& & -4 & -2 & -4 & -2 \\
\hline
& 8 & 4 & 8 & 4 & 0
\end{array}
$$

coefficients of depressed equation

This illustrates an important point: A rational root may be a multiple root! Applying the same technique to the resulting depressed equation, we see that $-\frac{1}{2}$ is once again a root.

$$
\begin{array}{r|rrrr}
-\frac{1}{2} & 8 & 4 & 8 & 4 \\
& & -4 & 0 & -4 \\
\hline
& 8 & 0 & 8 & 0
\end{array}
$$

coefficients of depressed equation

The final depressed equation

$$8x^2 + 8 = 0 \quad \text{or} \quad x^2 + 1 = 0$$

has the roots $\pm i$. Thus, the original equation has the roots

$$-\frac{1}{2}, \ -\frac{1}{2}, \ -\frac{1}{2}, \ i, \ \text{and} \ -i$$

PROGRESS CHECK 3
Find all roots of the polynomial

$$P(x) = 9x^4 - 3x^3 + 16x^2 - 6x - 4$$

Answer

$$\frac{2}{3}, \ -\frac{1}{3}, \ \pm \sqrt{2}i$$

EXAMPLE 4
Prove that $\sqrt{3}$ is not a rational number.

Solution
If we let $x = \sqrt{3}$, then $x^2 = 3$ or $x^2 - 3 = 0$. By the Rational Root Theorem, the only possible rational roots are $\pm 1, \pm 3$. Synthetic division can be used to show that none of these are roots. However, $\sqrt{3}$ is a root of $x^2 - 3 = 0$. Hence, $\sqrt{3}$ is not a rational number.

PROGRESS CHECK 4
Prove that $\sqrt[3]{2}$ is not a rational number.

A number that is a root of some polynomial equation with integer coefficients is said to be **algebraic.** The requirement that the coefficients be integers is critical since any real number a will satisfy the equation $x - a = 0$. We see that $\frac{2}{3}$ is algebraic since it is a root of the equation $3x - 2 = 0$; $\sqrt{2}$ is also algebraic since it satisfies the equation $x^2 - 2 = 0$.

Are all real numbers algebraic? To show that a real number a is *not* algebraic we must show that there is *no* polynomial equation with integer coefficients that has a as one of its roots. Although this appears to be an impossible task, Georg Cantor (1845–1918), in his brilliant work on infinite sets, provided an answer. There are indeed numbers which are not algebraic. We call such numbers **transcendental.** You are already familiar with two transcendental numbers: π and e. Thus, the numbers π and e are not roots of any polynomial equations with integer coefficients.

EXERCISE SET 10.4

In Exercises 1–12 use Descartes' Rule of Signs to analyze the nature of the roots of the given equation. List all possibilities.

1. $3x^4 - 2x^3 + 6x^2 + 5x - 2 = 0$

2. $2x^6 + 5x^5 + x^3 - 6 = 0$

3. $x^6 + 2x^4 + 4x^2 + 1 = 0$

4. $3x^3 - 2x + 2 = 0$

5. $x^5 - 4x^3 + 7x - 4 = 0$

6. $2x^3 - 5x^2 + 8x - 2 = 0$

7. $5x^3 + 2x^2 + 7x - 1 = 0$

8. $x^5 + 6x^4 - x^3 - 2x - 3 = 0$

9. $x^4 - 2x^3 + 5x^2 + 2 = 0$

10. $3x^4 - 2x^3 - 1 = 0$

11. $x^8 + 7x^3 + 3x - 5 = 0$

12. $x^7 + 3x^5 - x^3 - x + 2 = 0$

In Exercises 13–22 find all rational roots of the given equation.

13. $x^3 - 2x^2 - 5x + 6 = 0$

14. $3x^3 - x^2 - 3x + 1 = 0$

15. $6x^4 - 7x^3 - 13x^2 + 4x + 4 = 0$

16. $36x^4 - 15x^3 - 26x^2 + 3x + 2 = 0$

17. $5x^6 - x^5 - 5x^4 + 6x^3 - x^2 - 5x + 1 = 0$

18. $16x^4 - 16x^3 - 29x^2 + 32x - 6 = 0$

19. $4x^4 - x^3 + 5x^2 - 2x - 6 = 0$

20. $6x^4 + 2x^3 + 7x^2 + x + 2 = 0$

21. $2x^5 - 13x^4 + 26x^3 - 22x^2 + 24x - 9 = 0$

22. $8x^5 - 4x^4 + 6x^3 - 3x^2 - 2x + 1 = 0$

In Exercises 23–30 find all roots of the given equation.

23. $4x^4 + x^3 + x^2 + x - 3 = 0$

24. $x^4 + x^3 + x^2 + 3x - 6 = 0$

25. $5x^5 - 3x^4 - 10x^3 + 6x^2 - 40x + 24 = 0$

26. $12x^4 - 52x^3 + 75x^2 - 16x - 5 = 0$

27. $6x^4 - x^3 - 5x^2 + 2x = 0$

28. $2x^4 - \dfrac{3}{2}x^3 + \dfrac{11}{2}x^2 + \dfrac{23}{2}x + \dfrac{5}{2} = 0$

29. $2x^4 - x^3 - 28x^2 + 30x - 8 = 0$

30. $12x^4 + 4x^3 - 17x^2 + 6x = 0$

In Exercises 31–34 find the integer value(s) of k for which the given equation has rational roots, and find the roots. (*Hint:* Use synthetic division.)

31. $x^3 + kx^2 + kx + 2 = 0$

32. $x^4 - 4x^3 - kx^2 + 6kx + 9 = 0$

33. $x^4 - 3x^3 + kx^2 - 4x - 1 = 0$

34. $x^3 - 3kx^2 + k^2x + 4 = 0$

35. If $P(x)$ is a polynomial with real coefficients and has one variation in sign, prove that $P(x)$ has exactly one positive root.

36. If $P(x)$ is a polynomial with integer coefficients and the leading coefficient is $+1$ or -1, prove that the rational roots of $P(x)$ are all integers and are factors of the constant term.

37. Prove that $\sqrt{5}$ is not a rational number.

38. If p is a prime, prove that $\sqrt{p}$ is not a rational number.

39. Prove that if $P(x)$ is a polynomial with real coefficients and r is a positive root of $P(x)$, then the reduced equation

$$Q(x) = \frac{P(x)}{(x - r)}$$

has at least one fewer variation in sign than $P(x)$. (*Hint:* Assume the leading coefficient of $P(x)$ to be positive and use synthetic division to obtain $Q(x)$. Note that the coefficients of $Q(x)$ remain positive at least until there is a variation in sign in $P(x)$.)

40. Prove that if $P(x)$ is a polynomial with real coefficients then the number of positive roots is not greater than the number of variations in sign in $P(x)$. (*Hint:* Let $r_1, r_2, \ldots, r_k$ be the positive roots of $P(x)$, and let

$$P(x) = (x - r_1)(x - r_2) \cdots (x - r_k)Q(x)$$

Use the result of Exercise 39 to show that $Q(x)$ has at least k fewer variations in sign than does $P(x)$.)

41. Prove that if $r_1, r_2, \ldots, r_k$ are positive numbers, then

$$P(x) = (x - r_1)(x - r_2) \cdots (x - r_k)$$

has alternating signs. (*Hint:* Use the result of Exercise 40.)

42. Prove that the number of variations in sign of a polynomial with real coefficients is even if the first and last coefficients have the same sign and is odd if they are of opposite sign.

43. Prove that if the number of positive roots of the polynomial $P(x)$ with real coefficients is less than the number of variations in sign, then it is less by an even number. (*Hint:* Write $P(x)$ as a product of linear factors corresponding to the positive and negative roots, and of quadratic factors corresponding to complex roots. Apply the results of Exercises 41 and 42.)

44. Prove that the positive roots of $P(-x)$ correspond to the negative roots of $P(x)$, that is, if $a > 0$ is a root of $P(-x)$, then $-a$ is a root of $P(x)$.

TERMS AND SYMBOLS

polynomial function of
   degree n (p. 279)
polynomial equation of
   degree n (p. 280)
roots or zeros of a
   polynomial   (p. 280)
synthetic division   (p. 282)

complex conjugate z
   (p. 288)
root of multiplicity k
   (p. 292)
depressed equation
   (p. 292)

variation in sign   (p. 297)
algebraic numbers
   (p. 300)
transcendental numbers
   (p. 301)

KEY IDEAS FOR REVIEW

☐ Polynomial division results in a quotient and a remainder, both of which are polynomials. The degree of the remainder must be less than the degree of the divisor.

☐ Synthetic division is a quick way to divide a polynomial by a first-degree polynomial $x - r$, where r is a real constant.

☐ The following are the primary theorems concerning polynomials and their roots.

Remainder Theorem

If a polynomial $P(x)$ is divided by $x - r$, then the remainder is $P(r)$.

Factor Theorem

A polynomial $P(x)$ has a root r if and only if $x - r$ is a factor of $P(x)$.

Linear Factor Theorem

A polynomial $P(x)$ of degree $n \geq 1$ can be written as the product of n linear factors

$$P(x) = a(x - r_1)(x - r_2) \cdots (x - r_n)$$

where $r_1, r_2, \ldots, r_n$ are the complex roots of $P(x)$ and a is the leading coefficient of $P(x)$.

Number of Roots Theorem

If $P(x)$ is a polynomial of degree $n \geq 1$, then $P(x)$ has precisely n roots among the complex numbers.

Conjugate Roots Theorem

If $a + bi$, $b \neq 0$, is a root of the polynomial $P(x)$ with real coefficients, then $a - bi$ is also a root of $P(x)$.

Rational Root Theorem

If p/q is a rational root (in lowest terms) of the polynomial $P(x)$ with integer coefficients, then p is a factor of the constant term a_0 of $P(x)$ and q is a factor of the leading coefficient a_n of $P(x)$.

☐ If r is a real root of the polynomial $P(x)$, then the roots of the depressed equation are the other roots of $P(x)$. The depressed equation can be found by using synthetic division.

☐ Descartes' Rule of Signs tells us the maximum number of positive roots and the maximum number of negative roots of a polynomial $P(x)$ with real coefficients.

☐ If $P(x)$ has integer coefficients, then the Rational Root Theorem enables us to list all possible rational roots of $P(x)$. Synthetic division can then be used to test these potential rational roots, since r is a root if and only if the remainder is zero, that is, if and only if $P(r) = 0$.

PROGRESS TEST 10A

1. Find the quotient and remainder when $2x^4 - x^2 + 1$ is divided by $x^2 + 2$.
2. Use synthetic division to find the quotient and remainder when $3x^4 - x^3 - 2$ is divided by $x + 2$.
3. If $P(x) = x^3 - 2x^2 + 7x + 5$, use synthetic division to find $P(-2)$.
4. Determine the remainder when $4x^5 - 2x^4 - 5$ is divided by $x + 2$.
5. Use the Factor Theorem to show that $x - 3$ is a factor of

$$2x^4 - 9x^3 + 9x^2 + x - 3$$

In Problems 6-7 find a polynomial of lowest degree that has the indicated roots.

6. $-2, 1, 3$ 7. $-1, 1, 3 \pm \sqrt{2}$

In Problems 8-9 find the roots of the given equation.

8. $(x^2 + 1)(x - 2) = 0$ 9. $(x + 1)^2(x^2 - 3x - 2) = 0$

In Problems 10-12 find a polynomial that has the indicated roots and no others.

10. -3 of multiplicity 2; 1 of multiplicity 3
11. $-\frac{1}{4}$ of multiplicity 2; i, $-i$, and 1
12. $i, 1 + i$

In Problems 13-14 use the given root(s) to help in finding the remaining roots of the equation.

13. $4x^3 - 3x + 1 = 0$; -1 14. $x^4 - x^2 - 2x + 2 = 0$; 1

15. If $2 + i$ is a root of $x^3 - 6x^2 + 13x - 10 = 0$, write the equation as a product of linear and quadratic factors with real coefficients.

In Problems 16-17 determine the maximum number of roots of the given equation of the type indicated.

16. $2x^5 - 3x^4 + 1 = 0$; positive real roots
17. $3x^4 + 2x^3 - 2x^2 - 1 = 0$; negative real roots

In Problems 18-19 find all rational roots of the given equation.

18. $6x^3 - 17x^2 + 14x + 3 = 0$
19. $2x^5 - x^4 - 4x^3 + 2x^2 + 2x - 1 = 0$
20. Find all roots of the equation $3x^4 + 7x^3 - 3x^2 + 7x - 6 = 0$

PROGRESS TEST 10B

1. Find the quotient and remainder when $3x^5 + 2x^3 - x^2 - 2$ is divided by $2x^2 - x - 1$.
2. Use synthetic division to find the quotient and remainder when $-2x^3 + 3x^2 - 1$ is divided by $x - 1$.
3. If $P(x) = 2x^4 - 2x^3 + x - 4$, use synthetic division to find $P(-1)$.
4. Determine the remainder when $3x^4 - 5x^3 + 3x^2 + 4$ is divided by $x - 2$.
5. Use the Factor Theorem to show that $x + 2$ is a factor of $x^3 - 4x^2 - 9x + 6$.

In Problems 6-7 find a polynomial of lowest degree that has the indicated roots.

6. $-\frac{1}{2}, 1, 1, -1$ 7. $2, 1 \pm \sqrt{3}$

In Problems 8-9 find the roots of the given equation.

8. $(x^2 - 3x + 2)(x - 2)^2$ 9. $(x^2 + 3x - 1)(x - 2)(x + 3)^2$

In Problems 10-12 find a polynomial that has the indicated roots and no others.

10. $\frac{1}{2}$ of multiplicity 3; -2 of multiplicity 1
11. -3 of multiplicity 2; $1 + i$, $1 - i$
12. $3 \pm \sqrt{-1}$, -1 of multiplicity 2

In Problems 13–14 use the given root(s) to help in finding the remaining roots of the equation.

13. $x^3 - x^2 - 8x - 4 = 0$; -2 14. $x^4 - 3x^3 - 22x^2 + 68x - 40 = 0$

15. If $1 - i$ is a root of $2x^4 - x^3 - 4x^2 + 10x - 4 = 0$, write the equation as a product of linear and quadratic factors with real coefficients.

In Problems 16–17 determine the maximum number of roots, of the type indicated, of the given equation.

16. $3x^4 + 3x - 1 = 0$; positive real roots

17. $2x^4 + x^3 - 3x^2 + 2x + 1 = 0$; negative real roots

In Problems 18–19 find all rational roots of the given equation.

18. $3x^3 + 7x^2 - 4 = 0$ 19. $4x^4 - 4x^3 + x^2 - 4x - 3 = 0$

20. Find all roots of the equation $2x^4 - x^3 - 2x^2 + 2 = 0$.

CHAPTER ELEVEN
EXPONENTIAL AND LOGARITHMIC FUNCTIONS

The function concept can serve as a unifying idea in mathematics. Functions can be combined in a variety of ways, some of which you may anticipate (such as the addition and multiplication of two or more functions). We can also combine functions by applying one function to another to obtain a composite function.

At times it is also possible to define a function g which reverses the correspondence of a function f. An important pair of such inverse functions are the exponential and logarithmic functions. Many processes in nature are described by exponential functions. They are useful in chemistry, biology, and economics, as well as in mathematics and engineering. We will study applications of exponential functions in calculating such things as compound interest and the growth rate of bacteria in a culture medium.

Logarithms can be viewed as another way of writing exponents. Historically, logarithms have been used to simplify calculations; in fact, the slide rule, a device long used by engineers, is based on logarithmic scales. In today's world of inexpensive hand calculators, the need for logarithms is reduced. The section in this chapter on computing with logarithms will provide enough background to allow you to use this powerful tool, but omits some of the details found in older textbooks.

11.1
COMBINING FUNCTIONS; INVERSE FUNCTIONS

We can combine two functions such as

$$f(x) = x^2 \qquad g(x) = x - 1$$

by the usual operations of addition, subtraction, multiplication, and division. Using these functions f and g, we can form

$$(f + g)(x) = f(x) + g(x) = x^2 + x - 1$$
$$(f - g)(x) = f(x) - g(x) = x^2 - (x - 1) = x^2 - x + 1$$
$$(f \cdot g)(x) = f(x) \cdot g(x) = x^2(x - 1) = x^3 - x^2$$
$$\left(\frac{f}{g}\right)(x) = \frac{f(x)}{g(x)} = \frac{x^2}{x - 1}$$

In each case we have combined two functions f and g to form a new function. Note, however, that the domain of the new function need not be the same as the domain of the original functions. The function formed by division

$$\left(\frac{f}{g}\right)(x) = \frac{x^2}{x - 1}$$

has as its domain the set of all real numbers x except $x = 1$ since we cannot divide by 0. On the other hand, the functions $f(x) = x^2$ and $g(x) = x - 1$ are both defined at $x = 1$.

EXAMPLE 1
Given $f(x) = x - 4$, $g(x) = x^2 - 4$, find   (a) $(f + g)(x)$ (b) $(f - g)(x)$
(c) $(f \cdot g)(x)$ (d) $\left(\dfrac{f}{g}\right)(x)$ (e)   the domain of $\left(\dfrac{f}{g}\right)(x)$.

Solution
(a) $(f + g)(x) = f(x) + g(x) = x - 4 + x^2 - 4 = x^2 + x - 8$
(b) $(f - g)(x) = f(x) - g(x) = x - 4 - (x^2 - 4) = -x^2 + x$
(c) $(f \cdot g)(x) = f(x) \cdot g(x) = (x - 4)(x^2 - 4) = x^3 - 4x^2 - 4x + 16$
(d) $\left(\dfrac{f}{g}\right)(x) = \dfrac{f(x)}{g(x)} = \dfrac{x - 4}{x^2 - 4}$

(e)   The domain of $\left(\dfrac{f}{g}\right)(x)$ must exclude values of x for which $x^2 - 4 = 0$. Thus,

the domain of $\left(\dfrac{f}{g}\right)(x)$ consists of all real numbers except 2 and -2.

PROGRESS CHECK 1
Given $f(x) = 2x^2$, $g(x) = x^2 - 5x + 6$, find

(a) $(f + g)(x)$ (b) $(f - g)(x)$ (c) $(f \cdot g)(x)$ (d) $\left(\dfrac{f}{g}\right)(x)$

(e)   the domain of $\left(\dfrac{f}{g}\right)(x)$.

Answers
(a) $3x^2 - 5x + 6$ (b) $x^2 + 5x - 6$ (c) $2x^4 - 10x^3 + 12x^2$

(d) $\dfrac{2x^2}{x^2 - 5x + 6}$ (e) *All real numbers except 2 and 3.*

EXAMPLE 2

The treasurer of a corporation that manufactures tennis balls finds that the gross revenue R (in dollars) can be expressed as a function of the number of cans x (in millions) sold:

$$R(x) = -x^2 + 4x$$

Further, the total manufacturing cost C (in dollars) is given by

$$C(x) = \frac{5}{4}x + \frac{1}{2}$$

(a) Express the profit P in terms of the number of cans sold.
(b) Express the ratio of the profit P to the gross revenue R.

Solution

(a) Since

$$\text{profit} = \text{revenue} - \text{cost}$$

$$P(x) = R(x) - C(x)$$

$$= -x^2 + 4x - \left(\frac{5}{4}x + \frac{1}{2}\right)$$

$$= -x^2 + \frac{11}{4}x - \frac{1}{2}$$

(b) The ratio we seek is $\dfrac{\text{profit}}{\text{revenue}}$ or

$$\frac{P(x)}{R(x)} = \frac{-x^2 + \frac{11}{4}x - \frac{1}{2}}{-x^2 + 4x} = \frac{-4x^2 + 11x - 2}{-4x^2 + 16x}$$

PROGRESS CHECK 2

The Natural Fertilizer Company sets the price P (in dollars) per ton of fertilizer by

$$P(x) = \begin{cases} 200 - 10x, & 0 < x \le 5 \\ 150, & x > 5 \end{cases}$$

where x is the number of tons ordered.
(a) Express the gross revenue R as a function of the quantity x. (*Hint:* gross revenue = price × quantity.)
(b) Find the gross revenue when the demand x is 4 tons.

Answers

(a) $R(x) = \begin{cases} 200x - 10x^2, & 0 < x \le 5 \\ 150x, & x > 5 \end{cases}$

(b) $840

There is another, important way in which two functions f and g can be combined to form a new function. In Figure 11.1, the function f assigns the

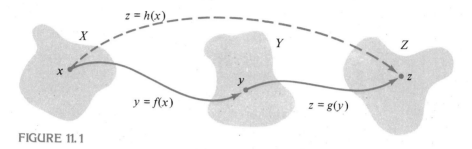

FIGURE 11.1

value y in set Y to x in set X; then, function g assigns the value z in set Z to y in Y. The net effect of this combination of f and g is a new function h, called the **composite function of g and f**, $g \circ f$, which assigns z in Z to x in X. We write this function as

$$h(x) = (g \circ f)(x) = g[f(x)]$$

which is read "g of f of x."

EXAMPLE 3

Given $f(x) = x^2$, $g(x) = x - 1$. Find the following.

(a) $f[g(3)]$

We begin by evaluating $g(3)$. Given

$$g(x) = x - 1$$
$$g(3) = 3 - 1 = 2$$

Therefore

$$f[g(3)] = f(2)$$

But since

then

$$f(x) = x^2$$
$$f(2) = 2^2 = 4$$

Thus $f[g(3)] = 4$.

(b) $g[f(3)]$

Beginning with $f(3)$, we have

$$f(3) = 3^2 = 9$$

Then we find that

$$g[f(3)] = g(9) = 9 - 1 = 8$$

(c) $f[g(x)]$

Since $g(x) = x - 1$, we make the substitution

$$f[g(x)] = f(x - 1) = (x - 1)^2 = x^2 - 2x + 1$$

(d) $g[f(x)]$

Since $f(x) = x^2$, we make the substitution

$$g[f(x)] = g(x^2) = x^2 - 1$$

Note that $f[g(x)] \neq g[f(x)]$.

PROGRESS CHECK 3

Given $f(x) = x^2 - 2x$, $g(x) = 3x$. Find the following.

(a) $f[g(-1)]$ (b) $g[f(-1)]$ (c) $f[g(x)]$

(d) $g[f(x)]$ (e) $(f \circ g)(2)$ (f) $(g \circ f)(2)$

Answers

(a) 15   (b) 9   (c) $9x^2 - 6x$ (d) $3x^2 - 6x$ (e) 24   (f) 0

Sometimes it is useful to think of a function as a composite of functions. For example, if

$$h(x) = (3x + 1)^4$$

we can let

$$f(x) = 3x + 1 \quad \text{and} \quad g(x) = x^4$$

Then

$$h(x) = (3x + 1)^4 = g(3x + 1) = g[f(x)]$$

Thus, $h(x) = g[f(x)]$ is a way of writing the function h as a composite of the functions f and g. In general, there may be many ways of writing a function as a composite.

EXAMPLE 4

Given

$$h(x) = \frac{\sqrt{x + 1}}{4}$$

Write the function h as a composite of two functions f and g.

Solution

(a) The form $\sqrt{x + 1}$ suggests that we let $g(x) = x + 1$. If we then let

$$f(x) = \frac{\sqrt{x}}{4}$$

we have

$$h(x) = \frac{\sqrt{x + 1}}{4} = f(x + 1) = f[g(x)]$$

(b) Alternatively, let $g(x) = \sqrt{x + 1}$. If

$$f(x) = \frac{x}{4}$$

we have

$$h(x) = \frac{\sqrt{x + 1}}{4} = f(\sqrt{x + 1}) = f[g(x)]$$

PROGRESS CHECK 4
Given

$$h(x) = \frac{1}{(2x - 5)^{25}}$$

Write the function h as a composite of two functions f and g.

Answer

$h(x) = f[g(x)]$, *where* $f(x) = \dfrac{1}{x^{25}}$, *and* $g(x) = 2x - 5$, *or*

$h(x) = f[g(x)]$, *where* $f(x) = x^{25}$, *and* $g(x) = \dfrac{1}{2x - 5}$

An element in the range of a function may correspond to more than one element in the domain of the function. In Figure 11.2 we see that y in Y

FIGURE 11.2

corresponds to both x_1 and x_2 in X. If we demand that every element in the domain be assigned to a *different* element of the range, then the function is called **one-to-one.** More formally,

A function f is one-to-one if $f(a) = f(b)$ only when $a = b$.

There is a simple means of determining if a function $y = f(x)$ is one-to-one by examining the graph of the function. In Figure 11.3a we see that a horizontal line meets the graph in more than one point. Thus, $f(x_1) = f(x_2)$ although $x_1 \neq x_2$; hence, the function is not one-to-one. On the other hand, no horizontal line meets the graph in Figure 11.3b in more than one point;

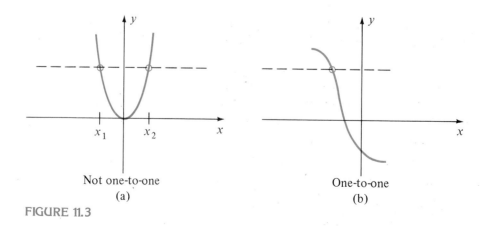

Not one-to-one
(a)

One-to-one
(b)

FIGURE 11.3

the graph thus determines a one-to-one function. In summary, we have the

Horizontal Line Test

If no horizontal line meets the graph of a function $y = f(x)$ in more than one point, then the function is one-to-one.

EXAMPLE 5
Which of the graphs in Figure 11.4 are graphs of one-to-one functions?

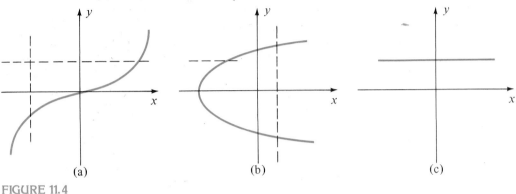

(a)   (b)   (c)

FIGURE 11.4

Solution
(a) No *vertical* line meets the graph in more than one point; hence, it is the graph of a function. No *horizontal* line meets the graph in more than one point; hence, it is the graph of a one-to-one function.
(b) No *horizontal* line meets the graph in more than one point. But *vertical* lines do meet the graph in more than one point. It is therefore not the graph of a function and consequently cannot be the graph of a one-to-one function.
(c) No *vertical* line meets the graph in more than one point; hence it is the graph of a function. But a *horizontal* line does meet the graph in more than one point. This is the graph of a function but not of a one-to-one function.

PROGRESS CHECK 5
Which of the graphs in Figure 11.5 are graphs of one-to-one functions?

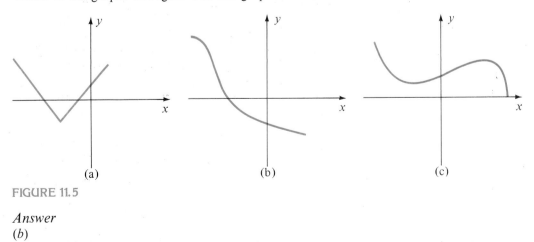

(a)   (b)   (c)

FIGURE 11.5

Answer
(b)

Suppose the function f in Figure 11.6a is a one-to-one function and that $y = f(x)$. Since f is one-to-one, we know that the correspondence is unique, that is, x in X is the *only* element of the domain for which $y = f(x)$. It is then

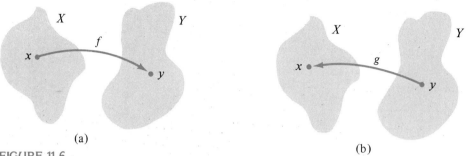

(a)

(b)

FIGURE 11.6

possible to define a function g (Figure 11.6b) with domain Y and range X which reverses the correspondence, that is,

$$g(y) = x \quad \text{for every } x \text{ in } X$$

If we substitute $y = f(x)$, we have

$$g[f(x)] = x \quad \text{for every } x \text{ in } X$$

The functions f and g in Figure 11.6 are said to be **inverse functions.** If we begin with x in X and successively apply the rules f and g, we see that

$$g[f(x)] = x$$

so that $g \circ f$ returns us to the starting point. Similarly, if we begin with y in Y and successively apply the rules g and f, we see that

$$f[g(y)] = y$$

so that $f \circ g$ returns us to the starting point. It is easy to show that the inverse of a one-to-one function is unique.

If f is a one-to-one function with domain X and range Y, then the function g with domain Y and range X and satisfying

$$g[f(x)] = x \quad \text{for every } x \text{ in } X$$
$$f[g(y)] = y \quad \text{for every } y \text{ in } Y$$

is called the **inverse function of f.**

Since the inverse (reciprocal) of a real number x ($\neq 0$) is $1/x$ or x^{-1}, it is natural to write the inverse of a function f as f^{-1}. Thus we have

$$f^{-1}[f(x)] = x \quad \text{for every } x \text{ in } X$$
$$f[f^{-1}(y)] = y \quad \text{for every } y \text{ in } Y$$

Graphically,

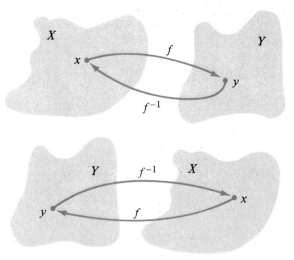

In the following sections we will study a very important class of inverse functions, the exponential and logarithmic functions. Always remember that we can define an inverse function of f only if f is one-to-one.

It is sometimes possible to find an inverse by algebraic methods. Here is a useful technique.

EXAMPLE 6

Find the inverse function of $f(x) = 2x - 3$.

Solution

TABLE 11.1

Inverse of f	Example
Step 1. Write $y = f(x)$.	*Step 1.* $y = 2x - 3$
Step 2. Solve for x.	*Step 2.* $y + 3 = 2x$ $x = \dfrac{y + 3}{2}$
Step 3. $f^{-1}(y)$ is the expression for x found in Step 2.	*Step 3.* $f^{-1}(y) = \dfrac{y + 3}{2}$
Step 4. To obtain $f^{-1}(x)$, replace y by x. (Note that y and x are simply symbols for a variable.)	*Step 4.* $f^{-1}(x) = \dfrac{x + 3}{2}$

PROGRESS CHECK 6

Given $f(x) = 3x + 5$, find f^{-1}.

Answer

$$f^{-1}(x) = \frac{x - 5}{3}$$

EXAMPLE 7

Given

$$h(x) = \frac{x - 3}{2}$$

Find   (a) $h^{-1}(x)$ (b) $h^{-1}(-1)$ (c) $h[h^{-1}(-1)]$ (d) $h^{-1}[h(-5)]$.

Solution

(a)   We find h^{-1} by following the procedure outlined in Example 6.

$$h(x) = \frac{x - 3}{2}$$

$$y = \frac{x - 3}{2} \qquad \text{Write } y = h(x).$$

$$x = 2y + 3 \qquad \text{Solve for } x.$$

$$h^{-1}(y) = 2y + 3$$

$$h^{-1}(x) = 2x + 3 \qquad \text{Replace } y \text{ by } x.$$

(b) $h^{-1}(-1) = 2(-1) + 3 = 1$

(c) $h[h^{-1}(-1)] = h(1) = \dfrac{1 - 3}{2} = -1$

(d) $h(-5) = \dfrac{-5 - 3}{2} = -4$

$h^{-1}[h(-5)] = h^{-1}(-4) = 2(-4) + 3 = -5$

Note that (c) and (d) illustrate the rules

$$h[h^{-1}(u)] = u$$

$$h^{-1}[h(x)] = x$$

PROGRESS CHECK 7

Given

$$g(x) = \frac{6 - x}{2}$$

Find   (a) $g^{-1}(x)$ (b) $g[g^{-1}(3)]$ (c) $g^{-1}[g(-2)]$.

Answers

(*a*) $g^{-1}(x) = 6 - 2x$ (*b*) 3   (*c*) -2

WARNING

(a) In general,

$$f^{-1}(x) \neq \frac{1}{f(x)}$$

If $f(x) = x - 1$, *don't* write

$$f^{-1}(x) = \frac{1}{x - 1}$$

Use the methods of this section to show that

$$f^{-1}(x) = x + 1$$

(b) The inverse function notation is *not* to be thought of as a power.

EXERCISE SET 11.1

Suppose $f(x) = x^2 + 1$, $g(x) = x - 2$. Compute the following.

1. (a) $(f + g)(2)$ (b) $(f + g)(x)$ (c) $(f - g)(3)$
2. (a) $(f - g)(x)$ (b) $(f \cdot g)(1)$ (c) $(f \cdot g)(x)$
3. (a) $\left(\frac{f}{g}\right)(-2)$ (b) $\left(\frac{f}{g}\right)(x)$ (c) $(f \cdot g)(2)$
4. (a) $(g \cdot f)(2)$ (b) $(g \cdot f)(-3)$ (c) $(g \cdot f)(x)$
5. The domain of (a) $\left(\frac{f}{g}\right)(x)$ (b) $(f \circ g)(x)$
6. The domain of (a) $(g \circ f)(x)$ (b) $\left(\frac{g}{f}\right)(x)$

Suppose $f(x) = 2x^2 - 1$, $g(x) = x + 1$. Compute the following.

7. (a) $(f + g)(-1)$ (b) $(f + g)(x)$ (c) $(f \cdot g)(x)$
8. (a) $(f - g)(x)$ (b) $(f \cdot g)(-2)$ (c) $(f - g)(2)$
9. (a) $\left(\frac{f}{g}\right)(-3)$ (b) $(g \cdot f)(-3)$ (c) $(f \cdot g)(1)$
10. (a) $(f \cdot g)(a)$ (b) $\left(\frac{f}{g}\right)(x)$ (c) $(g \cdot f)(x)$

Suppose $f(x) = 2x + 1$, $g(x) = 2x^2 + x$. Compute the following.

11. (a) $(f \circ g)(2)$ (b) $(f \circ g)(x)$ (c) $(g \circ f)(3)$
12. (a) $(g \circ f)(x)$ (b) $(f \circ g)(x + 1)$ (c) $(g \circ f)(x - 1)$
13. (a) $(f \circ f)(-2)$ (b) $(f \circ f)(x)$ (c) $(g \circ g)(2)$
14. (a) $(g \circ g)(x)$ (b) $(f \circ f)(x + 2)$ (c) $(g \circ g)(x - 3)$

Suppose $f(x) = x^2 + 4$, $g(x) = \sqrt{x + 2}$. Compute the following.

15. (a) $(f \circ g)(2)$ (b) $(f \circ g)(x)$ (c) $(g \circ f)(3)$
16. (a) $(g \circ f)(x)$ (b) $(f \circ f)(-1)$ (c) $(f \circ f)(x)$

17. The domain of (a) $(f \circ g)(x)$ (b) $(g \circ f)(x)$

18. The domain of (a) $(f \circ f)(x)$ (b) $(g \circ g)(x)$

In Exercises 19–22, compute $(f \circ g)(x)$ and $(g \circ f)(x)$.

19. $f(x) = x - 1,\ g(x) = x + 2$ 20. $f(x) = \sqrt{x + 1},\ g(x) = x + 2$

21. $f(x) = \dfrac{1}{x + 1},\ g(x) = \dfrac{1}{x - 1}$ 22. $f(x) = \dfrac{x + 1}{x - 1},\ g(x) = x$

In Exercises 23–34, write the given function $h(x)$ as a composite of two functions f and g so that $h(x) = (f \circ g)(x)$.

23. $h(x) = (3x + 2)^8$ 24. $h(x) = (x^3 + 2x^2 + 1)^{15}$

25. $h(x) = (x^3 - 2x^2)^{1/3}$ 26. $h(x) = \left(\dfrac{x^2 + 2x}{x^3 - 1}\right)^{3/2}$

27. $h(x) = (3x^2 + 1)^{20}$ 28. $h(x) = (3 - 2x^3)^{30}$

29. $h(x) = \sqrt{4 - x}$ 30. $h(x) = \sqrt{2x^2 - x + 2}$

31. $h(x) = (2 - 5x^2)^{-10}$ 32. $h(x) = \dfrac{1}{(3x^2 + 2x)^8}$

33. $h(x) = \sqrt{\dfrac{x - 2}{x + 5}}$ 34. $h(x) = (5x^3 + 4x^2 - 2x + 4)^{1/5}$

In Exercises 35–40, verify that $g = f^{-1}$ for the given functions f and g by showing that $f[g(x)] = x$ and $g[f(x)] = x$.

35. $f(x) = 2x + 4$ $g(x) = \dfrac{1}{2}x - 2$

36. $f(x) = 3x - 2$ $g(x) = \dfrac{1}{3}x + \dfrac{2}{3}$

37. $f(x) = 2 - 3x$ $g(x) = -\dfrac{1}{3}x + \dfrac{2}{3}$

38. $f(x) = x^3$ $g(x) = \sqrt[3]{x}$

39. $f(x) = \dfrac{1}{x}$ $g(x) = \dfrac{1}{x}$

40. $f(x) = \dfrac{1}{x - 2}$ $g(x) = \dfrac{1}{x} + 2$

Find the inverse function.

41. $f(x) = 2x + 3$ 42. $f(x) = 3x - 4$ 43. $f(x) = 3 - 2x$

44. $f(x) = \dfrac{1}{2}x + 1$ 45. $f(x) = \dfrac{1}{3}x - 5$ 46. $f(x) = 2 - \dfrac{1}{5}x$

47. $f(x) = x^3 + 1$ 48. $f(x) = \dfrac{1}{x + 1}$ 49. $f(x) = x^2,\ x \geq 0$

50. $f(x) = (x + 3)^2$

51. If $f(x) = \dfrac{1}{3}x + 2$, find (a) $f^{-1}(x)$ (b) $f^{-1}(2)$ (c) $(f \circ f^{-1})(2)$
     (d) $(f^{-1} \circ f)(3)$.

52. If $g(x) = 2x - 5$, find   (a) $g^{-1}(x)$ (b) $g^{-1}(3)$ (c) $(g \circ g^{-1})(2)$
    (d) $(g^{-1} \circ g)(0)$.

53. If $h(x) = 2 - 3x$, find   (a) $h^{-1}(x)$ (b) $h^{-1}(-2)$ (c) $(h \circ h^{-1})(3)$
    (d) $(h^{-1} \circ h)(-3)$.

54. If $F(x) = 4 - \frac{1}{3}x$, find   (a) $F^{-1}(x)$ (b) $F^{-1}(4)$ (c) $(F \circ F^{-1})(2)$
    (d) $(F^{-1} \circ F)(-1)$.

55. If $f(x) = x^3 - 2$, find   (a) $f^{-1}(x)$ (b) $f^{-1}(3)$ (c) $(f \circ f^{-1})(1)$
    (d) $(f^{-1} \circ f)(-2)$.

Use the horizontal line test to determine whether each of the following is a one-to-one function.

56. $f(x) = 2x - 1$ 57. $f(x) = 3 - 5x$

58. $f(x) = x^2 - 2x + 1$ 59. $f(x) = x^2 + 4x + 4$

60. $f(x) = -x^3 + 1$ 61. $f(x) = x^3 - 2$

62.
$$f(x) = \begin{cases} 2x, & x \le -1 \\ x^2, & -1 < x \le 0 \\ 3x - 1, & x > 0 \end{cases}$$

63.
$$f(x) = \begin{cases} x^2 - 4x + 4, & x \le 2 \\ x, & x > 2 \end{cases}$$

11.2
EXPONENTIAL FUNCTIONS

The function $f(x) = 2^x$ is very different from any of the functions we have worked with thus far. Previously, we defined functions by using the basic algebraic operations (addition, subtraction, multiplication, division, powers, and roots). However, $f(x) = 2^x$ has a variable in the exponent and doesn't fall into the class of algebraic functions. Rather, it is our first example of an **exponential function.**

> An exponential function has the form
> $$f(x) = a^x$$
> where $a > 0$, $a \ne 1$. The constant a is called the **base,** and the independent variable x may assume any real value.

The simplest way to become familiar with the exponential functions is to sketch their graphs.

EXAMPLE 1
Sketch the graph of $f(x) = 2^x$.

Solution
We let $y = 2^x$ and we form a table of values of x and y. We can now plot these points and sketch a smooth curve. See Figure 11.7.

x	y
-3	$\dfrac{1}{8}$
-2	$\dfrac{1}{4}$
-1	$\dfrac{1}{2}$
0	1
1	2
2	4
3	8

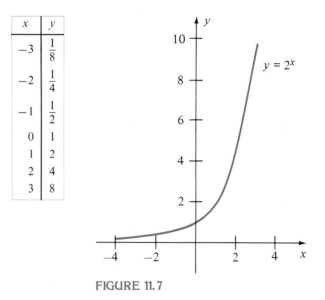

FIGURE 11.7

PROGRESS CHECK 1

Sketch the graphs of $f(x) = 2^x$ and $g(x) = 3^x$ on the same coordinate axes.

Answer

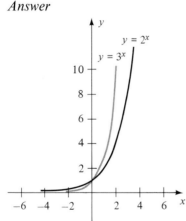

In a sense, we have cheated in our definition of $f(x) = 2^x$ and in sketching the graph in Figure 11.7. Since we have not explained the meaning of 2^x when x is irrational, we have no right to plot values such as $2^{\sqrt{2}}$. For our purposes, it will be adequate to think of $2^{\sqrt{2}}$ as the value we approach by taking successively closer approximations to $\sqrt{2}$, such as $2^{1.4}, 2^{1.41}, 2^{1.414}, \ldots$. A precise definition is given in more advanced mathematics courses where it is also shown that the laws of exponents hold for irrational exponents.

Let's look at $f(x) = a^x$ when $0 < a < 1$.

EXAMPLE 2

Sketch the graph of $f(x) = \left(\dfrac{1}{2}\right)^x = 2^{-x}$.

Solution
We form a table, plot points, and sketch the graph. See Figure 11.8.

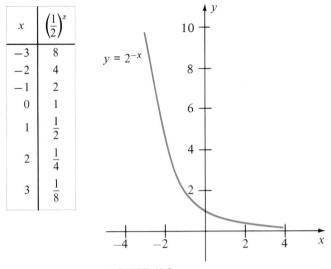

FIGURE 11.8

PROGRESS CHECK 2
Sketch the graphs of $f(x) = \left(\frac{1}{2}\right)^x$ and $g(x) = \left(\frac{1}{3}\right)^x$ on the same coordinate axes.

Answer

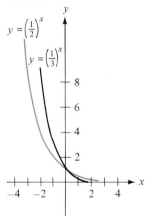

There are a number of interesting observations we can make by examining the graphs in Figures 11.7 and 11.8, and the answers to Progress Checks 1 and 2, always assuming that $a > 0$ and $a \neq 1$.

(a) Since $a^0 = 1$, the graph of $f(x) = a^x$ passes through the point $(0, 1)$ for any value of a.

(b) There are two basic types of exponential functions. The general shape of $f(x) = a^x$ when $a > 1$ can be seen in Figure 11.7 and the answer to Progress Check 1; the general shape when $0 < a < 1$ can be seen in Figure 11.8 and the answer to Progress Check 2.

(c) The domain of $f(x) = a^x$ consists of the set of all real numbers; the range is the set of all positive numbers.

(d) If $a > 1$, a^x is an increasing function; if $a < 1$, a^x is a decreasing function.

(e) If $a < b$, then $a^x < b^x$ for all $x > 0$ and $a^x > b^x$ for all $x < 0$. Notice in the answers to Progress Checks 1 and 2 that the curves exchange positions as they cross the y-axis, that is, at $x = 0$.

Since a^x is either increasing or decreasing, it never assumes the same value twice. (Recall that $a \neq 1$.) This leads to a useful conclusion.

$$\text{If } a^x = a^y, \text{ then } x = y$$

The graphs of a^x and b^x intersect only at $x = 0$. We can conclude that

$$\text{If } a^x = b^x \text{ for all } x \neq 0, \text{ then } a = b$$

EXAMPLE 3

Solve for x.

(a) $3^{10} = 3^{5x}$. Since $a^x = a^y$ implies $x = y$, we must have

$$10 = 5x$$
$$2 = x$$

(b) $2^7 = (x - 1)^7$. Since $a^x = b^x$ implies $a = b$, we must have

$$2 = x - 1$$
$$3 = x$$

PROGRESS CHECK 3

Solve for x.

(a) $2^8 = 2^{x+1}$ (b) $4^{2x+1} = 4^{11}$

Answers
(*a*) $x = 7$ (*b*) $x = 5$

There is an irrational number, denoted by the letter e, which plays an important role in mathematics. In calculus, we show that the expression

$$\left(1 + \frac{1}{m}\right)^m$$

gets closer and closer to the number e as m gets larger and larger. We can evaluate this expression for different values of m.

m	1	2	10	100	1000	10,000	100,000	1,000,000
$\left(1 + \dfrac{1}{m}\right)^m$	2.0	2.25	2.5937	2.7048	2.7169	2.7181	2.7182	2.71828

From this table we see that as m gets larger and larger the expression

$$\left(1 + \frac{1}{m}\right)^m$$

gets closer and closer to the number 2.71828 . . . which is an approximation to e.

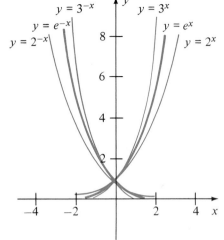

FIGURE 11.9

The graphs of $y = e^x$ and $y = e^{-x}$ are shown in Figure 11.9.

APPLICATIONS

Exponential functions occur in a wide variety of applied problems. We will look at problems dealing with population growth, such as the growth of bacteria in a culture medium; radioactive decay, such as determining the half-life of strontium-90; and interest earned when the interest rate is compounded.

The function Q defined by

$$Q(t) = q_0 e^{kt} \qquad (k > 0)$$

where the variable t represents time, is called an **exponential growth model;** k is a constant and t is the independent variable. We may think of Q as the quantity of a substance available at any given time t. Note that when $t = 0$ we have

$$Q(0) = q_0 e^0 = q_0$$

which says that q_0 is the initial quantity. (It is customary to use the subscript 0 to denote an initial value.) The constant k is called the **growth constant.**

EXAMPLE 4
The number of bacteria in a culture after t hours is described by the exponential growth model

$$Q(t) = 50 e^{0.7t}$$

(a) Find the initial number of bacteria, q_0, in the culture.
(b) How many bacteria are there in the culture after 10 hours?

Solution

(a) To find q_0 we need to evaluate $Q(t)$ at $t = 0$.

$$Q(0) = 50e^0 = 50 = q_0$$

Thus, initially there are 50 bacteria in the culture.

(b) The number of bacteria in the culture after 10 hours is given by $Q(10)$.

$$Q(10) = 50e^{0.7(10)} = 50e^7 = 50(1096.6) = 54{,}830$$

Thus, there are 54,830 bacteria after 10 hours. (The value $e^7 = 1096.6$ can be found by using Table I in the Appendix; it can also be found by using a calculator with a "y^x" key, with $y = e = 2.71828$ and $x = 7$.)

PROGRESS CHECK 4

The number of bacteria in a culture after t minutes is described by the exponential growth model $Q(t) = q_0e^{0.005t}$. If there were 100 bacteria present initially, how many bacteria will be present after one hour has elapsed?

Answer
135

The model defined by the function

$$Q(t) = q_0e^{-kt} \qquad (k > 0)$$

is called an **exponential decay model;** k is a constant, called the **decay constant,** and t is the independent variable.

EXAMPLE 5

A substance has a decay rate of 5% per hour. If 500 grams are present initially, how much of the substance remains after 4 hours?

Solution

The general equation of an exponential decay model is

$$Q(t) = q_0e^{-kt}$$

In our model, $q_0 = 500$ grams (since the quantity available initially is 500 grams) and $k = 0.05$ (since the decay rate is 5% per hour). After 4 hours

$$Q(4) = 500e^{-0.05(4)} = 500e^{-0.2} = 500(0.8187) = 409.4$$

($e^{-0.2} = 0.8187$ from Table I in the Appendix). Thus, there remain 409.4 grams of the substance.

PROGRESS CHECK 5

The number of grams Q of a certain radioactive substance present after t seconds is given by the exponential decay model $Q(t) = q_0e^{-0.4t}$. If 200 grams of the substance is present initially, find how much remains after 6 seconds.

Answer
18.1 grams

In Section 3.2 we studied simple interest as an application of linear equations. Recall that if the principal P is invested at a simple annual interest rate r for t years then the amount or sum that we will have on hand is given by

$$S = P + Prt$$

or

$$S = P(1 + rt)$$

In many business transactions the interest that is added to the principal at regular time intervals also earns interest. This is called the **compound interest process.**

The time period between successive additions of interest is known as the **conversion period.** Thus, if interest is compounded quarterly, the conversion period is three months; if interest is compounded semiannually, the conversion period is six months.

Suppose now that a principal P is invested at an annual interest rate r, compounded k times a year. Then each conversion period lasts $1/k$ years. Thus, the amount S_1 at the end of the first conversion period is

$$S_1 = P + P \cdot r \cdot \frac{1}{k} = P\left(1 + \frac{r}{k}\right)$$

The amount S_2 at the end of the second conversion period is

$$S_2 = S_1 + \text{interest earned by } S_1$$

$$S_2 = P\left(1 + \frac{r}{k}\right) + P\left(1 + \frac{r}{k}\right) \cdot r \cdot \frac{1}{k}$$

$$= \left[P\left(1 + \frac{r}{k}\right)\right]\left(1 + \frac{r}{k}\right)$$

or

$$S_2 = P\left(1 + \frac{r}{k}\right)^2$$

In this way we see that the amount S after n conversion periods is given by

$$S = P\left(1 + \frac{r}{k}\right)^n$$

which is usually written

$$S = P(1 + i)^n$$

where $i = r/k$. Table IV in the Appendix gives values of $(1 + i)^n$ for a number of values of i and n.

EXAMPLE 6
Suppose that $6000 is invested at an annual interest rate of 8% compounded quarterly. What is the value of the investment after 3 years?

Solution
We are given $P = 6000$, $r = 0.08$, $k = 4$, and $n = 12$ (since there are four conversion periods per year for three years). Thus,

$$i = \frac{r}{k} = \frac{0.08}{4} = 0.02$$

and

$$S = P(1 + i)^n$$
$$= 6000(1 + 0.02)^{12}$$

We refer to Table IV in the Appendix, with $i = 0.02$ and $n = 12$, and obtain

$$S = 6000(1.26824179) = 7609.45$$

Thus, the sum at the end of the three-year period is $7609.45.

PROGRESS CHECK 6

Suppose that $5000 is invested at an annual interest rate of 6% compounded semi-annually. What is the value of the investment after 12 years?

Answer

$10,163.97

EXERCISE SET 11.2

Sketch the graph of each given function.

1. $f(x) = 4^x$

2. $f(x) = 4^{-x}$

3. $f(x) = \left(\dfrac{1}{4}\right)^x$

4. $f(x) = \left(\dfrac{1}{4}\right)^{-x}$

5. $f(x) = \left(\dfrac{1}{2}\right)2^x$

6. $f(x) = \left(-\dfrac{1}{3}\right)2^x$

7. $f(x) = 2(3^x)$

8. $f(x) = -2(3^x)$

9. $f(x) = 2^{x+1}$

10. $f(x) = 2^{x-1}$

11. $f(x) = 2^{2x}$

12. $f(x) = 3^{-2x}$

13. $f(x) = e^{2x}$

14. $f(x) = e^{-2x}$

15. $f(x) = e^{x+1}$

16. $f(x) = e^{x-2}$

17. $f(x) = 40e^{0.20x}$

18. $f(x) = 50e^{-0.40x}$

Solve for x.

19. $2^x = 2^3$

20. $2^{x-1} = 2^4$

21. $2^{2x-1} = 2^5$

22. $3^{-x+1} = 3^4$

23. $3^x = 9^{x-2}$

24. $2^x = 8^{x+2}$

25. $2^{3x} = 4^{x+1}$

26. $3^{4x} = 9^{x-1}$

27. $e^x = e^3$

28. $e^{x-1} = e^3$

29. $e^{2x+1} = e^3$

30. $e^{-2x-3} = e^9$

31. $e^x = e^{2x+1}$

32. $e^{x-1} = 1$

Solve for a.

33. $(a + 1)^x = (2a - 1)^x$

34. $(2a + 1)^x = (a + 4)^x$

35. $(a + 1)^x = (2a)^x$

36. $(2a + 3)^x = (3a + 1)^x$

37. The number of bacteria in a culture after t hours is described by the exponential growth model $Q(t) = 200e^{0.25t}$.
    (a) What is the initial number of bacteria in the culture?
    (b) What is the growth constant?
    (c) Find the number of bacteria in the culture after 20 hours.
    (d) Use Table I in the Appendix to help you complete the following table.

t	1	4	8	10
Q				

38. The number of bacteria in a culture after t hours is described by the exponential growth model $Q(t) = q_0 e^{0.01t}$. If there were 500 bacteria present initially, how many bacteria will be present after 2 *days?*

39. At the beginning of 1975, the world population was approximately 4 billion. Suppose that the population is described by an exponential growth model, and that the rate of growth is 2% per year. Give the approximate world population in the year 2000.

40. The number of grams of potassium-42 present after t hours is given by the exponential decay model $Q(t) = q_0 e^{-0.055t}$. If 400 grams of the substance were present initially, how much remains after 10 hours?

41. A radioactive substance has a decay rate of 4% per hour. If 1000 grams are present initially, how much of the substance remains after 10 hours?

42. An investor purchases a $12,000 savings certificate paying 10% annual interest compounded semiannually. Find the amount received when the savings certificate is redeemed at the end of 8 years.

43. The parents of a newborn infant place $10,000 in an investment that pays 8% annual interest compounded quarterly. What sum is available at the end of 18 years to finance the child's college education?

44. A widow is offered a choice of two investments. Investment A pays 5% annual interest compounded semiannually, and investment B pays 6% compounded annually. Which investment will yield a greater return?

45. A firm intends to replace its present computer in 5 years. The treasurer suggests that $25,000 be set aside in an investment paying 6% compounded monthly. What sum will be available for the purchase of the new computer?

11.3
LOGARITHMIC FUNCTIONS

The two forms of the graph of $f(x) = a^x$ are shown in Figure 11.10. We have previously noted that the range of the function $f(x) = a^x$ is the set of all positive real numbers. When we combine this fact with the observation that $f(x) = a^x$ is either increasing or decreasing (since $a \neq 1$), we can conclude that the exponential function is a one-to-one function.

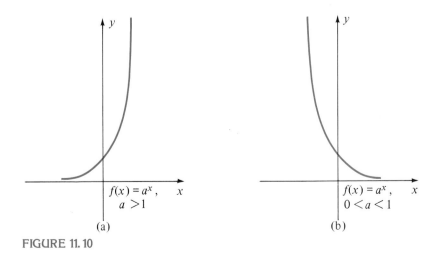

$f(x) = a^x$, $a > 1$

(a)

$f(x) = a^x$, $0 < a < 1$

(b)

FIGURE 11.10

In Figure 11.11a we see the function $f(x) = 2^x$ assigning values in the set Y for various values of x in the domain X. Since $f(x) = 2^x$ is a one-to-one

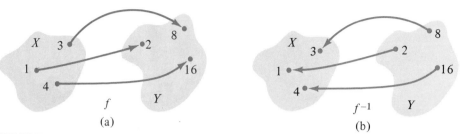

FIGURE 11.11

function, it makes sense to seek a function f^{-1} which will return the values of the range of f back to their origin as in Figure 11.11b. That is,

$$f \text{ maps 3 into 8}, \quad f^{-1} \text{ maps 8 into 3}$$

$$f \text{ maps 4 into 16}, \quad f^{-1} \text{ maps 16 into 4}$$

and so on. Since 2^x is always positive, we see that the domain of f^{-1} is the set of all positive real numbers. Its range is the set of all real numbers.

The function f^{-1} of Figure 11.11b has a special name, the **logarithmic function base 2,** which we write as $\log_2$. It is also possible to generalize and define the logarithmic function as the inverse of the exponential function with any base a.

Logarithmic Function Base a

$$y = \log_a x \quad \text{if and only if} \quad x = a^y$$

Examination of this definition shows that a *logarithm is an exponent.* If a is a positive number, then $\log_a x$ represents that power to which a must be raised to obtain x. Thus,

$$\log_2 16 = 4$$

since 2 must be raised to the fourth power to obtain 16. Similarly,

$$\log_{10} 100 = 2$$

since $10^2 = 100$. In short, when $a > 0$ and $a \neq 1$,

$$y = \log_a x \quad \text{and} \quad x = a^y$$

are equivalent statements.

The notation $\ln x$ is used to indicate logarithms to the base e. We call $\ln x$ the **natural logarithm of x.** Thus,

---

$$\ln x = \log_e x$$

EXAMPLE 1

Write in exponential form.

(a) $\log_3 9 = 2$ The exponential form is $3^2 = 9$.

(b) $\log_2 \dfrac{1}{8} = -3$ The exponential form is $2^{-3} = \dfrac{1}{8}$.

(c) $\log_{16} 4 = \dfrac{1}{2}$ The exponential form is $16^{1/2} = 4$.

(d) $\ln 7.39 = 2$ The exponential form is $e^2 = 7.39$.

(e) $\log_{10} 5 = 0.70$ The exponential form is $10^{0.70} = 5$.

PROGRESS CHECK 1

Write in exponential form.

(a) $\log_4 64 = 3$ (b) $\log_{10} \left(\dfrac{1}{10,000} \right) = -4$ (c) $\log_{25} 5 = \dfrac{1}{2}$

(d) $\ln 0.05 = -3$

Answers

(a) $4^3 = 64$ (b) $10^{-4} = \dfrac{1}{10,000}$ (c) $25^{1/2} = 5$ (d) $e^{-3} = 0.05$

EXAMPLE 2

Write in logarithmic form.

(a) $36 = 6^2$ The logarithmic form is $\log_6 36 = 2$.

(b) $7 = \sqrt{49}$ The logarithmic form is $\log_{49} 7 = \dfrac{1}{2}$.

(c) $\dfrac{1}{16} = 4^{-2}$ The logarithmic form is $\log_4 \dfrac{1}{16} = -2$.

(d) $0.1353 = e^{-2}$ The logarithmic form is $\ln 0.1353 = -2$.

PROGRESS CHECK 2

Write in logarithmic form.

(a) $64 = 8^2$ (b) $6 = 36^{1/2}$ (c) $\dfrac{1}{7} = 7^{-1}$ (d) $20.09 = e^3$

Answers

(a) $\log_8 64 = 2$ (b) $\log_{36} 6 = \dfrac{1}{2}$ (c) $\log_7 \dfrac{1}{7} = -1$ (d) $\ln 20.09 = 3$

Logarithmic equations can often be solved by changing to an equivalent exponential form.

EXAMPLE 3

Solve for x.

(a) $\log_3 x = -2$. The equivalent exponential form is

$$x = 3^{-2}$$

Thus,

$$x = \dfrac{1}{9}$$

(b) $\log_5 125 = x$. In exponential form we have

$$5^x = 125$$

Writing 125 in exponential form to the base 5, we have

$$5^x = 5^3$$

and since $a^x = a^y$ implies $x = y$, we conclude that

$$x = 3$$

(c) $\log_x 81 = 4$. The equivalent exponential form is

$$x^4 = 81 = 3^4$$

and thus

$$x = 3$$

(d) $\ln x = \dfrac{1}{2}$. The equivalent exponential form is

$$x = e^{1/2}$$

or

$$x = 1.65$$

which we obtain from Table I in the Appendix or by using a calculator with a "y^x" key.

PROGRESS CHECK 3
Solve for x.

(a) $\log_x 1000 = 3$ (b) $\log_2 x = 5$ (c) $x = \log_7 \dfrac{1}{49}$

Answers
(*a*) *10*   (*b*) *32*   (*c*) *−2*

If $f(x) = a^x$, then $f^{-1}(x) = \log_a x$. Recall that inverse functions have the property that

$$f[f^{-1}(x)] = x \quad \text{and} \quad f^{-1}[f(x)] = x$$

Substituting $f(x) = a^x$ and $f^{-1}(x) = \log_a x$, we have

$$f[f^{-1}(x)] = x \qquad f^{-1}[f(x)] = x$$
$$f(\log_a x) = x \qquad f^{-1}(a^x) \quad = x$$
$$a^{\log_a x} = x \qquad \log_a a^x \quad = x$$

The following two identities are useful in simplifying expressions and should be remembered.

$$a^{\log_a x} = x$$
$$\log_a a^x = x$$

Here is another pair of identities that can be verified by converting to the equivalent exponential form.

$$\log_a a = 1$$
$$\log_a 1 = 0$$

EXAMPLE 4

Evaluate.

(a) $8^{\log_8 5} = 5$

(b) $\log_{10} 10^{-3} = -3$

(c) $\log_7 7 = 1$

(d) $\log_4 1 = 0$

PROGRESS CHECK 4

Evaluate.

(a) $\log_3 3^4$ (b) $6^{\log_6 9}$ (c) $\log_5 1$ (d) $\log_8 8$

Answers

(a) 4   (b) 9   (c) 0   (d) 1

To graph $y = \log_a x$, convert to the equivalent exponential form $x = a^y$ and graph the second equation.

EXAMPLE 5

Sketch the graph of $f(x) = \log_2 x$.

Solution

To obtain the equivalent exponential equation, let

$$y = \log_2 x$$

Then solve for x.

$$x = 2^y$$

Now we form a table of values for $x = 2^y$.

y	-3	-2	-1	0	1	2	3
$x = 2^y$	$\dfrac{1}{8}$	$\dfrac{1}{4}$	$\dfrac{1}{2}$	1	2	4	8

We can now plot these points and sketch a smooth curve. See Figure 11.12.

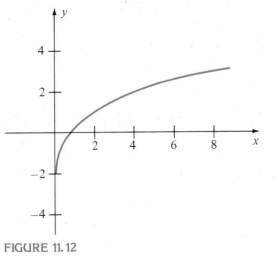

FIGURE 11.12

PROGRESS CHECK 5

Sketch the graphs of $y = \log_3 x$ and $y = \log_{1/3} x$ on the same coordinate axes.

Answer

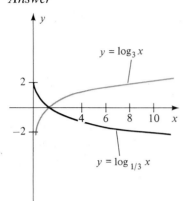

We can reach several conclusions by examining the graph in Figure 11.12 and the answer to Progress Check 5.

(a) The point $(1, 0)$ lies on the curve $y = \log_a x$ for any positive real number a. This is another way of saying $\log_a 1 = 0$.

(b) The domain of $f(x) = \log_a x$ is the set of all positive real numbers; the range is the set of all real numbers.

(c) When $a > 1$, $f(x) = \log_a x$ is an increasing function; when $0 < a < 1$, $f(x) = \log_a x$ is a decreasing function.

Since $\log_a x$ is either increasing or decreasing, the same value cannot be assumed more than once. Thus,

$$\text{If } \log_a x = \log_a y, \text{ then } x = y$$

Since the graphs of $\log_a x$ and $\log_b x$ intersect only at $x = 1$, we have:

$$\text{If } \log_a x = \log_b x \text{ and } x \neq 1, \text{ then } a = b$$

EXAMPLE 6

Solve for x.

(a) $\log_5(x + 1) = \log_5 25$

   $\qquad x + 1 = 25 \qquad\qquad$ If $\log_a x = \log_a y$, then $x = y$.

   $\qquad\qquad x = 24$

(b) $\log_{x-1} 31 = \log_5 31$

   $\qquad x - 1 = 5 \qquad\qquad$ If $\log_a x = \log_b x$ and $x \neq 1$, then $a = b$.

   $\qquad\qquad x = 6$

PROGRESS CHECK 6
Solve for x.

(a) $\log_2 x^2 = \log_2 9$ (b) $\log_7 14 = \log_{2x} 14$

Answers

(a) 3, −3   (b) $\dfrac{7}{2}$

The base 10 is used so frequently in work with logarithms that the notation $\log x$, with no base specified, is interpreted to mean $\log_{10} x$.

EXERCISE SET 11.3
Write each of the following in exponential form.

1. $\log_2 4 = 2$

2. $\log_5 125 = 3$

3. $\log_9 \dfrac{1}{81} = -2$

4. $\log_{64} 4 = \dfrac{1}{3}$

5. $\ln 20.09 = 3$

6. $\ln \dfrac{1}{73.9} = -2$

7. $\log_{10} 1000 = 3$

8. $\log_{10} \dfrac{1}{1000} = -3$

9. $\ln 1 = 0$

10. $\log_{10} 0.01 = -2$

11. $\log_3 \dfrac{1}{27} = -3$

12. $\log_{125} \dfrac{1}{5} = -\dfrac{1}{3}$

Write each of the following in logarithmic form.

13. $25 = 5^2$

14. $27 = 3^3$

15. $10,000 = 10^4$

16. $\dfrac{1}{100} = 10^{-2}$

17. $\dfrac{1}{8} = 2^{-3}$

18. $\dfrac{1}{27} = 3^{-3}$

19. $1 = 2^0$

20. $1 = e^0$

21. $6 = \sqrt{36}$

22. $2 = \sqrt[3]{8}$

23. $64 = 16^{3/2}$

24. $81 = 27^{4/3}$

25. $\dfrac{1}{3} = 27^{-1/3}$

26. $\dfrac{1}{2} = 16^{-1/4}$

Solve for x.

27. $\log_5 x = 2$

28. $\log_4 x = 2$

29. $\log_{16} x = \dfrac{1}{2}$

30. $\log_{25} x = -\dfrac{1}{2}$

31. $\log_{1/2} x = 3$

32. $\log_2 x = 1$

33. $\ln x = 2$

34. $\ln x = -3$

35. $\ln x = \dfrac{1}{3}$

36. $\ln x = -\dfrac{1}{2}$

37. $\log_4 64 = x$

38. $\log_5 \dfrac{1}{25} = x$

39. $\log_x 4 = \dfrac{1}{2}$

40. $\log_x \dfrac{1}{8} = -\dfrac{1}{3}$

41. $\log_3(x - 1) = 2$

42. $\log_5(x + 1) = 3$

43. $\log_3(x + 1) = \log_3 27$

44. $\log_2(x - 1) = \log_2 10$

45. $\log_{x+1} 24 = \log_3 24$

46. $\log_3 x^3 = \log_3 64$ 47. $\log_{x+1} 17 = \log_4 17$

48. $\log_{3x} 18 = \log_4 18$

Compute.

49. $3^{\log_3 6}$ 50. $2^{\log_2(2/3)}$ 51. $e^{\ln 2}$

52. $e^{\ln 1/2}$ 53. $\log_5 5^3$ 54. $\log_4 4^{-2}$

55. $\log_8 8^{1/2}$ 56. $\log_{64} 64^{-1/3}$ 57. $\log_7 49$

58. $\log_7 \sqrt{7}$ 59. $\log_5 5$ 60. $\ln e$

61. $\ln 1$ 62. $\log_4 1$ 63. $\log_3 \dfrac{1}{3}$

64. $\log_2 \dfrac{1}{4}$ 65. $\log_{16} 4$ 66. $\log_{36} \left(\dfrac{1}{6}\right)$

67. $\log 10{,}000$ 68. $\log 0.001$ 69. $\ln e^2$

70. $\ln e^{1/3}$ 71. $\ln e^{-2/3}$ 72. $\ln e^{-3}$

Sketch the graph of each given function.

73. $f(x) = \log_4 x$ 74. $f(x) = \log_{1/2} x$ 75. $f(x) = \log 2x$

76. $f(x) = \dfrac{1}{2}\log x$ 77. $f(x) = \ln \dfrac{x}{2}$ 78. $f(x) = \ln 3x$

79. $f(x) = \log_3(x - 1)$ 80. $f(x) = \log_3(x + 1)$

11.4
PROPERTIES OF LOGARITHMS

There are three fundamental properties of logarithms that have made them a powerful computational aid.

Property 1. $\log_a(x \cdot y) = \log_a x + \log_a y$

Property 2. $\log_a \left(\dfrac{x}{y}\right) = \log_a x - \log_a y$

Property 3. $\log_a x^n = n \log_a x$

These properties can be proved by using equivalent exponential forms. To prove the first property, $\log_a(x \cdot y) = \log_a x + \log_a y$, we let

$$\log_a x = u \quad \text{and} \quad \log_a y = v$$

Then the equivalent exponential forms are

$$a^u = x \quad \text{and} \quad a^v = y$$

Multiplying the left-hand and right-hand sides of these equations we have

$$a^u \cdot a^v = x \cdot y$$

or

$$a^{u+v} = x \cdot y$$

Substituting a^{u+v} for $x \cdot y$ in $\log_a(x \cdot y)$ we have

$$\log_a(x \cdot y) = \log_a(a^{u+v})$$

$$= u + v \qquad \text{Since } \log_a a^x = x.$$

Substituting for u and v,

$$\log_a(x \cdot y) = \log_a x + \log_a y$$

Properties 2 and 3 can be established in much the same way.

EXAMPLE 1

(a) $\log_{10}(225 \times 478) = \log_{10} 225 + \log_{10} 478$

(b) $\log_8\left(\dfrac{422}{735}\right) = \log_8 422 - \log_8 735$

(c) $\log_2(2^5) = 5 \log_2 2 = 5 \cdot 1 = 5$

(d) $\log_3(x \cdot y \cdot z) = \log_3 x + \log_3 y + \log_3 z$

(e) $\log_a\left(\dfrac{x \cdot y}{z}\right) = \log_a x + \log_a y - \log_a z$

PROGRESS CHECK 1

Write in terms of simpler logarithmic forms.

(a) $\log_4(1.47 \times 22.3)$ (b) $\log_5\left(\dfrac{149}{37.62}\right)$ (c) $\log_6(8)^4$

(d) $\log_a\left(\dfrac{m \cdot n}{p \cdot q}\right)$

Answers
(a) $\log_4 1.47 + \log_4 22.3$ (b) $\log_5 149 - \log_5 37.62$
(c) $4 \log_6 8$ (d) $\log_a m + \log_a n - \log_a p - \log_a q$

EXAMPLE 2

(a) $\log_a\left(\dfrac{x^2\sqrt{y}}{z^3}\right) = \log_a(x^2 y^{1/2}) - \log_a z^3$ Property 2

$$= \log_a x^2 + \log_a y^{1/2} - \log_a z^3 \qquad \text{Property 1}$$

$$= 2 \log_a x + \frac{1}{2} \log_a y - 3 \log_a z \qquad \text{Property 3}$$

(b) $\log_a \dfrac{x^{-1/2} y^{3/2}}{(z+1)^5} = \log_a \dfrac{y^{3/2}}{x^{1/2}(z+1)^5}$

$$= \log_a y^{3/2} - \log_a x^{1/2}(z+1)^5 \qquad \text{Property 2}$$

$$= \log_a y^{3/2} - \log_a x^{1/2} - \log_a(z+1)^5 \qquad \text{Property 1}$$

$$= \frac{3}{2} \log_a y - \frac{1}{2} \log_a x - 5 \log_a(z+1) \qquad \text{Property 3}$$

PROGRESS CHECK 2

Write in terms of simpler logarithmic forms.

(a) $\log_a \dfrac{x - 1}{\sqrt{x}}$ (b) $\log_a \dfrac{(x + 1)^5(x - 1)^4}{x^2}$

Answers

(a) $log_a(x - 1) - \dfrac{1}{2} log_a\, x$ (b) $5\, log_a(x + 1) + 4\, log_a(x - 1) - 2\, log_a\, x$

These rules will let you manipulate logarithmic forms more rapidly.

Rule	Example
1. Rewrite the expression so that each factor has a positive exponent.	$\log_a \dfrac{(x - 1)^{-2}(y + 2)^3}{(2x + 1)(y - 1)^{-4}}$
	$= \log_a \dfrac{(y + 2)^3(y - 1)^4}{(2x + 1)(x - 1)^2}$
2. Apply the rules for multiplication and division of logarithms (Property 1 and Property 2). Each factor in the numerator will yield a term with a plus sign. Each factor in the denominator will yield a term with a minus sign.	$= \log_a(y + 2)^3 + \log_a(y - 1)^4$ $- \log_a(2x + 1) - \log_a(x - 1)^2$

EXAMPLE 3

Simplify.

$$\log_a \sqrt{\dfrac{(x + 1)^3(x - 5)^{2/3}}{x - 2}}$$

Solution

$\log_a \sqrt{\dfrac{(x + 1)^3(x - 5)^{2/3}}{x - 2}}$

$= \dfrac{1}{2} \log_a \left[\dfrac{(x + 1)^3(x - 5)^{2/3}}{x - 2} \right]$ Property 3

$= \dfrac{1}{2}[\log_a(x + 1)^3 + \log_a(x - 5)^{2/3} - \log_a(x - 2)]$ Properties 1, 2

$= \dfrac{1}{2} \cdot 3 \log_a(x + 1) + \dfrac{1}{2} \cdot \dfrac{2}{3} \log_a(x - 5) - \dfrac{1}{2} \log_a(x - 2)$ Property 3

$= \dfrac{3}{2} \log_a(x + 1) + \dfrac{1}{3} \log_a(x - 5) - \dfrac{1}{2} \log_a(x - 2)$

PROGRESS CHECK 3

Simplify.

$$\log_a \left[\dfrac{(2x - 3)^{1/2}(y + 2)^{2/3}}{z^4} \right]$$

Answer

$\dfrac{1}{2} log_a(2x - 3) + \dfrac{2}{3} log_a(y + 2) - 4\, log_a\, z$

EXAMPLE 4

If $\log_a 1.5 = 0.37$, $\log_a 2 = 0.63$, and $\log_a 5 = 1.46$, find the following.

(a) $\log_a 7.5$

Since

$$7.5 = 1.5 \times 5$$

$$
\begin{aligned}
\log_a 7.5 &= \log_a(1.5 \times 5) \\
&= \log_a 1.5 + \log_a 5 \quad \text{Property 1} \\
&= 0.37 + 1.46 \quad\quad \text{Substitution} \\
&= 1.83
\end{aligned}
$$

(b) $\log_a(1.5)^3 \cdot \sqrt[5]{\dfrac{2}{5}}$

We write this as $\log_a(1.5)^3 + \log_a\left(\dfrac{2}{5}\right)^{1/5}$ Property 1

$$= 3\log_a 1.5 + \frac{1}{5}\log_a\left(\frac{2}{5}\right) \quad\quad \text{Property 3}$$

$$= 3\log_a 1.5 + \frac{1}{5}[\log_a 2 - \log_a 5] \quad \text{Property 2}$$

$$= 3(0.37) + \frac{1}{5}(0.63 - 1.46) \quad\quad \text{Substitution}$$

$$= 0.944$$

PROGRESS CHECK 4

If $\log_a 2 = 0.43$ and $\log_a 3 = 0.68$, find   (a) $\log_a 18$ (b) $\log_a \sqrt[3]{\dfrac{9}{2}}$.

Answers

(a) 1.79   (b) 0.31

WARNING

(a) *Don't* write

$$\log_a(x + y) = \log_a x + \log_a y$$

Property 1 tells us that

$$\log_a(x \cdot y) = \log_a x + \log_a y$$

Don't try to apply this property to $\log_a(x + y)$, which cannot be simplified.

(b) *Don't* write

$$\log_a x^n = (\log_a x)^n$$

By Property 3,

$$\log_a x^n = n\log_a x$$

We can also apply the properties of logarithms to combine terms involving logarithms.

EXAMPLE 5

Write as a single logarithm.

$$2 \log_a x - 3 \log_a(x + 1) + \log_a \sqrt{x - 1}$$

$$= \log_a x^2 - \log_a(x + 1)^3 + \log_a \sqrt{x - 1} \qquad \text{Property 3}$$

$$= \log_a x^2 \sqrt{x - 1} - \log_a(x + 1)^3 \qquad \text{Property 1}$$

$$= \log_a \frac{x^2 \sqrt{x - 1}}{(x + 1)^3} \qquad \text{Property 2}$$

PROGRESS CHECK 5

Write as a single logarithm.

$$\frac{1}{3}[\log_a(2x - 1) - \log_a(2x - 5)] + 4 \log_a x$$

Answer

$$\log_a x^4 \sqrt[3]{\frac{2x - 1}{2x - 5}}$$

WARNING

(a) *Don't* write

$$\frac{\log_a x}{\log_a y} = \log_a(x - y)$$

Property 2 tells us that

$$\log_a \left(\frac{x}{y}\right) = \log_a x - \log_a y$$

Don't try to apply this property to

$$\frac{\log_a x}{\log_a y}$$

which cannot be simplified.

(b) The expressions

and

$$\log_a x + \log_b x$$

$$\log_a x - \log_b x$$

cannot be simplified. Logarithms with different bases do not readily combine except in special cases.

EXERCISE SET 11.4

Find the error in each of the following.

1. $\log_2 12 = \log_2 3 - \log_2 4$

2. $\log_3 \left(\frac{7}{4}\right) = \log_3 7 + \log_3 4$

3. $\log_4(8.4 + 1.5) = \log_4 8.4 + \log_4 1.5$

4. $\log_3(7.6 - 4.2) = \log_3 7.6 - \log_3 4.2$

5. $\log_2 5^3 = (\log_2 5)^3$

6. $\log_3 \sqrt{5} = \sqrt{\log_3 5}$

7. $\ln 15 = \ln 5 - \ln 3$

8. $\ln \dfrac{8}{5} = \ln 8 + \ln 5$

9. $\ln(4 + 7) = \ln 4 + \ln 7$

10. $\ln(12.3 - 8.4) = \ln 12.3 - \ln 8.4$

11. $\ln 4^4 = (\ln 4)^4$

12. $\ln \sqrt[3]{5} = \sqrt[3]{\ln 5}$

Express each of the following in terms of simpler logarithmic forms.

13. $\log_{10}(120 \times 36)$

14. $\log_6\left(\dfrac{187}{39}\right)$

15. $\log_3(3^4)$

16. $\log_3(4^3)$

17. $\log_a(2xy)$

18. $\ln(4x \cdot y \cdot z)$

19. $\log_a\left(\dfrac{x}{yz}\right)$

20. $\ln\left(\dfrac{2x}{y}\right)$

21. $\ln x^5$

22. $\log_3 y^{2/3}$

23. $\log_a(x^2y^3)$

24. $\log_a(xy)^3$

25. $\log_a \sqrt{xy}$

26. $\log_a \sqrt[3]{xy^4}$

27. $\ln(x^2y^3z^4)$

28. $\log_a(xy^2z^2)$

29. $\ln(\sqrt{x} \sqrt[3]{y})$

30. $\ln \sqrt[3]{xy^2} \sqrt[4]{z}$

31. $\log_a\left(\dfrac{x^2y^3}{z^4}\right)$

32. $\ln \dfrac{x^4y^2}{z^{1/2}}$

33. $\ln\left(\dfrac{1}{a^2}\right)$

34. $\ln \sqrt{\dfrac{xz^2}{y}}$

If $\log_{10} 2 = 0.30$, $\log_{10} 3 = 0.47$, and $\log_{10} 5 = 0.70$, compute the following.

35. $\log_{10} 6$

36. $\log_{10} \dfrac{2}{3}$

37. $\log_{10} 9$

38. $\log_{10} \sqrt{5}$

39. $\log_{10} 12$

40. $\log_{10} 18$

41. $\log_{10} \dfrac{6}{5}$

42. $\log_{10} \dfrac{15}{2}$

43. $\log_{10} 0.30$

44. $\log_{10} \sqrt{7.5}$

45. $\log_{10} \dfrac{125}{36}$

46. $\log_{10} \sqrt[4]{30}$

Write each of the following as a single logarithm.

47. $2 \log x + \dfrac{1}{2} \log y$

48. $3 \log_a x - 2 \log_a z$

49. $\dfrac{1}{3} \ln x + \dfrac{1}{3} \ln y$

50. $\dfrac{1}{3} \ln x - \dfrac{2}{3} \ln y$

51. $\dfrac{1}{3} \log_a x + 2 \log_a y - \dfrac{3}{2} \log_a z$

52. $\dfrac{2}{3} \log_a x + \log_a y - 2 \log_a z$

53. $\dfrac{1}{2}(\log_a x + \log_a y)$

54. $\dfrac{2}{3}(4 \ln x - 5 \ln y)$

55. $\dfrac{1}{3}(2 \ln x + 4 \ln y) - 3 \ln z$

56. $\ln x - \dfrac{1}{2}(3 \ln x + 5 \ln y)$

57. $\dfrac{1}{2} \log_a(x - 1) - 2 \log_a(x + 1)$

58. $2 \log_a(x + 2) - \dfrac{1}{2}(\log_a y + \log_a z)$

59. $3 \log_a x - 2 \log_a(x - 1) + \dfrac{1}{2} \log_a \sqrt[3]{x + 1}$

60. $4 \ln(x - 1) + \dfrac{1}{2} \ln(x + 1) - 3 \ln y$

11.5
COMPUTING WITH LOGARITHMS

Properties 1 and 2 of logarithms allow us to replace multiplication and division by addition and subtraction; Property 3 permits us to replace exponentiation by multiplication. Logarithms can therefore be used to simplify complex calculations.

We will use 10 as the base for our computations with logarithms since 10 is the base of our number system. We call logarithms to the base 10 **common logarithms,** and it is customary to write log x rather than $\log_{10} x$. Logarithms to the base e are denoted by ln x, and are called **natural logarithms.**

We begin with the observation that any positive real number can be written as a product of a number c, $1 \le c < 10$, and a power of 10, say 10^k. This format is often referred to as **scientific notation.** Here are some examples.

$$643 = 6.43 \times 10^2 \qquad\qquad 4629 = 4.629 \times 10^3$$

$$754{,}000 = 7.54 \times 10^5 \qquad\qquad 1.76 = 1.76 \times 10^0$$

$$0.0423 = 4.23 \times 10^{-2} \qquad 0.0000926 = 9.26 \times 10^{-5}$$

Let's apply Property 1 of logarithms to the number 643 expressed in scientific notation. Thus,

$$643 = 6.43 \times 10^2$$

$$\log 643 = \log(6.43 \times 10^2)$$

$$= \log 6.43 + \log 10^2$$

$$= \log 6.43 + 2 \qquad\qquad \text{Since } \log_{10} 10^x = x.$$

In general,

> If x is a positive real number and $x = c \cdot 10^k$, then
>
> $$\log x = \log c + k$$

When $x = c \cdot 10^k$, with $1 \le c < 10$, the number log c is called the **mantissa** and the number k is called the **characteristic** of the number x. Since

$$x = c \cdot 10^k, \quad \text{where } 1 \le c < 10$$

and since the function $y = \log x$ is an increasing function, we see that

$$\log 1 \le \log c < \log 10$$

$$0 \le \log c < 1$$

We may conclude that the mantissa is always a number between 0 and 1.

Table II in the Appendix can be used to approximate the logarithm of any three-digit number between 1.00 and 9.99, at intervals of 0.01. Here is how it is used.

EXAMPLE 1
Find the following.

(a) log 73.5. Since $73.5 = 7.35 \times 10^1$, the characteristic of 7.35 is 1. Using Table II in the Appendix we find that log 7.35 is (approximately) 0.8663. Then

$$\log 73.5 = \log(7.35 \times 10^1) = \log 7.35 + \log 10^1$$
$$= 0.8663 + 1 = 1.8663$$

(b) log 2980. Since $2980 = 2.980 \times 10^3$, the characteristic of 2980 is 3. Using Table II we see that log 2.98 is (approximately) 0.4742. Then

$$\log 2980 = 0.4742 + 3 = 3.4742$$

(c) log 0.00451. Since $0.00451 = 4.51 \times 10^{-3}$, the characteristic of 0.00451 is -3. From Table II we have

$$\log 0.00451 = 0.6542 - 3$$

Here we have a positive mantissa and a negative characteristic. For reasons that will be clear later, we always leave the answer in this form.

PROGRESS CHECK 1
Find the following.

(a) log 69,700   (b) log 69.7   (c) log 0.000697   (d) log 0.697

Answers
(a) 4.8432   (b) 1.8432   (c) 0.8432 – 4   (d) 0.8432 – 1

WARNING
Note that $\log 0.00547 = 0.7380 - 3$. *Don't* write

$$\log 0.00547 = 0.7380 - 3 = -3.7380$$

since this is algebraically incorrect. In fact,

$$0.7380 - 3 = -2.2620$$

Table II in the Appendix can also be used in a reverse manner, that is, to find x if $\log x$ is known. Since the entries in the body of Table II are numbers between 0 and 1, we must first write the number $\log x$ in the form

$$\log x = m + c$$

where m is the mantissa, $0 \le m < 1$, and c the characteristic is an integer. (This is why we insisted in Example 1c that the number be left in the form $0.6542 - 3$. We can then identify the positive mantissa, 0.6542, and the integer characteristic, -3.)

EXAMPLE 2
Find x in the following.

(a) $\log x = 2.8351$. We look for the mantissa $c = 0.8351$ in the body of Table II in the Appendix and find that it corresponds to $\log 6.84$. Since the characteristic $k = 2$, we have

$$x = 6.84 \times 10^2 = 684$$

(b) $\log x = 0.4983 - 3$. From Table II,

$$0.4983 = \log 3.15$$

so that

$$x = 3.15 \times 10^{-3} = 0.00315$$

(c) $\log x = -6.6478$. We must rewrite this number before we can use the table since the mantissa m must be between 0 and 1. If we add and subtract 7, we have

$$\log x = 7 - 6.6478 - 7$$
$$= 0.3522 - 7$$

Now we may use Table II to find

$$0.3522 = \log 2.25$$

so that

$$x = 2.25 \times 10^{-7} = 0.000000225$$

PROGRESS CHECK 2
Find x in the following.

(a) $\log x = 3.8457$ (b) $\log x = 0.6201 - 2$ (c) $\log x = -2.0487$

Answers
(a) 7010   (b) 0.0417   (c) 0.00894

Inexpensive calculators have reduced the importance of logarithms as a computational device. Still, many calculators cannot, for example, handle $\sqrt[5]{14.2}$ directly. If you know how to compute with logarithms and combine this knowledge with a calculator that can handle logarithms, you will have enhanced the power of your calculator. Many additional applications of logarithms occur in more advanced mathematics, especially in calculus.

EXAMPLE 3
Approximate 478×0.0345 by using logarithms.

Solution
If $N = 478 \times 0.0345$, then

$$\log N = \log(478 \times 0.0345)$$
$$= \log 478 + \log 0.0345 \qquad \text{Property 1}$$

Using Table II in the Appendix and adding the logarithms,

$$\log 478 = 2.6794$$

$$\log 0.0345 = 0.5378 - 2$$

$$\overline{\log N = 3.2172 - 2}$$

$$= 1.2172$$

Looking in the body of Table II, we find that the mantissa 0.2172 does not appear. However, 0.2175 does appear and corresponds to log 1.65. Thus,

$$N \approx 1.65 \times 10^1 = 16.5$$

PROGRESS CHECK 3
Approximate 22.6×0.0023 by using logarithms.

Answer
0.052

EXAMPLE 4
Approximate by using logarithms.

$$\frac{\sqrt{47.4}}{(2.3)^3}$$

Solution
If

$$N = \frac{\sqrt{47.4}}{(2.3)^3}$$

then

$$\log N = \frac{1}{2} \log 47.4 - 3 \log 2.3$$

$$\frac{1}{2} \log 47.4 = \frac{1}{2}(1.6758) \qquad = \qquad 0.8379$$

$$\frac{3 \log 2.3 = 3(0.3617) \qquad = \qquad 1.0851}{\log N = 0.8379 - 1.0851 = -0.2472}$$

or

$$\log N = 0.7528 - 1$$

From the body of Table II in the Appendix

$$N \approx 5.66 \times 10^{-1} = 0.566$$

PROGRESS CHECK 4
Approximate by logarithms.

$$\frac{(4.64)^{3/2}}{\sqrt{7.42 \times 165}}$$

Answer
0.286

Problems in compound interest provide us with an opportunity to demonstrate the power of logarithms in computational work. In Section 11.2 we showed that the amount S available when a principal P is invested at an annual interest rate r compounded k times a year is given by

$$S = P(1 + i)^n$$

where $i = r/k$, and n is the number of conversion periods.

EXAMPLE 5

If $1000 is left on deposit at an interest rate of 8% per year compounded quarterly, how much money is in the account at the end of 6 years?

Solution

We have $P = 1000$, $r = 0.08$, $k = 4$, and $n = 24$ (since there are 24 quarters in 6 years). Thus,

$$S = P(1 + i)^n = 1000\left(1 + \frac{0.08}{4}\right)^{24}$$

$$= 1000(1 + 0.02)^{24} = 1000(1.02)^{24}$$

Then

$$\log S = \log 1000 + 24 \log 1.02$$

$$= 3 + 24(0.0086) = 3.2064$$

From the body of Table II in the Appendix,

$$S \approx 1.61 \times 10^3 = 1610$$

The account contains (approximately) $1610 at the end of 6 years.

PROGRESS CHECK 5

If $1000 is left on deposit at an interest rate of 6% per year compounded semiannually, approximately how much is in the account at the end of 4 years?

Answer
$1267

EXERCISE SET 11.5

Write each of the following in scientific notation.

1. 2725	2. 493	3. 0.0084
4. 0.000914	5. 716,000	6. 527,600,000
7. 296.2	8. 32.767	9. 965.01
10. 76.85	11. 3.45	12. 6.831

Compute the following logarithms using Tables II and III in the Appendix.

13. $\log 3.56$	14. $\log 4.84$	15. $\ln 3.2$
16. $\ln 16$	17. $\log 37.5$	18. $\log 85.3$
19. $\ln 4.7$	20. $\ln 60$	21. $\log 74$
22. $\log 4230$	23. $\log 48,200$	24. $\log 7,890,000$
25. $\log 0.342$	26. $\log 0.2$	27. $\log 0.00532$
28. $\ln 0.2$	29. $\log 0.8$	30. $\log 0.00000654$

Using Tables II and III in the Appendix, find x.

31. $\log x = 0.4014$ 32. $\log x = 0.8476$ 33. $\ln x = -0.5108$

34. $\ln x = 1.0647$ 35. $\log x = 2.7332$ 36. $\log x = 3.8142$

37. $\ln x = 2.0669$ 38. $\ln x = 4.2485$ 39. $\log x = 0.1903 - 2$

40. $\log x = 0.4099 - 1$ 41. $\log x = 0.7024 - 2$ 42. $\log x = 0.7832 - 4$

43. $\log x = 0.9320 - 2$ 44. $\log x = 0.9671 - 1$ 45. $\log x = -0.3118$

46. $\log x = -1.6599$ 47. $\log x = -2.9600$ 48. $\log x = -3.9004$

Using Table II in the Appendix, find the approximate value of x.

49. $\log x = 2.8144$ 50. $\log x = 1.7885$ 51. $\log x = 0.1830$

52. $\log x = 3.6614$ 53. $\log x = 2.9060$ 54. $\log x = 0.2100 - 1$

55. $\log x = 0.7366 - 2$ 56. $\log x = 0.9312 - 3$ 57. $\log x = -0.3200$

58. $\log x = -1.1940$ 59. $\log x = -2.4784$ 60. $\log x = -3.8425$

Find an approximate answer using logarithms.

61. $(320)(0.00321)$ 62. $(8780)(2.13)$

63. $(2.13)(48.2)(6.54)$ 64. $(378)(25.4)(0.0123)$

65. $\dfrac{679}{321}$ 66. $\dfrac{88.3}{97.2}$ 67. $\dfrac{(32.1)(2.78)}{(0.0474)}$

68. $\dfrac{73.4}{(4.25)(0.382)}$ 69. $(3.19)^4$ 70. $(27.4)^3$

71. $(42.3)^3(71.2)^2$ 72. $\dfrac{(87.3)^2(0.125)^3}{(17.3)^3}$ 73. $\sqrt[3]{(66.9)^4(0.781)^2}$

74. $\dfrac{\sqrt{7870}}{(46.3)^4}$ 75. $\dfrac{(7.28)^{2/3}}{\sqrt[3]{(87.3)(16.2)^4}}$ 76. $\dfrac{(32,870)(0.00125)}{(12.8)(124,000)}$

77. $\dfrac{\sqrt[3]{748}\,\sqrt[4]{(56.3)^3}}{\sqrt{76.3}}$ 78. $\dfrac{(14.7)^2}{\sqrt[3]{(87.3)^2}\,\sqrt{(0.00281)^3}}$

79. The period T (in seconds) of a simple pendulum of length L (in feet) is given by the formula

$$T = 2\pi\sqrt{\dfrac{L}{g}}$$

Using common logarithms, find the approximate value of T if $L = 4.72$ feet, $g = 32.2$, and $\pi = 3.14$.

80. Use logarithms to find the sum available when $6000 is invested for 8 years in a bank paying 7% interest per year compounded quarterly.

81. Use logarithms to find the sum available when $8000 is invested for 6 years in a bank paying 8% interest per year compounded monthly.

82. If $10,000 is invested at 7.8% interest per year compounded semiannually, what sum is available after 5 years?

83. Which of the following offers will yield a greater return, 8% annual interest compounded annually or 7.75% annual interest compounded quarterly?

84. Which of the following offers will yield a greater return, 9% annual interest compounded annually or 8.75% annual interest compounded quarterly?

85. The area of a triangle whose sides are a, b, and c is given by the formula

$$A = \sqrt{s(s-a)(s-b)(s-c)}$$

where $s = \frac{1}{2}(a+b+c)$. Use logarithms to find the approximate area of a triangle whose sides are 12.86 feet, 13.72 feet, and 20.3 feet.

11.6
EXPONENTIAL AND LOGARITHMIC EQUATIONS

The following approach will often help to solve exponential and logarithmic equations.

> To solve an exponential equation, take logarithms of both sides of the equation.
>
> To solve a logarithmic equation, convert to the equivalent exponential form.

At times it may be necessary to simplify before converting to an equivalent form. Here are some examples.

EXAMPLE 1
Solve $3^{2x-1} = 17$.

Solution
Taking logarithms to the base 10 of both sides of the equation,

$$\log 3^{2x-1} = \log 17$$

$$(2x - 1)\log 3 = \log 17 \qquad \text{Property 3}$$

$$2x - 1 = \frac{\log 17}{\log 3}$$

$$2x = 1 + \frac{\log 17}{\log 3}$$

$$x = \frac{1}{2} + \frac{\log 17}{2 \log 3}$$

If a numerical value is required, Table II in the Appendix can be used to approximate log 17 and log 3.

PROGRESS CHECK 1
Solve $2^{x+1} = 3^{2x-3}$.

Answer
$$\frac{\log 2 + 3 \log 3}{2 \log 3 - \log 2}$$

EXAMPLE 2

Solve $\log(2x + 8) = 1 + \log(x - 4)$.

Solution

If we rewrite the equation in the form

$$\log(2x + 8) - \log(x - 4) = 1$$

then we can apply Property 2 to form a single logarithm.

$$\log \frac{2x + 8}{x - 4} = 1$$

Now we convert to the equivalent exponential form.

$$\frac{2x + 8}{x - 4} = 10^1 = 10$$

$$2x + 8 = 10x - 40$$

$$-8x = -48$$

$$x = 6$$

PROGRESS CHECK 2

Solve $\log(x + 1) = 2 + \log(3x - 1)$.

Answer

$$\frac{101}{299}$$

EXAMPLE 3

Solve $\log_2 x = 3 - \log_2(x + 2)$.

Solution

Rewriting the equation with a single logarithm, we have

$\log_2 x + \log_2(x + 2) = 3$

$\log_2[x(x + 2)] = 3$ Why?

$x(x + 2) = 2^3 = 8$ Equivalent exponential form

$x^2 + 2x - 8 = 0$

$(x - 2)(x + 4) = 0$ Factoring

$x = 2$ or $x = -4$

The "solution" $x = -4$ must be rejected since the original equation contains $\log_2 x$ which requires x to be positive.

PROGRESS CHECK 3

Solve $\log_3(x - 8) = 2 - \log_3 x$.

Answer

$x = 9$

EXAMPLE 4

World population is increasing at an annual rate of 2.5%. If we assume an exponential growth model, in how many years will the population double?

Solution

The exponential growth model

$$Q(t) = q_0 e^{0.025t}$$

describes the population Q as a function of time t. Since the initial population is $Q(0) = q_0$, we seek the time t required for the population to double or become $2q_0$. We wish to solve the equation

$$Q(t) = 2q_0 = q_0 e^{0.025t}$$

for t. We then have

$$2q_0 = q_0 e^{0.025t}$$

$$2 = e^{0.025t} \qquad \text{Divide by } q_0.$$

$$\ln 2 = \ln e^{0.025t} \qquad \text{Take natural logs of both sides.}$$

$$\ln 2 = 0.025t \qquad \text{Since } \ln e^x = x.$$

$$t = \frac{\ln 2}{0.025} = \frac{0.6931}{0.025} = 27.7$$

or approximately 28 years.

PROGRESS CHECK 4

In a nuclear power plant accident, strontium-90 has been deposited outside the plant and nearby residents have been evacuated. The decay constant of strontium-90 is 2.77% per year. Public officials will not allow the residents to return to their homes until 90% of the radioactivity has disappeared. Assuming that no cleanup efforts are made, how long will it be before the nearby area is once again fit for human habitation?

Answer
83 years

EXERCISE SET 11.6

Solve for x.

1. $5^x = 18$	2. $2^x = 24$	3. $2^{x-1} = 7$
4. $3^{x-1} = 12$	5. $3^{2x} = 46$	6. $2^{2x-1} = 56$
7. $5^{2x-5} = 564$	8. $3^{3x-2} = 23.1$	9. $3^{x-1} = 2^{2x+1}$
10. $4^{2x-1} = 3^{2x+3}$	11. $2^{-x} = 15$	12. $3^{-x+2} = 103$
13. $4^{-2x+1} = 12$	14. $3^{-3x+2} = 2^{-x}$	15. $e^x = 18$
16. $e^{x-1} = 2.3$	17. $e^{2x+3} = 20$	18. $e^{-3x+2} = 40$

19. $\log x + \log 2 = 3$ 20. $\log x - \log 3 = 2$

21. $\log_x(3 - 5x) = 1$ 22. $\log_x(8 - 2x) = 2$

23. $\log x + \log(x - 3) = 1$ 24. $\log x + \log(x + 21) = 2$

25. $\log(3x + 1) - \log(x - 2) = 1$ 26. $\log(7x - 2) - \log(x - 2) = 1$

27. $\log_2 x = 4 - \log_2(x - 6)$ 28. $\log_2(x - 4) = 2 - \log_2 x$

29. $\log_2(x + 4) = 3 - \log_2(x - 2)$

30. Suppose that world population is increasing at an annual rate of 2%. If we assume an exponential growth model, in how many years will the population double?

31. Suppose that the population of a certain city is increasing at an annual rate of 3%. If we assume an exponential growth model, in how many years will the population triple?

32. The population P of a certain city t years from now is given by

$$P = 20,000e^{0.05t}$$

How many years from now will the population be 50,000?

33. Potassium-42 has a decay rate of approximately 5.5% per hour. Assuming an exponential decay model, in how many hours will the original quantity of potassium-42 have been halved?

34. Consider an exponential decay model given by

$$Q = q_0e^{-0.4t}$$

where t is in weeks. How many weeks does it take for Q to decay to one fourth of its original amount?

35. How long does it take an amount of money to double if it is invested at a rate of 8% per year compounded semiannually?

36. At what rate of annual interest, compounded semiannually, should a certain amount of money be invested so it will double in 8 years?

37. The number N of radios that an assembly-line worker can assemble daily after t days of training is given by

$$N = 60 - 60e^{-0.04t}$$

After how many days of training does the worker assemble 40 radios daily?

38. The quantity Q (in grams) of a radioactive substance that is present after t days of decay is given by

$$Q = 400e^{-kt}$$

If $Q = 300$ when $t = 3$, find k, the decay rate.

39. A person on an assembly line produces P items per day after t days of training. If

$$P = 400(1 - e^{-t})$$

how many days of training will it take this person to be able to produce 300 items per day?

40. Suppose that the number N of mopeds sold when x thousands of dollars are spent on advertising is given by

$$N = 4000 + 1000 \ln(x + 2)$$

How much advertising money must be spent to sell 6000 mopeds?

TERMS AND SYMBOLS

composite function (p. 310)
$g \circ f$ (p. 310)
$g[f(x)]$ (p. 310)
one-to-one function (p. 312)
horizontal line test (p. 313)
inverse function (p. 314)
$f^{-1}(x)$ (p. 314)
exponential function (p. 319)

a^x (p. 319)
base (p. 319)
e (p. 322)
exponential growth model (p. 323)
growth constant (p. 323)
exponential decay model (p. 324)
compound interest (p. 325)
conversion period (p. 325)

logarithmic function base a (p. 328)
$\log_a x$ (p. 328)
$\ln x$ (p. 328)
natural logarithm (p. 328)
common logarithm (p. 340)
scientific notation (p. 340)
mantissa (p. 340)
characteristic (p. 340)

KEY IDEAS FOR REVIEW

☐ A composite function is a function of a function.

☐ The inverse of a function, f^{-1}, reverses the correspondence defined by the function f. The domain of f becomes the range of f^{-1} and the range of f becomes the domain of f^{-1}.

☐ The inverse of a function f is defined only if f is a one-to-one function.

☐ An exponential function has a variable in the exponent and a constant base.

☐ The graph of the exponential function $f(x) = a^x$
  (a) passes through the points $(0, 1)$ and $(1, a)$ for any value of x.
  (b) is increasing when $a > 1$ and decreasing when $0 < a < 1$.

☐ The domain of the exponential function is the set of all real numbers; the range is the set of all positive numbers.

☐ If $a^x = a^y$, then $x = y$, assuming $a > 0$.

☐ If $a^x = b^x$ for all $x \neq 0$, then $a = b$, assuming $a, b > 0$.

☐ The logarithmic and exponential functions are inverse functions.

☐ $y = \log_a x$ if and only if $x = a^y$. This enables us to write logarithmic equations in exponential form and exponential equations in logarithmic form—a "trick" that is often useful in solving such equations.

☐ The following identities are useful in simplifying expressions and in solving equations.

$$a^{\log_a x} = x \qquad \log_a a = 1$$
$$\log_a a^x = x \qquad \log_a 1 = 0$$

☐ The graph of the logarithmic function $f(x) = \log_a x$
  (a) passes through the points $(1, 0)$ and $(a, 1)$ for any $a > 0$.
  (b) is increasing if $a > 1$ and decreasing if $0 < a < 1$.

☐ The domain of the logarithmic function is the set of all positive real numbers; the range is the set of all real numbers.

☐ If $\log_a x = \log_a y$, then $x = y$.

☐ If $\log_a x = \log_b x$ and $x \neq 1$, then $a = b$.

☐ The fundamental properties of logarithms are

  Property 1. $\log_a(x \cdot y) = \log_a x + \log_a y$
  Property 2. $\log_a\left(\dfrac{x}{y}\right) = \log_a x - \log_a y$
  Property 3. $\log_a x^n = n \log_a x$

☐ The fundamental properties of logarithms in conjunction with tables of logarithms are a powerful tool in performing calculations.

COMMON ERRORS

1. In general, $f^{-1}(x) \neq \dfrac{1}{f(x)}$.

2. *Don't* write

  or
  $$\log_a(x + y) = \log_a x + \log_a y$$
  $$\log_a(x - y) = \log_a x - \log_a y$$

  Logarithms are not distributive.

3. *Don't write*

$$\log_a x^n = (\log_a x)^n$$

Instead, note that

$$\log_a x^n = n \log_a x$$

4. *Don't write*

$$\frac{\log_a x}{\log_a y} = \log_a \left(\frac{x}{y}\right)$$

You cannot simplify

$$\frac{\log_a x}{\log_a y}$$

5. *Don't write*

$$\frac{\log_a x}{\log_a y} = \log_a(x - y)$$

Recall that

$$\log_a \left(\frac{x}{y}\right) = \log_a x - \log_a y$$

6. In calculating with logarithms, note that

$$\log 0.0362 = 0.5587 - 2$$

Don't write

$$\log 0.0362 = -2.5587$$

since these statements are not algebraically equivalent.

PROGRESS TEST 11A

1. If $f(x) = x - 1$ and $g(x) = -x^2 + x$, compute the following.
   (a) $\left(\frac{f}{g}\right)(2)$ (b) $(g \circ g)(-2)$

2. Verify that $f(x) = 2x + 4$ and $g(x) = \frac{x}{2} - 2$ are inverse functions.

3. Solve $2^{2x} = 8^{x-1}$ for x.

4. Solve $(2a + 1)^x = (3a - 1)^x$ for a.

5. Write $\log_{64} 8 = \frac{1}{2}$ in exponential form.

6. Write $27 = 9^{3/2}$ in logarithmic form.

7. Solve $\ln(x + 1) = -\frac{1}{4}$ for x.

8. Solve $\log_x 16 = 4$ for x.

9. Solve $\log_5 \frac{1}{125} = x - 1$ for x.

10. Compute $\log_3 3^5$.

11. Compute $\ln e^{-1/3}$.

12. Sketch the graph of $f(x) = \left(\dfrac{1}{3}\right)^x$.

13. Write $\log_5 \sqrt{\dfrac{(x-1)^3 y^2}{\sqrt{z}}}$ in terms of simpler logarithmic forms.

14. Write $\dfrac{1}{3}\log_a x - \dfrac{1}{2}\log_a y$ as a single logarithm.

15. Given $\log 4 = 0.60$, $\log 5 = 0.70$. Compute $\log 80$.

16. Solve $2^{3x-1} = 14$ for x.

17. Solve $\log(2x - 1) = 2 + \log(x - 2)$ for x.

PROGRESS TEST 11B

1. If $f(x) = x^2 + 3$ and $g(x) = 2x - 1$, compute the following.

   (a) $\left(\dfrac{g}{f}\right)(x)$ (b) $(f \circ f)(-1)$

2. Verify that $f(x) = -3x + 1$ and $g(x) = -\dfrac{1}{3}x + \dfrac{1}{3}$ are inverse functions.

3. Solve $\left(\dfrac{1}{2}\right)^x = \left(\dfrac{1}{4}\right)^{2x+1}$ for x.

4. Solve $(a + 3)^x = (2a - 5)^x$ for a.

5. Write $\log_2 \dfrac{1}{8} = -3$ in exponential form.

6. Write $16 = 64^{2/3}$ in logarithmic form.

7. Solve $2 \ln x^2 = \dfrac{1}{3}$ for x.

8. Solve $\log_x 27 = \dfrac{1}{3}$ for x.

9. Solve $\log_6 \dfrac{1}{36} = 3x + 1$ for x.

10. Compute $\log_9 3^{10}$.

11. Compute $\ln e^{5/2}$.

12. Sketch the graph of $f(x) = \log_3 x$.

13. Write $\log \dfrac{x^2(2y-1)^{1/2}}{y^3 z}$ in terms of simpler logarithmic forms.

14. Write $\log(x - 1) + 3\log x - \dfrac{1}{3}\log(x + 1)$ as a single logarithm.

15. Given $\log 2 = 0.30$, $\log 5 = 0.70$. Compute $\log \sqrt{10}$.

16. Solve $3^{1-x} = 10$ for x.

17. Solve $\log(x + 1) - 1 = \log(2x + 1)$ for x.

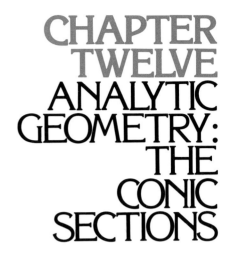

CHAPTER TWELVE
ANALYTIC GEOMETRY: THE CONIC SECTIONS

Analytic geometry combines the techniques of algebra with those of geometry. Analytic geometry enables us to apply algebraic methods and equations to the solution of problems in geometry and, conversely, to obtain geometric representations of algebraic equations.

We will first develop two simple but powerful devices: a formula for the distance between two points and a formula for the coordinates of the midpoint of a line segment. With these as tools, we will demonstrate the power of analytic geometry by proving a number of general theorems from plane geometry.

The graphs of second-degree equations in two variables are known as the conic sections. The power of the methods of analytic geometry are very well demonstrated in a study of the conic sections. We will see that (a) a geometric definition can be converted into an algebraic equation and (b) an algebraic equation can be classified by the type of graph it represents.

12.1
THE DISTANCE AND MIDPOINT FORMULAS

Our first objective will be to find an expression for the distance d between two points $P_1(x_1, y_1)$ and $P_2(x_2, y_2)$. In Figure 12.1 we have used the fact that the x-coordinate of a point is the distance from the y-axis to the point and the y-coordinate is the distance from the x-axis to the point. In Figure 12.2 we

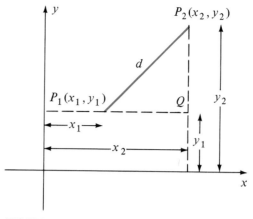

FIGURE 12.1

use the coordinates from Figure 12.1 to indicate the lengths of the sides P_1Q and P_2Q of right triangle P_1QP_2. Applying the Pythagorean theorem, we have

$$d^2 = (x_2 - x_1)^2 + (y_2 - y_1)^2$$

FIGURE 12.2

Since distance cannot be negative, we have

The distance d between two points $P_1(x_1, y_1)$ and $P_2(x_2, y_2)$ is given by
$$d = \sqrt{(x_2 - x_1)^2 + (y_2 - y_1)^2}$$

The student should verify that the formula is true regardless of the quadrants in which P_1 and P_2 may be located.

EXAMPLE 1
Find the distance between points $(3, -2)$ and $(-1, -5)$.

Solution

We let $(x_1, y_1) = (3, -2)$ and $(x_2, y_2) = (-1, -5)$. Substituting these values in the distance formula, we have

$$d = \sqrt{(x_2 - x_1)^2 + (y_2 - y_1)^2}$$
$$= \sqrt{(-1 - 3)^2 + [-5 - (-2)]^2}$$
$$= \sqrt{(-4)^2 + (-3)^2} = \sqrt{25} = 5$$

If we had let $(x_1, y_1) = (-1, -5)$ and $(x_2, y_2) = (3, -2)$, we would have obtained the same result for d. Verify this.

PROGRESS CHECK 1

Find the distance between the points.

(a) $(-4, 3), (-2, 1)$ (b) $(-6, -7), (3, 0)$

Answers

(a) $2\sqrt{2}$ (b) $\sqrt{130}$

Another useful expression that is easily obtained is the one for the midpoint of a line segment. In Figure 12.3, we let $P(x, y)$ be the midpoint of the

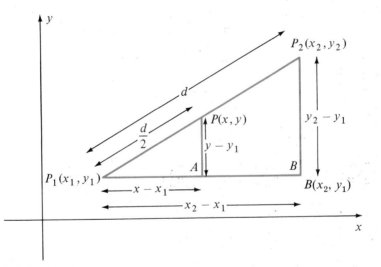

FIGURE 12.3

line segment whose endpoints are $P_1(x_1, y_1)$ and $P_2(x_2, y_2)$. Let d denote the length of P_1P_2. Since P is the midpoint of P_1P_2, the length of P_1P is $d/2$. The lines PA and P_2B are parallel, so triangles P_1AP and P_1BP_2 are similar. Since corresponding sides of similar triangles are in proportion, we can write

$$\frac{P_1P_2}{P_2B} = \frac{P_1P}{PA}$$

or

$$\frac{d}{y_2 - y_1} = \frac{\dfrac{d}{2}}{(y - y_1)}$$

This gives

$$\frac{d}{2}(y_2 - y_1) = d(y - y_1)$$

Canceling d we have

$$\frac{1}{2}(y_2 - y_1) = (y - y_1)$$

$$\frac{1}{2}y_2 - \frac{1}{2}y_1 = y - y_1$$

$$y = \frac{y_1 + y_2}{2}$$

Similarly, from

$$\frac{P_1 P_2}{P_1 B} = \frac{P_1 P}{P_1 A}$$

we obtain

$$\frac{d}{x_2 - x_1} = \frac{\dfrac{d}{2}}{x - x_1}$$

so that

$$x = \frac{x_1 + x_2}{2}$$

We have the following general result.

If $P(x, y)$ is the midpoint of the line segment whose endpoints are $P_1(x_1, y_1)$ and $P_2(x_2, y_2)$, then

$$x = \frac{x_1 + x_2}{2} \qquad y = \frac{y_1 + y_2}{2}$$

EXAMPLE 2
Find the midpoint of the line segment whose endpoints are $P_1(3, 4)$ and $P_2(-2, -6)$.

Solution
If $P(x, y)$ is the midpoint, then

$$x = \frac{x_1 + x_2}{2} = \frac{3 - 2}{2} = \frac{1}{2} \qquad y = \frac{y_1 + y_2}{2} = \frac{4 - 6}{2} = -1$$

Thus, the midpoint is $\left(\dfrac{1}{2}, -1\right)$.

PROGRESS CHECK 2
Find the midpoint of the line segment whose endpoints are the following.
(a) $(0, -4), (-2, -2)$ (b) $(-10, 4), (7, -5)$

Answers

(a) $(-1, -3)$ (b) $\left(-\dfrac{3}{2}, -\dfrac{1}{2}\right)$

The formulas for distance, midpoint of a line segment, and slope of a line are adequate to allow us to demonstrate the beauty and power of analytic geometry. With these tools, we can prove theorems from plane geometry by placing the figures on a rectangular coordinate system.

EXAMPLE 3
Prove that the line joining the midpoints of two sides of a triangle is parallel to the third side and equal to one half the third side.

Solution
We place the triangle OAB in a convenient location, namely, with one vertex at the origin and one side on the positive x-axis (Figure 12.4). If Q and R are the midpoints

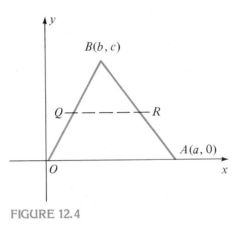

FIGURE 12.4

of OB and AB, then by the midpoint formula the coordinates of Q are

$$\left(\frac{b}{2}, \frac{c}{2}\right)$$

and the coordinates of R are

$$\left(\frac{a+b}{2}, \frac{c}{2}\right)$$

We see that the line joining

$$Q\left(\frac{b}{2}, \frac{c}{2}\right) \quad \text{and} \quad R\left(\frac{a+b}{2}, \frac{c}{2}\right)$$

has slope of 0 since the difference of the y coordinates is

$$\frac{c}{2} - \frac{c}{2} = 0$$

But side OA also has slope 0, which proves that QR is parallel to OA.

Applying the distance formula to QR we have

$$d = \sqrt{(x_2 - x_1)^2 + (y_2 - y_1)^2}$$

$$= \sqrt{\left(\frac{a+b}{2} - \frac{b}{2}\right)^2 + \left(\frac{c}{2} - \frac{c}{2}\right)^2}$$

$$= \sqrt{\left(\frac{a}{2}\right)^2} = \frac{a}{2}$$

Since OA has length a, we have shown that QR is one half of OA.

PROGRESS CHECK 3
Prove that the midpoint of the hypotenuse of a right triangle is equidistant from all three vertices.

Answer
Hint: Place the triangle so that two legs coincide with the positive x- and y-axes. Find the coordinates of the midpoint of the hypotenuse by the midpoint formula. Finally, compute the distance from the midpoint to each vertex by the distance formula.

EXERCISE SET 12.1
In each of the following find the distance between the given points.

1. $(5, 4), (2, 1)$
2. $(-4, 5), (-2, 3)$
3. $(-1, -5), (-5, -1)$
4. $(2, -4), (3, -1)$
5. $(3, 1), (-4, 5)$
6. $(2, 4), (-3, 3)$
7. $(-2, 4), (-4, -2)$
8. $(-3, 0), (2, -4)$
9. $\left(-\frac{1}{2}, 3\right), \left(-1, -\frac{3}{4}\right)$
10. $(3, 0), (0, 4)$
11. $(2, -4), (0, -1)$
12. $\left(\frac{2}{3}, \frac{3}{2}\right), (-2, -4)$

Find the midpoint of the line segment whose endpoints are the given pair of points.

13. $(2, 6), (3, 4)$
14. $(1, 1), (-2, 5)$
15. $(2, 0), (0, 5)$
16. $(-3, 0), (-5, 2)$
17. $(-2, 1), (-5, -3)$
18. $(2, 3), (-1, 3)$
19. $(0, -4), (0, 3)$
20. $(1, -3), (3, 2)$
21. $(-1, 3), (-1, 6)$
22. $(3, 2), (0, 0)$
23. $(1, -1), (-1, 1)$
24. $(2, 4), (2, -4)$

25. Show that the medians from the equal angles of an isosceles triangle are of equal length. (*Hint:* Place the triangle so that its vertices are at the points $A(-a, 0)$, $B(a, 0)$, and $C(0, b)$.)
26. Show that the midpoints of the sides of a rectangle are the vertices of a rhombus (a quadrilateral with four equal sides). (*Hint:* Place the rectangle so that its vertices are at the points $(0, 0), (a, 0), (0, b)$, and (a, b).)
27. Show that a triangle with two equal medians is isosceles.
28. Show that the sum of the squares of the lengths of the medians of a triangle equals three fourths the sum of the squares of the lengths of the sides. (*Hint:* Place the triangle so that its vertices are the points $(-a, 0), (b, 0)$, and $(0, c)$.)
29. Show that the diagonals of a rectangle are equal in length. (*Hint:* Place the rectangle so that its vertices are at the points $(0, 0), (a, 0), (0, b)$, and (a, b).)

The definition of symmetry given below provides a precise means for testing for a property that you have always dealt with intuitively. In Figure 12.5a we have sketched the graph of the equation $x = y^2 - 3$ which appears to be symmetric about the x-axis; that is, each point above the x-axis appears

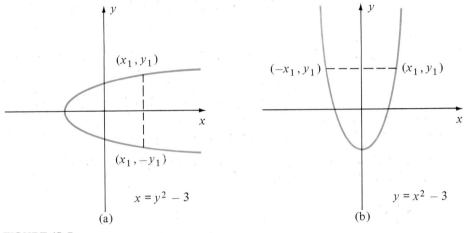

FIGURE 12.5

to be reflected to a point below the x-axis. Thus, if (x_1, y_1) is a point on the upper portion of the curve then $(x_1, -y_1)$ must lie on the lower portion of the curve. Similarly, using the graph of $y = x^2 - 3$ in Figure 12.5b, we can argue that symmetry about the y-axis occurs if, for every point (x_1, y_1) on the curve, $(-x_1, y_1)$ also lies on the curve. In summary,

If the equation of a curve is unchanged when x is replaced by $-x$, then the curve is **symmetric with respect to the y-axis.**

If the equation of a curve is unchanged when y is replaced by $-y$, then the curve is **symmetric with respect to the x-axis.**

EXAMPLE 1
Without sketching the graph, determine symmetry with respect to the x- and y-axes.

(a) $x^2 + 4y^2 - y = 1$

Replacing x by $-x$,

$$(-x)^2 + 4y^2 - y = 1$$
$$x^2 + 4y^2 - y = 1$$

Since the equation is unaltered, the curve is symmetric with respect to the y-axis. Now, replacing y by $-y$,

$$x^2 + 4(-y)^2 - (-y) = 1$$
$$x^2 + 4y^2 + y = 1$$

is *not* the same equation. Thus, the curve is not symmetric with respect to the x-axis.

(b) $xy = 5$

Replacing x by $-x$, we have $-xy = 5$, which is not the same equation. Replacing y by $-y$, we have $xy = -5$. Thus, the curve is not symmetric with respect to either axis.

PROGRESS CHECK 1
Without graphing, determine symmetry with respect to the coordinate axes.

(a) $x^2 - y^2 = 1$ (b) $x + y = 10$ (c) $y = \dfrac{1}{x^2 + 1}$

Answers
(a)   Symmetric with respect to both x- and y-axes.
(b)   Not symmetric with respect to either axis.
(c)   Symmetric with respect to the y-axis.

There is another type of symmetry that is easy to verify—symmetry with respect to the origin. In Figure 12.6 we have sketched the graph of $y = x^3$

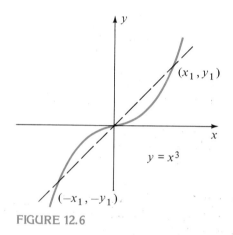

FIGURE 12.6

which appears to be symmetric with respect to the origin. We see from the graph that if (x_1, y_1) lies on the curve, so does $(-x_1, -y_1)$. We conclude

If the equation of a curve is unchanged when x is replaced by $-x$ and y is replaced by $-y$, then the curve is **symmetric with respect to the origin.**

EXAMPLE 2
Determine symmetry with respect to the origin.

(a) $y = x^3 - 1$

Replacing x by $-x$ and y by $-y$ we have

$$-y = (-x)^3 - 1$$
$$-y = -x^3 - 1$$
$$y = x^3 + 1$$

Since the equation is altered, it is not symmetric with respect to the origin.

(b) $y^2 = \dfrac{x^2 + 1}{x^2 - 1}$

Replacing x by $-x$ and y by $-y$,

$$(-y)^2 = \dfrac{(-x)^2 + 1}{(-x)^2 - 1}$$

$$y^2 = \dfrac{x^2 + 1}{x^2 - 1}$$

The equation is unchanged and we conclude that the curve is symmetric with respect to the origin.

PROGRESS CHECK 2

Determine symmetry with respect to the origin.

(a) $x^2 + y^2 = 1$ (b) $y^2 = x - 1$ (c) $y = x + \dfrac{1}{x}$

Answers
(a) *Symmetric with respect to the origin.*
(b) *Not symmetric with respect to the origin.*
(c) *Symmetric with respect to the origin.*

Note that in Example 2b and Progress Check 2a, the given curves are symmetric with respect to *both* the x- and y-axes, as well as the origin. In fact, we have the following general rule.

A curve that is symmetric with respect to both coordinate axes is also symmetric with respect to the origin. The converse, however, is not true.

The curve $y = x^3$ in Figure 12.6 is an example of one that is symmetric with respect to the origin, but not with respect to the coordinate axes.

EXERCISE SET 12.2

Without graphing, determine whether each given curve is symmetric with respect to the x-axis, the y-axis, the origin, or none.

1. $y = x + 3$
2. $x - 2y = 5$
3. $2x - 5y = 0$
4. $3x + 2y = 5$
5. $y = 4x^2$
6. $x = 2y^2$
7. $y^2 = x - 4$
8. $x^2 - y = 2$
9. $y = 4 - 9x^2$
10. $y = x(x - 4)$
11. $y = 1 + x^3$
12. $y^2 = 1 + x^3$
13. $y = (x - 2)^2$
14. $y^2 = (x - 2)^2$
15. $y^2 = x^2 - 9$
16. $y^2 x + 2x = 4$
17. $y + yx^2 = x$
18. $y^2 x + 2x^2 = 4x^2 y$
19. $y^3 = x^2 - 9$
20. $y^3 = x^3 + 9$
21. $y = \dfrac{1}{x^2}$
22. $y = \dfrac{x^2 + 4}{x^2 - 4}$
23. $y = \dfrac{1}{x^2 + 1}$
24. $y^2 = \dfrac{x^2 + 1}{x^2 - 1}$

25. $4y^2 - x^2 = 1$ 26. $4x^2 + 9y^2 = 36$ 27. $9y^2 - 4x^2 = 36$

28. $4 + x^2y = x^2$ 29. $y^2 = \dfrac{1}{x^2}$ 30. $y = \dfrac{1}{x^3}$

31. $y = x + \dfrac{1}{x^2}$ 32. $y^2 = \dfrac{x}{x - 1}$ 33. $xy = 4$

34. $x^2y = 4$

12.3
THE CIRCLE

The conic sections provide us with an outstanding opportunity to demonstrate the double-edged power of analytic geometry. We will see that a geometric figure defined as a set of points can often be described analytically by an algebraic equation; conversely, we can start with an algebraic equation and use graphing procedures to study the properties of the curve.

First, let's see how the term conic section originates. If we pass a plane through a cone at various angles, the intersections are called **conic sections.** Figure 12.7 shows the four conic sections: the circle, parabola, ellipse, and hyperbola.

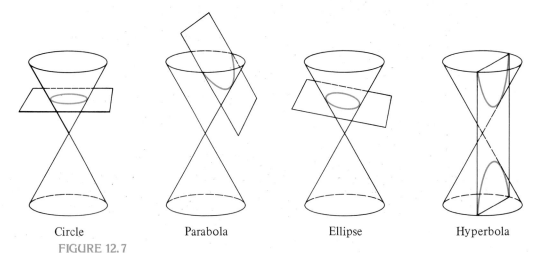

Circle   Parabola   Ellipse   Hyperbola

FIGURE 12.7

Let's begin with the geometric definition of a circle.

A **circle** is the set of all points in a plane that are at a given distance from a fixed point. The fixed point is called the **center** of the circle and the given distance is called the **radius.**

Using the methods of analytic geometry, we place the center at a point (h, k) as in Figure 12.8. If $P(x, y)$ is a point on the circle, then the distance from P to the center (h, k) must be equal to the radius r. By the distance formula

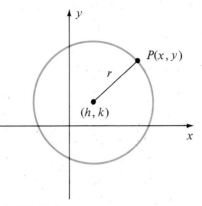

FIGURE 12.8

$$\sqrt{(x - h)^2 + (y - k)^2} = r$$

or

$$(x - h)^2 + (y - k)^2 = r^2$$

Since $P(x, y)$ is any point on the circle, we say that

$$(x - h)^2 + (y - k)^2 = r^2$$

is the **standard form of the equation of the circle** with center (h, k) and radius r.

EXAMPLE 1

Write the equation of the circle with center $(2, -5)$ and radius 3.

Solution

Substituting $h = 2$, $k = -5$, and $r = 3$ in the equation

$$(x - h)^2 + (y - k)^2 = r^2$$

yields

$$(x - 2)^2 + (y + 5)^2 = 9$$

PROGRESS CHECK 1

Write the equation of the circle with center $(-4, -6)$ and radius 5.

Answer

$(x + 4)^2 + (y + 6)^2 = 25$

EXAMPLE 2

Find the center and radius of the circle whose equation is $(x + 1)^2 + (y - 3)^2 = 4$.

Solution

Since the standard form is

$$(x - h)^2 + (y - k)^2 = r^2$$

we must have

$$x - h = x + 1, \qquad y - k = y - 3, \qquad r^2 = 4$$

Solving, we find that

$$h = -1, \quad k = 3, \quad r = 2$$

Thus, the center is at $(-1, 3)$ and the radius is 2.

PROGRESS CHECK 2

Find the center and radius of the circle whose equation is $(x - \frac{1}{2})^2 + (y + 5)^2 = 15$.

Answer
center $(\frac{1}{2}, -5)$, radius $\sqrt{15}$

It is also possible to begin with the equation of a circle in the **general form**

$$Ax^2 + Ay^2 + Dx + Ey + F = 0, \quad A \neq 0$$

and to rewrite the equation in standard form. The process involves completing the square in each variable.

EXAMPLE 3

Write in standard form the equation of the circle

$$2x^2 + 2y^2 - 12x + 16y - 22 = 0$$

Solution

Dividing by the coefficient of x^2 and y^2, we have

$$x^2 + y^2 - 6x + 8y = 11$$

We rewrite the equation by regrouping as

$$(x^2 - 6x) + (y^2 + 8y) = 11$$

and complete the square in both x and y.

$$(x^2 - 6x + 9) + (y^2 + 8y + 16) = 11 + 9 + 16$$
$$(x - 3)^2 + (y + 4)^2 = 36$$

which is the standard form of the equation of the circle with center $(3, -4)$ and radius 6.

PROGRESS CHECK 3

Write in standard form and determine the center and radius of the circle

$$4x^2 + 4y^2 - 8x + 4y = 103$$

Answer
$(x - 1)^2 + \left(y + \dfrac{1}{2}\right)^2 = 27$, *center $\left(1, -\dfrac{1}{2}\right)$, radius $\sqrt{27}$*

EXAMPLE 4

Write in standard form

$$x^2 + y^2 - 6x + 13 = 0$$

Solution

Regrouping, we have

$$(x^2 - 6x) + y^2 = -13$$

We now complete the square in x and y.

$$(x^2 - 6x + 9) + y^2 = -13 + 9$$
$$(x - 3)^2 + y^2 = -4$$

Since $r^2 = -4$ is an impossible situation, the graph of the equation is not a circle. Note that the left-hand side of the equation in standard form is a sum of squares and is therefore nonnegative, while the right-hand side is negative. Thus, there are no real values of x and y which satisfy the equation. This is an example of an equation that does not have a graph.

PROGRESS CHECK 4
Write in standard form and analyze the graph of $x^2 + y^2 - 12y + 36 = 0$.

Answer
Standard form is $x^2 + (y - 6)^2 = 0$. The equation is that of a "circle" with center $(0, 6)$ and radius of 0. The "circle" is actually the point $(0, 6)$.

EXERCISE SET 12.3
In each of the following write the equation of the circle with center at (h, k) and radius r.

1. $(h, k) = (2, 3)$ $r = 2$
2. $(h, k) = (-3, 0)$ $r = 3$
3. $(h, k) = (-2, -3)$ $r = \sqrt{5}$
4. $(h, k) = (2, -4)$ $r = 4$
5. $(h, k) = (0, 0)$ $r = 3$
6. $(h, k) = (0, -3)$ $r = 2$
7. $(h, k) = (-1, 4)$ $r = 2\sqrt{2}$
8. $(h, k) = (2, 2)$ $r = 2$

In each of the following find the center and radius of the circle with the given equation.

9. $(x - 2)^2 + (y - 3)^2 = 16$
10. $(x + 2)^2 + y^2 = 9$
11. $(x - 2)^2 + (y + 2)^2 = 4$
12. $\left(x + \dfrac{1}{2}\right)^2 + (y - 2)^2 = 8$
13. $(x + 4)^2 + \left(y + \dfrac{3}{2}\right)^2 = 18$
14. $x^2 + (y - 2)^2 = 4$
15. $\left(x - \dfrac{1}{3}\right)^2 + y^2 = -\dfrac{1}{9}$
16. $x^2 + \left(y - \dfrac{1}{2}\right)^2 = 3$

Write the equation of each given circle in standard form and determine the center and radius, if possible.

17. $x^2 + y^2 + 4x - 8y + 4 = 0$
18. $x^2 + y^2 - 2x + 6y - 15 = 0$
19. $2x^2 + 2y^2 - 6x - 10y + 6 = 0$
20. $2x^2 + 2y^2 + 8x - 12y - 8 = 0$
21. $2x^2 + 2y^2 - 4x - 5 = 0$
22. $4x^2 + y^2 - 2y + 7 = 0$
23. $3x^2 + 3y^2 - 12x + 18y + 15 = 0$
24. $4x^2 + 4y^2 + 4x + 4y - 4 = 0$

Write each equation in standard form and determine the nature of the graph.

25. $x^2 + y^2 - 6x + 8y + 7 = 0$
26. $x^2 + y^2 + 4x + 6y + 5 = 0$
27. $x^2 + y^2 + 3x - 5y + 7 = 0$
28. $x^2 + y^2 - 4x - 6y - 13 = 0$
29. $2x^2 + 2y^2 - 12x - 4 = 0$
30. $2x^2 + 2y^2 + 4x - 4y + 25 = 0$
31. $2x^2 + 2y^2 - 6x - 4y - 2 = 0$
32. $2x^2 + 2y^2 - 10y + 6 = 0$

33. $3x^2 + 3y^2 + 12x - 4y - 20 = 0$ 34. $x^2 + y^2 + x + y = 0$

35. $4x^2 + 4y^2 + 12x - 20y + 38 = 0$ 36. $4x^2 + 4y^2 - 12x - 36 = 0$

12.4
THE PARABOLA

We begin our study of the parabola with the geometric definition.

> A **parabola** is the set of all points that are equidistant from a given point and a given line.

The given point is called the **focus** and the given line is called the **directrix** of the parabola. Thus, in Figure 12.9, all points P are equidistant from F and line L. In other words, $PF = PQ$ for any point P on the parabola.

We can apply the methods of analytic geometry to find the equation of a parabola. We first consider parabolas that are conveniently located. Place the given line L parallel to the x-axis and the given point F on the y-axis with the origin halfway between the point and the line (Figure 12.9). Then F has

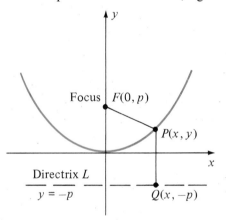

FIGURE 12.9

coordinates $(0, p)$ and the equation of L is $y = -p$. If $P(x, y)$ is a point on the parabola, then the distance from P to the given point F must equal the distance from P to the given line L. Using the distance formula,

$$PF = PQ$$
$$\sqrt{(x - 0)^2 + (y - p)^2} = \sqrt{(x - x)^2 + (y + p)^2}$$

Squaring both sides

$$x^2 + y^2 - 2py + p^2 = y^2 + 2py + p^2$$
$$x^2 = 4py$$

If we treat $4p$ as a constant and substitute $a = 4p$, we have the general equation of a parabola.

$$x^2 = ay$$

We see that the parabola is symmetric about a line called the **axis of symmetry** of the parabola (or simply the **axis**) and opens from a point on the axis called the **vertex.** The graph of the equation $x^2 = ay$ (Figure 12.9) has its vertex at the origin and the y-axis is the axis of the parabola. If a parabola has the x-axis as the axis of symmetry, and the vertex is at the origin, the equation of the parabola is

$$y^2 = ax$$

EXAMPLE 1
Sketch the graph of each of the following.

(a) $x^2 = -2y$ (b) $y^2 = \dfrac{x}{2}$

Solution
We form tables of values giving points on the graphs and draw smooth curves. See Figure 12.10.

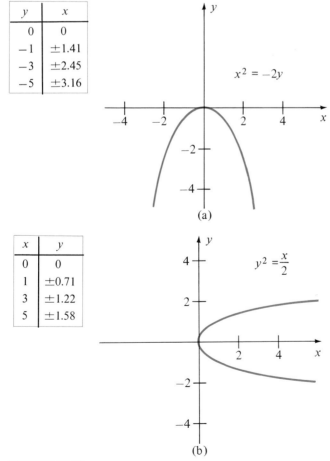

y	x
0	0
-1	±1.41
-3	±2.45
-5	±3.16

$x^2 = -2y$

(a)

x	y
0	0
1	±0.71
3	±1.22
5	±1.58

$y^2 = \dfrac{x}{2}$

(b)

FIGURE 12. 10

PROGRESS CHECK 1

Sketch the graph of

(a) $x^2 = 3y$ (b) $y^2 = -2x$

Answers

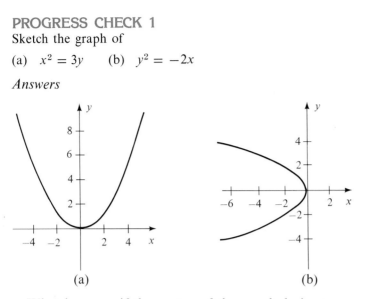

(a)   (b)

What happens if the vertex of the parabola is at some point (h, k) other than the origin? Again, the result will depend upon whether the axis of the parabola is parallel to the x-axis or to the y-axis.

The **standard forms of the equations of the parabolas** with vertex at (h, k) are

$$(x - h)^2 = a(y - k) \qquad \text{The axis is } x = h.$$
$$(y - k)^2 = a(x - h) \qquad \text{The axis is } y = k.$$

Note that if the vertex is at the origin, then $h = k = 0$, and we arrive at the equations we derived previously, $x^2 = ay$ and $y^2 = ax$.

EXAMPLE 2

Sketch the graph of the equation $(y - 3)^2 = -2(x + 2)$.

Solution

The equation is the standard form of a parabola, with vertex at $(-2, 3)$ and axis of symmetry $y = 3$. See Figure 12.11.

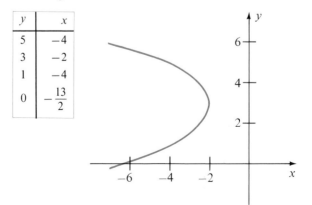

y	x
5	-4
3	-2
1	-4
0	$-\dfrac{13}{2}$

FIGURE 12.11

PROGRESS CHECK 2

Sketch the graph of the equation $(x + 1)^2 = 2(y + 2)$. Locate the vertex and the axis of symmetry.

Answer

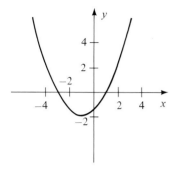

vertex: $(-1, -2)$, *axis of symmetry:* $x = -1$

From the graphs in Figures 12.10 and 12.11 and the answers to Progress Checks 1 and 2, we can make the following observations.

The graph of a parabola whose equation is

$$(x - h)^2 = a(y - k)$$

opens upward if $a > 0$, opens downward if $a < 0$, and the axis of symmetry is $x = h$. See Figures 12.12a and 12.12b.

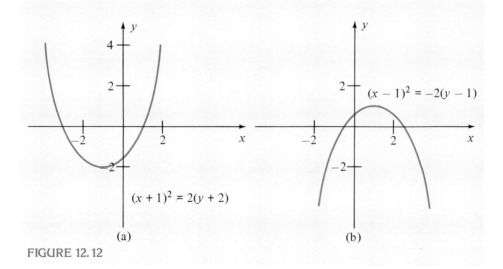

(a)   (b)

FIGURE 12.12

The graph of a parabola whose equation is

$$(y - k)^2 = a(x - h)$$

opens to the right if $a > 0$, opens to the left if $a < 0$, and the axis of symmetry is $y = k$. See Figures 12.12c and 12.12d.

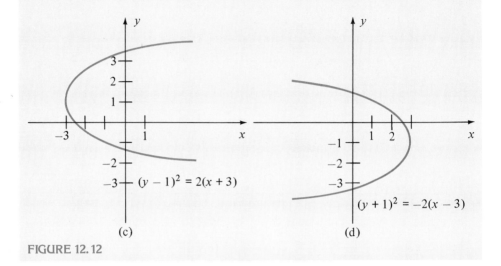

(c)

(d)

FIGURE 12.12

EXAMPLE 3
Determine the vertex, axis, and direction of the graph of the parabola

$$\left(x - \frac{1}{2}\right)^2 = -\frac{1}{2}(y + 4)$$

Solution

Comparison of the equation with the standard form

$$(x - h)^2 = a(y - k)$$

yields $h = \frac{1}{2}, k = -4, a = -\frac{1}{2}$. The axis of the parabola is always found by setting the square term equal to 0.

$$\left(x - \frac{1}{2}\right)^2 = 0$$

$$x = \frac{1}{2}$$

Thus, the vertex is $(h, k) = \left(\frac{1}{2}, -4\right)$, the axis is $x = \frac{1}{2}$, and the parabola opens downward since $a < 0$.

PROGRESS CHECK 3
Determine the vertex, axis, and direction of the graph of the parabola

$$(y + 1)^2 = 4\left(x - \frac{1}{3}\right)$$

Answer

vertex: $\left(\frac{1}{3}, -1\right)$, axis: $y = -1$, opens to the right

Any second-degree equation in x and y that has a square term in one variable but only first-degree terms in the other, represents a parabola. We can put such an equation in standard form by completing the square.

EXAMPLE 4

Determine the vertex, axis, and direction of opening of the parabola

$$2y^2 - 12y + x + 19 = 0$$

Solution

First, we complete the square in y.

$$2y^2 - 12y + x + 19 = 0$$
$$2(y^2 - 6y) = -x - 19$$
$$2(y^2 - 6y + 9) = -x - 19 + 18$$
$$2(y - 3)^2 = -x - 1 = -(x + 1)$$
$$(y - 3)^2 = -\frac{1}{2}(x + 1)$$

With the equation in standard form, we see that $(h, k) = (-1, 3)$ is the vertex, $y = 3$ is the axis, and the curve opens to the left.

PROGRESS CHECK 4

Write the equation of the parabola $x^2 + 4x + y + 9 = 0$ in standard form. Determine the vertex, axis, and direction of opening.

Answer

$(x + 2)^2 = -(y + 5)$, vertex: $(-2, -5)$, axis: $x = -2$, opens downward

EXERCISE SET 12.4

In each of the following sketch the graph of the given equation.

1. $x^2 = 4y$ 2. $x^2 = -4y$ 3. $y^2 = 2x$

4. $y^2 = -\frac{3}{2}x$ 5. $x^2 = y$ 6. $y^2 = x$

7. $x^2 + 5y = 0$ 8. $2y^2 - 3x = 0$

9. $(x - 2)^2 = 2(y + 1)$ 10. $(x + 4)^2 = 3(y - 2)$

11. $(y - 1)^2 = 3(x - 2)$ 12. $(y - 2)^2 = -2(x + 1)$

13. $(x + 4)^2 = -\frac{1}{2}(y + 2)$ 14. $(y - 1)^2 = -3(x - 2)$

15. $y^2 = -2(x + 1)$ 16. $x^2 = \frac{1}{2}(y - 3)$

Determine the vertex, axis, and direction of each given parabola.

17. $x^2 - 2x - 3y + 7 = 0$ 18. $x^2 + 4x + 2y - 2 = 0$

19. $y^2 - 8y + 2x + 12 = 0$ 20. $y^2 + 6y - 3x + 12 = 0$

21. $x^2 - x + 3y + 1 = 0$

22. $y^2 + 2y - 4x - 3 = 0$

23. $y^2 - 10y - 3x + 24 = 0$

24. $x^2 + 2x - 5y - 19 = 0$

25. $x^2 - 3x - 3y + 1 = 0$

26. $y^2 + 4y + x + 3 = 0$

27. $y^2 + 6y + \dfrac{1}{2}x + 7 = 0$

28. $x^2 + 2x - 3y + 19 = 0$

29. $x^2 + 2x + 2y + 3 = 0$

30. $y^2 - 6y + 2x + 17 = 0$

12.5
THE ELLIPSE AND HYPERBOLA

Here is the geometric definition of an ellipse.

> An **ellipse** is the set of all points the sum of whose distances from two fixed points is a constant.

The two fixed points are called the **foci** of the ellipse. The ellipse is in standard position if the two fixed points are on either the x-axis or the y-axis and are equidistant from the origin. Thus, if F_1 and F_2 are the foci of the ellipse in Figure 12.13 and P and Q are points on the ellipse, then $F_1P + F_2P = c$ and $F_1Q + F_2Q = c$, where c is a constant.

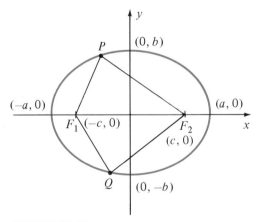

FIGURE 12.13

The equation of an ellipse in standard position is

$$\frac{x^2}{a^2} + \frac{y^2}{b^2} = 1$$

This is called the **standard form of the equation of an ellipse.**

If we let $x = 0$ in the standard form, we find $y = \pm b$; if we let $y = 0$, we find $x = \pm a$. Thus, the ellipse whose equation is

$$\frac{x^2}{a^2} + \frac{y^2}{b^2} = 1$$

has intercepts $(\pm a, 0)$ and $(0, \pm b)$. See Figure 12.13.

EXAMPLE 1

Find the intercepts and sketch the graph of the ellipse whose equation is

$$\frac{x^2}{16} + \frac{y^2}{9} = 1$$

Solution

The intercepts are found by setting $x = 0$ and then $y = 0$ and solving. Thus, the intercepts are $(\pm 4, 0)$ and $(0, \pm 3)$. The graph is then easily sketched (Figure 12.14).

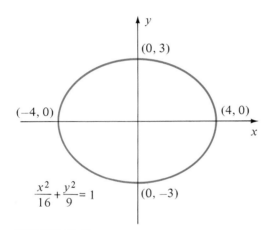

FIGURE 12.14

PROGRESS CHECK 1

Find the intercepts and sketch the graph of

$$\frac{x^2}{9} + \frac{y^2}{16} = 1$$

Answer
$(\pm 3, 0), (0, \pm 4)$

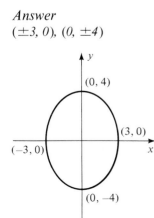

EXAMPLE 2

Write the equation of the ellipse in standard form and determine the intercepts.

(a) $4x^2 + 3y^2 = 12$

Dividing by 12 to make the right-hand side equal to 1, we have

$$\frac{x^2}{3} + \frac{y^2}{4} = 1$$

The x-intercepts are $(\pm\sqrt{3}, 0)$; the y-intercepts are $(0, \pm 2)$

(b) $9x^2 + y^2 = 10$

Dividing by 10 we have

$$\frac{9x^2}{10} + \frac{y^2}{10} = 1$$

But this is *not* standard form. However, if we write

$$\frac{9x^2}{10} = \frac{x^2}{\frac{10}{9}}$$

then

$$\frac{x^2}{\frac{10}{9}} + \frac{y^2}{10} = 1$$

is the standard form of an ellipse. The intercepts are

$$\left(\frac{\pm\sqrt{10}}{3}, 0\right) \quad \text{and} \quad (0, \pm\sqrt{10})$$

PROGRESS CHECK 2

Find the standard form and determine the intercepts of the ellipse.

(a) $2x^2 + 3y^2 = 6$ (b) $3x^2 + y^2 = 5$

Answers

(a) $\dfrac{x^2}{3} + \dfrac{y^2}{2} = 1$ $(\pm\sqrt{3}, 0), (0, \pm\sqrt{2})$

(b) $\dfrac{x^2}{\frac{5}{3}} + \dfrac{y^2}{5} = 1$ $\left(\dfrac{\pm\sqrt{15}}{3}, 0\right), (0, \pm\sqrt{5})$

The hyperbola is the last of the conic sections we will study.

> A **hyperbola** is the set of all points the difference of whose distances from two fixed points is a constant.

The two fixed points are called the **foci** of the hyperbola. The hyperbola is in standard position if the two fixed points are on either the x-axis or the y-axis and are equidistant from the origin. Here are the equations of the hyperbolas in standard position.

$$\frac{x^2}{a^2} - \frac{y^2}{b^2} = 1 \qquad \text{Foci on the } x\text{-axis} \qquad (1)$$

$$\frac{y^2}{a^2} - \frac{x^2}{b^2} = 1 \qquad \text{Foci on the } y\text{-axis} \qquad (2)$$

These equations are called the **standard forms of the equations of a hyperbola.**

Letting $y = 0$, we see that the x-intercepts of the graph of Equation (1) are $\pm a$. Letting $x = 0$, we find there are no y-intercepts since the equation $y^2 = -b^2$ has no real roots. (See Figure 12.15.) Similarly, the graph of Equation (2) has y-intercepts of $\pm a$ and no x-intercepts.

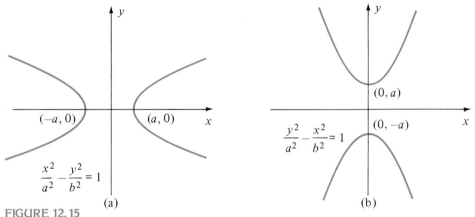

FIGURE 12. 15

EXAMPLE 3
Find the intercepts and sketch the graph of the equation.

(a) $\dfrac{x^2}{9} - \dfrac{y^2}{4} = 1$

When $y = 0$, we have $x^2 = 9$ or $x = \pm 3$. The intercepts are $(3, 0)$ and $(-3, 0)$. With the assistance of a few plotted points, we can sketch the graph (Figure 12.16).

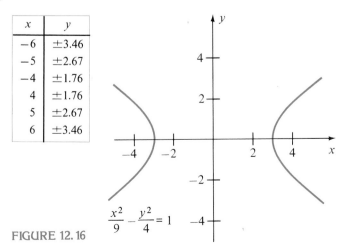

x	y
-6	± 3.46
-5	± 2.67
-4	± 1.76
4	± 1.76
5	± 2.67
6	± 3.46

FIGURE 12. 16

(b) $\dfrac{y^2}{4} - \dfrac{x^2}{3} = 1$

When $x = 0$, we have $y^2 = 4$ or $y = \pm 2$. The intercepts are $(0, 2)$ and $(0, -2)$. Plotting a few points, we can sketch the graph (Figure 12.17).

x	y
-4	± 5.03
-2	± 3.05
2	± 3.05
4	± 5.03

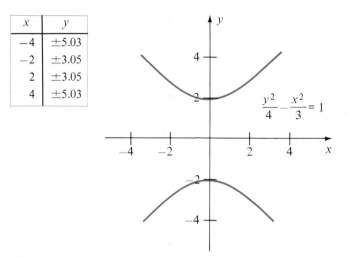

FIGURE 12. 17

PROGRESS CHECK 3

Find the intercepts and sketch the graph.

(a) $\dfrac{x^2}{16} - \dfrac{y^2}{9} = 1$ (b) $\dfrac{y^2}{16} - \dfrac{x^2}{9} = 1$

Answers
(a) *Intercepts are* $(4, 0)$ *and* $(-4, 0)$.   (b) *Intercepts are* $(0, 4)$ *and* $(0, -4)$.

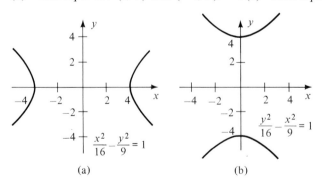

(a)   (b)

EXAMPLE 4

Write the equation of the hyperbola in standard form and determine the intercepts.

(a) $9y^2 - 4x^2 = 36$

Dividing by 36 to produce a 1 on the right-hand side, we have

$$\frac{y^2}{4} - \frac{x^2}{9} = 1$$

The y-intercepts are $(0, \pm 2)$. There are no x-intercepts.

(b) $9x^2 - 5y^2 = 10$

Dividing by 10 we have

$$\frac{9x^2}{10} - \frac{y^2}{2} = 1$$

Rewriting in standard form, the equation becomes

$$\frac{x^2}{\frac{10}{9}} - \frac{y^2}{2} = 1$$

The x-intercepts are

$$\left(\frac{\pm\sqrt{10}}{3}, 0\right)$$

There are no y-intercepts.

PROGRESS CHECK 4

Write the equation of the hyperbola in standard form and determine the intercepts.

(a) $2x^2 - 5y^2 = 6$ (b) $4y^2 - x^2 = 5$

Answers

(a) $\dfrac{x^2}{3} - \dfrac{y^2}{\frac{6}{5}} = 1;\ (\pm\sqrt{3}, 0)$ (b) $\dfrac{y^2}{\frac{5}{4}} - \dfrac{x^2}{5} = 1;\ \left(0, \dfrac{\pm\sqrt{5}}{2}\right)$

There is a way of sketching the graph of a hyperbola without the need for plotting points. Given the equation of the hyperbola

$$\frac{x^2}{a^2} - \frac{y^2}{b^2} = 1$$

in standard form, we plot the four points $(a, \pm b)$, $(-a, \pm b)$ as in Figure 12.18 and draw the diagonals of the rectangle formed by the four points. The hyperbola opens from the intercepts $(\pm a, 0)$ and *approaches the lines formed*

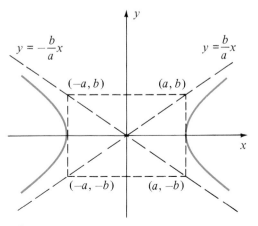

FIGURE 12.18

by the diagonals of the rectangle. We call these lines the **asymptotes** of the hyperbola. Since one asymptote passes through the points $(0, 0)$ and (a, b), its equation is

$$y = \frac{b}{a}x$$

The equation of the other asymptote is found to be

$$y = -\frac{b}{a}x$$

Of course, a similar argument can be made about the standard form

$$\frac{y^2}{a^2} - \frac{x^2}{b^2} = 1$$

In this case, the four points $(\pm b, \pm a)$ determine the rectangle and the equations of the asymptotes are

$$y = \pm\frac{a}{b}x$$

EXAMPLE 5

Using asymptotes, sketch the graph of the equation

$$\frac{y^2}{4} - \frac{x^2}{9} = 1$$

Solution

The points $(\pm 3, \pm 2)$ form the vertices of the rectangle. See Figure 12.19.

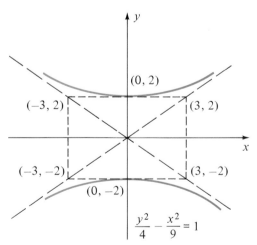

FIGURE 12.19

Using the fact that $(0, \pm 2)$ are intercepts, we can sketch the graph opening from these points and approaching the asymptotes.

PROGRESS CHECK 5

Using asymptotes, sketch the graph of the equation $\dfrac{x^2}{9} - \dfrac{y^2}{9} = 1$.

Answer

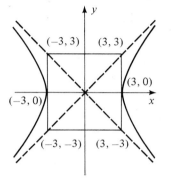

EXERCISE SET 11.5

In each of the following find the intercepts and sketch the graph of the ellipse.

1. $\dfrac{x^2}{25} + \dfrac{y^2}{4} = 1$

2. $\dfrac{x^2}{4} + \dfrac{y^2}{16} = 1$

3. $\dfrac{x^2}{9} + \dfrac{y^2}{4} = 1$

4. $\dfrac{x^2}{12} + \dfrac{y^2}{18} = 1$

5. $\dfrac{x^2}{16} + \dfrac{y^2}{25} = 1$

6. $\dfrac{x^2}{1} + \dfrac{y^2}{3} = 1$

7. $\dfrac{x^2}{20} + \dfrac{y^2}{10} = 1$

8. $\dfrac{x^2}{6} + \dfrac{y^2}{24} = 1$

In each of the following write the equation of the ellipse in standard form and determine the intercepts.

9. $4x^2 + 9y^2 = 36$

10. $16x^2 + 9y^2 = 144$

11. $4x^2 + 16y^2 = 16$

12. $25x^2 + 4y^2 = 100$

13. $4x^2 + 16y^2 = 4$

14. $8x^2 + 4y^2 = 32$

15. $8x^2 + 6y^2 = 24$

16. $5x^2 + 6y^2 = 50$

17. $36x^2 + 8y^2 = 9$

18. $5x^2 + 4y^2 = 45$

In each of the following find the intercepts and sketch the graph of the hyperbola.

19. $\dfrac{x^2}{25} - \dfrac{y^2}{16} = -1$

20. $\dfrac{y^2}{9} - \dfrac{x^2}{4} = 1$

21. $\dfrac{x^2}{36} - \dfrac{y^2}{1} = 1$

22. $\dfrac{y^2}{49} - \dfrac{x^2}{25} = 1$ 23. $\dfrac{x^2}{6} - \dfrac{y^2}{8} = 1$ 24. $\dfrac{y^2}{8} - \dfrac{x^2}{10} = -1$

25. $\dfrac{x^2}{12} - \dfrac{y^2}{2} = 1$ 26. $\dfrac{y^2}{6} - \dfrac{x^2}{5} = 1$

In each of the following write the equation of the hyperbola in standard form and determine the intercepts.

27. $16x^2 - y^2 = 64$ 28. $4x^2 - 25y^2 = 100$

29. $4y^2 - 4x^2 = 1$ 30. $2x^2 - 3y^2 = 6$

31. $4x^2 - 5y^2 = 20$ 32. $25y^2 - 16x^2 = 400$

33. $4y^2 - 16x^2 = 64$ 34. $35x^2 - 9y^2 = 45$

35. $8x^2 - 4y^2 = 32$ 36. $4y^2 - 36x^2 = 9$

Using asymptotes, sketch the graph of each given hyperbola.

37. $\dfrac{x^2}{16} - \dfrac{y^2}{4} = 1$ 38. $\dfrac{y^2}{4} - \dfrac{x^2}{25} = 1$

39. $\dfrac{y^2}{4} - \dfrac{x^2}{4} = 1$ 40. $\dfrac{x^2}{4} - \dfrac{y^2}{16} = 1$

41. $16x^2 - 4y^2 = 144$ 42. $16y^2 - 25x^2 = 400$

43. $9y^2 - 9x^2 = 1$ 44. $25x^2 - 9y^2 = 225$

45. $\dfrac{x^2}{25} - \dfrac{y^2}{36} = 1$ 46. $y^2 - 4x^2 = 4$

47. $y^2 - x^2 = 1$ 48. $\dfrac{x^2}{36} - \dfrac{y^2}{36} = 1$

12.6
IDENTIFYING THE CONIC SECTIONS

Each of the conic sections we have studied in this chapter has one or more axes of symmetry. We studied the circle and parabola when their axes of symmetry were the coordinate axes or lines parallel to them. Although our study of the ellipse and hyperbola was restricted to those that have the coordinate axes as their axes of symmetry, the same method of completing the square allows us to transform the general equation of the conic section

$$Ax^2 + Cy^2 + Dx + Ey + F = 0$$

into standard form. This transformation is very helpful in sketching the graph of the conic. It is easy, also, to identify the conic section from the general equation (see Table 12.1).

TABLE 12.1

$Ax^2 + Cy^2 + Dx$ $+ Ey + F = 0$	Conic Section	Remarks
$A = 0$ or $C = 0$	Parabola	Second degree in one variable, first degree in the other.
$A = C$ $(\neq 0)$	Circle	Coefficients A and C are the same. *Caution:* Complete the square and check that $r > 0$.
$A \neq C$ $AC > 0$	Ellipse	A and C are unequal but have the same sign. *Caution:* Complete the square and check that the right-hand side is a positive constant.
$A \neq C$ $AC < 0$	Hyperbola	A and C have opposite signs.

EXAMPLE 1

Identify the conic section.

(a) $3x^2 + 3y^2 - 2y = 4$

Since the coefficients of x^2 and y^2 are the same, the graph will be a circle if the standard form yields $r > 0$. Completing the square, we have

$$3x^2 + 3\left(y - \frac{1}{3}\right)^2 = \frac{13}{3}$$

which is the equation of a circle.

(b) $3x^2 - 9y^2 + 2x - 4y = 7$

Since the coefficients of x^2 and y^2 are of opposite sign, the graph is a hyperbola.

(c) $2x^2 + 5y^2 - 7x + 3y - 4 = 0$

The coefficients of x^2 and y^2 are unequal but of like sign, so the graph is an ellipse. (Verify that the right-hand side is positive.)

(d) $3y^2 - 4x + 17y = -10$

The graph is a parabola since the equation is of the second degree in y and of first degree in x.

PROGRESS CHECK 1

Identify the conic section.

(a) $\dfrac{x^2}{5} - 3y^2 - 2x + 2y - 4 = 0$

(b) $x^2 - 2y - 3x = 2$ (c) $x^2 + y^2 - 4x - 6y = -11$

(d) $4x^2 + 3y^2 + 6x - 10 = 0$

Answers

(a) *hyperbola*   (b) *parabola*   (c) *circle*   (d) *ellipse*

A summary of the characteristics of the conic sections is given in Table 12.2.

TABLE 12.2

Curve and Standard Equation	Characteristics	Example
Circle $(x - h)^2 + (y - k)^2 = r^2$	Center: (h, k) Radius: r	$(x - 2)^2 + (y + 4)^2 = 25$ Center: $(2, -4)$ Radius: 5
Parabola $(x - h)^2 = a(y - k)$	Vertex: (h, k) Axis: $x = h$ $a > 0$: Opens up $a < 0$: Opens down	$(x + 1)^2 = 2(y - 3)$ Vertex: $(-1, 3)$ Axis: $x = -1$ Opens up
$(y - k)^2 = a(x - h)$	Vertex: (h, k) Axis: $y = k$ $a < 0$: Opens left $a > 0$: Opens right	$(y + 4)^2 = -3(x + 5)$ Vertex: $(-5, -4)$ Axis: $y = -4$ Opens left
Ellipse $\dfrac{x^2}{a^2} + \dfrac{y^2}{b^2} = 1$	Intercepts: $(\pm a, 0), (0, \pm b)$	$\dfrac{x^2}{4} + \dfrac{y^2}{6} = 1$ Intercepts: $(\pm 2, 0), (0, \pm\sqrt{6})$
Hyperbola $\dfrac{x^2}{a^2} - \dfrac{y^2}{b^2} = 1$	Intercepts: $(\pm a, 0)$ Asymptotes: $y = \pm\dfrac{b}{a}x$ Opens to left and right	$\dfrac{x^2}{4} - \dfrac{y^2}{9} = 1$ Intercepts: $(\pm 2, 0)$ Asymptotes: $y = \pm\dfrac{3}{2}x$ Opens to left and right
$\dfrac{y^2}{a^2} - \dfrac{x^2}{b^2} = 1$	Intercepts: $(0, \pm a)$ Asymptotes: $y = \pm\dfrac{a}{b}x$ Opens up and down	$\dfrac{y^2}{9} - \dfrac{x^2}{4} = 1$ Intercepts: $(0, \pm 3)$ Asymptotes: $y = \pm\dfrac{3}{2}x$ Opens up and down

EXERCISE SET 12.6

Identify the conic section.

1. $2x^2 + y - x + 3 = 0$

2. $4y^2 - x^2 + 2x - 3y + 5 = 0$

3. $4x^2 + 4y^2 - 2x + 3y - 4 = 0$

4. $3x^2 + 6y^2 - 2x + 8 = 0$

5. $36x^2 - 4y^2 + x - y + 2 = 0$

6. $x^2 + y^2 - 6x + 4y + 13 = 0$

7. $16x^2 + 4y^2 - 2y + 3 = 0$

8. $2y^2 - 3x + y + 4 = 0$

9. $x^2 + y^2 - 4x - 2y + 8 = 0$

10. $x^2 + y^2 - 2x - 2y + 6 = 0$

11. $4x^2 + 9y^2 - x + 2 = 0$

12. $3x^2 + 3y^2 - 3x + y = 0$

13. $4x^2 - 9y^2 + 2x + y + 3 = 0$

14. $x^2 + y^2 + 6x - 2y + 10 = 0$

15. $x^2 + y^2 - 4x + 4 = 0$ 16. $2x^2 + 3x - 5y^2 + 4y - 6 = 0$

17. $4x^2 + y^2 - 2x + y + 4 = 0$ 18. $x^2 - \frac{1}{2}y^2 + 2x - y + 3 = 0$

19. $4x^2 + 4y^2 - x + 2y - 1 = 0$ 20. $x^2 + y^2 + 6x - 6y + 18 = 0$

21. $4x^2 + y + 2x - 3 = 0$ 22. $y^2 - \frac{1}{4}x^2 + 2x + 6 = 0$

23. $x^2 + y^2 + 4x - 2y + 7 = 0$ 24. $y^2 + 2x - \frac{1}{2}x + 3 = 0$

25. $x^2 + y^2 - 2x - 10y - 26 = 0$ 26. $3x^2 + 2y^2 - y + 2 = 0$

27. $\frac{1}{2}x^2 + y - x - 3 = 0$ 28. $3x^2 - 2y^2 + 2x - 5y + 5 = 0$

29. $2x^2 + y^2 - 3x + 2y - 5 = 0$ 30. $x^2 - 2x - y + 1 = 0$

TERMS AND SYMBOLS

distance formula   (p. 354)
midpoint formula   (p. 356)
symmetry   (p. 359)
conic sections (p. 362)
circle   (p. 362)
center   (p. 362)
radius   (p. 362)
standard equation of a
   circle   (p. 363)
general equation of a
   circle   (p. 364)

parabola   (p. 366)
focus of a parabola
   (p. 366)
directrix   (p. 366)
axis of a parabola
   (p. 367)
vertex   (p. 367)
standard equation of a
   parabola   (p. 368)
ellipse   (p. 372)
foci of an ellipse   (p. 372)

standard equation of an
   ellipse   (p. 372)
hyperbola   (p. 374)
foci of a hyperbola
   (p. 374)
standard equations of a
   hyperbola   (p. 375)
asymptotes of a
   hyperbola   (p. 378)
general equation of the
   conic sections   (p. 380)

KEY IDEAS FOR REVIEW

☐ If $P_1(x_1, y_1)$ and $P_2(x_2, y_2)$ are any two points, then the distance d between the points is given by the formula

$$d = \sqrt{(x_2 - x_1)^2 + (y_2 - y_1)^2}$$

and the midpoint Q has coordinates

$$\left(\frac{x_1 + x_2}{2}, \frac{y_1 + y_2}{2} \right)$$

☐ Theorems from plane geometry can be proven using the methods of analytic geometry. In general, place the given geometric figure in a convenient position relative to the origin and axes. The distance formula, the midpoint formula, and the computation of slope are the tools to apply in proving the theorem.

☐ Symmetry about a line or a point means that the curve is its own reflection about that line or that point. The graph of an equation in x and y will be symmetric with respect to the
   (a) x-axis if the equation is unchanged when y is replaced by $-y$.
   (b) y-axis if the equation is unchanged when x is replaced by $-x$.
   (c)   origin if the equation is unchanged when both x and y are replaced by $-x$ and $-y$, respectively.

☐ The conic sections are the circle, parabola, ellipse, and hyperbola. In some special cases, these reduce to a point, a line, two lines, or no graph. The conic sections represent the possible intersections of a cone and a plane.

☐ Each conic section has a geometric definition as a set of points satisfying certain given conditions.

☐ A second-degree equation in x and y can be written in standard form by completing the square in the variables. It is much easier to sketch the graph when the equation is in standard form than it is when the equation is in general form.

☐ It is often possible to distinguish the various conic sections even when the equation is given in the general form.

COMMON ERRORS

1. When completing the square, be careful to balance both sides of the equation.

2. The first step in writing the equation

$$4x^2 + 25y^2 = 9$$

in standard form is to divide both sides of the equation by 9. The result,

$$\frac{4x^2}{9} + \frac{25y^2}{9} = 1$$

is *not* in standard form. You must rewrite this as

$$\frac{x^2}{\frac{9}{4}} + \frac{y^2}{\frac{9}{25}} = 1$$

to obtain standard form and to determine the intercepts $\left(\pm\frac{3}{2}, 0\right)$, $\left(0, \pm\frac{3}{5}\right)$.

3. The graph of the equation $3y^2 - 4x - 6 = 0$ is *not* a hyperbola. If only one variable appears to the second degree, the equation is that of a parabola.

4. The safest way to find the intercepts is to let one variable equal 0 and solve for the other variable. If you attempt to memorize the various forms, you might conclude that the intercepts of

$$\frac{y^2}{16} - \frac{x^2}{9} = 1$$

are $(\pm 4, 0)$. However, when $x = 0$ we see that $y^2 = 16$ or $y = \pm 4$ and the intercepts are $(0, \pm 4)$.

5. When analyzing the type of conic section from the general form of the second-degree equation, remember that the circle and ellipse have degenerate cases. The equation

$$2x^2 + y^2 - 8x + 6y + 21 = 0$$

is equivalent to

$$2(x - 2)^2 + (y + 3)^2 = -4$$

which is impossible. There are no points we can graph that will satisfy this equation.

PROGRESS TEST 12A

1. Find the distance between the points $P_1(-3, 4)$ and $P_2(4, -3)$.

2. Find the midpoint of the line segment whose endpoints are $P_1\left(\frac{1}{2}, -1\right)$ and $P_2\left(2, \frac{3}{4}\right)$.

3. Given the points $A(1, -2)$, $B(5, -1)$, $C(2, 7)$, and $D(6, 8)$, show that AC is equal and parallel to BD.

4. Show that $A(-1, 7)$, $B(-3, 2)$, and $C(4, 5)$ are the coordinates of an isosceles triangle.

5. Without sketching, determine symmetry with respect to the x-axis, y-axis, and origin:
$$3x^2 - 2x - 4y^2 = 6$$

6. Without sketching, determine symmetry with respect to the x-axis, y-axis, and origin:
$$y^2 = \frac{2x}{x^2 - 1}$$

7. Find the center and radius of the circle having the equation $x^2 - 8x + y^2 + 6y + 15 = 0$. Sketch.

8. Find the vertex and axis of symmetry of the parabola whose equation is $4y^2 - 4y + 12x = -13$. Sketch.

9. Find the intercepts and asymptotes of the hyperbola whose equation is $4y^2 - \frac{x^2}{9} = 1$. Sketch.

10. Find the equation of the circle having center $\left(\frac{2}{3}, -3\right)$ and radius $\sqrt{3}$.

11. Find the intercepts and vertex and sketch the graph of the parabola whose equation is $y = x^2 - x - 6$.

12. Find the intercepts and sketch the graph of the equation
$$\frac{x^2}{36} + \frac{y^2}{9} = 1$$

13. Identify the conic section whose equation is
$$3x^2 + 2x - 7y^2 + 3y - 14 = 0$$

14. Identify the conic section whose equation is
$$y^2 - 3y - 5x = 20$$

15. Identify the conic section whose equation is
$$x^2 - 6x + y^2 = 0$$

PROGRESS TEST 12B

1. Find the distance between the points $P_1(6, -7)$ and $P_2(2, -5)$.

2. Find the midpoint of the line segment whose endpoints are $P_1\left(-\frac{3}{2}, -\frac{1}{2}\right)$ and $P_2(-2, 1)$.

3. The point $(3, 2)$ is the midpoint of a line segment having the point $(4, -1)$ as an endpoint. Find the other endpoint.

4. Show that the points $A(-8, 4)$, $B(5, 3)$, and $C(2, -2)$ are the vertices of a right triangle.

5. Without sketching, determine symmetry with respect to the x-axis, y-axis, and origin:

$$y = 3x^3 - 8x$$

6. Without sketching, determine symmetry with respect to the x-axis, y-axis, and origin:

$$y = \frac{1}{4 - x^2}$$

7. Find the center and radius of the circle having the equation $x^2 + x + y^2 - 6y = -9$. Sketch.

8. Find the vertex and axis of symmetry of the parabola whose equation is $16x^2 - 8x + 32y + 65 = 0$. Sketch.

9. Find the intercepts and asymptotes of the hyperbola whose equation is $3x^2 - 8y^2 = 2$. Sketch.

10. Find the equation of the circle having center $\left(-1, -\frac{1}{2}\right)$ and radius $\sqrt{5}$.

11. Find the intercepts and vertex and sketch the graph of the parabola whose equation is $y = -x^2 + 16x - 14$.

12. Find the intercepts and sketch the graph of the equation $4x^2 + y^2 = 9$.

13. Identify the conic section whose equation is

$$x^2 - 4x + y^2 - 6y + 12 = 0$$

14. Identify the conic section whose equation is

$$x^2 + 2x - 5y + 4 = 0$$

15. Identify the conic section whose equation is

$$x^2 + 2y^2 - 4x + 3 = 0$$

CHAPTER THIRTEEN
SYSTEMS OF EQUATIONS

Many problems in business and engineering result in systems of equations or inequalities. In fact, systems of linear equations and inequalities occur with such frequency that mathematicians and computer scientists have devoted considerable energy to devising methods for their solution. With the aid of large-scale computers it is possible to solve systems involving thousands of equations or inequalities, a task that previous generations would not have dared tackle.

In this chapter we will study some of the basic methods for solving systems of linear equations. We will conclude with the method of Gaussian elimination, which can be extended to systems of linear equations of any size, thereby providing the basic concepts needed in the study of advanced methods for solving linear systems. We will also demonstrate that the same techniques can be applied to systems of equations that include at least one second-degree equation.

13.1
SYSTEMS OF LINEAR EQUATIONS

A pile of nine coins consists of nickels and quarters. If the value of the coins is $1.25, how many of each type of coin are there?

This type of word problem was handled in earlier chapters by using one variable. Now we tackle this problem by using two variables. If we let

$$x = \text{the number of nickels}$$

and

$$y = \text{the number of quarters}$$

then

$$x + y = 9$$
$$5x + 25y = 125$$

We call this a **system of linear equations** or simply a **linear system** and we seek values of x and y that satisfy *both* equations. A pair of values $x = r$, $y = s$ that satisfy both equations is called a **solution** of the system. Thus,

$$x = 5, \quad y = 4$$

is a solution because substituting in the equations of the system gives

$$5 + 4 \overset{\checkmark}{=} 9$$
$$5(5) + 25(4) \overset{\checkmark}{=} 125$$

However,

$$x = 6, \quad y = 3$$

is not a solution, since

$$6 + 3 \overset{\checkmark}{=} 9$$
$$5(6) + 25(3) \neq 125$$

We will devote most of this chapter to methods of solving linear systems.

SOLVING BY GRAPHING

We know that the graph of a linear equation is a straight line. Returning to the linear system

$$x + y = 9$$
$$5x + 25y = 125$$

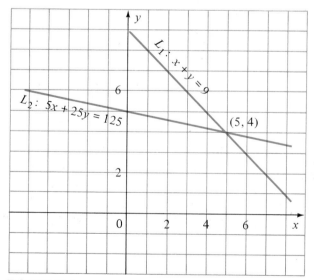

FIGURE 13.1

we can graph both lines on the same set of coordinate axes, as in Figure 13.1. The coordinates of every point on line L_1 satisfy the equation $x + y = 9$, while the coordinates of every point on line L_2 satisfy the equation $5x + 25y = 125$. Therefore the coordinates $(5, 4)$ of the point where the lines intersect must satisfy *both* equations, and hence these coordinates are a solution of the linear system. In general,

If the straight lines obtained from a system of linear equations intersect, then the coordinates of the point of intersection are a solution of the system.

EXAMPLE 1
Solve by graphing.

$$2x - y = -5$$
$$x + 2y = 0$$

Solution
We graph both lines (see Figure 13.2). The point of intersection is at $(-2, 1)$. Verify that $x = -2$, $y = 1$ is a solution of the system.

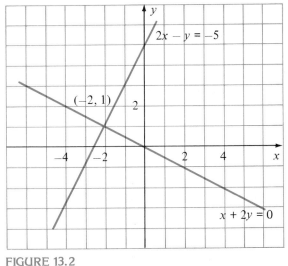

FIGURE 13.2

PROGRESS CHECK 1
Solve by graphing.

$$2x - 3y = 12$$
$$3x - 2y = 13$$

Answer
$x = 3, y = -2$

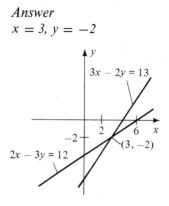

When we graph two linear equations on the same set of coordinate axes, there are three possibilities that can occur.

(a) The two lines intersect in a point.

(b) The two lines are parallel (do not intersect).

(c) The two lines are really the same line.

The following example illustrates the latter two cases.

EXAMPLE 2

Solve by graphing.

(a) $6x - 2y = 9$ (b) $2x + 3y = 5$

     $3x - y = 12$ $6x + 9y = 15$

Solution

(a) See Figure 13.3a. The system results in parallel lines. Therefore, since the lines do not intersect, there is no solution, or we may say that the solution consists of the null set. A system with no solution is said to be **inconsistent.**

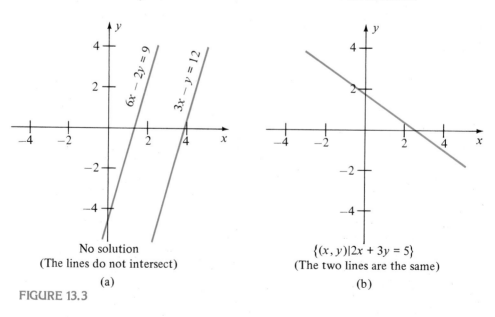

No solution
(The lines do not intersect)
(a)

$\{(x, y)|2x + 3y = 5\}$
(The two lines are the same)
(b)

FIGURE 13.3

(b)   See Figure 13.3b. The system yields only one line because the two equations are equivalent. Since any point on the line is a solution, we can say that the solution set is

$$\{(x, y) \mid 2x + 3y = 5\}$$

PROGRESS CHECK 2

Solve by graphing.

(a) $3x - 2y = -2$ (b) $5x + y = -4$

$\frac{3}{2}x - y = -1$ $10x + 2y = 4$

Answers

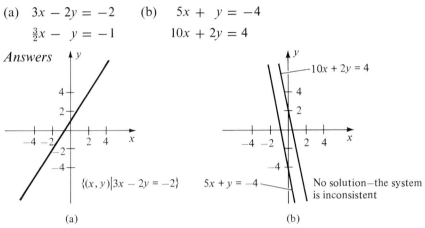

(a)   (b)

The method of graphing has severe limitations since the accuracy of the solution depends on the accuracy of the graph. The algebraic methods that follow avoid this limitation.

SOLVING BY SUBSTITUTION

Let's demonstrate the method of substitution using the system

$$x - 2y = -10$$
$$2x + 3y = 8$$

It is easy to solve the first equation for x.

$$x - 2y = -10$$
$$x = 2y - 10$$

We now *substitute* this expression for x in the second equation.

$$2x + 3y = 8$$
$$2(2y - 10) + 3y = 8$$

This last equation is a linear equation in one variable, which we know how to solve.

$$2(2y - 10) + 3y = 8$$
$$4y - 20 + 3y = 8$$
$$7y = 28$$

$$y = 4$$

We can now substitute $y = 4$ in either equation to find x.

$$x - 2y = -10$$
$$x - 2(4) = -10$$
$$x = -2$$

You can verify that $x = -2$, $y = 4$ is a solution of the linear system.

What happens if the lines represented by the linear system are parallel or reduce to the same line? Let's apply the method of substitution to these cases.

TABLE 13.1

Parallel lines	Same line
$x - 3y = 5$ $2x - 6y = 20$	$x + 4y = 10$ $-2x - 8y = -20$
Solving the first equation for x,	Solving the first equation for x,
$x = 5 + 3y$	$x = 10 - 4y$
and substituting in the second equation,	and substituting in the second equation,
$2(5 + 3y) - 6y = 20$	$-2(10 - 4y) - 8y = -20$
$10 + 6y - 6y = 20$	$-20 + 8y - 8y = -20$
$10 = 20$	$-20 = -20$
Impossible!	Always true!
The system is satisfied by *no* values of x and y.	All solutions to one equation will also satisfy the other equation.

We see that the method of substitution signals us when there is no solution. In general,

The method of substitution will
(a) provide a unique value for x and for y if the lines intersect.
(b) result in a contradiction if the lines are parallel.
(c) result in an identity if there is only one line.

EXAMPLE 3
Solve by substitution.

$$3x - 2y = -19$$
$$2x + 3y = -4$$

Solution
From the first equation we have

$$3x - 2y = -19$$
$$3x + 19 = 2y$$
$$\frac{3x + 19}{2} = y$$

Substituting in the second equation,

$$2x + 3y = -4$$

$$2x + 3\left(\frac{3x + 19}{2}\right) = -4$$

$$4x + 3(3x + 19) = -8$$

$$4x + 9x + 57 = -8$$

$$13x = -65$$

$$x = -5$$

Substituting $x = -5$ in either equation, we have

$$y = 2$$

Verify that $x = -5$, $y = 2$ is a solution of the system of equations.

PROGRESS CHECK 3

Solve by substitution.

(a) $2x - y = 10$ (b) $4x + 3y = 3$ (c) $3x - y = 7$
     $3x - 2y = 14$ $-3x - 2y = -1$ $-9x + 3y = -22$

Answers
(a) $x = 6$, $y = 2$ (b) $x = -3$, $y = 5$ (c) *no solution*

WARNING

The expression for x or y obtained from an equation *must not be substituted into the same equation.* From the first equation of the linear system

$$x + 2y = -1$$
$$3x + y = 2$$

we obtain

$$x = -1 - 2y$$

Substituting (incorrectly) into the same equation would result in

$$(-1 - 2y) + 2y = -1$$
$$-1 = -1$$

The substitution $x = -1 - 2y$ must be made in the *second* equation.

EXERCISE SET 13.1

1. Which of the following is a solution of the linear system?

$$x + 2y = 9$$
$$2x - y = 1$$

(a) $x = 3$, $y = 2$ (b) $x = -3$, $y = 2$ (c) $x = 2$, $y = 3$
(d) $x = -3$, $y = -2$ (e) none of these

2. Which of the following is a solution of the linear system?

$$2x - 3y = 8$$
$$3x - 2y = 2$$

(a) $x = 4, \quad y = 2$ (b) $x = -2, \quad y = 4$

(c) $x = -4, \quad y = -2$ (d) $x = -2, \quad y = -4$

(e) none of these

Solve the given linear system by graphing.

3. $x + y = 1$ 4. $x - y = 1$

$x - y = 3$ $x + y = 5$

5. $x + 2y = -5$ 6. $x + y = 5$

$x - 3y = 0$ $3x + 3y = 10$

7. $3x - y = 4$ 8. $x + 2y = 8$

$6x - 2y = -8$ $3x - 4y = 4$

9. $x + 3y = -2$ 10. $2x - y = -1$

$3x - 5y = 8$ $3x - y = -1$

11. $2x + 5y = -6$ 12. $-3x + y = 11$

$4x + 10y = -12$ $4x - 2y = -16$

13. $2x + 3y = -8$ 14. $3x + 2y = 10$

$x - y = 11$ $-9x - 6y = 8$

Solve the given linear system by the method of substitution and check your answer.

15. $x + y = -2$ 16. $-x - 2y = 4$

$x - y = 4$ $x + 3y = -7$

17. $x + 3y = 3$ 18. $x - 4y = -7$

$x - 5y = -1$ $2x - 8y = -4$

19. $x + 6y = 4$ 20. $3x - y = -9$

$3x - 4y = 1$ $2x - y = -7$

21. $2x + 2y = 3$ 22. $8x - 2y = -3$

$4x + 4y = 6$ $3x - 5y = 1$

23. $3x + 3y = 9$ 24. $2x + 3y = -2$

$2x + 2y = -6$ $-3x - 5y = 4$

25. $3x - y = 18$ 26. $2x + y = 6$

$\frac{3}{2}x - \frac{1}{2}y = 9$ $x + \frac{1}{2}y = 3$

SOLVING BY ELIMINATION

If we require answers to a system of linear equations accurate to, say, five decimal places, then the estimates obtained by graphing won't suffice. The method of substitution provides exact answers but is not suited for a computer.

The **method of elimination** overcomes these difficulties. The strategy of the method is to replace the original system of equations by a simpler, equivalent system that has the same solution set. The following operations are used to obtain an equivalent system.

(a) An equation may be multiplied by a nonzero constant.

(b) An equation may be replaced by the sum of itself and another equation.

TABLE 13.2

Solving by elimination	Example
	$\begin{cases} 3x + y = 7 \\ 2x - 4y = 14 \end{cases}$
Step 1. Multiply each equation by a constant so that the coefficients of either x or y will differ only in sign.	*Step 1.* Multiply first equation by 4 and second equation by 1 so that the coefficients of y will be 4 and -4. $\begin{cases} 12x + 4y = 28 \\ -2x - 4y = 14 \end{cases}$
Step 2. Add the equations. The resulting equation will contain (at most) one variable.	*Step 2.* $\begin{aligned} 12x + 4y &= 28 \\ 2x - 4y &= 14 \\ \hline 14x &= 42 \end{aligned}$
Step 3. Solve the resulting linear equation in one variable.	*Step 3.* $$14x = 42$$ $$x = 3$$
Step 4. Substitute in either of the *original* equations to solve for the second variable.	*Step 4.* Substituting $x = 3$ in the first equation of the original system, $$3x + y = 7$$ $$3(3) + y = 7$$ $$9 + y = 7$$ $$y = -2$$
Step 5. Check in both equations.	*Step 5.* $3x + y = 7 \qquad\qquad 2x - 4y = 14$ $3(3) + (-2) \overset{?}{=} 7 \qquad 2(3) - 4(-2) \overset{?}{=} 14$ $9 - 2 \overset{?}{=} 7 \qquad\qquad 6 + 8 \overset{?}{=} 14$ $7 \overset{\checkmark}{=} 7 \qquad\qquad\qquad 14 \overset{\checkmark}{=} 14$

Note that in Step 2 we have "eliminated" y, which is why we call this the method of elimination.

WARNING ───
When multiplying an equation by a constant, be sure to multiply each term of both sides of the equation by the constant.

EXAMPLE 1

Solve by elimination.

$$2x + 5y = -5$$
$$3x - 2y = -17$$

Solution

We choose to eliminate x. If we multiply the first equation by 3 and the second equation by -2, we will have coefficients of 6 and -6 as desired.

$$6x + 15y = -15$$

Adding,
$$-6x + 4y = 34$$
$$\overline{19y = 19}$$

$$y = 1$$

Substituting in the first equation,

$$2x + 5y = -5$$
$$2x + 5(1) = -5$$
$$2x = -10$$

$$x = -5$$

Check that this solution satisfies both equations.

PROGRESS CHECK 1

Solve by elimination.

(a) $\quad 2x - y = 2$ (b) $\quad 5x + 2y = 5$
$\quad -4x - 2y = 8$ $\quad -2x + 3y = -21$

Answers

(a) $\quad x = -\dfrac{1}{2}, \quad y = -3$ (b) $\quad x = 3, \quad y = -5$

In Section 13.1 we saw that the graphs of the equations

$$6x - 2y = 9$$
$$3x - y = 12$$

are parallel lines and that the system of equations has no solution. If we attempt to solve this system by elimination, we can multiply the second equation by -2 to eliminate y.

$$6x - 2y = 9$$
$$-6x + 2y = -24$$
$$\overline{0x + 0y = -15} \quad \text{Impossible!}$$

We see that the elimination method signals us when there is no solution, by yielding an impossibility. In general,

If the elimination method results in an equation of the form

$$0x + 0y = c$$

where $c \neq 0$, then there is no solution of the original system of linear equations, and the system is inconsistent.

We also saw in Section 13.1 that the system of equations

$$2x + 3y = 5$$
$$6x + 9y = 15$$

is really two forms of the same equation. If we eliminate x by multiplying the first equation by -3, we have

$$-6x - 9y = -15$$
$$\underline{6x + 9y = 15}$$
$$0x + 0y = 0 \qquad \text{Always true.}$$

Again, the elimination method has signaled us that any point on the line is a solution. In general,

If the elimination method results in an equation of the form

$$0x + 0y = 0$$

then the equations in the original system are equivalent.

You may have noticed that the special cases are signaled in essentially the same way by both the method of elimination and the method of substitution. The case of parallel lines (no solution) is signaled by a contradiction; the case of the same line (equivalent equations) is indicated by an identity.

EXAMPLE 2
Solve by elimination.

(a) $6x - 2y = 7$ (b) $5x + 6y = 4$
     $3x - y = 16$ $-10x - 12y = -8$

Solution
(a) If we multiply the second equation by -2, we have

$$6x - 2y = 7$$
$$\underline{-6x + 2y = -32}$$
$$0x + 0y = 25$$

We can conclude that the system of equations represents a pair of parallel lines and that there are no solutions.

(b) Multiplying the first equation by 2, we have

$$10x + 12y = 8$$
$$-10x - 12y = -8$$
$$\overline{\quad 0x + \quad 0y = 0\quad}$$

We can conclude that the equations are equivalent and that the solution set is given by $\{(x, y)\,|\,5x + 6y = 4\}$.

PROGRESS CHECK 2
Solve by elimination.

(a) $x - y = 2$ (b) $4x + 6y = 3$
      $3x - 3y = -6$ $-2x - 3y = -\frac{3}{2}$

Answers
(*a*) *no solution*   (*b*) $\{(x, y)\,|\,4x + 6y = 3\}$

EXERCISE SET 13.2
Solve by elimination and check.

1. $x + y = -1$
    $x - y = 3$

2. $x - 2y = 8$
    $2x + y = 1$

3. $x + 4y = -1$
    $2x - 4y = 4$

4. $2x - 2y = 4$
    $x - y = 8$

5. $x + 2y = 6$
    $2x + 4y = 12$

6. $x - 2y = 4$
    $2x + y = 3$

7. $x + 3y = 2$
    $3x - 5y = -6$

8. $x + 2y = 0$
    $5x - y = 22$

9. $2x + 2y = 6$
    $3x + 3y = 6$

10. $2x + y = 2$
    $3x - y = 8$

11. $x + 2y = 1$
    $5x + 2y = 13$

12. $x - 4y = -7$
    $2x + 3y = -8$

13. $x - 3y = 9$
    $x + 5y = 11$

14. $2x - 3y = 8$
    $4x - 6y = 16$

15. $x - y = 3$
    $3x + 2y = 14$

16. $4x + y = 3$
    $2x - y = 3$

17. $2x + y = 7$
    $3x - 2y = 0$

18. $2x + 3y = 4$
    $4x + 6y = 6$

19. $2x + 3y = 13$
    $3x - 4y = 1$

20. $3x - 2y = 5$
    $\frac{3}{2}x - y = \frac{5}{2}$

In Section 13.1 we saw that many of the word problems that we previously solved by using one variable can be recast as a system of linear equations. There are, in addition, many word problems that are difficult to handle with one variable but are easily formulated by using two variables.

EXAMPLE 1
If 3 sulfa pills and 4 penicillin pills cost 69 cents while 5 sulfa pills and 2 penicillin pills cost 73 cents, what is the cost of each kind of pill?

Solution
Using two variables, we let

$$x = \text{the cost of each sulfa pill}$$
$$y = \text{the cost of each penicillin pill}$$

Then

$$3x + 4y = 69$$
$$5x + 2y = 73$$

Multiplying the second equation by -2 and adding to eliminate y,

$$3x + 4y = 69$$
$$\underline{-10x - 4y = -146}$$
$$-7x \qquad = -77$$

$$x = 11$$

Substituting in the first equation,

$$3(11) + 4y = 69$$
$$33 + 4y = 69$$
$$4y = 36$$

$$y = 9$$

Thus, each sulfa pill costs 11 cents and each penicillin pill costs 9 cents. (Could you have set up this problem using only one variable? Unlikely!)

PROGRESS CHECK 1
If 2 pounds of rib steak and 6 pounds of hamburger meat costs $12.30 while 3 pounds of rib steak and 2 pounds of hamburger meat costs $9.70, what is the cost per pound of each type of meat?

Answer
steak: $2.40,   hamburger: $1.25

EXAMPLE 2
Swimming downstream, a swimmer can cover 2 kilometers in 15 minutes, while the return trip upstream requires 20 minutes. What is the rate of the swimmer and of the current? (The rate of the swimmer is the speed at which he would swim if there were no current.)

Solution
Let

$$x = \text{rate of the swimmer (in km per hour)}$$
$$y = \text{rate of the current (in km per hour)}$$

When swimming downstream, the rate of the current is added to the rate of the swimmer so that $x + y$ is the rate downstream. Similarly, $x - y$ is the rate while swimming upstream. We display the information we have, expressing time in hours.

	Rate	$\times$	Time	$=$	Distance
Downstream	$x + y$		$\dfrac{1}{4}$		$\dfrac{1}{4}(x + y)$
Upstream	$x - y$		$\dfrac{1}{3}$		$\dfrac{1}{3}(x - y)$

Since distance upstream = distance downstream = 2 kilometers,

$$\tfrac{1}{4}(x + y) = 2$$
$$\tfrac{1}{3}(x - y) = 2$$

or, equivalently,

$$x + y = 8$$
$$x - y = 6$$

Solving, we have

$$x = 7 \qquad \text{Rate of the swimmer}$$
$$y = 1 \qquad \text{Rate of the current}$$

Thus, the rate of the swimmer is 7 kilometers per hour and the rate of the current is 1 kilometer per hour. (The student is urged to verify the solution.)

PROGRESS CHECK 2
Rowed downstream, a boat can travel a distance of 16 miles in 2 hours. If the return trip upstream requires 8 hours, what is the rate of the boat in still water and what is the rate of the current?

Answer
boat: 5 miles per hour,   current: 3 miles per hour

EXAMPLE 3
The sum of a two-digit number and its units digit is 64, while the sum of the number and its tens digit is 62. Find the number.

Solution
The basic idea in solving digit problems is to note that if we let

$$t = \text{tens digit}$$

and

$$u = \text{units digit}$$

then

$$10t + u = \text{the two-digit number}$$

Our word problem then translates into

$$(10t + u) + u = 64 \quad \text{or} \quad 10t + 2u = 64$$

and

$$(10t + u) + t = 62 \quad \text{or} \quad 11t + \ u = 62$$

Solving, we find that $t = 5$, $u = 7$ (verify) and the number we seek is 57.

PROGRESS CHECK 3

The sum of twice the tens digit and the units digit of a two-digit number is 16, while the sum of the tens digit and twice the units digit is 14. Find the number.

Answer
64

EXAMPLE 4

A landscaping firm prepares two plans for a homeowner. Plan A uses 8 hemlocks and 12 junipers at a cost of $520. If Plan B uses 6 hemlocks and 15 junipers at a cost of $510, what is the cost of each hemlock and of each juniper?

Solution

Let

$$x = \text{the cost of each hemlock}$$

and

$$y = \text{the cost of each juniper}$$

Then Plan A results in the equation

$$8x + 12y = 520$$

while Plan B yields the equation

$$6x + 15y = 510$$

Solving, we find that $x = 35$ and $y = 20$ so that each hemlock costs $35 and each juniper costs $20.

PROGRESS CHECK 4

A manufacturer of children's toys is producing two types of model airplanes. Model A requires 7 minutes on the jigsaw and 10 minutes on the lathe, while model B requires 15 minutes on the jigsaw and 6 minutes on the lathe. If the jigsaw and lathe are each used 15 hours per day, what is the daily production of each type of model airplane?

Answer
75 of model A and 25 of model B

APPLICATIONS IN BUSINESS AND ECONOMICS: BREAK-EVEN ANALYSIS

One of the problems faced by a manufacturer is that of determining the **level of production,** that is, how many units of the product should be manufactured during a given fixed time period such as a day, a week, or a month. Suppose that

$$C = 400 + 2x \tag{1}$$

is the total cost (in thousands of dollars) of producing x units of the product and that

$$R = 4x \qquad (2)$$

is the total revenue (in thousands of dollars) when x units of the product are sold. This would happen, for example, if after setting up production at a cost of \$400,000, there is an additional cost of \$2000 to make each unit [Equation (1)] and a revenue of \$4000 is earned from the sale of each unit [Equation (2)]. If all the units that are manufactured are sold, then the total profit P is the difference between total revenue and total cost.

$$P = R - C$$
$$= 4x - (400 + 2x)$$
$$= 2x - 400$$

The value of x for which $R = C$, so that the profit is zero, is called the **break-even point.** When that many units of the product have been produced and sold the manufacturer neither makes money nor loses money. To find the break-even point we set $R = C$. Using Equations (1) and (2) we obtain

$$400 + 2x = 4x$$

or

$$400 = 2x$$

$$x = \frac{400}{2} = 200$$

Thus, the break-even point is 200 units.

The break-even point can also be obtained graphically as follows. Observe that Equations (1) and (2) are linear equations and therefore equations of straight lines. The break-even point is the x-coordinate of the point where the

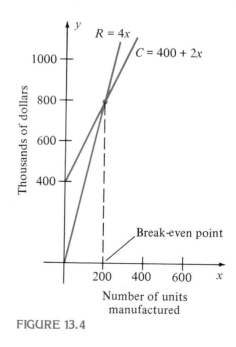

FIGURE 13.4

lines given by Equations (1) and (2) intersect. Figure 13.4 shows the lines and their point of intersection (200, 800). When 200 units of the product are made, the cost ($800,000) is exactly equal to the revenue, and the profit is $0. If $x > 200$, then $R > C$, so that the manufacturer is making a profit. If $x < 200$, $R < C$ and the manufacturer is losing money.

EXAMPLE 5
A steel producer finds that when x million tons of steel are made, the total cost and revenue are given (in millions of dollars) by

$$C = 20 + 0.4x$$

$$R = 0.8x$$

(a) Find the break-even point graphically.
(b) What is the total revenue at the break-even point?

Solution
(a) See Figure 13.5. The break-even point is $x = 50$ million tons.

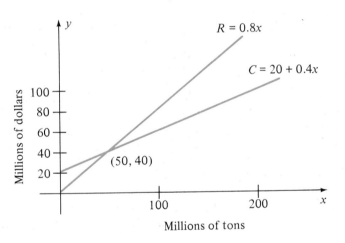

FIGURE 13.5

(b) When the level of production is 50 million tons, the total revenue is

$$R = 0.8x$$

$$= (0.8)(50)$$

$$= 40 \text{ million dollars}$$

PROGRESS CHECK 5
A producer of photographic developer finds that the total weekly cost of producing x liters of developer is given (in dollars) by $C = 550 + 0.40x$. The manufacturer sells the product at $0.50 per liter.
(a) What is the total revenue received when x liters of developer are sold?
(b) Find the break-even point graphically.
(c) What is the total revenue received at the break-even point?

Answers
(a) $R = 0.50x$ (b) 5500 liters   (c) $2750

APPLICATIONS IN BUSINESS AND ECONOMICS: SUPPLY AND DEMAND

A manufacturer of a product is free to set any price p (in dollars) for each unit of the product. Of course, if the price is too high, not enough people will buy the product; if the price is too low, so many people will rush to buy the product that the producer will not be able to satisfy demand. Thus, in setting price, the manufacturer must take into consideration the demand for the product.

Let S be the number of units that the manufacturer is willing to supply at the price p. S is called the **supply.** Generally, the value of S will increase as p increases; that is, the manufacturer is willing to supply more of the product as the price p increases. Let D be the number of units of the product that consumers are willing to buy at the price p. D is called the **demand.** Generally, the value of D will decrease as p increases; that is, consumers are willing to buy fewer units of the product as the price rises. For example, suppose that S and D are given by

$$S = 2p + 3 \tag{3}$$

$$D = -p + 12 \tag{4}$$

Equations (3) and (4) are linear equations, so they are equations of straight lines (see Figure 13.6). The price at which supply S and demand D are equal is called the **equilibrium price.** At this price, every unit that is supplied is purchased. Thus, there is neither a surplus nor a shortage. In Figure 13.6, the equilibrium price is $p = 3$. At this price, the number of units supplied equals the number of units demanded and is given by substituting in Equation (3): $2(3) + 3 = 9$. This value can also be obtained by finding the ordinate at the point of intersection in Figure 13.6.

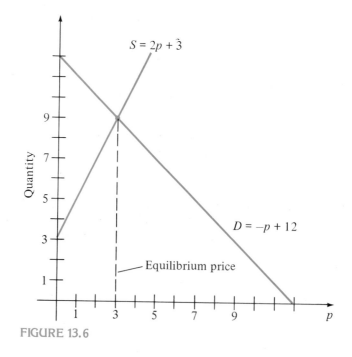

FIGURE 13.6

If we are in an economic system in which there is pure competition, then the law of supply and demand states that the selling price of a product will be its equilibrium price. That is, if the selling price were higher than the equilibrium price, then the consumers' reduced demand would leave the manufacturer with an unsold surplus. This would force the manufacturer to reduce the selling price. If the selling price is below the equilibrium price, then the increased demand would cause a shortage of the product. This would lead the manufacturer to raise the selling price. Of course, in actual practice, the marketplace does not operate under pure competition since manufacturers consult with each other on selling prices, governments try to influence selling prices, and many other factors are present. In addition, deeper mathematical analysis of economic systems requires the use of more sophisticated equations.

EXAMPLE 6

Suppose that the supply and demand for ball-point pens are given by

$$S = p + 5$$
$$D = -p + 7$$

(a)   Find the equilibrium price.
(b)   Find the number of pens sold at that price.

Solution

(a)   Figure 13.7 illustrates the graphical solution. Thus, the equilibrium price is $p = 1$. (Algebraic methods will, of course, yield the same solution.)

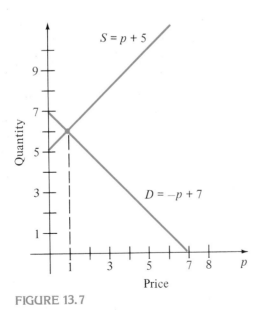

FIGURE 13.7

(b)   When $p = 1$, the number of pens sold is $S = 1 + 5 = 6$, the y-coordinate of the point of intersection.

PROGRESS CHECK 6

Suppose that the supply and demand for radios are given by

$$S = 3p + 120$$
$$D = -p + 200$$

(a) Find the equilibrium price.
(b) Find the number of radios sold at this price.

Answers

(a) 20   (b) 180

EXERCISE SET 13.3

1. A pile of 40 coins consists of nickels and dimes. If the total value of the coins is $2.75, how many of each type of coin are there?

2. An automatic vending machine in the post office, which charges no more than a clerk, provides a packet of 27 ten-cent and twenty-cent stamps worth $3.00. How many of each type of stamp are there?

3. A photography store sells sampler A, consisting of 6 rolls of color film and 4 rolls of black and white film for $21.00. It also sells sampler B consisting of 4 rolls of color film and 6 rolls of black and white film for $19.00. What is the cost per roll of each type of film?

4. A hardware store sells power pack A, consisting of 4 D cells and 2 C cells, for $1.70, and power pack B, consisting of 6 D cells and 4 C cells, for $2.80. What is the price of each cell?

5. A fund is planning to invest $6000 in two types of bonds, A and B. Bond A is safer than bond B and pays a dividend of 8%, while bond B pays a dividend of 10%. If the total return of both investments is $520, how much was invested in each type of bond?

6. A trash removal company carries waste material in two sizes of sealed containers weighing 4 and 3 kilograms, respectively. On a certain trip there are a total of 30 containers weighing 100 kilograms. How many of each type of container are there?

7. A paper firm makes rolls of paper 12″ wide and 15″ wide by cutting a sheet that is 180″ wide. Suppose that a total of 14 rolls of paper are to be cut without any waste. How many of each type of roll will be made?

8. An animal-feed producer mixes two types of grain, A and B. Each unit of grain A contains 2 grams of fat and 80 calories, while each unit of grain B contains 3 grams of fat and 60 calories. If the producer wants the final product to provide 18 grams of fat and 480 calories, how much of each type of grain should be used?

9. A supermarket mixes coffee that sells for $1.20 per pound with coffee selling for $1.80 per pound to obtain 24 pounds of coffee selling for $1.60 per pound. How much of each type of coffee should be used?

10. An airplane flying against the wind covers a distance of 3000 kilometers in 6 hours, and the return trip with the aid of the wind takes 5 hours. What is the speed of the airplane in still air and what is the speed of the wind?

11. A cyclist who is traveling against the wind can cover a distance of 45 miles in 4 hours. The return trip with the aid of the wind takes 3 hours. What is the speed of the bicycle in still air and what is the speed of the wind?

12. The sum of a two-digit number and its units digit is 20 while the sum of the number and its tens digit is 16. Find the number.

13. The sum of the digits of a two-digit number is 7. If the digits are reversed, the resulting number exceeds the given number by 9. Find the number.

14. The sum of three times the tens digit and the units digit of a two-digit number is 14, while the sum of the tens digit and twice the units digit is 18. Find the number.

15. A health food shop mixes nuts and raisins into a snack pack. How many pounds of nuts, selling for $2 per pound, and how many pounds of raisins, selling for $1.50 per pound, must be mixed to produce a 50-pound mixture selling for $1.80 per pound?

16. A movie theater charges $3 admission for an adult and $1.50 for a child. If 600 tickets were sold and the total revenue received was $1350, how many tickets of each type were sold?

17. A moped dealer selling a model A and a model B moped has $18,000 in inventory. The profit on selling a model A moped is 12%, while the profit on a model B moped is 18%. If the profit on the entire stock would be 16%, how much was invested in each type of model?

18. The cost of sending a telegram is determined as follows. There is a flat charge for the first 10 words and a uniform rate for each additional word. Suppose that an 18-word telegram costs $1.94, while a 22-word telegram costs $2.16. Find the cost of the first 10 words and the rate for each additional word.

19. A certain epidemic disease is treated by a combination of the drugs Epiline I and Epiline II. Suppose that each unit of Epiline I contains 1 milligram of factor X and 2 milligrams of factor Y, while each unit of Epiline II contains 2 milligrams of factor X and 3 milligrams of factor Y. Successful treatment of the disease calls for 13 milligrams of factor X and 22 milligrams of factor Y. How many milligrams of Epiline I and Epiline II should be administered to a patient?

20. **(Break-even analysis)** An animal-feed manufacturer finds that the weekly cost of making x kilograms of feed is given (in dollars) by $C = 2000 + 0.50x$, while the revenue received from selling the feed is given by $R = 0.75x$.
    (a) Find the break-even point graphically.
    (b) What is the total revenue at the break-even point?

21. **(Break-even analysis)** A small manufacturer of a new solar device finds that the annual cost of making x units is given (in dollars) by $C = 24,000 + 55x$. Each device sells for $95.
    (a) What is the total revenue received when x devices are sold?
    (b) Find the break-even point graphically.
    (c) What is the total revenue received at the break-even point?

22. **(Supply and demand)** A manufacturer of calculators finds that the supply and demand are given by

$$S = 4p + 12$$
$$D = -2p + 18$$

    (a) Find the equilibrium price.
    (b) Find the number of calculators sold at this price.

23. **(Supply and demand)** A manufacturer of mopeds finds that the supply and demand are given by

$$S = 2p + 10$$
$$D = -p + 19$$

    (a) Find the equilibrium price.
    (b) Find the number of mopeds sold at this price.

13.4
SYSTEMS OF LINEAR EQUATIONS IN THREE UNKNOWNS

The method of substitution and the method of elimination can both be applied to systems of linear equations in three unknowns and, more generally, to systems of linear equations in any number of unknowns. There is yet another method, known as **Gaussian elimination,** which is ideally suited for computers and which we will now apply to linear systems in three unknowns.

The objective of Gaussian elimination is to transform a given linear system into triangular form, such as

$$3x - y + 3z = -11$$
$$2y + z = 2$$
$$2z = -4$$

A linear system is in **triangular form** when the only nonzero coefficient of x appears in the first equation, the only nonzero coefficients of y appear in the first and second equations, and so on. Note that when a linear system is in triangular form, the last equation must involve only one variable and immediately yields the value of an unknown. In our example, we see that

$$2z = -4$$

$$z = -2$$

Substituting $z = -2$ in the second equation yields

$$2y + (-2) = 2$$

$$y = 2$$

Finally, substituting $z = -2$, $y = 2$ in the first equation yields
$$3x - (2) + 3(-2) = -11$$
$$3x = -3$$

$$x = -1$$

This process of **back-substitution** thus allows us to solve a linear system quickly when it is in triangular form.

The challenge, then, is to find a means of transforming a linear system into triangular form. In Section 13.2 we listed several operations which may be used to obtain an equivalent linear system. We now offer (without proof) a more complete list of operations that transform a system of linear equations into an equivalent system.

(a)   Interchange any two equations.

(b)   Multiply an equation by a nonzero constant.

(c)   Replace an equation by the sum of itself plus a constant times another equation.

Using these operations we can now demonstrate the method of Gaussian elimination.

EXAMPLE 1

Solve the linear system

$$2y - z = -5$$
$$x - 2y + 2z = 9$$
$$2x + y - z = -2$$

Solution

Gaussian elimination	Example
Step 1. (a) If necessary, interchange equations to obtain a nonzero coefficient for x in the first equation.	*Step 1.* (a) Interchanging the first two equations yields $x - 2y + 2z = 9$ $2y - z = -5$ $2x + y - z = -2$
(b) Replace the second equation by the sum of itself and an appropriate multiple of the first equation which will result in a zero coefficient for x. (c) Replace the third equation by the sum of itself and an appropriate multiple of the first equation which will result in a zero coefficient for x.	(b) The coefficient of x in the second equation is already 0. (c) Replace the third equation by the sum of itself and -2 times the first equation. $x - 2y + 2z = 9$ $2y - z = -5$ $5y - 5z = -20$
Step 2. Apply the procedures of Step 1 to the second and third equations.	*Step 2.* Replace the third equation by the sum of itself and $-\frac{5}{2}$ times the second equation. $x - 2y + 2z = 9$ $2y - z = -5$ $-\frac{5}{2}z = -\frac{15}{2}$
Step 3. The system is now in triangular form. The solution is obtained by back-substitution.	*Step 3.* From the third equation, $$-\frac{5}{2}z = -\frac{15}{2}$$ $$z = 3$$ Substituting this value for z in the second equation, $$2y - (3) = -5$$ $$y = -1$$ Substituting for y and for z in the first equation $$x - 2(-1) + 2(3) = 9$$ $$x + 8 = 9$$ $$x = 1$$ The solution is $x = 1$, $y = -1$, $z = 3$.

PROGRESS CHECK 1
Solve by Gaussian elimination.

(a) $2x - 4y + 2z = 1$

$3x + y + 3z = 5$

$x - y - 2z = -8$

(b) $-2x + 3y - 12z = -17$

$3x - y - 15z = 11$

$-x + 5y + 3z = -9$

Answers

(a) $x = -\dfrac{3}{2}$, $y = \dfrac{1}{2}$, $z = 3$ (b) $x = 5$, $y = -1$, $z = \dfrac{1}{3}$

EXERCISE SET 13.4
Solve by Gaussian elimination and check.

1. $x + 2y + 3z = -6$

    $2x - 3y - 4z = 15$

    $3x + 4y + 5z = -8$

2. $2x + 3y + 4z = -12$

    $x - 2y + z = -5$

    $3x + y + 2z = 1$

3. $x + y + z = 1$

    $x + y - 2z = 3$

    $2x + y + z = 2$

4. $2x - y + z = 3$

    $x - 3y + z = 4$

    $-5x - 2z = -5$

5. $x + y + z = 2$

    $x - y + 2z = 3$

    $3x + 5y + 2z = 6$

6. $x + y + z = 0$

    $x + y = 3$

    $y + z = 1$

7. $x + 2y + z = 7$

    $x + 2y + 3z = 11$

    $2x + y + 4z = 12$

8. $4x + 2y - z = 5$

    $3x + 3y + 6z = 1$

    $5x + y - 8z = 8$

9. $x + y + z = 2$

    $x + 2y + z = 3$

    $x + y - z = 2$

10. $x + y - z = 2$

    $x + 2y + z = 3$

    $x + y + 4z = 3$

11. $2x + y + 3z = 8$

    $-x + y + z = 10$

    $x + y + z = 12$

12. $2x - y + z = 2$

    $3x + y + 2z = 3$

    $x + y - z = -1$

13. $x + 2y - 2z = 8$

    $5y - z = 6$

    $-2x + y + 3z = -2$

14. $x - 2y + z = -5$

    $2x + z = -10$

    $y - z = 15$

15. $3x - y + 4z = 4$

    $x + y + z = 8$

    $2x - y - 2z = 10$

16. $2x - y + z = 4$

    $3x + y - z = 8$

    $x - y - z = 6$

17. $x + y - 2z = 6$
    $x - y \quad\ = -12$
    $y - z = 8$

18. $x + 2y - 3z = 10$
    $5x + y - z = -12$
    $3x - 3y + 5z = 15$

19. $2x - y + 2z = 6$
    $2x - y + 3z = -5$
    $x + y \quad\ = -4$

20. $x - y - z = 3$
    $y + z = 5$
    $x + y - z = 8$

21. A special low-calorie diet consists of dishes A, B, and C. Each unit of A has 2 grams of fat, 1 gram of carbohydrate, and 3 grams of protein. Each unit of B has 1 gram of fat, 2 grams of carbohydrate, and 1 gram of protein. Each unit of C has 1 gram of fat, 2 grams of carbohydrate, and 3 grams of protein. The diet must provide exactly 10 grams of fat, 14 grams of carbohydrate, and 18 grams of protein. How much of each dish should be used?

22. A furniture manufacturer makes chairs, coffee tables, and dining-room tables. Each chair requires 2 minutes of sanding, 2 minutes of staining, and 4 minutes of varnishing. Each coffee table requires 5 minutes of sanding, 4 minutes of staining, and 3 minutes of varnishing. Each dining-room table requires 5 minutes of sanding, 4 minutes of staining, and 6 minutes of varnishing. The sanding and varnishing benches are each available 6 hours per day, and the staining bench is available 5 hours per day. How many of each type of furniture should be made?

23. A manufacturer produces 12", 16", and 19" television sets that require assembly, testing, and packing. The 12" sets each require 45 minutes to assemble, 30 minutes to test, and 10 minutes to package. The 16" sets each require 1 hour to assemble, 45 minutes to test, and 15 minutes to package. The 19" sets each require $1\frac{1}{2}$ hours to assemble, 1 hour to test, and 15 minutes to package. If the assembly line operates for $17\frac{3}{4}$ hours per day, the test facility is available for $12\frac{1}{2}$ hours per day, and the packing equipment is available for $3\frac{3}{4}$ hours per day, how many of each type of set can be produced?

13.5
SYSTEMS INVOLVING
NONLINEAR EQUATIONS

The algebraic methods of substitution and elimination can also be applied to systems of equations that involve at least one second-degree equation. It is also a good idea to consider the graph of each equation since this tells you the maximum number of points of intersection or solutions of the system. Here are some examples.

EXAMPLE 1
Solve the system of equations.

$$x^2 + y^2 = 25$$
$$x + y = -1$$

Solution
From the second equation we have

$$y = -1 - x$$

Substituting for y in the first equation,

$$x^2 + (-1 - x)^2 = 25$$
$$x^2 + 1 + 2x + x^2 = 25$$
$$2x^2 + 2x - 24 = 0$$
$$x^2 + x - 12 = 0$$
$$(x + 4)(x - 3) = 0$$

which yields $x = -4$ and $x = 3$. Substituting these values for x in the equation $x + y = -1$ we obtain the corresponding values of y.

$$
\begin{array}{ll}
x + y = -1 & \quad x + y = -1 \\
-4 + y = -1 & \quad 3 + y = -1 \\
y = 3 & \quad y = -4
\end{array}
$$

Thus, $x = -4$, $y = 3$ and $x = 3$, $y = -4$ are solutions of the system of equations.

Note that the equations represent a circle and a line. The algebraic solution tells us that they intersect in two points. The student is urged to verify the solution by sketching the graphs of the equations.

PROGRESS CHECK 1
Solve the system of equations.

(a) $x^2 + 3y^2 = 12$ (b) $x^2 + y^2 = 34$
    $x + 3y = 6$ $x - y = 2$

Answers
(*a*) $x = 3$, $y = 1$ and $x = 0$, $y = 2$
(*b*) $x = -3$, $y = -5$ and $x = 5$, $y = 3$

EXAMPLE 2
Solve the system.

$$3x^2 + 8y^2 = 21$$
$$x^2 + 4y^2 = 10$$

Solution
We can employ the method of elimination to obtain an equation that has just one variable. If we multiply the second equation by -3 and add the resulting equations we have

$$
\begin{array}{r}
3x^2 + 8y^2 = 21 \\
-3x^2 - 12y^2 = -30 \\
\hline
-4y^2 = -9
\end{array}
$$

$$y^2 = \frac{9}{4}$$

$$y = \pm\frac{3}{2}$$

We can now substitute for y in either of the original equations. Using the second equation,

$$x^2 + 4y^2 = 10 \qquad\qquad x^2 + 4y^2 = 10$$

$$x^2 + 4\left(\frac{3}{2}\right)^2 = 10 \qquad\qquad x^2 + 4\left(-\frac{3}{2}\right)^2 = 10$$

$$x^2 + 9 = 10 \qquad\qquad x^2 + 9 = 10$$

$$x^2 = 1 \qquad\qquad x^2 = 1$$

$$x = \pm 1 \qquad\qquad x = \pm 1$$

We then have four solutions: $x = 1, y = \frac{3}{2}$; $x = -1, y = \frac{3}{2}$; $x = 1, y = -\frac{3}{2}$; $x = -1$, $y = -\frac{3}{2}$.

   Note that the equations represent two ellipses; the algebraic solution tells us that the ellipses intersect in four points. The student is urged to sketch the graphs.

PROGRESS CHECK 2
Solve the system.

$$2x^2 + y^2 = 5$$
$$2x^2 - 3y^2 = 3$$

Answer

$$x = \frac{3}{2}, \quad y = \frac{\sqrt{2}}{2}; \qquad x = -\frac{3}{2}, \quad y = \frac{\sqrt{2}}{2}; \qquad x = \frac{3}{2}, \quad y = -\frac{\sqrt{2}}{2};$$

$$x = -\frac{3}{2}, \quad y = -\frac{\sqrt{2}}{2}$$

EXAMPLE 3
Solve the system.

$$4x - y = 7$$
$$x^2 - y = 3$$

Solution
By subtracting the first equation from the second we obtain

$$x^2 - 4x = -4$$
$$x^2 - 4x + 4 = 0$$
$$(x - 2)(x - 2) = 0$$

which yields $x = 2$. Substituting $x = 2$ in the first equation,

$$4(2) - y = 7$$
$$y = 1$$

Thus, $x = 2, y = 1$ is a solution of the system.

   Note that the equations represent a line and a parabola. Since our algebraic techniques yield just one solution, the line is tangent to the parabola at the point (2, 1).

PROGRESS CHECK 3
Find the real solutions of the system.

$$x^2 - 4x + y^2 - 4y = 1$$
$$x^2 - 4x \qquad + y = -5$$

Answer
$x = 2$, $y = -1$ (*The parabola is tangent to the circle.*)

EXERCISE SET 13.5

Find all solutions of each of the following systems.

1. $x^2 + y^2 = 13$
$2x - y = 4$

2. $x^2 + 4y^2 = 32$
$x + 2y = 0$

3. $y^2 - x = 0$
$y - 4x = -3$

4. $xy = -4$
$4x - y = 8$

5. $x^2 - 2x + y^2 = 3$
$2x + y = 4$

6. $4x^2 + y^2 = 4$
$x - y = 3$

7. $xy = 1$
$x - y + 1 = 0$

8. $x^2 - y^2 = 3$
$x^2 + y^2 = 5$

9. $4x^2 + 9y^2 = 72$
$4x - 3y^2 = 0$

10. $x^2 + y^2 + 2y = 9$
$y - 2x = 4$

11. $2y^2 - x^2 = -1$
$4y^2 + x^2 = 25$

12. $x^2 + 4y^2 = 25$
$4x^2 + y^2 = 25$

13. $25y^2 - 16x^2 = 400$
$9y^2 - 4x^2 = 36$

14. $16y^2 + 5x^2 = 26$
$25y^2 - 4x^2 = 17$

15. $y^2 - 8x^2 = 9$
$y^2 + 3x^2 = 31$

16. $4y^2 + 3x^2 = 24$
$3y^2 - 2x^2 = 35$

17. $x^2 - 3xy - 2y^2 - 2 = 0$
$x - y - 2 = 0$

18. $3x^2 + 8y^2 = 21$
$x^2 + 4y^2 = 10$

19. The sum of the squares of the sides of a rectangle is 100 square meters. If the area of the rectangle is 48 square meters, find the length of each side of the rectangle.

20. Find the dimensions of a rectangle with an area of 30 square feet and perimeter of 22 feet.

21. Find two numbers such that their product is 20 and their sum is 9.

22. Find two numbers such that the sum of their squares is 30 and their sum is 10.

TERMS AND SYMBOLS

system of linear
equations   (p. 388)
linear system   (p. 388)
solution of a system of
linear equations
(p. 388)
solving by graphing
(p. 389)
inconsistent system
(p. 390)

method of substitution
(p. 391)
method of elimination
(p. 395)
equivalent equations
(p. 397)
level of production   (p. 401)
revenue and cost   (p. 402)
break-even point   (p. 402)

supply   (p. 404)
demand   (p. 404)
equilibrium price   (p. 404)
Gaussian elimination
(p. 408)
triangular form   (p. 408)
back-substitution   (p. 408)
nonlinear system   (p. 411)

KEY IDEAS FOR REVIEW

☐ The graph of a pair of linear equations in two variables is two straight lines that may either (a) intersect in a point, (b) be parallel, or (c) be the same line. If the two straight lines intersect, the coordinates of the point of intersection are a solution of the system of linear equations. If the lines do not intersect, the system is inconsistent.

☐ The method of substitution involves solving an equation for one variable and substituting the result into another equation.

☐ The method of elimination involves multiplying an equation by a nonzero constant so that when it is added to a second equation a variable drops out.

☐ When using the method of elimination, it is possible to detect the special cases in which the lines are parallel or reduce to the same line.

☐ It is often easier and more natural to set up word problems by using two or more variables.

☐ Gaussian elimination is a systematic way of transforming a linear system to triangular form. A system of equations in triangular form is easily solved by back-substitution.

☐ Many of the methods used in solving linear systems are applicable in solving nonlinear systems. Nonlinear systems often have more than one solution.

COMMON ERRORS

1. When multiplying an equation by a constant, you must multiply each term of both sides of the equation by the constant.

2. When using the method of substitution, the expression obtained for a variable from one equation must be substituted into a *different* equation.

3. Don't be frustrated if the method of substitution or method of elimination doesn't yield a solution to a linear system. A linear system need not have a solution, and these methods will call your attention to this situation.

4. Given the linear system

$$x + 3y - 2z = 5$$
$$2x - 2y + 5z = -4$$
$$-3x + y - 6z = 13$$

the variable x is eliminated from the *second* equation by replacing the *second* equation by the sum of itself plus -2 times the first equation. This yields

$$x + 3y - 2z = 5$$
$$-8y + 9z = -14$$
$$-3x + y - 6z = 13$$

Remember that the triangular form requires zero coefficients for x in all equations other than the first. This will help you to avoid the common error of replacing the wrong equation.

PROGRESS TEST 13A

1. Solve by graphing

$$2x + 3y = 2$$
$$-4x - 5y = -3$$

2. Solve by substitution.

$$-x + 6y = -11$$
$$2x + 5y = 5$$

3. Solve by substitution.

$$2x - 4y = -14$$
$$-x - 6y = -5$$

4. Solve by elimination.

$$x + y = 1$$
$$3x - 6y = -3$$

5. Solve by elimination.

$$2x - 5y = 15$$
$$-4x + 10y = 7$$

6. Solve by any method.

$$4x - 3y = 0$$
$$6x + y = -\tfrac{11}{6}$$

7. Solve by Gaussian elimination.

$$-3x - y + z = 12$$
$$2x + 5y - 2z = 1$$
$$-x + 4y + 2z = 8$$

8. The sum of a two-digit number and its tens digit is 49. If we reverse the digits of the number, the resulting number is 9 more than the original number. Find the number.

9. An airplane flying with a tail wind can complete a journey of 3500 kilometers between two cities in 5 hours. When flying back, the plane travels the same distance in 7 hours. What is the speed of the plane in still air and what is the wind speed?

10. Two bottles of Brand A aspirin and 3 bottles of Brand B aspirin cost $2.80. The cost of 4 bottles of Brand A aspirin and 2 bottles of Brand B is $3.20. What is the cost per bottle of Brand A and Brand B aspirin?

11. An auto repair shop finds that the monthly expenditures (in dollars) is given by $C = 4025 + 9x$, where x is the total number of hours worked by all employees. If the revenue received (in dollars) is given by $R = 16x$, find the break-even point and the total revenue received at that point.

12. A manufacturer of faucets finds that the supply S and demand D are given by

$$S = 3p + 2$$
$$D = -2p + 17$$

Find the equilibrium price and the number of faucets sold at this price.

PROGRESS TEST 13B

1. Solve by graphing.

$$3x - y = -17$$
$$x + 2y = -1$$

2. Solve by substitution.

$$2x + y = 4$$
$$3x - 2y = -15$$

3. Solve by substitution.

$$3x + 6y = -1$$
$$6x - 3y = 3$$

4. Solve by substitution.

$$4x - 4y = -3$$
$$-2x + 8y = 3$$

5. Solve by elimination.

$$7x - 2y = 3$$
$$-21x + 6y = -9$$

6. Solve by any method.

$$3x + y = -11$$
$$-2x + 2y = 2$$

7. Solve by Gaussian elimination.

$$x + 2y - 3z = -2$$
$$-x + 3y + 6z = 15$$
$$2x + 3y + 6z = 0$$

8. The sum of the digits of a two-digit number is 10 and the units digit exceeds the tens digit by 4. Find the number.
9. A motorboat can travel 60 kilometers downstream in 3 hours, and the return trip requires 4 hours. What is the rate of the boat in still water and what is the rate of the current?
10. If a dozen pencils and 5 pens cost 96 cents, while 8 pencils and 10 pens cost $1.44, what is the cost of each pencil and of each pen?
11. A school cafeteria manager finds that the weekly cost of operation is $1375 plus $1.25 for every meal served. If the average meal produces a revenue of $2.50, find the break-even point.
12. Suppose the supply and demand for a particular tennis racket is given by

$$S = 5p + 1$$
$$D = -2p + 43$$

Find the equilibrium price and the number of rackets sold at this price.

CHAPTER FOURTEEN
MATRICES AND DETERMINANTS

The material on matrices and determinants presented in this chapter serves as an introduction to linear algebra, a mathematical subject with wide application in other branches of mathematics as well as in the sciences, business, economics, and the social sciences.

Our study of matrices and determinants will concentrate on their application to the problem of solving systems of linear equations. We will see that the method of Gaussian elimination studied in Chapter Thirteen can be neatly implemented using matrices. Determinants will provide us with an additional technique, known as Cramer's rule, for the solution of special linear systems.

It should be emphasized that this material is a very brief introduction to matrices and determinants. Their properties and applications are both extensive and important.

14.1
MATRICES AND LINEAR SYSTEMS

In Chapter Thirteen we studied methods for solving a system of linear equations such as

$$2x + 3y = -7$$
$$3x - y = 17$$

This system can be displayed by a **matrix,** which is simply a rectangular array of $m \times n$ real numbers arranged in m horizontal **rows** and n vertical **columns.** The numbers are called the **entries** or **elements** of the matrix and are enclosed within brackets. Thus,

$$A = \begin{bmatrix} 2 & 3 & -7 \\ 3 & -1 & 17 \end{bmatrix} \begin{matrix} \leftarrow \\ \leftarrow \end{matrix} \text{rows}$$

columns

is a matrix consisting of 2 rows and 3 columns, whose entries are obtained from the two given equations. In general, a matrix of m rows and n columns is said to be of **dimension** $m \times n$. The matrix A is seen to be of dimension 2×3.

EXAMPLE 1

(a) $A = \begin{bmatrix} -1 & 4 \\ 0.5 & -2 \end{bmatrix}$ is a 2×2 matrix and is called a **square matrix** since $m = n$.

(b) $B = \begin{bmatrix} 4 & -5 \\ -2 & 1 \\ 3 & 0 \end{bmatrix}$ is a 3×2 matrix.

(c) $C = [-8 \quad 6 \quad 1]$ and $D = \begin{bmatrix} 2 \\ -4 \end{bmatrix}$ are of dimension 1×3 and 2×1, respectively.

PROGRESS CHECK 1

Determine the dimension of each matrix.

(a) $A = \begin{bmatrix} -0.25 & -5 & 6 \\ 10 & 0.1 & -3 \\ -2 & 4 & 6 \end{bmatrix}$

(b) $B = [2 \quad 3]$ (c) $C = \begin{bmatrix} -1 & 2 \\ 20 & 0 \\ 4 & -3 \end{bmatrix}$

Answers
(a) 3×3 (b) 1×2 (c) 3×2

There is a convenient way of denoting a general $m \times n$ matrix, using "double subscripts."

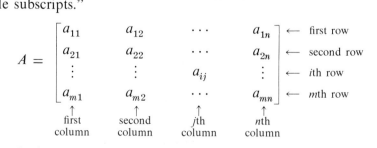

$$A = \begin{bmatrix} a_{11} & a_{12} & \cdots & a_{1n} \\ a_{21} & a_{22} & \cdots & a_{2n} \\ \vdots & \vdots & a_{ij} & \vdots \\ a_{m1} & a_{m2} & \cdots & a_{mn} \end{bmatrix} \begin{matrix} \leftarrow \text{first row} \\ \leftarrow \text{second row} \\ \leftarrow i\text{th row} \\ \leftarrow m\text{th row} \end{matrix}$$

first column   second column jth column nth column

Thus, a_{ij} is the entry in the ith row and jth column of the matrix A.

EXAMPLE 2

Let

$$A = \begin{bmatrix} 3 & -2 & 4 & 5 \\ 9 & 1 & 2 & 0 \\ -3 & 2 & -4 & 8 \end{bmatrix}$$

Then A is of dimension 3×4 and $a_{12} = -2$, $a_{23} = 2$, $a_{31} = -3$, $a_{33} = -4$, and $a_{34} = 8$.

PROGRESS CHECK 2

Let

$$B = \begin{bmatrix} 4 & 8 & 1 \\ 2 & -5 & 3 \\ -8 & 6 & -4 \\ 0 & 1 & -1 \end{bmatrix}$$

Find   (a) b_{11} (b) b_{23} (c) b_{31} (d) b_{42}.

Answers

(a) 4   (b) 3   (c) −8   (d) 1

If we begin with the system of linear equations

$$2x + 3y = -7$$
$$3x - y = 17$$

the matrix

$$\begin{bmatrix} 2 & 3 \\ 3 & -1 \end{bmatrix}$$

formed from the coefficients of the unknowns is called the **coefficient matrix.**
The matrix

$$\begin{bmatrix} 2 & 3 & -7 \\ 3 & -1 & 17 \end{bmatrix}$$

which includes the right-hand side of the system, is called the **augmented matrix.**

EXAMPLE 3

Write a system of linear equations that corresponds to the augmented matrix

$$\begin{bmatrix} -5 & 2 & -1 & 15 \\ 0 & -2 & 1 & -7 \\ \frac{1}{2} & 1 & -1 & 3 \end{bmatrix}$$

Solution

We attach the unknown x to the first column, the unknown y to the second column, and the unknown z to the third column. The resulting system is

$$-5x + 2y - z = 15$$
$$-2y + z = -7$$
$$\tfrac{1}{2}x + y - z = 3$$

PROGRESS CHECK 3

Write a system of linear equations that corresponds to the augmented matrix

$$\begin{bmatrix} 3 & -2 & 5 & -1 \\ 2 & 0 & \frac{1}{3} & 2 \\ -1 & 4 & 2 & 5 \end{bmatrix}$$

Answer

$3x - 2y + 5z = -1$

$2x \qquad + \frac{1}{3}z = 2$

$-x + 4y + 2z = 5$

' Now that we have seen how a matrix can be used to represent a system of linear equations, we next proceed to show how routine operations on that matrix can yield the solution of the system. These "matrix methods" are simply a clever streamlining of the methods studied in Chapter Thirteen.

In Section 13.4 we used three elementary operations to transform a system of linear equations into triangular form. When applying the same procedures to a matrix, we speak of rows, columns, and elements instead of equations, variables, and coefficients. The three elementary operations of Section 13.4 which yielded an equivalent system now become the elementary row operations.

> The following **elementary row operations** transform an augmented matrix into an equivalent system.
> (1) Interchange any two rows.
> (2) Multiply each element of any row by a constant k ($\neq 0$).
> (3) Replace each element of a given row by the sum of itself plus k times the corresponding element of any other row.

The method of Gaussian elimination introduced in Section 13.4 can now be restated in terms of matrices. By use of elementary row operations we seek to transform an augmented matrix into a matrix for which $a_{ij} = 0$ when $i > j$. The resulting matrix will have the following appearance for a system of three linear equations in three unknowns.

Since this matrix represents a linear system in triangular form, back-substitution will provide a solution of the original system. We will illustrate the process with an example.

EXAMPLE 4

Solve the system.

$$x - y + 4z = 4$$
$$2x + 2y - z = 2$$
$$3x - 2y + 3z = -3$$

Solution

TABLE 14.1

Gaussian Elimination	Example
Step 1. Form the augmented matrix.	*Step 1.* The augmented matrix is $$\begin{bmatrix} 1 & -1 & 4 & 4 \\ 2 & 2 & -1 & 2 \\ 3 & -2 & 3 & -3 \end{bmatrix}$$
Step 2. If necessary, interchange rows to make sure that a_{11}, the first element of the first row, is nonzero.	*Step 2.* We see that $a_{11} = 1 \neq 0$.
Step 3. Arrange to have 0 as the first element of every row below row 1. This is done by replacing row 2, row 3, and so on, by the sum of itself and an appropriate multiple of row 1. We will call row 1 the **pivot row.**	*Step 3.* Replacing row 2 by the sum of itself and (-2) times row 1, and replacing row 3 by the sum of itself and (-3) times row 1 yields $$\begin{bmatrix} 1 & -1 & 4 & 4 \\ 0 & 4 & -9 & -6 \\ 0 & 1 & -9 & -15 \end{bmatrix}$$
Step 4. Repeat the process defined by Steps 2 and 3, allowing row 2, row 3, and so on, to play the role of the first row. Thus, row 2, row 3, and so on, serve as the pivot rows.	*Step 4.* Replacing row 3 by the sum of itself and $(-\frac{1}{4})$ times row 2 yields $$\begin{bmatrix} 1 & -1 & 4 & 4 \\ 0 & 4 & -9 & -6 \\ 0 & 0 & -\frac{27}{4} & -\frac{27}{2} \end{bmatrix}$$
Step 5. Solve the triangular system by back-substitution.	*Step 5.* The third row of the final matrix yields $$-\frac{27}{4}z = -\frac{27}{2}$$ $$\boxed{z = 2}$$ From the second row of the final matrix we have $$4y - 9z = -6$$ $$4y - 9(2) = -6$$ $$\boxed{y = 3}$$ Substituting $y = 3$, $z = 2$, we obtain from the first row of the matrix $$x - y + 4z = 4$$ $$x - 3 + 4(2) = 4$$ $$\boxed{x = -1}$$ The solution is $x = -1$, $y = 3$, $z = 2$.

PROGRESS CHECK 4

Solve the linear system by matrix methods.

$$
\begin{aligned}
2x + 4y - z &= 0 \\
x - 2y - 2z &= 2 \\
-5x - 8y + 3z &= -2
\end{aligned}
$$

Answer

$x = 6, \ y = -2, \ z = 4$

Note that we have described the process of Gaussian elimination in a manner that will apply to any augmented matrix of dimension $n \times (n + 1)$; that is, Gaussian elimination may be used on any system of n linear equations in n unknowns that has a unique solution. (We will not deal with the situation in which the solution is not unique in this brief introduction to matrices.) It is also permissible to perform these elementary row operations in clever ways to simplify the arithmetic. For instance, you may wish to interchange rows, or to multiply a row by a constant to obtain a pivot row such that $a_{ii} = 1$. We will illustrate these ideas with an example.

EXAMPLE 5

Solve by matrix methods.

$$
\begin{aligned}
2y + 3z &= 4 \\
4x + y + 8z + 15w &= -14 \\
x - y + 2z &= 9 \\
-x - 2y - 3z - 6w &= 10
\end{aligned}
$$

Solution

We begin with the augmented matrix and perform a sequence of elementary row operations which will lead to a matrix representing a linear system in triangular form.

$$
\begin{bmatrix}
0 & 2 & 3 & 0 & 4 \\
4 & 1 & 8 & 15 & -14 \\
1 & -1 & 2 & 0 & 9 \\
-1 & -2 & -3 & -6 & 10
\end{bmatrix}
$$

Augmented matrix
Note that $a_{11} = 0$.
Interchange rows 1 and 3 so that $a_{11} = 1$.

$$
\begin{bmatrix}
1 & -1 & 2 & 0 & 9 \\
4 & 1 & 8 & 15 & -14 \\
0 & 2 & 3 & 0 & 4 \\
-1 & -2 & -3 & -6 & 10
\end{bmatrix}
$$

Replace row 2 by the sum of itself and (-4) times row 1.

Replace row 4 by the sum of itself plus row 1.

$$
\begin{bmatrix}
1 & -1 & 2 & 0 & 9 \\
0 & 5 & 0 & 15 & -50 \\
0 & 2 & 3 & 0 & 4 \\
0 & -3 & -1 & -6 & 19
\end{bmatrix}
$$

Multiply row 2 by $\frac{1}{5}$ so that $a_{22} = 1$.

$$\begin{bmatrix} 1 & -1 & 2 & 0 & 9 \\ 0 & 1 & 0 & 3 & -10 \\ 0 & 2 & 3 & 0 & 4 \\ 0 & -3 & -1 & -6 & 19 \end{bmatrix}$$

Replace row 3 by the sum of
itself and (-2) times row 2.
Replace row 4 by the sum of
itself plus 3 times row 2.

$$\begin{bmatrix} 1 & -1 & 2 & 0 & 9 \\ 0 & 1 & 0 & 3 & -10 \\ 0 & 0 & 3 & -6 & 24 \\ 0 & 0 & -1 & 3 & -11 \end{bmatrix}$$

Interchange rows 3 and 4.

$$\begin{bmatrix} 1 & -1 & 2 & 0 & 9 \\ 0 & 1 & 0 & 3 & -10 \\ 0 & 0 & -1 & 3 & -11 \\ 0 & 0 & 3 & -6 & 24 \end{bmatrix}$$

Replace row 4 by the sum of
itself and 3 times row 3.

$$\begin{bmatrix} 1 & -1 & 2 & 0 & 9 \\ 0 & 1 & 0 & 3 & -10 \\ 0 & 0 & -1 & 3 & -11 \\ 0 & 0 & 0 & 3 & -9 \end{bmatrix}$$

The last row of the matrix indicates that

$$3w = -9$$

$$w = -3$$

The remaining variables are found by back-substitution.

Third row of final matrix	Second row of final matrix	First row of final matrix
$-z + 3w = -11$	$y + 3w = -10$	$x - y + 2z = 9$
$-z + 3(-3) = -11$	$y + 3(-3) = -10$	$x - (-1) + 2(2) = 9$
$z = 2$	$y = -1$	$x = 4$

The solution is $x = 4$, $y = -1$, $z = 2$, $w = -3$.

PROGRESS CHECK 5
Solve by matrix methods.

$$2x \qquad - 3z + 2w = -7$$
$$2y + 2z - 3w = 1$$
$$-2x + y - 2z + w = -9$$
$$4x - 3y \qquad + 5w = 6$$

Answer

$x = \dfrac{1}{2}$, $y = -3$, $z = 2$, $w = -1$

EXERCISE SET 14.1

Give the dimension of each matrix.

1. $\begin{bmatrix} 3 & -1 \\ 2 & 4 \end{bmatrix}$

2. $[1 \quad 2 \quad 3 \quad -1]$

3. $\begin{bmatrix} 4 & 2 & 3 \\ 5 & -1 & 4 \\ 2 & 3 & 6 \\ -8 & -1 & 2 \end{bmatrix}$

4. $\begin{bmatrix} -1 \\ 3 \\ 2 \end{bmatrix}$

5. $\begin{bmatrix} 4 & 2 & 1 \\ 3 & 1 & 5 \\ -4 & -2 & 3 \end{bmatrix}$

6. $\begin{bmatrix} 3 & -1 & 2 & 6 \\ 2 & 8 & 4 & 1 \end{bmatrix}$

7. If

$$A = \begin{bmatrix} 3 & -4 & -2 & 5 \\ 8 & 7 & 6 & 2 \\ 1 & 0 & 9 & -3 \end{bmatrix}$$

find (a) a_{12} (b) a_{22} (c) a_{23} (d) a_{34}.

8. If

$$B = \begin{bmatrix} -5 & 6 & 8 \\ 4 & 1 & 3 \\ 0 & 2 & -6 \\ -3 & 9 & 7 \end{bmatrix}$$

find (a) b_{13} (b) b_{21} (c) b_{33} (d) b_{42}.

For each given linear system write the coefficient matrix and the augmented matrix.

9. $3x - 2y = 12$
    $5x + y = -8$

10. $3x - 4y = 15$
    $4x - 3y = 12$

11. $\frac{1}{2}x + y + z = 4$
    $2x - y - 4z = 6$
    $4x + 2y - 3z = 8$

12. $2x + 3y - 4z = 10$
    $-3x + y \quad\quad = 12$
    $5x - 2y + z = -8$

Write the linear system whose augmented matrix is given.

13. $\begin{bmatrix} \frac{3}{2} & 6 & -1 \\ 4 & 5 & 3 \end{bmatrix}$

14. $\begin{bmatrix} 4 & 0 & 2 \\ -7 & 8 & 3 \end{bmatrix}$

15. $\begin{bmatrix} 1 & 1 & 3 & -4 \\ -3 & 4 & 0 & 8 \\ 2 & 0 & 7 & 6 \end{bmatrix}$

16. $\begin{bmatrix} 4 & 8 & 3 & 12 \\ 1 & -5 & 3 & -14 \\ 0 & 2 & 7 & 18 \end{bmatrix}$

Decide whether the given augmented matrix represents a linear system in triangular form.

17. $\begin{bmatrix} 1 & 0 & 2 & 4 \\ 0 & 1 & -3 & 5 \\ 0 & 0 & 1 & 6 \end{bmatrix}$

18. $\begin{bmatrix} 1 & 0 & -3 & -4 \\ 0 & 1 & 2 & 5 \\ 0 & 0 & 1 & 2 \end{bmatrix}$

19. $\begin{bmatrix} -2 & 0 & 3 & -1 \\ 0 & 1 & 4 & 2 \\ 0 & 0 & 1 & 5 \end{bmatrix}$

20. $\begin{bmatrix} 1 & 0 & 4 & 2 \\ 2 & -1 & -1 & 5 \\ 0 & 0 & 1 & 6 \end{bmatrix}$

In each of the following, the augmented matrix of a linear system has been transformed by elementary row operations to the given matrix. Find a solution of the original linear system.

21. $\begin{bmatrix} 1 & 2 & 0 & 3 \\ 0 & 1 & -2 & 4 \\ 0 & 0 & 1 & 2 \end{bmatrix}$

22. $\begin{bmatrix} 1 & 0 & 2 & -1 \\ 0 & 1 & 3 & 2 \\ 0 & 0 & 1 & 5 \end{bmatrix}$

23. $\begin{bmatrix} 1 & -2 & 1 & 3 \\ 0 & 1 & 3 & 2 \\ 0 & 0 & 1 & -4 \end{bmatrix}$

24. $\begin{bmatrix} 1 & -4 & 2 & -4 \\ 0 & 1 & 3 & -2 \\ 0 & 0 & 1 & 5 \end{bmatrix}$

Transform each given augmented matrix so that the corresponding linear system is in triangular form.

25. $\begin{bmatrix} 2 & -4 & -2 & 1 \\ 2 & -2 & -4 & 4 \\ 2 & 4 & 2 & 2 \end{bmatrix}$

26. $\begin{bmatrix} 1 & -2 & 0 & 1 \\ 2 & -3 & -1 & 3 \\ 1 & 3 & 2 & 4 \end{bmatrix}$

27. $\begin{bmatrix} 1 & 2 & 1 & 0 \\ 1 & -3 & 2 & 1 \\ 1 & 4 & 3 & 2 \end{bmatrix}$

28. $\begin{bmatrix} 1 & 3 & 2 & 10 \\ -2 & 2 & 2 & 8 \\ -1 & 1 & 3 & 9 \end{bmatrix}$

Solve each given linear system by matrix methods.

29. $\begin{aligned} x - 2y &= -4 \\ 2x + 3y &= 13 \end{aligned}$

30. $\begin{aligned} 2x + y &= -1 \\ 3x - y &= -7 \end{aligned}$

31. $\begin{aligned} x + y + z &= 4 \\ 2x - y + 3z &= 14 \\ x + 2y + z &= 3 \end{aligned}$

32. $\begin{aligned} x - y + z &= -5 \\ 3x + y + 2z &= -5 \\ 2x - y - z &= -2 \end{aligned}$

33. $\begin{aligned} 2x + y - z &= 9 \\ x - 2y + 2z &= -3 \\ 3x + 3y + 4z &= 11 \end{aligned}$

34. $\begin{aligned} 2x + y - z &= -2 \\ -2x - 2y + 3z &= 2 \\ 3x + y - z &= -4 \end{aligned}$

35. $\begin{aligned} -x - y + 2z &= 9 \\ x + 2y - 2z &= -7 \\ 2x - y + z &= -9 \end{aligned}$

36. $\begin{aligned} 4x + y - z &= -1 \\ x - y + 2z &= 3 \\ -x + 2y - z &= 0 \end{aligned}$

37. $\begin{aligned} x + y - z + 2w &= 0 \\ 2x + y \quad\;\; - w &= -2 \\ 3x \quad\;\; + 2z \quad &= -3 \\ -x + 2y \quad\;\; + 3w &= 1 \end{aligned}$

38. $\begin{aligned} 2x + y \quad\;\; - 3w &= -7 \\ 3x \quad\;\; + 2z + w &= 0 \\ -x + 2y \quad\;\; + 3w &= 10 \\ -2x - 3y + 2z - w &= 7 \end{aligned}$

14.2
DETERMINANTS

In this section we will define a determinant and will develop manipulative skills for evaluating determinants. In the next section we will show you that determinants have important applications and can be used to solve linear systems.

Associated with every square matrix A is a number called the **determinant** of A denoted by $|A|$. If A is a 2×2 matrix,

$$A = \begin{bmatrix} a_{11} & a_{12} \\ a_{21} & a_{22} \end{bmatrix}$$

we define $|A|$ by the rule

$$|A| = \begin{vmatrix} a_{11} & a_{12} \\ a_{21} & a_{22} \end{vmatrix} = a_{11}a_{22} - a_{21}a_{12}$$

and call this a **determinant of second order.**

EXAMPLE 1
Compute the real number represented by

$$\begin{vmatrix} 4 & -5 \\ 3 & 1 \end{vmatrix}$$

Solution
We apply the rule for a determinant of second order.

$$\begin{vmatrix} 4 & -5 \\ 3 & 1 \end{vmatrix} = a_{11}a_{22} - a_{21}a_{22} = (4)(1) - (3)(-5) = 19$$

PROGRESS CHECK 1
Compute the real number represented by each determinant.

(a) $\begin{vmatrix} -6 & 2 \\ -1 & -2 \end{vmatrix}$ (b) $\begin{vmatrix} \frac{1}{2} & -4 \\ \frac{1}{4} & -2 \end{vmatrix}$

Answers
(a) 14   (b) 0

To simplify matters, when we want to compute the number represented by a determinant we will say "evaluate the determinant." This is not technically correct, however, since a determinant *is* a real number.

The rule for evaluating a **determinant of order 3** is

$$\begin{vmatrix} a_{11} & a_{12} & a_{13} \\ a_{21} & a_{22} & a_{23} \\ a_{31} & a_{32} & a_{33} \end{vmatrix} = \begin{aligned} & a_{11}a_{22}a_{33} - a_{11}a_{32}a_{23} - a_{12}a_{21}a_{33} \\ & + a_{12}a_{31}a_{23} + a_{13}a_{21}a_{32} - a_{13}a_{31}a_{22} \end{aligned}$$

The situation becomes even more cumbersome for determinants of higher order! Fortunately, we don't have to memorize this rule; instead, we shall see that it is possible to evaluate a determinant of order 3 by reducing the problem to that of evaluating a number of determinants of order 2.

The **minor of an element** a_{ij} is the determinant of the matrix remaining after deleting the row and column in which the element appears. Given the matrix.

$$\begin{bmatrix} 4 & 0 & -2 \\ 1 & -6 & 7 \\ -3 & 2 & 5 \end{bmatrix}$$

the minor of the element 7 in row 2, column 3, is

$$\begin{vmatrix} 4 & 0 & -2 \\ 1 & 6 & 7 \\ -3 & 2 & 5 \end{vmatrix} = \begin{vmatrix} 4 & 0 \\ -3 & 2 \end{vmatrix} = 8 - 0 = 8$$

The **cofactor of the element** a_{ij} is the minor of the element a_{ij} multiplied by $(-1)^{i+j}$. Since $(-1)^{i+j}$ is $+1$ if $i + j$ is even and is -1 if $i + j$ is odd, we see that the cofactor is the minor with a sign attached. The cofactor attaches the sign to the minor according to this checkerboard pattern:

$$
\begin{array}{ccccc}
+ & - & + & - & \cdots \\
- & + & - & + & \cdots \\
+ & - & + & - & \cdots \\
- & + & - & + & \cdots
\end{array}
$$

EXAMPLE 2

Find the cofactor of each element in the first row of the matrix

$$\begin{bmatrix} -2 & 0 & 12 \\ -4 & 5 & 3 \\ 7 & 8 & -6 \end{bmatrix}$$

Solution

The cofactor of -2 is

$$(-1)^{1+1} \begin{vmatrix} -2 & 0 & 12 \\ -4 & 5 & 3 \\ 7 & 8 & -6 \end{vmatrix} = \begin{vmatrix} 5 & 3 \\ 8 & -6 \end{vmatrix}$$

$$= -30 - 24 = -54$$

The cofactor of 0 is

$$(-1)^{1+2} \begin{vmatrix} -2 & 0 & 12 \\ -4 & 5 & 3 \\ 7 & 8 & -6 \end{vmatrix} = -\begin{vmatrix} -4 & 3 \\ 7 & -6 \end{vmatrix}$$

$$= -(24 - 21) = -3$$

The cofactor of 12 is

$$(-1)^{1+3} \begin{vmatrix} -2 & 0 & 12 \\ -4 & 5 & 3 \\ 7 & 8 & -6 \end{vmatrix} = \begin{vmatrix} -4 & 5 \\ 7 & 8 \end{vmatrix}$$

$$= -32 - 35 = -67$$

PROGRESS CHECK 2
Find the cofactor of each element in the second column of the matrix.

$$\begin{bmatrix} 16 & -9 & 3 \\ -5 & 2 & 0 \\ -3 & 4 & -1 \end{bmatrix}$$

Answer
Cofactor of −9 is −5; cofactor of 2 is −7; cofactor of 4 is −15.

The cofactor is the key to the process of evaluating determinants of order 3 or higher.

To evaluate a determinant, form the sum of the products obtained by multiplying each element of any row or any column by its cofactor. This process is called **expansion by cofactors.**

Let's illustrate the process with an example.

EXAMPLE 3
Evaluate the determinant by expansion by cofactors.

$$\begin{vmatrix} -2 & 7 & 2 \\ 6 & -6 & 0 \\ 4 & 10 & -3 \end{vmatrix}$$

Solution
TABLE 14.2

Expansion by Cofactors	Example
Step 1. Choose a row or column about which to expand. In general, a row or column containing zeros will simplify the work.	*Step 1.* We will expand about column 3.
Step 2. Expand about the cofactors of the chosen row or column, by multiplying each element of the row or column by its cofactor.	*Step 2.* The expansion about column 3 is $$(2)(-1)^{1+3}\begin{vmatrix} 6 & -6 \\ 4 & 10 \end{vmatrix}$$ $$+ (0)(-1)^{2+3}\begin{vmatrix} -2 & 7 \\ 4 & 10 \end{vmatrix}$$ $$+ (-3)(-1)^{3+3}\begin{vmatrix} -2 & 7 \\ 6 & -6 \end{vmatrix}$$
Step 3. Evaluate the cofactors and form their sum.	*Step 3.* Using the rule for evaluating a determinant of order 2, we have $$(2)(1)[(6)(10) - (4)(-6)] + 0$$ $$+ (-3)(1)[(-2)(-6) - (6)(7)]$$ $$= 2(60 + 24) - 3(12 - 42)$$ $$= 258$$

Note that the expansion by the cofactors of *any* row or *any* column will produce the same result. This important property of determinants can be used to simplify the arithmetic. The best choice of a row or column about which to expand is one that has the most zero elements. The reason for this is that if an element is zero, the element times its cofactor will be zero, so we don't have to evaluate that cofactor.

PROGRESS CHECK 3
Evaluate the determinant of Example 3 by expanding about the second row.

Answer
258

EXAMPLE 4
Verify the rule for evaluating a determinant of order 3.

$$\begin{vmatrix} a_{11} & a_{12} & a_{13} \\ a_{21} & a_{22} & a_{23} \\ a_{31} & a_{32} & a_{33} \end{vmatrix} = \begin{aligned} & a_{11}a_{22}a_{33} - a_{11}a_{32}a_{23} - a_{12}a_{21}a_{33} \\ & + a_{12}a_{31}a_{23} + a_{13}a_{21}a_{32} - a_{13}a_{31}a_{22} \end{aligned}$$

Solution
Expanding about the first row we have

$$\begin{vmatrix} a_{11} & a_{12} & a_{13} \\ a_{21} & a_{22} & a_{23} \\ a_{31} & a_{32} & a_{33} \end{vmatrix} = a_{11}\begin{vmatrix} a_{22} & a_{23} \\ a_{32} & a_{33} \end{vmatrix} - a_{12}\begin{vmatrix} a_{21} & a_{23} \\ a_{31} & a_{33} \end{vmatrix} + a_{13}\begin{vmatrix} a_{21} & a_{22} \\ a_{31} & a_{32} \end{vmatrix}$$

$$= a_{11}(a_{22}a_{33} - a_{32}a_{23}) - a_{12}(a_{21}a_{33} - a_{31}a_{23}) + a_{13}(a_{21}a_{32} - a_{31}a_{22})$$

$$= a_{11}a_{22}a_{33} - a_{11}a_{32}a_{23} - a_{12}a_{21}a_{33} + a_{12}a_{31}a_{23} + a_{13}a_{21}a_{32} - a_{13}a_{31}a_{22}$$

PROGRESS CHECK 4
Show that the determinant

$$\begin{vmatrix} a & b & c \\ a & b & c \\ d & e & f \end{vmatrix} = 0$$

The process of expanding by cofactors works for determinants of any order. If we apply the method to a determinant of order 4, we will produce determinants of order 3; applying the method again will result in determinants of order 2.

EXAMPLE 5
Evaluate.

$$\begin{vmatrix} -3 & 5 & 0 & -1 \\ 1 & 2 & 3 & -3 \\ 0 & 4 & -6 & 0 \\ 0 & -2 & 1 & 2 \end{vmatrix}$$

Solution

Expanding about the cofactors of the first column,

$$
\begin{vmatrix} -3 & 5 & 0 & -1 \\ 1 & 2 & 3 & -3 \\ 0 & 4 & -6 & 0 \\ 0 & -2 & 1 & 2 \end{vmatrix} = -3 \begin{vmatrix} 2 & 3 & -3 \\ 4 & -6 & 0 \\ -2 & 1 & 2 \end{vmatrix} - 1 \begin{vmatrix} 5 & 0 & -1 \\ 4 & -6 & 0 \\ -2 & 1 & 2 \end{vmatrix}
$$

Each determinant of order 3 can then be evaluated, using the above method. The student is urged to verify that

$$
-3 \begin{vmatrix} 2 & 3 & -3 \\ 4 & -6 & 0 \\ -2 & 1 & 2 \end{vmatrix} = (-3)(-24) = 72 \qquad -1 \begin{vmatrix} 5 & 0 & -1 \\ 4 & -6 & 0 \\ -2 & 1 & 2 \end{vmatrix} = (-1)(-52) = 52
$$

and the original determinant has the value $72 + 52 = 124$.

PROGRESS CHECK 5

Evaluate.

$$
\begin{vmatrix} 0 & -1 & 0 & 2 \\ 3 & 0 & 4 & 0 \\ 0 & 5 & 0 & -3 \\ 1 & 0 & 1 & 0 \end{vmatrix}
$$

Answer

7

EXERCISE SET 14.2

In Exercises 1–4 let

$$
A = \begin{bmatrix} 3 & -1 & 2 \\ 4 & 1 & -3 \\ 5 & -2 & 0 \end{bmatrix}
$$

1. Compute the minor of each of the following elements.
   (a) a_{11} (b) a_{23} (c) a_{31} (d) a_{33}
2. Compute the minor of each of the following elements.
   (a) a_{12} (b) a_{22} (c) a_{23} (d) a_{32}
3. Compute the cofactor of each of the following elements.
   (a) a_{11} (b) a_{23} (c) a_{31} (d) a_{33}
4. Compute the cofactor of each of the following elements.
   (a) a_{12} (b) a_{22} (c) a_{23} (d) a_{32}

In Exercises 5–8 let

$$
A = \begin{bmatrix} -1 & 0 & 3 \\ -2 & 4 & 5 \\ -3 & -4 & 2 \end{bmatrix}
$$

5. Compute the minor of each of the following elements.
   (a) a_{12} (b) a_{21} (c) a_{23} (d) a_{31}
6. Compute the minor of each of the following elements.
   (a) a_{11} (b) a_{13} (c) a_{22} (d) a_{33}

7. Compute the cofactor of each of the following elements.
    (a) a_{12} (b) a_{21} (c) a_{23} (d) a_{31}
8. Compute the cofactor of each of the following elements.
    (a) a_{11} (b) a_{13} (c) a_{22} (d) a_{33}

In Exercises 9–14 let

$$A = \begin{bmatrix} 3 & -2 & 5 \\ 0 & 1 & 2 \\ -4 & 3 & 4 \end{bmatrix}$$

Mark each of the following true (T) or false (F).
9. The cofactor of a_{11} is 2.
10. The cofactor of a_{22} is 32.
11. The cofactor of a_{13} is -4.
12. The cofactor of a_{13} is 3.
13. The cofactor of a_{23} is -1.
14. The cofactor of a_{33} is 3.

Evaluate the given determinant.

15. $\begin{vmatrix} 2 & -3 \\ 4 & 5 \end{vmatrix}$

16. $\begin{vmatrix} 3 & 4 \\ -1 & 2 \end{vmatrix}$

17. $\begin{vmatrix} -4 & 1 \\ 0 & 2 \end{vmatrix}$

18. $\begin{vmatrix} 2 & 2 \\ 3 & 3 \end{vmatrix}$

19. $\begin{vmatrix} 0 & 0 \\ 1 & 3 \end{vmatrix}$

20. $\begin{vmatrix} -4 & -1 \\ -2 & 3 \end{vmatrix}$

21. $\begin{vmatrix} 4 & -2 & 5 \\ 5 & 2 & 0 \\ 2 & 0 & 4 \end{vmatrix}$

22. $\begin{vmatrix} 4 & 1 & 2 \\ 0 & 2 & 3 \\ 0 & 0 & -4 \end{vmatrix}$

23. $\begin{vmatrix} -1 & 2 & 0 \\ 3 & 4 & 1 \\ 6 & 5 & 2 \end{vmatrix}$

24. $\begin{vmatrix} -1 & 3 & 2 \\ 0 & 7 & 7 \\ 2 & 1 & 3 \end{vmatrix}$

25. $\begin{vmatrix} 2 & 2 & 4 \\ 3 & 8 & 1 \\ 1 & 1 & 2 \end{vmatrix}$

26. $\begin{vmatrix} 0 & 1 & 3 \\ 2 & 5 & -1 \\ 4 & 2 & -2 \end{vmatrix}$

27. $\begin{vmatrix} 3 & 2 & 1 & 0 \\ -1 & -3 & -1 & 0 \\ 0 & 0 & 2 & 2 \\ 4 & 1 & 3 & 3 \end{vmatrix}$

28. $\begin{vmatrix} -1 & 2 & 4 & 0 \\ 3 & -2 & -3 & 0 \\ 0 & 4 & 2 & 5 \\ 0 & -3 & 1 & 4 \end{vmatrix}$

29. $\begin{vmatrix} 2 & -3 & 2 & -4 \\ 0 & 4 & -1 & 9 \\ 0 & 1 & 2 & 0 \\ 0 & 1 & 3 & -1 \end{vmatrix}$

30. $\begin{vmatrix} 1 & 1 & 0 & 1 \\ 0 & -1 & 4 & -1 \\ -2 & 3 & 1 & -4 \\ 0 & 2 & 0 & 2 \end{vmatrix}$

14.3
CRAMER'S RULE

Determinants provide a convenient way of expressing formulas in many areas of mathematics, particularly in geometry. One of the best known uses of determinants is in solving systems of linear equations, a procedure known as **Cramer's rule.**

In Chapter Thirteen we solved systems of linear equations by the method of elimination. Let's apply this method to the general system of two equations in two unknowns:

$$a_{11}x + a_{12}y = c_1 \qquad (1)$$

$$a_{21}x + a_{22}y = c_2 \qquad (2)$$

If we multiply Equation (1) by a_{22} and Equation (2) by $-a_{12}$ and add, we will eliminate y.

$$a_{11}a_{22}x + a_{12}a_{22}y = c_1a_{22}$$
$$\underline{-a_{21}a_{12}x - a_{12}a_{22}y = -c_2a_{12}}$$
$$a_{11}a_{22}x - a_{21}a_{12}x = c_1a_{22} - c_2a_{12}$$

Thus,

$$x(a_{11}a_{22} - a_{21}a_{12}) = c_1a_{22} - c_2a_{12}$$

or

$$x = \frac{c_1a_{22} - c_2a_{12}}{a_{11}a_{22} - a_{21}a_{12}}$$

Similarly, multiplying Equation (1) by a_{21} and Equation (2) by $-a_{11}$ and adding, we can eliminate x and solve for y.

$$y = \frac{c_2a_{11} - c_1a_{21}}{a_{11}a_{22} - a_{21}a_{12}}$$

The denominators in the expressions for x and y are identical and can be written as a determinant.

$$|D| = \begin{vmatrix} a_{11} & a_{12} \\ a_{21} & a_{22} \end{vmatrix}$$

If we apply this same idea to the numerators, we have

$$x = \frac{\begin{vmatrix} c_1 & a_{12} \\ c_2 & a_{22} \end{vmatrix}}{|D|} \qquad y = \frac{\begin{vmatrix} a_{11} & c_1 \\ a_{21} & c_2 \end{vmatrix}}{|D|}, \qquad |D| \neq 0$$

What we have arrived at is Cramer's rule, which is a means of expressing the solution of a system of linear equations in determinant form.

The following example outlines the steps for using Cramer's rule.

EXAMPLE 1

Solve by Cramer's rule.

$$3x - y = 9$$
$$x + 2y = -4$$

Solution

TABLE 14.3

Cramer's Rule	Example
Step 1. Form the determinant $|D|$ from the coefficients of x and y.	*Step 1.* $$|D| = \begin{vmatrix} 3 & -1 \\ 1 & 2 \end{vmatrix}$$
Step 2. The numerator for x is obtained from the determinant $|D|$ by replacing the column of coefficients of x by the column of constants obtained from the right-hand sides of the equations.	*Step 2.* $$x = \frac{\begin{vmatrix} 9 & -1 \\ -4 & 2 \end{vmatrix}}{|D|}$$
Step 3. The numerator for y is obtained from the determinant $|D|$ by replacing the column of coefficients of y by the column of constants.	*Step 3.* $$y = \frac{\begin{vmatrix} 3 & 9 \\ 1 & -4 \end{vmatrix}}{|D|}$$
Step 4. Evaluate the determinants to obtain the solution. If $|D| = 0$, Cramer's rule cannot be used.	*Step 4.* $|D| = 6 + 1 = 7$ $$x = \frac{18 - 4}{7} = \frac{14}{7} = 2$$ $$y = \frac{-12 - 9}{7} = -\frac{21}{7} = -3$$

PROGRESS CHECK 1

Solve by Cramer's rule.

$$2x + 3y = -4$$
$$3x + 4y = -7$$

Answer
$x = -5, y = 2$

The steps outlined in Example 1 can be applied to any systems of linear equations in which the number of equations is the same as the number of variables and in which $|D| \neq 0$. Here is an example with three equations and three unknowns.

EXAMPLE 2

Solve by Cramer's rule.

$$3x \quad\quad + 2z = -2$$
$$2x - y \quad\quad = 0$$
$$2y + 6z = -1$$

Solution

We form the determinant of coefficients:

$$|D| = \begin{vmatrix} 3 & 0 & 2 \\ 2 & -1 & 0 \\ 0 & 2 & 6 \end{vmatrix}$$

Then

$$x = \frac{|D_1|}{|D|} \qquad y = \frac{|D_2|}{|D|} \qquad z = \frac{|D_3|}{|D|}$$

where $|D_1|$ is obtained from $|D|$ by replacing its first column by the column of constants, $|D_2|$ is obtained from $|D|$ by replacing its second column by the column of constants, and $|D_3|$ is obtained from $|D|$ by replacing its third column by the column of constants. Thus,

$$x = \frac{\begin{vmatrix} -2 & 0 & 2 \\ 0 & -1 & 0 \\ -1 & 2 & 6 \end{vmatrix}}{|D|} \qquad y = \frac{\begin{vmatrix} 3 & -2 & 2 \\ 2 & 0 & 0 \\ 0 & -1 & 6 \end{vmatrix}}{|D|} \qquad z = \frac{\begin{vmatrix} 3 & 0 & -2 \\ 2 & -1 & 0 \\ 0 & 2 & -1 \end{vmatrix}}{|D|}$$

Expanding by cofactors we calculate (verify) $|D| = -10$, $|D_1| = 10$, $|D_2| = 20$, and $|D_3| = -5$, obtaining

$$x = \frac{10}{-10} = -1, \qquad y = \frac{20}{-10} = -2, \qquad z = \frac{-5}{-10} = \frac{1}{2}$$

PROGRESS CHECK 2

Solve by Cramer's rule.

$$3x \qquad - z = 1$$
$$-6x + 2y \qquad = -5$$
$$-4y + 3z = 5$$

Answer

$$x = \frac{2}{3}, \quad y = -\frac{1}{2}, \quad z = 1$$

WARNING

(a) The system of linear equations must be written in standard form before using Cramer's rule.

(b) If $|D| = 0$, Cramer's rule cannot be used.

EXERCISE SET 14.3

If possible, solve each given linear system by Cramer's rule.

1. $x - y + 2z = 10$
    $2x + y - 2z = -4$
    $3x + y + z = 7$

2. $x + 2y + z = 9$
    $x - y + 2z = 2$
    $x + y - z = 2$

3. $x - y + z = 2$
    $2x + y - 3z = 10$
    $x - y + 2z = 2$

4. $2x - 2y - 2z = 2$
    $-2x + 2y + 2z = -2$
    $3x + y - z = 3$

5. $2x + y = 2$
    $3x - y + 5z = 3$
    $x + y - z = 4$

6. $x - 2y + 2z = -7$
    $2x + 2y - z = -2$
    $3x + y + z = -7$

7. $x - y + 3z = 6$
    $-x + 2y - 2z = -10$
    $2x + 3y - z = 4$

8. $2x + 4y - 3z = -10$
    $x - 2y + z = 7$
    $3x - 2y - 3z = -5$

9. $2x + y + 5z = -5$
    $x - y + z = 0$
    $5x + y - 2z = 3$

10. $x + 2y + z = 0$
    $3x - y + 2z = 18$
    $5x + 3y = 6$

TERMS AND SYMBOLS

matrix   (p. 419)
row   (p. 420)
column   (p. 420)
entry   (p. 420)
element   (p. 420)
dimension   (p. 420)
square matrix   (p. 420)

a_{ij} (p. 420)
coefficient matrix   (p. 421)
augmented matrix
   (p. 421)
elementary row
   operations   (p. 422)
pivot row   (p. 423)

determinant   (p. 428)
minor   (p. 429)
cofactor   (p. 429)
expansion by cofactors
   (p. 430)
Cramer's rule   (p. 433)

KEY IDEAS FOR REVIEW

☐ A matrix is a rectangular array of numbers.

☐ Systems of linear equations can be conveniently handled in matrix notation. By dropping the names of the variables, matrix notation focuses on the coefficients and the right-hand side of the system. The elementary row operations are then seen to be a restatement of those operations which produce equivalent systems of equations.

☐ Matrices have much broader use than simply solving linear systems. In more advanced mathematics courses (such as linear algebra) properties of matrices are studied and are applied to problems in many disciplines.

☐ The rule for evaluating a determinant of order 2 is

$$\begin{vmatrix} a & b \\ c & d \end{vmatrix} = ad - bc$$

For determinants of higher order, the method of expansion by cofactors may be used to reduce the problem to that of evaluating determinants of order 2.

☐ When expanding by cofactors, choose the row or column which contains the greatest number of zeros. This will ease the arithmetic burden.

☐ When expanding by cofactors, remember to attach the proper sign to each minor.

☐ Cramer's rule expresses the value of each unknown of a system of linear equations as a quotient of determinants.

COMMON ERRORS

1. After the first stage of Gaussian elimination, the elements in the first column, except a_{11}, are all zero. Subsequent elementary row operations must *not* use row 1. This will insure that the zeros will remain, that is, that $a_{i1} = 0$ for all $i > 1$.

2. Remember that the cofactor of an element a_{ij} is a determinant multiplied by $(-1)^{i+j}$. When evaluating a determinant by expanding by cofactors of a row or column, remember that the multipliers of the minors will alternate in sign.

3. Do not use Cramer's rule if the determinant of the coefficients is 0.

PROGRESS TEST 14A

1. Given the matrix
$$A = \begin{bmatrix} -2 & 0.5 \\ 1.1 & -3 \\ 14 & 7 \end{bmatrix}$$
    (a) Determine the dimension of matrix A.
    (b) Find a_{22}.

2. Write a system of linear equations that corresponds to the augmented matrix
$$\begin{bmatrix} 3 & -5 & 4 \\ 4 & 2 & -1 \end{bmatrix}$$

3. Solve the system of linear equations that corresponds to the matrix
$$\begin{bmatrix} 2 & 1 & 3 & -6 \\ 0 & -1 & 2 & -3 \\ 0 & 0 & 1 & -2 \end{bmatrix}$$

4. Given the matrix
$$B = \begin{bmatrix} 2 & 0 & 4 & -3 \\ 4 & 3 & -2 & -8 \\ -1 & 1 & 5 & 4 \end{bmatrix}$$
    (a) Replace row 2 of matrix B by the sum of itself and a multiple of row 1 so that $b_{21} = 0$.
    (b) Replace row 3 of matrix B by the sum of itself and a multiple of row 1 so that $b_{31} = 0$.

5. Solve by matrix methods.
$$3x - y = -17$$
$$2x + 3y = -4$$

6. Solve by matrix methods.
$$x + 3y + 2z = 0$$
$$-2x \qquad + 3z = -12$$
$$2x - 6y - z = 6$$

7. Find the value of the determinant
$$\begin{vmatrix} 2 & 2 \\ 4 & -1 \end{vmatrix}$$

8. Find the cofactor of each element in the first column of the determinant

$$\begin{vmatrix} 1 & 2 & -1 \\ 3 & 0 & -3 \\ 4 & 5 & 1 \end{vmatrix}$$

9. Evaluate by expanding by cofactors.

$$\begin{vmatrix} 4 & -1 & 2 \\ -3 & 0 & 5 \\ -2 & 1 & 6 \end{vmatrix}$$

10. Evaluate by expanding by cofactors.

$$\begin{vmatrix} 3 & 0 & -4 & 5 \\ -4 & 6 & -2 & 1 \\ 0 & 2 & -1 & 0 \\ 8 & 4 & 1 & -1 \end{vmatrix}$$

11. Solve by Cramer's rule.

$$9x - 3y = 7$$
$$x + 8y = -2$$

12. Solve by Cramer's rule.

$$2x + 2y - z = -3$$
$$x \quad\quad + 2z = 13$$
$$3x - 2y - z = 8$$

PROGRESS TEST 14B

1. Given the matrix

$$A = \begin{bmatrix} 4 & -1.5 & 0.6 \\ 1.2 & 15 & -3 \end{bmatrix}$$

  (a) Determine the dimension of matrix A.
  (b) Find a_{13}.

2. Write a system of linear equations that corresponds to the augmented matrix.

$$\begin{bmatrix} 2.6 & 1.5 & -13 \\ 0.2 & -3.7 & 7 \end{bmatrix}$$

3. Solve the system of linear equations that corresponds to the matrix

$$\begin{bmatrix} 2 & 1 & -4 & 13 \\ 0 & -3 & 2 & 0 \\ 0 & 0 & 4 & -6 \end{bmatrix}$$

4. Given the matrix

$$B = \begin{bmatrix} -1 & 2 & 4 & 0 \\ 3 & 0 & -4 & -3 \\ -\frac{1}{2} & -5 & -2 & 7 \end{bmatrix}$$

  (a) Replace row 2 of matrix B by the sum of itself and a multiple of row 1 so that $b_{21} = 0$.

(b) Replace row 3 of matrix B by the sum of itself and a multiple of row 1 so that $b_{31} = 0$.

5. Solve by matrix methods.

$$\tfrac{1}{2}x + y = 0$$
$$x - y = 10$$

6. Solve by matrix methods.

$$-2x + y - z = -3$$
$$x + 2y = -1$$
$$3x + 4y + 2z = -9$$

7. Find the value of the determinant

$$\begin{vmatrix} -3 & 4 \\ \tfrac{1}{2} & -2 \end{vmatrix}$$

8. Find the cofactor of each element in the second row of the determinant

$$\begin{vmatrix} 4 & 0 & 2 \\ 2 & 1 & -5 \\ -2 & 3 & -1 \end{vmatrix}$$

9. Evaluate by expanding by cofactors.

$$\begin{vmatrix} -1 & 4 & 5 \\ 0 & -2 & 4 \\ 2 & -3 & 0 \end{vmatrix}$$

10. Evaluate by expanding by cofactors.

$$\begin{vmatrix} -1 & 3 & 4 & -2 \\ 5 & 2 & -1 & 0 \\ 2 & 1 & 1 & 0 \\ -4 & 0 & 3 & 1 \end{vmatrix}$$

11. Solve by Cramer's rule.

$$3x - y = 2$$
$$2x + 3y = 5$$

12. Solve by Cramer's rule.

$$x + 2y + 4z = 3$$
$$3x - 2z = 8$$
$$-2x + y = -7$$

CHAPTER FIFTEEN
TOPICS IN ALGEBRA

This chapter presents several topics in algebra that are somewhat independent of the flow of ideas in the earlier chapters of this book. Some of these topics, such as sequences, deal with functions whose domain is the set of natural numbers. An important reason for studying sequences and series is that the underlying concepts can be used as an introduction to calculus.

The binomial theorem gives us a way to expand the expression $(a + b)^n$. Those students who proceed to a study of calculus will find this theorem used when they begin to study the derivative.

The theory of permutations and combinations enables us to count the ways in which we can arrange or select a subset of a set of objects and is necessary background to a study of probability theory.

15.1
ARITHMETIC PROGRESSIONS

Can you see a pattern or relationship that describes this string of numbers?

$$1, 4, 9, 16, 25, \ldots$$

If we rewrite this string as

$$1^2, 2^2, 3^2, 4^2, 5^2, \ldots$$

it is clear that these are the squares of successive natural numbers. Each number in the string is called a **term.** We could write the nth term of the list as a function a such that

$$a(n) = n^2$$

where n is a natural number. Such a string of numbers is called an **infinite sequence** since the list is infinitely long.

An infinite sequence is a function whose domain is the set of all natural numbers.

The range of the function a is

$$a(1), \ a(2), \ a(3), \ldots$$

which we write as

$$a_1, \ a_2, \ a_3, \ldots$$

That is, we indicate a sequence by using subscript notation rather than function notation. We say that a_1 is the first term of the sequence, a_2 is the second term, and so on, and write the nth term as a_n.

If a string of numbers satisfies the definition of a sequence except that it terminates, then it is called a **finite sequence.**

EXAMPLE 1
Write the first four terms of a sequence whose nth term is

$$a_n = \frac{n}{n + 1}$$

Solution
To find a_1, we substitute $n = 1$ in the formula

$$a_1 = \frac{1}{1 + 1} = \frac{1}{2}$$

Similarly, we have

$$a_2 = \frac{2}{2 + 1} = \frac{2}{3}, \quad a_3 = \frac{3}{3 + 1} = \frac{3}{4}, \quad a_4 = \frac{4}{4 + 1} = \frac{4}{5}$$

PROGRESS CHECK 1
Write the first four terms of a sequence whose nth term is

$$a_n = \frac{n - 1}{n^2}$$

Answer

$0, \ \dfrac{1}{4}, \ \dfrac{2}{9}, \ \dfrac{3}{16}$

Now let's try to find a pattern or relationship for the sequence

$$2, \ 5, \ 8, \ 11, \ldots$$

You may notice that the nth term can be written as $a_n = 3n - 1$. But there is a simpler way to describe this sequence. Each term after the first can be obtained by adding 3 to the preceding term.

$$a_1 = 2, \quad a_2 = a_1 + 3, \quad a_3 = a_2 + 3, \ldots$$

A sequence in which each successive term is obtained by adding a fixed number to the previous term is called an **arithmetic progression** or **arithmetic sequence.**

> In an arithmetic progression, there is a number d such that
> $$a_n = a_{n-1} + d$$
> for all $n > 1$. The number d is called the **common difference.**

The nth term of the sequence

$$2, 5, 8, 11, \ldots$$

can be written as

$$a_n = a_{n-1} + 3 \quad \text{with} \quad a_1 = 2$$

This is an arithmetic progression with first term equal to 2 and common difference 3. The formula $a_n = a_{n-1} + 3$ is said to be a **recursive formula** since it defines the nth term by means of preceding terms. Beginning with $a_1 = 2$, the formula is used "recursively" (over and over) to obtain a_2, then a_3, then a_4, and so on.

EXAMPLE 2
Which of the following are arithmetic progressions?

(a) $5, 7, 9, 11, \ldots$

Since each term can be obtained from the preceding by adding 2, this is an arithmetic progression with first term equal to 5 and a common difference of 2.

(b) $4, 8, 11, 13, \ldots$

This is not an arithmetic progression since there is not a common difference between terms. The difference between the first and second term is 4, while that between the next two terms is 3.

(c) $3, -1, -5, -9, \ldots$

The difference between terms is -4, that is,

$$a_n = a_{n-1} - 4$$

This is an arithmetic progression with first term equal to 3 and a common difference of -4.

(d) $1, \dfrac{3}{2}, 2, \dfrac{5}{2}, \ldots$

This is an arithmetic progression with a common difference of $\frac{1}{2}$.

PROGRESS CHECK 2

Which of the following are arithmetic progressions?

(a)   16, 17, 16, 17, ...   (b) $-2, -1, 0, 1, ...$

(c) $6, \dfrac{9}{2}, 3, \dfrac{3}{2}, ...$ (d)   2, 4, 8, 12, ...

Answer

(b) *and* (c)

EXAMPLE 3

Write the first four terms of an arithmetic progression whose first term is -4 and whose common difference is -3.

Solution

The arithmetic progression is defined by

$$a_n = a_{n-1} - 3, \quad a_1 = -4$$

which leads to the terms

$$a_1 = -4, \quad a_2 = -7, \quad a_3 = -10, \quad a_4 = -13$$

PROGRESS CHECK 3

Write the first four terms of an arithmetic progression whose first term is 4 and whose common difference is $-\frac{1}{3}$.

Answer

$4, \dfrac{11}{3}, \dfrac{10}{3}, 3$

For a given arithmetic progression it's easy to find a formula for the *n*th term, a_n, in terms of *n* and the first term a_1. Since

$$a_2 = a_1 + d$$

and

$$a_3 = a_2 + d$$

we see that

$$a_3 = (a_1 + d) + d = a_1 + 2d$$

Similarly, we can show that

$$a_4 = a_3 + d = (a_1 + 2d) + d = a_1 + 3d$$

In general,

> The *n*th term a_n of an arithmetic progression is given by
> $$a_n = a_1 + (n - 1)d$$

EXAMPLE 4

Find the 7th term of the arithmetic progression whose first term is 2 and whose common difference is 4.

Solution
We substitute $n = 7$, $a_1 = 2$, $d = 4$ in the formula

$$a_n = a_1 + (n - 1)d$$

obtaining

$$a_7 = 2 + (7 - 1)4 = 2 + 24 = 26$$

PROGRESS CHECK 4
Find the 16th term of the arithmetic progression whose first term is -5 and common difference is $\frac{1}{2}$.

Answer

$$a_{16} = \frac{5}{2}$$

EXAMPLE 5
Find the 25th term of an arithmetic progression whose first and 20th terms are -7 and 31, respectively.

Solution
We can apply the given information to find d.

$$a_n = a_1 + (n - 1)d$$
$$a_{20} = a_1 + (20 - 1)d$$
$$31 = -7 + 19d$$
$$d = 2$$

Now we use the formula for a_n to find a_{25}.

$$a_n = a_1 + (n - 1)d$$
$$a_{25} = -7 + (25 - 1)2$$
$$a_{25} = 41$$

PROGRESS CHECK 5
Find the 60th term of an arithmetic progression whose first and 10th terms are 3 and $-\frac{3}{2}$, respectively.

Answer

$$-\frac{53}{2}$$

In many applications of sequences, we wish to *add* the terms. A sum of the terms of a sequence is called a **series.** When we are dealing with an arithmetic progression, the associated series is called an **arithmetic series.**
We denote the sum of the first n terms of the arithmetic progression $a_1, a_2, a_3, \ldots$ by S_n.

$$S_n = a_1 + a_2 + a_3 + \cdots + a_{n-2} + a_{n-1} + a_n$$

Since an arithmetic progression has a common difference d, we may write

$$S_n = a_1 + (a_1 + d) + (a_1 + 2d) + \cdots + (a_n - 2d) + (a_n - d) + a_n \quad (1)$$

Rewriting Equation (1) in reverse order, we have

$$S_n = a_n + (a_n - d) + (a_n - 2d) + \cdots + (a_1 + 2d) + (a_1 + d) + a_1 \quad (2)$$

Summing the corresponding sides of Equations (1) and (2),

$$2S_n = (a_1 + a_n) + (a_1 + a_n) + (a_1 + a_n) + \cdots \qquad \text{Repeated } n \text{ times}$$

$$= n(a_1 + a_n)$$

Thus,

$$S_n = \frac{n}{2}(a_1 + a_n)$$

Since $a_n = a_1 + (n - 1)d$, we see that

$$S_n = \frac{n}{2}(a_1 + a_n)$$

$$= \frac{n}{2}[a_1 + a_1 + (n - 1)d]$$

$$= \frac{n}{2}[2a_1 + (n - 1)d]$$

We now have two useful formulas.

$$S_n = \frac{n}{2}(a_1 + a_n)$$

$$S_n = \frac{n}{2}[2a_1 + (n - 1)d]$$

EXAMPLE 6

Find the sum of the first 30 terms of an arithmetic sequence whose first term is -20 and whose common difference is 3.

Solution

We know that $n = 30$, $a_1 = -20$, and $d = 3$. Substituting in

$$S_n = \frac{n}{2}[2a_1 + (n - 1)d]$$

$$S_{30} = \frac{30}{2}[2(-20) + (30 - 1)3]$$

$$= 15(-40 + 87)$$

$$= 705$$

PROGRESS CHECK 6

Find the sum of the first 10 terms of an arithmetic sequence whose first term is 2 and whose common difference is $-\frac{1}{2}$.

Answer

$$-\frac{5}{2}$$

EXAMPLE 7

The first term of an arithmetic series is 2, the last term is 58, and the sum is 450. Find the number of terms and the common difference.

Solution

We have $a_1 = 2$, $a_n = 58$, and $S_n = 450$. Substituting in

$$S_n = \frac{n}{2}(a_1 + a_n)$$

we have

$$450 = \frac{n}{2}(2 + 58)$$

$$900 = 60n$$

$$n = 15$$

Now we substitute in

$$a_n = a_1 + (n - 1)d$$

$$58 = 2 + 14d$$

$$56 = 14d$$

$$d = 4$$

PROGRESS CHECK 7

The first term of an arithmetic series is 6, the last term is 1, and the sum is 77/2. Find the number of terms and the common difference.

Answer

$n = 11, \quad d = -\dfrac{1}{2}$

EXERCISE SET 15.1

In each of the following write the first four terms of the sequence whose nth term is given as a_n.

1. $a_n = 2n$

2. $a_n = 2n + 1$

3. $a_n = 2n + 3$

4. $a_n = 2n - 2$

5. $a_n = 3n - 1$

6. $a_n = \dfrac{2n + 5}{2}$

7. $a_n = \dfrac{n + 3}{3}$

8. $a_n = \dfrac{n - 1}{2n}$

9. $a_n = \dfrac{n^2 + n}{n + 1}$

10. $a_n = \dfrac{n^2 - 1}{n^2 + 1}$

11. $a_n = \dfrac{n^2}{2n + 1}$

12. $a_n = \dfrac{2n + 1}{n^2}$

Which of the following are arithmetic progressions?

13. $3, 6, 9, 12, \ldots$

14. $4, \dfrac{11}{2}, 7, \dfrac{17}{2}, \ldots$

15. $1, -2, -6, -10, \ldots$

16. $-\dfrac{1}{2}, -1, -\dfrac{3}{2}, -2, \ldots$

17. $0, \dfrac{1}{4}, \dfrac{1}{2}, \dfrac{3}{4}, \ldots$

18. $2, 0, 1, 4, \ldots$

19. $-3, -\dfrac{8}{3}, -\dfrac{7}{3}, -2, \ldots$

20. $2, 0, -2, -4, \ldots$

21. $-1, 2, 6, 11, \ldots$ 22. $18, 15, 12, 9, \ldots$

23. $-1, 2, 5, 8, \ldots$ 24. $12, 8, 4, 1, \ldots$

Write the first four terms of an arithmetic progression whose first term is a_1 and whose common difference is d.

25. $a_1 = 2, d = 4$ 26. $a_1 = -2, d = -5$

27. $a_1 = 3, d = -\dfrac{1}{2}$ 28. $a_1 = \dfrac{1}{2}, d = 2$

29. $a_1 = -4, d = 4$ 30. $a_1 = 17, d = \dfrac{3}{2}$

31. $a_1 = 21, d = -4$ 32. $a_1 = -7, d = 2$

33. $a_1 = \dfrac{1}{3}, d = -\dfrac{1}{3}$ 34. $a_1 = 6, d = \dfrac{5}{2}$

Find the specified term of the arithmetic progression whose first term is a_1 and whose common difference is d.

35. $a_1 = 4, d = 3$; 8th term   36. $a_1 = -3, d = \dfrac{1}{4}$; 14th term

37. $a_1 = 14, d = -2$; 12th term   38. $a_1 = 6, d = -\dfrac{1}{3}$; 9th term

Given two terms of an arithmetic progression, find the specified term.

39. $a_1 = -2, a_{20} = -2$; 24th term

40. $a_1 = \dfrac{1}{2}, a_{12} = 6$; 30th term

41. $a_1 = 0, a_{61} = 20$; 20th term

42. $a_1 = 23, a_{15} = -19$; 6th term

43. $a_1 = -\dfrac{1}{4}, a_{41} = 10$; 22nd term

44. $a_1 = -3, a_{18} = 65$; 30th term

Find the sum of the specified number of terms of an arithmetic progression whose first term is a_1 and whose common difference is d.

45. $a_1 = 3, d = 2$; 20 terms   46. $a_1 = -4, d = \dfrac{1}{2}$; 24 terms

47. $a_1 = \dfrac{1}{2}, d = -2$; 12 terms   48. $a_1 = -3, d = -\dfrac{1}{3}$; 18 terms

49. $a_1 = 82, d = -2$; 40 terms   50. $a_1 = 6, d = 4$; 16 terms

51. How many terms of the arithmetic progression $2, 4, 6, 8. \ldots$ add up to 930?

52. How many terms of the arithmetic progression $44, 41, 38, 35, \ldots$ add up to 305?

53. The first term of an arithmetic series is 3, the last term is 90, and the sum is 1395. Find the number of terms and the common difference.

54. The first term of an arithmetic series is -3, the last term is $\dfrac{5}{2}$, and the sum is -3. Find the number of terms and the common difference.

55. The first term of an arithmetic series is $\dfrac{1}{2}$, the last term is $\dfrac{7}{4}$, and the sum is $\dfrac{27}{4}$. Find the number of terms and the common difference.

56. The first term of an arithmetic series is 20, the last term is -14, and the sum is 54. Find the number of terms and the common difference.
57. Find the sum of the first 16 terms of an arithmetic progression whose 4th and 10th terms are $-\frac{5}{4}$ and $\frac{1}{4}$, respectively.
58. Find the sum of the first 12 terms of an arithmetic progression whose 3rd and 6th terms are 9 and 18, respectively.

15.2
GEOMETRIC PROGRESSIONS

The sequence

$$3, 6, 12, 24, 48$$

has a distinct pattern. Each term after the first is obtained by multiplying the preceding one by 2. Thus, we could rewrite the sequence as

$$3, 3 \cdot 2, (3 \cdot 2) \cdot 2, (3 \cdot 2 \cdot 2) \cdot 2, \ldots$$

Such a sequence is called a **geometric progression** or **geometric sequence.** Each successive term is found by multiplying the previous term by a fixed number.

In a geometric progression, there is a number r such that

$$a_n = ra_{n-1} \quad \text{for all } n > 1$$

The constant r is called the **common ratio.** The common ratio r can be found by dividing any term by the preceding one.

$$r = \frac{a_k}{a_{k-1}}$$

EXAMPLE 1
If the sequence is a geometric progression, find the common ratio.

(a) $2, -4, 8, -16, \ldots$

Since each term can be obtained by multiplying the preceding one by -2, this is a geometric progression with common ratio of -2.

(b) $1, 2, 6, 24, \ldots$

The ratio between successive terms is not constant. This is not a geometric progression.

(c) $\frac{1}{4}, \frac{1}{8}, \frac{1}{16}, \frac{1}{32}, \ldots$

This is a geometric progression with common ratio of $\frac{1}{2}$.

PROGRESS CHECK 1
For each geometric progression, find the common ratio.

(a) $3, -9, 27, -81, \ldots$ (b) $4, 1, -2, -5, \ldots$

(c) $6, 2, \frac{2}{3}, \frac{2}{9}, \ldots$ (d) $4, 16, 48, 96, \ldots$

Answers

Sequence (a) is a geometric progression with $r = -3$.

Sequence (c) is a geometric progression with $r = \dfrac{1}{3}$.

Let's look at successive terms of a geometric progression whose first term is a_1 and whose common ratio is r. We have

$$a_2 = ra_1$$
$$a_3 = ra_2 = r(ra_1) = r^2 a_1$$
$$a_4 = ra_3 = r(r^2 a_1) = r^3 a_1$$

The pattern suggests that

> The nth term of a geometric progression is given by
> $$a_n = a_1 r^{n-1}$$

EXAMPLE 2

Find the 7th term of the geometric progression $-4, -2, -1, \ldots$.

Solution

Since

$$r = \frac{a_k}{a_{k-1}}$$

we see that

$$r = \frac{-1}{-2} = \frac{1}{2}$$

Substituting $a_1 = -4$, $r = \frac{1}{2}$, and $n = 7$, we have

$$a_n = a_1 r^{n-1}$$
$$a_7 = (-4)\left(\frac{1}{2}\right)^{7-1} = (-4)\left(\frac{1}{2}\right)^6$$
$$= (-4)\left(\frac{1}{64}\right) = -\frac{1}{16}$$

PROGRESS CHECK 2

Find the 6th term of the geometric progression $2, -6, 18, \ldots$.

Answer

-486

In a geometric progression, the terms between the first and last terms are called **geometric means.** We will illustrate the method of calculating such means.

EXAMPLE 3

Insert three geometric means between 3 and 48.

Solution

The geometric progression must look like this.

$$3, a_2, a_3, a_4, 48$$

Thus, $a_1 = 3$, $a_5 = 48$, and $n = 5$. Substituting in

$$a_n = a_1 r^{n-1}$$
$$48 = 3r^4$$
$$r^4 = 16$$
$$r = \pm 2$$

There are two geometric progressions with three geometric means between 3 and 48.

$$3, 6, 12, 24, 48 \qquad r = 2$$
$$3, -6, 12, -24, 48 \qquad r = -2$$

PROGRESS CHECK 3

Insert two geometric means between 5 and $\dfrac{8}{25}$.

Answer

$5, 2, \dfrac{4}{5}, \dfrac{8}{25}$

If $a_1, a_2, \ldots a_n$ is a geometric progression, then the corresponding series

$$S_n = a_1 + a_2 + \cdots + a_n \qquad (1)$$

is called a **geometric series.** Since each term of the series can be rewritten as $a_k = a_1 r^{k-1}$, we can rewrite Equation (1) as

$$S_n = a_1 + a_1 r + a_1 r^2 + \cdots + a_1 r^{n-2} + a_1 r^{n-1} \qquad (2)$$

Multiplying each term in Equation (2) by r we have

$$rS_n = a_1 r + a_1 r^2 + a_1 r^3 + \cdots + a_1 r^{n-1} + a_1 r^n \qquad (3)$$

Subtracting Equation (2) from Equation (3) produces

$$rS_n - S_n = a_1 r^n - a_1$$
$$(r - 1)S_n = a_1(r^n - 1) \qquad \text{Factoring}$$
$$S_n = \frac{a_1(r^n - 1)}{r - 1} \qquad \text{Dividing by } r - 1$$

Changing the signs in both the numerator and the denominator gives us the following equation for the sum of n terms.

Geometric Series

$$S_n = \frac{a_1(1 - r^n)}{1 - r}$$

EXAMPLE 4

Find the sum of the first six terms of the geometric progression whose first three terms are 12, 6, 3.

Solution

The common ratio can be found by dividing any term by the preceding term:

$$r = \frac{a_k}{a_{k-1}} = \frac{6}{12} = \frac{1}{2}$$

Substituting $a_1 = 12$, $r = \frac{1}{2}$, $n = 6$ in the formula for S_n, we have

$$S_n = \frac{a_1(1 - r^n)}{1 - r}$$

$$= \frac{12\left[1 - \left(\frac{1}{2}\right)^6\right]}{1 - \frac{1}{2}} = \frac{12\left(1 - \frac{1}{64}\right)}{\frac{1}{2}}$$

$$= 24\left(\frac{63}{64}\right) = \frac{189}{8}$$

PROGRESS CHECK 4

Find the sum of the first five terms of the geometric progression whose first three terms are 2, $-\frac{4}{3}$, $\frac{8}{9}$.

Answer

$\dfrac{110}{81}$

EXAMPLE 5

A father promises to give each child 2 cents on the first evening, 4 cents on the second evening, and to continue doubling the amount each evening for a total of 8 days. How much will each child receive on the last evening? How much will each child have received in total after 8 evenings?

Solution

The daily payout forms a geometric progression 2, 4, 8, ... with $a_1 = 2$ and $r = 2$. The term a_8 is given by substituting in

$$a_n = a_1 r^{n-1}$$

$$a_8 = a_1 r^{8-1} = 2 \cdot 2^7 = 256$$

Thus, each child receives $2.56 on the last evening. The total received is given by

$$S_n = \frac{a_1(1 - r^n)}{1 - r}$$

$$S_8 = \frac{a_1(1 - r^8)}{1 - r} = \frac{2(1 - 2^8)}{1 - 2}$$

$$= \frac{2(1 - 256)}{-1} = 510$$

Each child receives a total of $5.10 after eight evenings.

PROGRESS CHECK 5

A ball is dropped from a height of 64 feet. On each bounce, it rebounds half the height it fell. How high is the ball at the top of the fifth bounce? What is the total distance the ball has traveled?

Answer
2 feet, 186 feet

We now want to focus on a geometric series for which $|r| < 1$, say

$$\frac{1}{2} + \frac{1}{4} + \frac{1}{8} + \cdots + \frac{1}{2^n} + \cdots$$

To see how the sum increases as n increases, let's form a table of values of S_n.

n	1	2	3	4	5	6	7	8	9
S_n	0.500	0.750	0.875	0.938	0.969	0.984	0.992	0.996	0.998

We begin to suspect that S_n gets closer and closer to 1 as n increases. To see that this is really so, let's look at the formula

$$S_n = \frac{a_1(1 - r^n)}{1 - r}$$

when $|r| < 1$. When a number r that is less than 1 in absolute value is raised to higher and higher positive integer powers, the value of r^n gets smaller and smaller. Thus, the term r^n can be made as small as we like by choosing n sufficiently large. Since we are dealing with an infinite series, we say that "r^n approaches 0 as n approaches infinity." We then replace r^n with 0 in the formula and denote the sum by S.

Sum of an Infinite Geometric Series

$$S = \frac{a_1}{1 - r} \quad \text{when} \quad |r| < 1$$

Applying this formula to the preceding series, we see that

$$S = \frac{\dfrac{1}{2}}{1 - \dfrac{1}{2}} = 1$$

which justifies the conjecture resulting from the examination of the table above.

It is appropriate to remark that the ideas we have used in deriving the formula for the sum of an infinite geometric series have led us to the very border of the beginning concepts of calculus.

EXAMPLE 6

Evaluate the sum $\dfrac{3}{2} + 1 + \dfrac{2}{3} + \dfrac{4}{9} + \cdots$.

Solution

The common ratio $r = \frac{2}{3}$. The sum of an infinite geometric series with $|r| < 1$ is given by

$$S = \frac{a_1}{1 - r} = \frac{\dfrac{3}{2}}{1 - \dfrac{2}{3}} = \frac{9}{2}$$

PROGRESS CHECK 6

Evaluate the sum $4 - 1 + \dfrac{1}{4} - \dfrac{1}{16} + \cdots$.

Answer

$\dfrac{16}{5}$

EXERCISE SET 15.2

In each of the following determine if the given sequence is a geometric progression. If it is, find the common ratio.

1. 3, 6, 12, 24, ...

2. −4, 12, −36, 108, ...

3. −2, 4, 12, −36, ...

4. 27, 18, 12, 8, ...

5. $-4, 3, -\dfrac{9}{4}, \dfrac{27}{16}, \ldots$

6. $3, -1, \dfrac{1}{2}, -\dfrac{1}{4}, \ldots$

7. 1.2, 0.24, 0.048, 0.0096, ...

8. $\dfrac{2}{3}, 1, \dfrac{3}{2}, \dfrac{9}{4}, \ldots$

9. $\dfrac{1}{4}, \dfrac{1}{2}, 2, 8, \ldots$

10. $\dfrac{1}{2}, \dfrac{3}{2}, \dfrac{9}{4}, \dfrac{27}{8}, \ldots$

Write the first four terms of the geometric progression whose first term is a_1 and whose common ratio is r.

11. $a_1 = 3, r = 3$

12. $a_1 = -4, r = 2$

13. $a_1 = 4, r = \dfrac{1}{2}$

14. $a_1 = 16, r = -\dfrac{3}{2}$

15. $a_1 = \dfrac{1}{2}, r = 4$

16. $a_1 = \dfrac{3}{2}, r = -\dfrac{2}{3}$

17. $a_1 = -3, r = 2$

18. $a_1 = 3, r = -\dfrac{2}{3}$

19. If $a_1 = 3$ and $r = -2$, find a_8.

20. If $a_1 = 18$ and $r = -\dfrac{1}{2}$, find a_6.

21. Given the sequence $24, -6, \dfrac{3}{2}, \ldots$, find a_5.

22. Given the sequence 16, 8, 4, ... , find a_7.

23. Given the sequence 15, -10, $\dfrac{20}{3}$, ... , find a_6.

24. Given the sequence 10, 15, $\dfrac{45}{2}$, ... , find a_6.

25. If $a_1 = 3$ and $a_5 = \dfrac{1}{27}$, find a_7.

26. If $a_1 = 2$ and $a_6 = \dfrac{1}{16}$, find a_3.

27. If $a_1 = \dfrac{16}{81}$ and $a_6 = \dfrac{3}{2}$, find a_8.

28. If $a_4 = \dfrac{1}{4}$ and $a_7 = 1$, find r.

29. If $a_2 = 4$ and $a_8 = 256$, find r.
30. If $a_3 = 3$ and $a_6 = -81$, find a_8.

31. If $a_1 = \dfrac{1}{2}$, $r = 2$, and $a_n = 32$, find n.

32. If $a_1 = -2$, $r = 3$, and $a_n = 162$, find n.
33. Insert two geometric means between 3 and 96.
34. Insert two geometric means between -3 and 192.

35. Insert two geometric means between 1 and $\dfrac{1}{64}$.

36. Insert three geometric means between $\dfrac{2}{3}$ and $\dfrac{32}{243}$.

37. Find the sum of the first seven terms of the geometric progression whose first three terms are 3, 1, $\dfrac{1}{3}$.

38. Find the sum of the first six terms of the geometric progression whose first three terms are $\dfrac{1}{3}$, 1, 3.

39. Find the sum of the first five terms of the geometric progression whose first three terms are -3, $\dfrac{6}{5}$, $-\dfrac{12}{25}$.

40. Find the sum of the first six terms of the geometric progression whose first three terms are 2, $\dfrac{4}{3}$, $\dfrac{8}{9}$.

41. If $a_1 = 4$ and $r = 2$, find S_8.

42. If $a_1 = -\dfrac{1}{2}$ and $r = -3$, find S_{10}.

43. If $a_1 = 2$ and $a_4 = \dfrac{-54}{8}$, find S_5.

44. If $a_1 = 64$ and $a_7 = 1$, find S_6.

45. A Christmas Club calls for savings of $5 in January, and twice as much on each successive month as in the previous month. How much money will have been saved by the end of November?

46. A city has a population of 20,000 people in 1980. If the population increases 5% per year, what will the population be in 1990?

47. A city has a population of 30,000 in 1980. If the population increases 25% every ten years, what will the population be in the year 2010?

48. For good behavior a child is offered a reward consisting of 1 cent on the first day, 2 cents on the second day, 4 cents on the third day, and so on. If the child behaves properly for two weeks, what is the total amount that the child will receive?

Evaluate the sum of each geometric series.

49. $1 + \dfrac{1}{2} + \dfrac{1}{4} + \dfrac{1}{8} + \cdots$

50. $\dfrac{4}{5} + \dfrac{1}{5} + \dfrac{1}{20} + \dfrac{1}{80} + \cdots$

51. $1 - \dfrac{1}{3} + \dfrac{1}{9} - \dfrac{1}{27} + \cdots$

52. $\dfrac{1}{2} - \dfrac{1}{4} + \dfrac{1}{8} - \dfrac{1}{16} + \cdots$

53. $2 + \dfrac{1}{2} + \dfrac{1}{8} + \dfrac{1}{32} + \cdots$

54. $1 + 0.1 + 0.01 + 0.001 + \cdots$

55. $0.5 + (0.5)^2 + (0.5)^3 + (0.5)^4 + \cdots$

56. $\dfrac{2}{5} + \dfrac{4}{25} + \dfrac{8}{125} + \dfrac{16}{625} + \cdots$

57. $\dfrac{1}{3} - \dfrac{2}{9} + \dfrac{4}{27} - \dfrac{8}{81} + \cdots$

15.3
THE BINOMIAL THEOREM

By sequential multiplication by $(a + b)$ you may verify that

$$(a + b)^1 = a + b$$
$$(a + b)^2 = a^2 + 2ab + b^2$$
$$(a + b)^3 = a^3 + 3a^2b + 3ab^2 + b^3$$
$$(a + b)^4 = a^4 + 4a^3b + 6a^2b^2 + 4ab^3 + b^4$$
$$(a + b)^5 = a^5 + 5a^4b + 10a^3b^2 + 10a^2b^3 + 5ab^4 + b^5$$

The expression on the right-hand side of the equation is called the **expansion** of the left-hand side. If we were to predict the form of the expansion of $(a + b)^n$, where n is a natural number, the preceding examples would lead us to conclude that it has the following properties.

(a) The expansion has $n + 1$ terms.

(b) The first term is a^n and the last term is b^n.

(c) The sum of the exponents of a and b in each term is n.

(d) In each successive term after the first, the power of a decreases by 1 and the power of b increases by 1.

(e) The coefficients may be obtained from the following array, which is known as **Pascal's triangle.** Each number, with the exception of those at the ends of a row,

is the sum of the two nearest numbers in the row above. The numbers at the ends of a row are always 1.

Note that the entries in the Pascal triangle correspond to the coefficients of the expansions at the beginning of this section. The Pascal triangle, however, is not a convenient means for determining the coefficients of the expansion when n is large. Here is an alternate method.

(e′) The coefficient of any term (after the first) can be found by this rule: In the preceding term, multiply the coefficient by the exponent of a and then divide by one more than the exponent of b.

EXAMPLE 1
Write the expansion of $(a + b)^6$.

Solution
From Property (b) we know that the first term is a^6. Thus,

$$(a + b)^6 = a^6 + \cdots$$

From Property (e′) the next coefficient is

$$\frac{1 \cdot 6}{1} = 6$$

(since the coefficient of b is 0). By Property (d) the term will be of the form a^5b. We have

$$(a + b)^6 = a^6 + 6a^5b + \cdots$$

Applying Property (e′) again the next coefficient is

$$\frac{6 \cdot 5}{2} = 15$$

and the term will be of the form a^4b^2. Thus,

$$(a + b)^6 = a^6 + 6a^5b + 15a^4b^2 + \cdots$$

Continuing in this manner we see that

$$(a + b)^6 = a^6 + 6a^5b + 15a^4b^2 + 20a^3b^3 + 15a^2b^4 + 6ab^5 + b^6$$

PROGRESS CHECK 1
Write the first five terms in the expansion of $(a + b)^{10}$.

Answer
$a^{10} + 10a^9b + 45a^8b^2 + 120a^7b^3 + 210a^6b^4$

The expansion of $(a + b)^n$ that we have described is called the **binomial theorem** or **binomial formula** and can be written as follows.

$$(a + b)^n = a^n + \frac{n}{1}a^{n-1}b + \frac{n(n - 1)}{1 \cdot 2}a^{n-2}b^2 + \frac{n(n - 1)(n - 2)}{1 \cdot 2 \cdot 3}a^{n-3}b^3$$

$$+ \cdots + \frac{n(n - 1)(n - 2) \cdots (n - r + 1)}{1 \cdot 2 \cdot 3 \cdots r}a^{n-r}b^r + \cdots + b^n$$

EXAMPLE 2
Find the expansion of $(2x - 1)^4$.

Solution
Let $a = 2x$, $b = -1$ and apply the binomial formula.

$$(2x - 1)^4 = (2x)^4 + \frac{4}{1}(2x)^3(-1) + \frac{4 \cdot 3}{1 \cdot 2}(2x)^2(-1)^2 + \frac{4 \cdot 3 \cdot 2}{1 \cdot 2 \cdot 3}(2x)(-1)^3 + (-1)^4$$

$$= 16x^4 - 32x^3 + 24x^2 - 8x + 1$$

PROGRESS CHECK 2
Find the expansion of $(x^2 - 2)^4$.

Answer
$x^8 - 8x^6 + 24x^4 - 32x^2 + 16$

You may have noticed that the denominator of the coefficient in the binomial formula is always a product of the first n natural numbers. We use the symbol **$n!$**, which is read as **n factorial,** to indicate this type of product. For example,

$$4! = 4 \cdot 3 \cdot 2 \cdot 1 = 24$$

$$6! = 6 \cdot 5 \cdot 4 \cdot 3 \cdot 2 \cdot 1 = 720$$

and

$$n! = n(n - 1)(n - 2) \cdots 4 \cdot 3 \cdot 2 \cdot 1$$

Since

$$(n - 1)! = (n - 1)(n - 2)(n - 3) \cdots 4 \cdot 3 \cdot 2 \cdot 1$$

we see that

$$n! = n(n - 1)!$$

Thus, $8! = 8 \cdot 7!$ and $15! = 15 \cdot 14!$ It should follow that $1! = 1 \cdot 0! = 1$ which leads us to *define* 0! as

$$0! = 1$$

EXAMPLE 3
Evaluate.

(a) $\dfrac{5!}{3!}$

Since $5! = 5 \cdot 4 \cdot 3!$, we may write

$$\frac{5!}{3!} = \frac{5 \cdot 4 \cdot 3!}{3!} = 5 \cdot 4 = 20$$

(b) $\dfrac{9!}{8!} = \dfrac{9 \cdot 8!}{8!} = 9$

(c) $\dfrac{10!4!}{12!} = \dfrac{10!4!}{12 \cdot 11 \cdot 10!} = \dfrac{4!}{12 \cdot 11} = \dfrac{4 \cdot 3 \cdot 2 \cdot 1}{12 \cdot 11}$

$$= \frac{4 \cdot 3 \cdot 2 \cdot 1}{4 \cdot 3 \cdot 11} = \frac{2}{11}$$

(d) $\dfrac{n!}{(n-2)!} = \dfrac{n(n-1)(n-2)!}{(n-2)!} = n(n-1) = n^2 - n$

(e) $\dfrac{(2-2)!}{3!} = \dfrac{0!}{3\cdot2} = \dfrac{1}{6}$

PROGRESS CHECK 3
Evaluate.

(a) $\dfrac{12!}{10!}$ (b) $\dfrac{6!}{4!2!}$ (c) $\dfrac{10!8!}{9!7!}$ (d) $\dfrac{n!(n-1)!}{(n+1)!(n-2)!}$ (e) $\dfrac{8!}{6!(3-3)!}$

Answers

(a) *132*   (b) *15*   (c) *80*   (d) $\dfrac{n-1}{n+1}$ (e) *56*

Here is what the binomial formula looks like in factorial notation.

$$(a+b)^n = a^n + \frac{n!}{1!(n-1)!}a^{n-1}b + \frac{n!}{2!(n-2)!}a^{n-2}b^2$$

$$+ \frac{n!}{3!(n-3)!}a^{n-3}b^3 + \cdots + \frac{n!}{r!(n-r)!}a^{n-r}b^r$$

$$+ \cdots + b^n$$

We can conclude with a formula for any term of a binomial expansion.

> The kth term of the expansion of $(a+b)^n$ has the form
>
> $$\frac{n!}{(k-1)!(n-k+1)!}a^{n-k+1}b^{k-1}$$

EXAMPLE 4
Find the fourth term in the expansion of $(x+2)^5$.

Solution
Using the formula for the kth term with $n = 5$ and $k = 4$, we have

$$(k-1)! = 3! \quad \text{and} \quad (n-k+1)! = 2!$$

Thus, the term is

$$\frac{5!}{3!2!}x^2(2)^3 = 80x^2$$

PROGRESS CHECK 4
Find the third term in the expansion of

$$\left(\frac{x}{2} - 1\right)^8$$

Answer

$\dfrac{7}{16}x^6$

EXAMPLE 5
Find the term in the expansion of $(x^2 - y^2)^6$ which involves y^8.

Solution
Since $y^8 = (-y^2)^4$, we seek that term which involves b^4 in the expansion of $(a + b)^6$. This term must have the form $a^2 b^4$ because $n = 6 = 2 + 4$. Setting $k = 5$ since the exponent of $b = 4 = k - 1$, we find the coefficient to be

$$\frac{n!}{(k-1)!(n-k+1)!} = \frac{6!}{4!2!} = 15$$

Thus, the term is $15(x^2)^2(-y^2)^4 = 15x^4 y^8$.

PROGRESS CHECK 5
Find the term in the expansion of $(x^3 - \sqrt{2})^5$ which involves x^6.

Answer
$-20\sqrt{2}x^6$

EXERCISE SET 15.3
Expand and simplify.

1. $(3x + 2y)^5$ 2. $(2a - 3b)^6$ 3. $(4x - y)^4$

4. $\left(3 + \frac{1}{2}x\right)^4$ 5. $(2 - xy)^5$ 6. $(3a^2 + b)^4$

7. $(a^2 b + 3)^4$ 8. $(x - y)^7$ 9. $(a - 2b)^8$

10. $\left(\frac{x}{y} + y\right)^6$ 11. $\left(\frac{1}{3}x + 2\right)^3$ 12. $\left(\frac{x}{y} + \frac{y}{x}\right)^5$

Find the first four terms in the given expansion and simplify.

13. $(2 + x)^{10}$ 14. $(x - 3)^{12}$ 15. $(3 - 2a)^9$

16. $(a^2 + b^2)^{11}$ 17. $(2x - 3y)^{14}$ 18. $\left(a - \frac{1}{a^2}\right)^8$

19. $(2x - yz)^{13}$ 20. $\left(x - \frac{1}{y}\right)^{15}$

Evaluate.

21. $5!$ 22. $7!$ 23. $\dfrac{12!}{11!}$

24. $\dfrac{13!}{12!}$ 25. $\dfrac{11!}{8!}$ 26. $\dfrac{7!}{9!}$

27. $\dfrac{10!}{6!}$ 28. $\dfrac{9!}{6!}$ 29. $\dfrac{6!}{3!}$

30. $\dfrac{8!}{5!3!}$ 31. $\dfrac{10!}{6!4!}$ 32. $\dfrac{(n+1)!}{(n-1)!}$

In each expansion find only the term specified.

33. The fourth term in $(2x - 4)^7$.

34. The third term in $(4a + 3b)^{11}$.

35. The fifth term in $\left(\dfrac{1}{2}x - y\right)^{12}$.

36. The sixth term in $(3x - 2y)^{10}$.

37. The fifth term in $\left(\dfrac{1}{x} - 2\right)^9$.

38. The next to last term in $(a + 4b)^5$.

39. The middle term in $(x - 3y)^6$.

40. The middle term in $\left(2a + \dfrac{1}{2}b\right)^6$.

41. The term involving x^4 in $(3x + 4y)^7$.

42. The term involving x^6 in $(2x^2 - 1)^9$.

43. The term involving x^6 in $(2x^3 - 1)^9$.

44. The term involving x^{10} in $\left(x^2 + \dfrac{1}{x}\right)^8$.

45. The term involving x^{12} in $\left(x^3 + \dfrac{1}{2}\right)^7$.

46. The term involving x^{-4} in $\left(x + \dfrac{1}{x^2}\right)^8$.

47. Evaluate $(1.3)^6$ to four decimal places by writing it as $(1 + 0.3)^6$ and using the binomial formula.

48. Using the method of Exercise 47, evaluate
    (a) $(3.4)^4$ (b) $(48)^5$ (*Hint:* $48 = 50 - 2$.)

15.4

COUNTING: PERMUTATIONS AND COMBINATIONS

How many arrangements can be made using the letters a, b, and c, two at a time? One way to solve this problem is to list all the possible arrangements.

$$ab \qquad ac \qquad bc$$
$$ba \qquad ca \qquad cb$$

Here is another way to view this problem. Each arrangement consists of a choice of candidates to fill two positions.

☐   ☐

Any one of three candidates (a, b, or c) can be assigned to the first position; once a candidate is assigned to the first position, any one of the two remaining candidates can be assigned to the second position. Since each candidate in the second position can be associated with each candidate in the first position, the *product*

$$3 \cdot 2 = 6$$

yields the total number of arrangements. This simple example illustrates a very important principle.

Counting Principle

If one event can occur in m different ways, and, after it has happened in one of these ways, a second event can occur in n different ways, then both events can occur in mn different ways.

Note that the order or sequence of events is significant since each arrangement is counted as one of the "mn different ways."

EXAMPLE 1
In how many ways can 5 students be seated in a row of 5 seats?

Solution
We have 5 positions to be filled. Any one of the 5 students may occupy the first position, after which any one of the remaining 4 students can occupy the next position. Reapplying the counting principle to the other positions, we see that the number of arrangements is

$$5 \cdot 4 \cdot 3 \cdot 2 \cdot 1 = 120$$

PROGRESS CHECK 1
How many different four-digit numbers can be formed using the digits 2, 4, 6, 8? (Don't repeat any of the digits.)

Answer
24

EXAMPLE 2
How many different three-letter arrangements can be made using the letters A, B, C, X, Y, Z (a) if no letter may be repeated in an arrangement; (b) if letters may be repeated?

Solution
(a) We need to fill 3 positions. Any one of the 6 letters may occupy the first position; then, any one of the *remaining* 5 letters may occupy the second position (since repetitions are not allowed). Thus, the total number of arrangements is $6 \cdot 5 \cdot 4 = 120$.
(b) Any one of the 6 letters may fill any of the 3 positions (since repetitions are allowed). The total number of arrangements is $6 \cdot 6 \cdot 6 = 216$.

PROGRESS CHECK 2
The positions of president, secretary, and treasurer are to be filled from a class of 15 students. In how many ways can these positions be filled if no student can hold more than one position?

Answer
2730

Each arrangement that can be made by using all or some of the elements of a set of objects without repetition is called a **permutation.** The phrase "without repetition" means that the permutations of the letters a, b, c taken three at a time include bac but exclude aab.

We will use the notation $P(n, r)$ to indicate the number of permutations of n distinct objects taken r at a time. (There are a number of other notations in common use: nPr, P_r^n, ^nP_r, $P_{n,r}$.) If r is equal to n, then by using the counting principle, we see that

$$P(n, n) = n(n - 1)(n - 2) \cdots 2 \cdot 1$$

since any one of the n objects may fill the first position, any one of the remaining $(n - 1)$ objects may fill the second position, and so on. Using factorial notation,

$$P(n, n) = n!$$

Let's try to calculate $P(n, r)$, that is, the number of permutations of n distinct objects taken r at a time, when r is less than n. We may think of this as the number of ways of filling r positions with n candidates. Once again, we may fill the first position with any one of the n candidates, the second position with any one of the remaining $(n - 1)$ candidates, and so on, so that

$$P(n, r) = \underbrace{n(n - 1)(n - 2) \cdots}_{r \text{ factors}}$$

We may write this as

$$P(n, r) = n(n - 1)(n - 2) \cdots (n - r + 1)$$

since $(n - r + 1)$ will be the rth factor. If we multiply the right-hand side of the equation by

$$\frac{(n - r)!}{(n - r)!} = 1$$

we have

$$P(n, r) = \frac{n(n - 1)(n - 2) \cdots (n - r + 1)(n - r)(n - r - 1) \cdots 2 \cdot 1}{(n - r)!}$$

or

$$P(n, r) = \frac{n!}{(n - r)!}$$

EXAMPLE 3
Compute.

(a) $P(5, 5) = \dfrac{5!}{(5 - 5)!} = \dfrac{5!}{0!} = \dfrac{5 \cdot 4 \cdot 3 \cdot 2 \cdot 1}{1} = 120$

(b) $P(5, 2) = \dfrac{5!}{(5 - 2)!} = \dfrac{5!}{3!} = \dfrac{5 \cdot 4 \cdot 3!}{3!} = 20$

(c) $\dfrac{P(6, 2)}{3!} = \dfrac{6!}{3!(6 - 2)!} = \dfrac{6!}{3!4!} = \dfrac{6 \cdot 5 \cdot 4!}{3 \cdot 2 \cdot 4!} = 5$

PROGRESS CHECK 3
Compute.

(a) $P(4, 4)$ (b) $P(6, 3)$ (c) $\dfrac{2P(6, 4)}{2!}$

Answers
(a) 24   (b) 120   (c) 360

EXAMPLE 4
How many different arrangements can be made by taking five of the letters of the word *relation?*

Solution
Since the word *relation* has 8 different letters, we are seeking the number of permutations of 8 objects taken 5 at a time or $P(8, 5)$. Thus,

$$P(n, r) = \frac{n!}{(n - r)!}$$

$$P(8, 5) = \frac{8!}{(8 - 5)!} = \frac{8!}{3!} = 6720$$

PROGRESS CHECK 4
There is space on a bookshelf for displaying 4 books. If there are 6 different novels available, how many arrangements can be made?

Answer
360

EXAMPLE 5
How many arrangements can be made using all of the letters of the word *quartz* if the vowels are always to remain adjacent to each other?

Solution
If we treat the vowel pair *ua* as a unit, then there are 5 "letters" (*q, ua, r, t, z*) which can be arranged in $P(5, 5)$ ways. But the vowels can themselves be arranged in $P(2, 2)$ ways. By the counting principle, the total number of arrangements is

$$P(5, 5) \cdot P(2, 2)$$

Since $P(5, 5) = 120$ and $P(2, 2) = 2$, the total number of arrangements is 240.

PROGRESS CHECK 5
A bookshelf is to be used to display 5 new textbooks. There are 7 mathematics textbooks and 4 biology textbooks available. If we wish to put 3 mathematics books and 2 biology books on display, how many arrangements can be made if the books in each discipline must be kept together?

Answer
5040

Let's take another look at the arrangements of the letters *a*, *b*, and *c* taken two at a time:

ab	*ac*	*bc*
ba	*ca*	*cb*

Now let's ask a different question. In how many ways can we *select* two letters from the letters a, b, and c? We are now asked to disregard the order in which the letters are chosen. The result is then

$$ab \qquad ac \qquad bc$$

In general, a set of r objects chosen from a set of n objects is called a **combination.** We denote the number of combinations of r objects chosen from n objects by $C(n, r)$. [Other notations in common use include nCr, C_r^n, ^nC_r, $C_{n,r}$, and $\binom{n}{r}$.]

EXAMPLE 6
List the combinations of the letters a, b, c, d taken three at a time.

Solution
The combinations are seen to be

$$abc \quad abd \quad acd \quad bcd$$

PROGRESS CHECK 6
List the combinations of the letters a, b, c, d taken two at a time.

Answer
ab, ac, ad, bc, bd, cd

Here is a rule that will prove helpful in determining whether a problem calls for the number of permutations or the number of combinations.

If we are interested in calculating the number of arrangements in which different ordering of the same objects are counted, we use permutations.

If we are interested in calculating the number of ways of selecting objects in which the order of the selected objects doesn't matter, we use combinations.

For example, suppose we want to determine the number of different four-card hands that can be dealt from a deck of 52 cards. Since a hand consisting of four cards is the same hand regardless of the order of the cards, we must use combinations.

Let's find a formula for $C(n, r)$. There are three combinations of the letters a, b, c taken two at a time, namely,

$$ab, \quad ac, \quad \text{and} \quad bc$$

so that $C(3, 2) = 3$. Now, each of these combinations can be arranged in 2! ways to yield the total list of permutations

$$ab, \quad ba, \quad ac, \quad ca, \quad bc, \quad cb$$

Thus, $P(3, 2) = 6 = 2!C(3, 2)$.

In general, each of the $C(n, r)$ combinations can be permuted in $r!$ ways so that by the counting principle the total number of permutations is

$P(n, r) = r!C(n, r)$ or

$$C(n, r) = \frac{P(n, r)}{r!} = \frac{n!}{r!(n - r)!}$$

EXAMPLE 7
Compute.

(a) $C(5, 2) = \dfrac{5!}{2!(5 - 2)!} = \dfrac{5!}{2!3!} = \dfrac{5 \cdot 4 \cdot 3!}{2 \cdot 3!} = 10$

(b) $C(4, 4) = \dfrac{4!}{4!(4 - 4)!} = \dfrac{4!}{4!0!} = 1$

(c) $\dfrac{P(6, 3)}{C(6, 3)}$

$$P(6, 3) = \frac{6!}{(6 - 3)!} = \frac{6!}{3!} = 6 \cdot 5 \cdot 4 = 120$$

$$C(6, 3) = \frac{6!}{3!(6 - 3)!} = \frac{6!}{3!3!} = \frac{6 \cdot 5 \cdot 4}{3 \cdot 2 \cdot 1} = 20$$

Thus,
$$\frac{P(6, 3)}{C(6, 3)} = \frac{120}{20} = 6$$

PROGRESS CHECK 7
Compute.

(a) $C(6, 2)$ (b) $C(10, 10)$ (c) $\dfrac{P(3, 2)}{3!C(5, 4)}$

Answers

(a)   *15*   (b)   *1*   (c) $\dfrac{1}{5}$

EXAMPLE 8
In how many ways can a committee of 4 be selected from a group of 10 people?

Solution
If A, B, C, and D constitute a committee, is the arrangement B, A, C, D a different committee? Of course not—which says that the order doesn't matter. We are therefore interested in computing $C(10, 4)$.

$$C(10, 4) = \frac{10!}{4!6!} = \frac{10 \cdot 9 \cdot 8 \cdot 7}{4 \cdot 3 \cdot 2 \cdot 1} = 210$$

PROGRESS CHECK 8
In how many ways can a five-card hand be dealt from a deck of 52 cards?

Answer
2,598,960

EXAMPLE 9
In how many ways can a committee of 3 girls and 2 boys be selected from a class of 8 girls and 7 boys?

Solution
The girls can be selected in $C(8, 3)$ ways and the boys can be selected in $C(7, 2)$ ways. By the counting principle, each choice of boys can be associated with each choice of girls.

$$C(8, 3) \cdot C(7, 2) = \frac{8!}{3!5!} \cdot \frac{7!}{2!5!} = (56)(21) = 1176$$

PROGRESS CHECK 9
From 5 representatives of District A and 8 representatives of District B, in how many ways can 4 be chosen if only 1 representative from District A is to be included?

Answer
280

EXAMPLE 10
A bookstore has 12 French books and 9 German books. In how many ways can a group of 6 books, consisting of 4 French books and 2 German books, be placed on a shelf?

Solution
The French books can be selected in $C(12, 4)$ ways and the German books in $C(9, 2)$ ways. The 6 books can then be selected in $C(12, 4) \cdot C(9, 2)$ ways. Each *selection* of 6 books can then be arranged on the shelf in $P(6, 6)$ ways so that the total number of arrangements is

$$C(12, 4) \cdot C(9, 2) \cdot P(6, 6) = \frac{12!}{4!8!} \cdot \frac{9!}{2!7!} \cdot \frac{6!}{(6-6)!} = 495 \cdot 36 \cdot 720 = 12,830,400$$

PROGRESS CHECK 10
From 6 different consonants and 4 different vowels, how many 5-letter words can be made consisting of 3 consonants and 2 vowels? (Assume every arrangement is a "word.")

Answer
14,400

EXERCISE SET 15.4
1. How many different five-digit numbers can be formed using the digits 1, 3, 4, 6, 8?
2. How many different ways are there to arrange the letters in the word *study?*
3. An employee identification number consists of two letters of the alphabet followed by a sequence of three digits selected from the digits 2, 3, 5, 6, 8, and 9. If repetitions are allowed, how many different identification numbers are possible?
4. In a psychological experiment, a subject has to arrange a cube, square, triangle, and rhombus in a row. How many different arrangements are possible?
5. A coin is tossed eight times and the result of each toss is recorded. How many different sequences of heads and tails are possible?
6. A die (from a pair of dice) is tossed four times and the result of each toss is recorded. How many different sequences are possible?
7. A concert is to consist of 3 guitar pieces, 2 vocal numbers, and 2 jazz selections. In how many ways can the program be arranged?

Compute.

8. $P(6, 6)$ 9. $P(6, 5)$ 10. $P(4, 2)$

11. $P(8, 3)$ 12. $P(5, 2)$ 13. $P(10, 2)$

14. $P(8, 4)$ 15. $\dfrac{P(9, 3)}{3!}$ 16. $\dfrac{4P(12, 3)}{2!}$

17. $P(3, 1)$ 18. $\dfrac{P(7, 3)}{2!}$ 19. $\dfrac{P(10, 4)}{4!}$

20. Find the number of ways in which 5 men and 5 women can be seated in a row
     (a) if any person may sit next to any other person.
     (b) if a man must always sit next to a woman.
21. Find the number of distinguishable permutations of the letters of the word
     money.
22. Find the number of distinguishable permutations of the letters of the word
     goose.
23. Find the number of distinguishable permutations of the letters of the word
     needed.
24. How many permutations of the letters a, b, e, g, h, k, and m are there when taken
     (a) two at a time?
     (b) three at a time?
25. How many three-letter labels of new chemical products can be formed from the
     letters a, b, c, d, f, g, l, and m?
26. Find the number of distinguishable permutations that can be formed from the
     letters of the word *Mississippi* taken four at a time.
27. A family consisting of a mother, father, and three children is having a picture
     taken. If all 5 people are arranged in a row, how many different photographs
     can be taken?
28. List all the combinations of the numbers 4, 3, 5, 8, and 9 taken three at a time.

Compute.

29. $C(9, 3)$ 30. $C(7, 3)$ 31. $C(10, 2)$
32. $C(7, 1)$ 33. $C(7, 7)$ 34. $C(5, 4)$
35. $C(n, n - 1)$ 36. $C(n, n - 2)$ 37. $C(n + 1, n - 1)$
38. In how many ways can a committee of 2 faculty members and 3 students be
     selected from 8 faculty members and 10 students?
39. In how many ways can a basketball team of 5 players be selected from among 15
     candidates?
40. In how many ways can a four-card hand be dealt from a deck of 52 cards?
41. How many three-letter moped plates on a local campus can be formed
     (a) if no letters can be repeated?
     (b) if letters can be repeated?
42. In a certain city each police car is staffed by two officers: one male and one
     female. A police captain, who needs to staff 8 cars, has 15 male officers and 12
     female officers available. How many different teams can be formed?
43. How many different 10-card hands with 4 aces can be dealt from a deck of 52
     cards?
44. A car manufacturer makes three different models, each of which is available in
     five different colors and with two different engines. How many cars must a
     dealer stock in the showroom to display the full line?
45. A penny, nickel, dime, quarter, half-dollar, and silver dollar are to be arranged

in a row. How many different arrangements can be formed if the penny and dime must always be next to each other?

46. An automobile manufacturer who is planning an advertising campaign is considering seven newspapers, two magazines, three radio stations, and four television stations. In how many ways can five advertisements be placed
    (a) if all five are to be in newspapers?
    (b) if two are to be in newspapers, two on radio, and one on television?

47. In a certain police station there are 12 prisoners and 10 police officers. How many possible line-ups consisting of 4 prisoners and 3 officers can be formed?

48. The notation $\binom{n}{r}$ is often used in place of $C(n, r)$. Show that $\binom{n}{r} = \binom{n}{n - r}$.

49. How many different 10-card hands with 6 red cards and 4 black cards can be dealt from a deck of 52 cards?

50. A bin contains 12 transistors, 7 of which are defective. In how many ways can four transistors be chosen so that
    (a) all four are defective?
    (b) two are good and two are defective?
    (c) all four are good?
    (d) three are defective and one is good?

TERMS AND SYMBOLS

term (p. 442)

infinite sequence (p. 442)

a_n (p. 442)

finite sequence (p. 442)

arithmetic progression (p. 443)

arithmetic sequence (p. 443)

common difference (p. 443)

recursive formula (p. 443)

series (p. 445)

arithmetic series (p. 445)

geometric progression (p. 449)

geometric sequence (p. 449)

common ratio (p. 449)

geometric means (p. 450)

geometric series (p. 451)

infinite geometric series (p. 453)

expansion of $(a + b)^n$ (p. 456)

Pascal's triangle (p. 456)

binomial theorem (p. 457)

binomial formula (p. 457)

factorial (p. 458)

$n!$ (p. 458)

permutation (p. 462)

$P(n, r)$ (p. 463)

combination (p. 465)

$C(n, r)$ (p. 465)

KEY IDEAS FOR REVIEW

☐ A sequence is a function whose domain is restricted to the set of natural numbers. We generally write a sequence by using subscript notation, that is, a_n replaces $a(n)$.

☐ An arithmetic progression has a common difference d between terms. We can define an arithmetic progression recursively by writing $a_n = a_{n-1} + d$ and specifying a_1.

☐ A geometric progression has a common ratio r between terms. We can define a geometric progression recursively by writing $a_n = ra_{n-1}$ and specifying a_1.

☐ The formulas for the nth term of an arithmetic or geometric progression are

$$a_n = a_1 + (n - 1)d \quad \text{Arithmetic}$$

$$a_n = a_1 r^{n-1} \quad \text{Geometric}$$

☐ A series is the sum of the terms of a sequence.

☐ The formulas for the sums S_n of the first n terms of arithmetic and geometric progressions are

$$S_n = \frac{n}{2}(a_1 + a_n) \qquad \text{Arithmetic}$$

$$S_n = \frac{n}{2}[2a_1 + (n-1)d] \qquad \text{Arithmetic}$$

$$S_n = \frac{a_1(1 - r^n)}{1 - r} \qquad \text{Geometric}$$

☐ If the common ratio r satisfies $-1 < r < 1$, then the infinite geometric series has the sum S given by

$$S = \frac{a_1}{1 - r}$$

☐ The notation $n!$ indicates the product of the natural numbers 1 through n.

$$n! = n(n-1)(n-2)\cdots 2 \cdot 1 \quad \text{for} \quad n \geq 1$$

$$0! = 1$$

☐ The binomial formula provides the terms of the expansion of $(a + b)^n$.

$$(a + b)^n = a^n + \frac{n!}{1!(n-1)!}a^{n-1}b + \frac{n!}{2!(n-2)!}a^{n-2}b^2$$

$$+ \frac{n!}{3!(n-3)!}a^{n-3}b^3 + \cdots + \frac{n!}{r!(n-r)!}a^{n-r}b^r$$

$$+ \cdots + b^n$$

☐ Permutations involve arrangements of order of objects; thus, abc and bac are distinct permutations of the letters a, b, c.

☐ Combinations involve selection of objects; the order is not significant. If we are selecting three letters from a box containing the letters a, b, c, d, then abc and bac are the same combination.

☐ The formulas for counting permutations and combinations of n objects taken r at a time are

$$P(n, r) = \frac{n!}{(n-r)!} \qquad \text{Permutations}$$

$$C(n, r) = \frac{n!}{r!(n-r)!} \qquad \text{Combinations}$$

COMMON ERRORS

1. The sum of the first n terms of a geometric progression is given by

$$S_n = \frac{a_1(1 - r^n)}{1 - r}$$

Don't write

$$S_n = \frac{a_1(1 - r)^n}{1 - r}$$

2. The formula for the sum of the terms of an infinite geometric progression can only be used if $|r| < 1$.

PROGRESS TEST 15A

1. Write the first four terms of the sequence whose nth term is

$$a_n = \frac{n + 2}{n^2 - 2}$$

2. Find the 8th term of the progression $3, \frac{5}{2}, 2, \ldots$.
3. Find the 16th term of the arithmetic progression whose first and 30th terms are 2 and 60, respectively.
4. Find the sum of the first 12 terms of an arithmetic series whose first term is -1 and whose common difference is $\frac{1}{2}$.
5. The first term of an arithmetic series is -5, the last term is 35, and the sum is 165. Find the number of terms and the common difference.
6. Find the 7th term of the progression $16, 8, 4, \ldots$.
7. Find the sum of the first six terms of the geometric progression whose first term is 6 and whose common ratio is $\frac{1}{2}$.
8. Find the sum of the infinite geometric series $5 + \frac{5}{2} + \frac{5}{4} \cdots$.
9. Expand and simplify $\left(x - \frac{y}{2}\right)^4$.
10. Evaluate $\dfrac{9!6!}{10!5!}$.
11. Five paintings are to be hung on a wall in a side-by-side arrangement. If seven paintings are available, how many arrangements are possible?
12. A row in a stamp album provides space for 5 postage stamps. There are 6 airmail and 5 regular postage stamps available. How many arrangements can be made consisting of 3 airmail and 2 regular postage stamps if the stamps of each type must be kept together?
13. In how many ways can a committee of 5 be selected from a group of 10 people if the past chairperson must serve on the committee?
14. An army task requires 2 privates, 2 sergeants, and a major. If 5 privates, 4 sergeants, and 3 majors are available, in how many ways can a team be formed?

PROGRESS TEST 15B

1. Write the first four terms of the sequence whose nth term is

$$a_n = n^2 + \frac{1}{n}$$

2. Find the 11th term of the progression $4, \frac{17}{4}, \frac{9}{2}, \ldots$.
3. Find the 40th term of the arithmetic progression whose first and 21st terms are 5 and 15, respectively.
4. Find the sum of the first 15 terms of an arithmetic series whose first term is 4 and whose common difference is $-\frac{1}{4}$.
5. The first term of an arithmetic series is -3, the last term is -39, and the sum is 525. Find the number of terms and the common difference.
6. Find the 6th term of the progression $1, 3, 9, \ldots$.
7. Find the sum of the first five terms of the geometric progression whose first term is 8 and whose common ratio is $-\frac{1}{2}$.
8. Find the sum of the infinite geometric series $3 - 2 + \frac{4}{3} + \cdots$.
9. Expand and simplify $\left(\frac{x^2}{2} - y\right)^4$.

10. Simplify $\dfrac{7!(n-2)!}{6!n!}$.

11. There is a space available for three books on a shelf. If there are six books from which to select, how many arrangements can be made?

12. A store clerk wants to display 3 symphonic and 3 jazz records on a shelf. There are 6 symphonic and 5 jazz records available. How many arrangements of the records are possible if the records of each type must be kept together?

13. How many different committees of 6 persons can be chosen if 9 people are available?

14. The capitals A, B, C, consonants m, n, p, r, and vowels a, e, i, o, u can be used for making words. If each word must begin with a capital and must contain 3 consonants and 2 vowels, how many words can be made?

APPENDIX/TABLES

TABLE I Exponentials and Their Reciprocals

x	e^x	e^{-x}	x	e^x	e^{-x}
0.00	1.0000	1.0000	1.4	4.0552	0.2466
0.01	1.0101	0.9900	1.5	4.4817	0.2231
0.02	1.0202	0.9802	1.6	4.9530	0.2019
0.03	1.0305	0.9704	1.7	5.4739	0.1827
0.04	1.0408	0.9608	1.8	6.0496	0.1653
0.05	1.0513	0.9512	1.9	6.6859	0.1496
0.06	1.0618	0.9418	2.0	7.3891	0.1353
0.07	1.0725	0.9324	2.1	8.1662	0.1225
0.08	1.0833	0.9231	2.2	9.0250	0.1108
0.09	1.0942	0.9139	2.3	9.9742	0.1003
0.10	1.1052	0.9048	2.4	11.023	0.0907
0.11	1.1163	0.8958	2.5	12.182	0.0821
0.12	1.1275	0.8869	2.6	13.464	0.0743
0.13	1.1388	0.8781	2.7	14.880	0.0672
0.14	1.1503	0.8694	2.8	16.445	0.0608
0.15	1.1618	0.8607	2.9	18.174	0.0550
0.16	1.1735	0.8521	3.0	20.086	0.0498
0.17	1.1853	0.8437	3.1	22.198	0.0450
0.18	1.1972	0.8353	3.2	24.533	0.0408
0.19	1.2092	0.8270	3.3	27.113	0.0369
0.20	1.2214	0.8187	3.4	29.964	0.0334
0.21	1.2337	0.8106	3.5	33.115	0.0302
0.22	1.2461	0.8025	3.6	36.598	0.0273
0.23	1.2586	0.7945	3.7	40.447	0.0247
0.24	1.2712	0.7866	3.8	44.701	0.0224
0.25	1.2840	0.7788	3.9	49.402	0.0202
0.26	1.2969	0.7711	4.0	54.598	0.0183
0.27	1.3100	0.7634	4.1	60.340	0.0166
0.28	1.3231	0.7558	4.2	66.686	0.0150
0.29	1.3364	0.7483	4.3	73.700	0.0136
0.30	1.3499	0.7408	4.4	81.451	0.0123
0.35	1.4191	0.7047	4.5	90.017	0.0111
0.40	1.4918	0.6703	4.6	99.484	0.0101
0.45	1.5683	0.6376	4.7	109.95	0.0091
0.50	1.6487	0.6065	4.8	121.51	0.0082
0.55	1.7333	0.5769	4.9	134.29	0.0074
0.60	1.8221	0.5488	5	148.41	0.0067
0.65	1.9155	0.5220	6	403.43	0.0025
0.70	2.0138	0.4966	7	1,096.6	0.0009
0.75	2.1170	0.4724	8	2,981.0	0.0003
0.80	2.2255	0.4493	9	8,103.1	0.0001
0.85	2.3396	0.4274	10	22,026	0.00005
0.90	2.4596	0.4066	11	59,874	0.00002
0.95	2.5857	0.3867	12	162,754	0.000006
1.0	2.7183	0.3679	13	442,413	0.000002
1.1	3.0042	0.3329	14	1,202,604	0.0000008
1.2	3.3201	0.3012	15	3,269,017	0.0000003
1.3	3.6693	0.2725			

TABLE II Common Logarithms

N	0	1	2	3	4	5	6	7	8	9
1.0	.0000	.0043	.0086	.0128	.0170	.0212	.0253	.0294	.0334	.0374
1.1	.0414	.0453	.0492	.0531	.0569	.0607	.0645	.0682	.0719	.0755
1.2	.0792	.0828	.0864	.0899	.0934	.0969	.1004	.1038	.1072	.1106
1.3	.1139	.1173	.1206	.1239	.1271	.1303	.1335	.1367	.1399	.1430
1.4	.1461	.1492	.1523	.1553	.1584	.1614	.1644	.1673	.1703	.1732
1.5	.1761	.1790	.1818	.1847	.1875	.1903	.1931	.1959	.1987	.2014
1.6	.2041	.2068	.2095	.2122	.2148	.2175	.2201	.2227	.2253	.2279
1.7	.2304	.2330	.2355	.2380	.2405	.2430	.2455	.2480	.2504	.2529
1.8	.2553	.2577	.2601	.2625	.2648	.2672	.2695	.2718	.2742	.2765
1.9	.2788	.2810	.2833	.2856	.2878	.2900	.2923	.2945	.2967	.2989
2.0	.3010	.3032	.3054	.3075	.3096	.3118	.3139	.3160	.3181	.3201
2.1	.3222	.3243	.3263	.3284	.3304	.3324	.3345	.3365	.3385	.3404
2.2	.3424	.3444	.3464	.3483	.3502	.3522	.3541	.3560	.3579	.3598
2.3	.3617	.3636	.3655	.3674	.3692	.3711	.3729	.3747	.3766	.3784
2.4	.3802	.3820	.3838	.3856	.3874	.3892	.3909	.3927	.3945	.3692
2.5	.3979	.3997	.4014	.4031	.4048	.4065	.4082	.4099	.4116	.4133
2.6	.4150	.4166	.4183	.4200	.4216	.4232	.4249	.4265	.4281	.4298
2.7	.4314	.4330	.4346	.4362	.4378	.4393	.4409	.4425	.4440	.4456
2.8	.4472	.4487	.4502	.4518	.4533	.4548	.4564	.4579	.4594	.4609
2.9	.4624	.4639	.4654	.4669	.4683	.4698	.4713	.4728	.4742	.4757
3.0	.4771	.4786	.4800	.4814	.4829	.4843	.4857	.4871	.4886	.4900
3.1	.4914	.4928	.4942	.4955	.4969	.4983	.4997	.5011	.5024	.5038
3.2	.5051	.5065	.5079	.5092	.5105	.5119	.5132	.5145	.5159	.5172
3.3	.5185	.5198	.5211	.5224	.5237	.5250	.5263	.5276	.5289	.5302
3.4	.5315	.5328	.5340	.5353	.5366	.5378	.5391	.5403	.5416	.5428
3.5	.5441	.5453	.5465	.5478	.5490	.5502	.5514	.5527	.5539	.5551
3.6	.5563	.5575	.5587	.5599	.5611	.5623	.5635	.5647	.5658	.5670
3.7	.5682	.5694	.5705	.5717	.5729	.5740	.5752	.5763	.5775	.5786
3.8	.5798	.5809	.5821	.5832	.5843	.5855	.5866	.5877	.5888	.5899
3.9	.5911	.5922	.5933	.5944	.5955	.5966	.5977	.5988	.5999	.6010
4.0	.6021	.6031	.6042	.6053	.6064	.6075	.6085	.6096	.6107	.6117
4.1	.6128	.6138	.6149	.6160	.6170	.6180	.6191	.6201	.6212	.6222
4.2	.6232	.6243	.6253	.6263	.6274	.6284	.6294	.6304	.6314	.6325
4.3	.6335	.6345	.6355	.6365	.6375	.6385	.6395	.6405	.6415	.6425
4.4	.6435	.6444	.6454	.6464	.6474	.6484	.6493	.6503	.6513	.6522
4.5	.6532	.6542	.6551	.6561	.6571	.6580	.6590	.6599	.6609	.6618
4.6	.6628	.6637	.6646	.6656	.6665	.6675	.6684	.6693	.6702	.6712
4.7	.6721	.6730	.6739	.6749	.6758	.6767	.6776	.6785	.6794	.6803
4.8	.6812	.6821	.6830	.6839	.6848	.6857	.6866	.6875	.6884	.6893
4.9	.6902	.6911	.6920	.6928	.6937	.6946	.6955	.6964	.6972	.6981
5.0	.6990	.6998	.7007	.7016	.7024	.7033	.7042	.7050	.7059	.7067
5.1	.7076	.7084	.7093	.7101	.7110	.7118	.7126	.7135	.7143	.7152
5.2	.7160	.7168	.7177	.7185	.7193	.7202	.7210	.7218	.7226	.7235
5.3	.7243	.7251	.7259	.7267	.7275	.7284	.7292	.7300	.7308	.7316
5.4	.7324	.7332	.7340	.7348	.7356	.7364	.7372	.7380	.7388	.7396

TABLE II *(continued)*

N	0	1	2	3	4	5	6	7	8	9
5.5	.7404	.7412	.7419	.7427	.7435	.7443	.7451	.7459	.7466	.7474
5.6	.7482	.7490	.7497	.7505	.7513	.7520	.7528	.7536	.7543	.7551
5.7	.7559	.7566	.7574	.7582	.7589	.7597	.7604	.7612	.7619	.7627
5.8	.7634	.7642	.7649	.7657	.7664	.7672	.7679	.7686	.7694	.7701
5.9	.7709	.7716	.7723	.7731	.7738	.7745	.7752	.7760	.7767	.7774
6.0	.7782	.7789	.7796	.7803	.7810	.7818	.7825	.7832	.7839	.7846
6.1	.7853	.7860	.7868	.7875	.7882	.7889	.7896	.7903	.7910	.7917
6.2	.7924	.7931	.7938	.7945	.7952	.7959	.7966	.7973	.7980	.7987
6.3	.7993	.8000	.8007	.8014	.8021	.8028	.8035	.8041	.8048	.8055
6.4	.8062	.8069	.8075	.8082	.8089	.8096	.8102	.8109	.8116	.8122
6.5	.8129	.8136	.8142	.8149	.8156	.8162	.8169	.8176	.8182	.8189
6.6	.8195	.8202	.8209	.8215	.8222	.8228	.8235	.8241	.8248	.8254
6.7	.8261	.8267	.8274	.8280	.8287	.8293	.8299	.8306	.8312	.8319
6.8	.8325	.8331	.8338	.8344	.8351	.8357	.8363	.8370	.8376	.8382
6.9	.8388	.8395	.8401	.8407	.8414	.8420	.8426	.8432	.8439	.8445
7.0	.8451	.8457	.8463	.8470	.8476	.8482	.8488	.8494	.8500	.8506
7.1	.8513	.8519	.8525	.8531	.8537	.8543	.8549	.8555	.8561	.8567
7.2	.8573	.8579	.8585	.8591	.8597	.8603	.8609	.8615	.8621	.8627
7.3	.8633	.8639	.8645	.8651	.8657	.8663	.8669	.8675	.8681	.8686
7.4	.8692	.8698	.8704	.8710	.8716	.8722	.8727	.8733	.8739	.8745
7.5	.8751	.8756	.8762	.8768	.8774	.8779	.8785	.8791	.8797	.8802
7.6	.8808	.8814	.8820	.8825	.8831	.8837	.8842	.8848	.8854	.8859
7.7	.8865	.8871	.8876	.8882	.8887	.8893	.8899	.8904	.8910	.8915
7.8	.8921	.8927	.8932	.8938	.8943	.8949	.8954	.8960	.8965	.8971
7.9	.8976	.8982	.8987	.8993	.8998	.9004	.9009	.9015	.9020	.9025
8.0	.9031	.9036	.9042	.9047	.9053	.9058	.9063	.9069	.9074	.9079
8.1	.9085	.9090	.9096	.9101	.9106	.9112	.9117	.9122	.9128	.9133
8.2	.9138	.9143	.9149	.9154	.9159	.9165	.9170	.9175	.9180	.9186
8.3	.9191	.9196	.9201	.9206	.9212	.9217	.9222	.9227	.9232	.9238
8.4	.9243	.9248	.9253	.9258	.9263	.9269	.9274	.9279	.9284	.9289
8.5	.9294	.9299	.9304	.9309	.9315	.9320	.9325	.9330	.9335	.9340
8.6	.9345	.9350	.9355	.9360	.9365	.9370	.9375	.9380	.9385	.9390
8.7	.9395	.9400	.9405	.9410	.9415	.9420	.9425	.9430	.9435	.9440
8.8	.9445	.9450	.9455	.9460	.9465	.9469	.9474	.9479	.9484	.9489
8.9	.9494	.9499	.9504	.9509	.9513	.9518	.9523	.9528	.9533	.9538
9.0	.9542	.9547	.9552	.9557	.9562	.9566	.9571	.9576	.9581	.9586
9.1	.9590	.9595	.9600	.9605	.9609	.9614	.9619	.9624	.9628	.9633
9.2	.9638	.9643	.9647	.9652	.9657	.9661	.9666	.9671	.9675	.9680
9.3	.9685	.9689	.9694	.9699	.9703	.9708	.9713	.9717	.9722	.9727
9.4	.9731	.9736	.9741	.9745	.9750	.9754	.9759	.9763	.9768	.9773
9.5	.9777	.9782	.9786	.9791	.9795	.9800	.9805	.9809	.9814	.9818
9.6	.9823	.9827	.9832	.9836	.9841	.9845	.9850	.9854	.9859	.9863
9.7	.9868	.9872	.9877	.9881	.9886	.9890	.9894	.9899	.9903	.9908
9.8	.9912	.9917	.9921	.9926	.9930	.9934	.9939	.9943	.9948	.9952
9.9	.9956	.9961	.9965	.9969	.9974	.9978	.9983	.9987	.9991	.9996

TABLE III Natural Logarithms

N	$\ln N$	N	$\ln N$	N	$\ln N$
		4.5	1.5041	9.0	2.1972
0.1	−2.3026	4.6	1.5261	9.1	2.2083
0.2	−1.6094	4.7	1.5476	9.2	2.2192
0.3	−1.2040	4.8	1.5686	9.3	2.2300
0.4	−0.9163	4.9	1.5892	9.4	2.2407
0.5	−0.6931	5.0	1.6094	9.5	2.2513
0.6	−0.5108	5.1	1.6292	9.6	2.2618
0.7	−0.3567	5.2	1.6487	9.7	2.2721
0.8	−0.2231	5.3	1.6677	9.8	2.2824
0.9	−0.1054	5.4	1.6864	9.9	2.2925
1.0	0.0000	5.5	1.7047	10	2.3026
1.1	0.0953	5.6	1.7228	11	2.3979
1.2	0.1823	5.7	1.7405	12	2.4849
1.3	0.2624	5.8	1.7579	13	2.5649
1.4	0.3365	5.9	1.7750	14	2.6391
1.5	0.4055	6.0	1.7918	15	2.7081
1.6	0.4700	6.1	1.8083	16	2.7726
1.7	0.5306	6.2	1.8245	17	2.8332
1.8	0.5878	6.3	1.8405	18	2.8904
1.9	0.6419	6.4	1.8563	19	2.9444
2.0	0.6931	6.5	1.8718	20	2.9957
2.1	0.7419	6.6	1.8871	25	3.2189
2.2	0.7885	6.7	1.9021	30	3.4012
2.3	0.8329	6.8	1.9169	35	3.5553
2.4	0.8755	6.9	1.9315	40	3.6889
2.5	0.9163	7.0	1.9459	45	3.8067
2.6	0.9555	7.1	1.9601	50	3.9120
2.7	0.9933	7.2	1.9741	55	4.0073
2.8	1.0296	7.3	1.9879	60	4.0943
2.9	1.0647	7.4	2.0015	65	4.1744
3.0	1.0986	7.5	2.0149	70	4.2485
3.1	1.1314	7.6	2.0281	75	4.3175
3.2	1.1632	7.7	2.0412	80	4.3820
3.3	1.1939	7.8	2.0541	85	4.4427
3.4	1.2238	7.9	2.0669	90	4.4998
3.5	1.2528	8.0	2.0794	95	4.5539
3.6	1.2809	8.1	2.0919	100	4.6052
3.7	1.3083	8.2	2.1041		
3.8	1.3350	8.3	2.1163		
3.9	1.3610	8.4	2.1282		
4.0	1.3863	8.5	2.1401		
4.1	1.4110	8.6	2.1518		
4.2	1.4351	8.7	2.1633		
4.3	1.4586	8.8	2.1748		
4.4	1.4816	8.9	2.1861		

TABLE IV Interest Rates

	$i = \frac{1}{2}\%$				$i = 1\%$				$i = 1\frac{1}{2}\%$		
n	$(1+i)^n$	n	$(1+i)^n$	n	$(1+i)^n$	n	$(1+i)^n$	n	$(1+i)^n$	n	$(1+i)^n$
1	1.0050 0000	51	1.2896 4194	1	1.0100 0000	51	1.6610 7814	1	1.0150 0000	51	2.1368 2106
2	1.0100 2500	52	1.2960 9015	2	1.0201 0000	52	1.6776 8892	2	1.0302 2500	52	2.1688 7337
3	1.0150 7513	53	1.3025 7060	3	1.0303 0100	53	1.6944 6581	3	1.0456 7838	53	2.2014 0647
4	1.0201 5050	54	1.3090 8346	4	1.0406 0401	54	1.7114 1047	4	1.0613 6355	54	2.2344 2757
5	1.0252 5125	55	1.3156 2887	5	1.0510 1005	55	1.7285 2457	5	1.0772 8400	55	2.2679 4398
6	1.0303 7751	56	1.3222 0702	6	1.0615 2015	56	1.7458 0982	6	1.0934 4326	56	2.3019 6314
7	1.0355 2940	57	1.3288 1805	7	1.0721 3535	57	1.7632 6792	7	1.1098 4491	57	2.3364 9259
8	1.0407 0704	58	1.3354 6214	8	1.0828 5671	58	1.7809 0060	8	1.1264 9259	58	2.3715 3998
9	1.0459 1058	59	1.3421 3946	9	1.0936 8527	59	1.7987 0960	9	1.1433 8998	59	2.4071 1308
10	1.0511 4013	60	1.3488 5015	10	1.1046 2213	60	1.8166 9670	10	1.1605 4083	60	2.4432 1978
11	1.0563 9583	61	1.3555 9440	11	1.1156 6835	61	1.8348 6367	11	1.1779 4894	61	2.4798 6807
12	1.0616 7781	62	1.3623 7238	12	1.1268 2503	62	1.8532 1230	12	1.1956 1817	62	2.5170 6609
13	1.0669 8620	63	1.3691 8424	13	1.1380 9328	63	1.8717 4443	13	1.2135 5244	63	2.5548 2208
14	1.0723 2113	64	1.3760 3016	14	1.1494 7421	64	1.8904 6187	14	1.2317 5573	64	2.5931 4442
15	1.0776 8274	65	1.3829 1031	15	1.1609 6896	65	1.9093 6649	15	1.2502 3207	65	2.6320 4158
16	1.0830 7115	66	1.3898 2486	16	1.1725 7864	66	1.9284 6015	16	1.2689 8555	66	2.6715 2221
17	1.0884 8651	67	1.3967 7399	17	1.1843 0443	67	1.9477 4475	17	1.2880 2033	67	2.7115 9504
18	1.0939 2894	68	1.4037 5785	18	1.1961 4748	68	1.9672 2220	18	1.3073 4064	68	2.7522 6896
19	1.0993 9858	69	1.4107 7664	19	1.2081 0895	69	1.9868 9442	19	1.3269 5075	69	2.7935 5300
20	1.1048 9558	70	1.4178 3053	20	1.2201 9004	70	2.0067 6337	20	1.3468 5501	70	2.8354 5629
21	1.1104 2006	71	1.4249 1968	21	1.2323 9194	71	2.0268 3100	21	1.3670 5783	71	2.8779 8814
22	1.1159 7216	72	1.4320 4428	22	1.2447 1586	72	2.0470 9931	22	1.3875 6370	72	2.9211 5796
23	1.1215 5202	73	1.4392 0450	23	1.2571 6302	73	2.0675 7031	23	1.4083 7715	73	2.9649 7533
24	1.1271 5978	74	1.4464 0052	24	1.2697 3465	74	2.0882 4601	24	1.4295 0281	74	3.0094 4996
25	1.1327 9558	75	1.4536 3252	25	1.2824 3200	75	2.1091 2847	25	1.4509 4535	75	3.0545 9171
26	1.1384 5955	76	1.4609 0069	26	1.2952 5631	76	2.1302 1975	26	1.4727 0953	76	3.1004 1059
27	1.1441 5185	77	1.4682 0519	27	1.3082 0888	77	2.1515 2195	27	1.4948 0018	77	3.1469 1674
28	1.1498 7261	78	1.4755 4622	28	1.3212 9097	78	2.1730 3717	28	1.5172 2218	78	3.1941 2050
29	1.1556 2197	79	1.4829 2395	29	1.3345 0388	79	2.1947 6754	29	1.5399 8051	79	3.2420 3230
30	1.1614 0008	80	1.4903 3857	30	1.3478 4892	80	2.2167 1522	30	1.5630 8022	80	3.2906 6279
31	1.1672 0708	81	1.4977 9026	31	1.3613 2740	81	2.2388 8237	31	1.5865 2642	81	3.3400 2273
32	1.1730 4312	82	1.5052 7921	32	1.3749 4068	82	2.2612 7119	32	1.6103 2432	82	3.3901 2307
33	1.1789 0833	83	1.5128 0561	33	1.3886 9009	83	2.2838 8390	33	1.6344 7918	83	3.4409 7492
34	1.1848 0288	84	1.5203 6964	34	1.4025 7699	84	2.3067 2274	34	1.6589 9637	84	3.4925 8954
35	1.1907 2689	85	1.5279 7148	35	1.4166 0276	85	2.3297 8997	35	1.6838 8132	85	3.5449 7838
36	1.1966 8052	86	1.5356 1134	36	1.4307 6878	86	2.3530 8787	36	1.7091 3954	86	3.5981 5306
37	1.2026 6393	87	1.5432 8940	37	1.4450 7647	87	2.3766 1875	37	1.7347 7663	87	3.6521 2535
38	1.2086 7725	88	1.5510 0585	38	1.4595 2724	88	2.4003 8494	38	1.7607 9828	88	3.7069 0723
39	1.2147 2063	89	1.5587 6087	39	1.4741 2251	89	2.4243 8879	39	1.7872 1025	89	3.7625 1084
40	1.2207 9424	90	1.5665 5468	40	1.4888 6373	90	2.4486 3267	40	1.8140 1841	90	3.8189 4851
41	1.2268 9821	91	1.5743 8745	41	1.5037 5237	91	2.4731 1900	41	1.8412 2868	91	3.8762 3273
42	1.2330 3270	92	1.5822 5939	42	1.5187 8989	92	2.4978 5019	42	1.8688 4712	92	3.9343 7622
43	1.2391 9786	93	1.5901 7069	43	1.5339 7779	93	2.5228 2869	43	1.8968 7982	93	3.9933 9187
44	1.2453 9385	94	1.5981 2154	44	1.5493 1757	94	2.5480 5698	44	1.9253 3302	94	4.0532 9275
45	1.2516 2082	95	1.6061 1215	45	1.5648 1075	95	2.5735 3755	45	1.9542 1301	95	4.1140 9214
46	1.2578 7892	96	1.6141 4271	46	1.5804 5885	96	2.5992 7293	46	1.9835 2621	96	4.1758 0352
47	1.2641 6832	97	1.6222 1342	47	1.5962 6344	97	2.6252 6565	47	2.0132 7910	97	4.2384 4057
48	1.2704 8916	98	1.6303 2449	48	1.6122 2608	98	2.6515 1831	48	2.0434 7829	98	4.3020 1718
49	1.2768 4161	99	1.6384 7611	49	1.6283 4834	99	2.6780 3349	49	2.0741 3046	99	4.3665 4744
50	1.2832 2581	100	1.6466 6849	50	1.6446 3182	100	2.7048 1383	50	2.1052 4242	100	4.4320 4565

(continued)

TABLE IV (*continued*)

	$i = 2\%$				$i = 2\frac{1}{2}\%$				$i = 3\%$
n	$(1 + i)^n$	n	$(1 + i)^n$	n	$(1 + i)^n$	n	$(1 + i)^n$	n	$(1 + i)^n$
1	1.0200 0000	51	2.7454 1979	1	1.0250 0000	51	3.5230 3644	1	1.0300 0000
2	1.0404 0000	52	2.8003 2819	2	1.0506 2500	52	3.6111 1235	2	1.0609 0000
3	1.0612 0800	53	2.8563 3475	3	1.0768 9063	53	3.7013 9016	3	1.0927 2700
4	1.0824 3216	54	2.9134 6144	4	1.1038 1289	54	3.7939 2491	4	1.1255 0881
5	1.1040 8080	55	2.9717 3067	5	1.1314 0821	55	3.8887 7303	5	1.1592 7407
6	1.1261 6242	56	3.0311 6529	6	1.1596 9342	56	3.9859 9236	6	1.1940 5230
7	1.1486 8567	57	3.0917 8859	7	1.1886 8575	57	4.0856 4217	7	1.2298 7387
8	1.1716 5938	58	3.1536 2436	8	1.2184 0290	58	4.1877 8322	8	1.2667 7008
9	1.1950 9257	59	3.2166 9685	9	1.2488 6297	59	4.2924 7780	9	1.3047 7318
10	1.2189 9442	60	3.2810 3079	10	1.2800 8454	60	4.3997 8975	10	1.3439 1638
11	1.2433 7431	61	3.3466 5140	11	1.3120 8666	61	4.5097 8449	11	1.3842 3387
12	1.2682 4179	62	3.4135 8443	12	1.3448 8882	62	4.6225 2910	12	1.4257 6089
13	1.2936 0663	63	3.4818 5612	13	1.3785 1104	63	4.7380 9233	13	1.4685 3371
14	1.3194 7876	64	3.5514 9324	14	1.4129 7382	64	4.8565 4464	14	1.5125 8972
15	1.3458 6834	65	3.6225 2311	15	1.4482 9817	65	4.9779 5826	15	1.5579 6742
16	1.3727 8571	66	3.6949 7357	16	1.4845 0562	66	5.1024 0721	16	1.6047 0644
17	1.4002 4142	67	3.7688 7304	17	1.5216 1826	67	5.2299 6739	17	1.6528 4763
18	1.4282 4625	68	3.8442 5050	18	1.5596 5872	68	5.3607 1658	18	1.7024 3306
19	1.4568 1117	69	3.9211 3551	19	1.5986 5019	69	5.4947 3449	19	1.7535 0605
20	1.4859 4740	70	3.9995 5822	20	1.6386 1644	70	5.6321 0286	20	1.8061 1123
21	1.5156 6634	71	4.0795 4939	21	1.6795 8185	71	5.7729 0543	21	1.8602 9457
22	1.5459 7967	72	4.1611 4038	22	1.7215 7140	72	5.9172 2806	22	1.9161 0341
23	1.5768 9926	73	4.2443 6318	23	1.7646 1068	73	6.0651 5876	23	1.9735 8651
24	1.6084 3725	74	4.3292 5045	24	1.8087 2595	74	6.2167 8773	24	2.0327 9411
25	1.6406 0599	75	4.4158 3546	25	1.8539 4410	75	6.3722 0743	25	2.0937 7793
26	1.6734 1811	76	4.5041 5216	26	1.9002 9270	76	6.5315 1261	26	2.1565 9127
27	1.7068 8648	77	4.5942 3521	27	1.9478 0002	77	6.6948 0043	27	2.2212 8901
28	1.7410 2421	78	4.6861 1991	28	1.9964 9502	78	6.8621 7044	28	2.2879 2768
29	1.7758 4469	79	4.7798 4231	29	2.0464 0739	79	7.0337 2470	29	2.3565 6551
30	1.8113 6158	80	4.8754 3916	30	2.0975 6758	80	7.2095 6782	30	2.4272 6247
31	1.8475 8882	81	4.9729 4794	31	2.1500 0677	81	7.3898 0701	31	2.5000 8035
32	1.8845 4059	82	5.0724 0690	32	2.2037 5694	82	7.5745 5219	32	2.5750 8276
33	1.9222 3140	83	5.1738 5504	33	2.2588 5086	83	7.7639 1599	33	2.6523 3524
34	1.9606 7603	84	5.2773 3214	34	2.3153 2213	84	7.9580 1389	34	2.7319 0530
35	1.9998 8955	85	5.3828 7878	35	2.3732 0519	85	8.1569 6424	35	2.8138 6245
36	2.0398 8734	86	5.4905 3636	36	2.4325 3532	86	8.3608 8834	36	2.8982 7833
37	2.0806 8509	87	5.6003 4708	37	2.4933 4870	87	8.5699 1055	37	2.9852 2668
38	2.1222 9879	88	5.7123 5402	38	2.5556 8242	88	8.7841 5832	38	3.0747 8348
39	2.1647 4477	89	5.8266 0110	39	2.6195 7448	89	9.0037 6228	39	3.1670 2698
40	2.2080 3966	90	5.9431 3313	40	2.6850 6384	90	9.2288 5633	40	3.2620 3779
41	2.2522 0046	91	6.0619 9579	41	2.7521 9043	91	9.4595 7774	41	3.3598 9893
42	2.2972 4447	92	6.1832 3570	42	2.8209 9520	92	9.6960 6718	42	3.4606 9589
43	2.3431 8936	93	6.3069 0042	43	2.8915 2008	93	9.9384 6886	43	3.5645 1677
44	2.3900 5314	94	6.4330 3843	44	2.9638 0808	94	10.1869 3058	44	3.6714 5227
45	2.4378 5421	95	6.5616 9920	45	3.0379 0328	95	10.4416 0385	45	3.7815 9584
46	2.4866 1129	96	6.6929 3318	46	3.1138 5086	96	10.7026 4395	46	3.8950 4372
47	2.5363 4352	97	6.8267 9184	47	3.1916 9713	97	10.9702 1004	47	4.0118 9503
48	2.5870 7039	98	6.9633 2768	48	3.2714 8956	98	11.2444 6530	48	4.1322 5188
49	2.6388 1179	99	7.1025 9423	49	3.3532 7680	99	11.5255 7693	49	4.2562 1944
50	2.6915 8803	100	7.2446 4612	50	3.4371 0872	100	11.8137 1635	50	4.3839 0602

TABLE IV (*continued*)

n	$(1+i)^n$	n	$(1+i)^n$	n	$(1+i)^n$	n	$(1+i)^n$	n	$(1+i)^n$
	$i=4\%$		$i=5\%$		$i=6\%$		$i=7\%$		$i=8\%$
1	1.0400 0000	1	1.0500 0000	1	1.0600 0000	1	1.0700 0000	1	1.0800 0000
2	1.0816 0000	2	1.1025 0000	2	1.1236 0000	2	1.1449 0000	2	1.1664 0000
3	1.1248 6400	3	1.1576 2500	3	1.1910 1600	3	1.2250 4300	3	1.2597 1200
4	1.1698 5856	4	1.2155 0625	4	1.2624 7696	4	1.3107 9601	4	1.3604 8896
5	1.2166 5290	5	1.2762 8156	5	1.3382 2558	5	1.4025 5173	5	1.4693 2808
6	1.2653 1902	6	1.3400 9564	6	1.4185 1911	6	1.5007 3035	6	1.5868 7432
7	1.3159 3178	7	1.4071 0042	7	1.5036 3026	7	1.6057 8148	7	1.7138 2427
8	1.3685 6905	8	1.4774 5544	8	1.5938 4807	8	1.7181 8618	8	1.8509 3021
9	1.4233 1181	9	1.5513 2822	9	1.6894 7896	9	1.8384 5921	9	1.9990 0463
10	1.4802 4428	10	1.6288 9463	10	1.7908 4770	10	1.9671 5136	10	2.1589 2500
11	1.5394 5406	11	1.7103 3936	11	1.8982 9856	11	2.1048 5195	11	2.3316 3900
12	1.6010 3222	12	1.7958 5633	12	2.0121 9647	12	2.2521 9159	12	2.5181 7012
13	1.6650 7351	13	1.8856 4914	13	2.1329 2826	13	2.4098 4500	13	2.7196 2373
14	1.7316 7645	14	1.9799 3160	14	2.2609 0396	14	2.5785 3415	14	2.9371 9362
15	1.8009 4351	15	2.0789 2818	15	2.3965 5819	15	2.7590 3154	15	3.1721 6911
16	1.8729 8125	16	2.1828 7459	16	2.5403 5168	16	2.9521 6375	16	3.4259 4264
17	1.9479 0050	17	2.2920 1832	17	2.6927 7279	17	3.1588 1521	17	3.7000 1805
18	2.0258 1652	18	2.4066 1923	18	2.8543 3915	18	3.3799 3228	18	3.9960 1950
19	2.1068 4918	19	2.5269 5020	19	3.0255 9950	19	3.6165 2754	19	4.3157 0106
20	2.1911 2314	20	2.6532 9771	20	3.2071 3547	20	3.8696 8446	20	4.6609 5714
21	2.2787 6807	21	2.7859 6259	21	3.3995 6360	21	4.1405 6237	21	5.0338 3372
22	2.3699 1879	22	2.9252 6072	22	3.6035 3742	22	4.4304 0174	22	5.4365 4041
23	2.4647 1554	23	3.0715 2376	23	3.8197 4966	23	4.7405 2986	23	5.8714 6365
24	2.5633 0416	24	3.2250 9994	24	4.0489 3464	24	5.0723 6695	24	6.3411 8074
25	2.6658 3633	25	3.3863 5494	25	4.2918 7072	25	5.4274 3264	25	6.8484 7520
26	2.7724 6978	26	3.5556 7269	26	4.5493 8296	26	5.8073 5292	26	7.3963 5321
27	2.8833 6858	27	3.7334 5632	27	4.8223 4594	27	6.2138 6763	27	7.9880 6147
28	2.9987 0332	28	3.9201 2914	28	5.1116 8670	28	6.6488 3836	28	8.6271 0639
29	3.1186 5145	29	4.1161 3560	29	5.4183 8790	29	7.1142 5705	29	9.3172 7490
30	3.2433 9751	30	4.3219 4238	30	5.7434 9117	30	7.6122 5504	30	10.0626 5689
31	3.3731 3341	31	4.5380 3949	31	6.0881 0064	31	8.1451 1290	31	10.8676 6944
32	3.5080 5875	32	4.7649 4147	32	6.4533 8668	32	8.7152 7080	32	11.7370 8300
33	3.6483 8110	33	5.0031 8854	33	6.8405 8988	33	9.3253 3975	33	12.6760 4964
34	3.7943 1634	34	5.2533 4797	34	7.2510 2528	34	9.9781 1354	34	13.6901 3361
35	3.9460 8899	35	5.5160 1537	35	7.6860 8679	35	10.6765 8148	35	14.7853 4429
36	4.1039 3255	36	5.7918 1614	36	8.1472 5200	36	11.4239 4219	36	15.9681 7184
37	4.2680 8986	37	6.0814 0694	37	8.6360 8712	37	12.2236 1814	37	17.2456 2558
38	4.4388 1345	38	6.3854 7729	38	9.1542 5235	38	13.0792 7141	38	18.6252 7563
39	4.6163 6599	39	6.7047 5115	39	9.7035 0749	39	13.9948 2041	39	20.1152 9768
40	4.8010 2063	40	7.0399 8871	40	10.2857 1794	40	14.9744 5784	40	21.7245 2150
41	4.9930 6145	41	7.3919 8815	41	10.9028 6101	41	16.0226 6989	41	23.4624 8322
42	5.1927 8391	42	7.7615 8756	42	11.5570 3267	42	17.1442 5678	42	25.3394 8187
43	5.4004 9527	43	8.1496 6693	43	12.2504 5463	43	18.3443 5475	43	27.3666 4042
44	5.6165 1508	44	8.5571 5028	44	12.9854 8191	44	19.6284 5959	44	29.5559 7166
45	5.8411 7568	45	8.9850 0779	45	13.7646 1083	45	21.0024 5176	45	31.9204 4939
46	6.0748 2271	46	9.4342 5818	46	14.5904 8748	46	22.4726 2338	46	34.4740 8534
47	6.3178 1562	47	9.9059 7109	47	15.4659 1673	47	24.0457 0702	47	37.2320 1217
48	6.5705 2824	48	10.4012 6965	48	16.3938 7173	48	25.7289 0651	48	40.2105 7314
49	6.8333 4937	49	10.9213 3313	49	17.3775 0403	49	27.5299 2997	49	43.4274 1899
50	7.1066 8335	50	11.4673 9979	50	18.4201 5427	50	29.4570 2506	50	46.9016 1251

ANSWERS TO ODD-NUMBERED EXERCISES AND PROGRESS TESTS

CHAPTER ONE

EXERCISE SET 1.1, page 5

1. a, b, c, d   3. a, c   5. a, b, c, d   7. a, c, d
9. c, e   11. c, e   13. F   15. F
17. T   19. T   21. T   23. F
25. F

27.

29. (a) 10   (b) -20 (c) 20   (d) -5 31. 0
33. -4 35. -5 37. 3   39. 0
41. 5

43.

EXERCISE SET 1.2, page 14

1. 4   3. 27   5. 0   7. 8/33
9. 10/3   11. 4/3   13. 3/2   15. 12
17. 7   19. 25   21. 17/12   23. 5/12
25. 8/23   27. 20/39   29. 1/5, 0.2   31. 131/200, 0.655
33. 6/125, 0.048   35. 6/5, 1.2   37. 1/500, 0.002

39. 5%	41. 42.5%	43. 628%	45. 60%
47. 225%	49. 28.57%	51. 21	53. 42
55. 0.2	57. $58	59. $6480	61. $15.18
63. $115,200	65. 8%		

EXERCISE SET 1.3, page 18

1. T	3. T	5. 11	7. 2
9. 64	11. 1/11	13. 10	15. 5
17. 33	19. 1/2	21. 12	23. 8/3
25. 5/4	27. 1/3	29. 18.84	31. 9.4

33. (a) $2160   (b) $2480   (c) $2080   (d) $2106.67

EXERCISE SET 1.4, page 22

1. 8	3. -7	5. 2	7. -2
9. 3	11. 2	13. 8	15. -5
17. 1	19. 2	21. -3	23. 10
25. $-1/2$	27. 4	29. -3	31. $-3/5$
33. 18	35. $-9/2$	37. 6	39. 0
41. 0	43. 3	45. -4	47. -12
49. $2x - 3y$	51. x/y	53. $-4/x$	55. -7
57. 2/3	59. 1	61. -8	63. 3
65. 12°C	67. $1200	69. $y - x$	

EXERCISE SET 1.5, page 27

1. commutative (addition)   3. commutative (multiplication)
5. distributive   7. distributive
9. associative (addition)   11. commutative (addition)
13. associative (multiplication)
15. commutative and associative (multiplication)
17. commutative and associative (addition)
19. $1 - 2 \neq 2 - 1;\ 1 - (3 - 2) \neq (1 - 3) - 2$
21. $2(a + 2) = 2a + 4$ 23. $(a - b)2 = 2a - 2b$
25. $(2a + 3) + a = 3(a + 1)$ 27. $4 - x$
29. $-x - 5$ 31. $12x$ 33. $6abc$
35. $6 + 8a - 4b$ 37. $-\dfrac{1}{2x}$ 39. $-\dfrac{4}{3}xy$
41. $2ab$ 43. x 45. $8a$
47. $2x + 8y$ 49. $3x - 2y - 5z$
51. 1.31   53. 2.01

EXERCISE SET 1.6, page 34

1. 2	3. 1.5	5. -2	7. 1
9. 4	11. 2	13. 1/5	15. -1
17. 6	19. -7	21. $-2/3$	23. 1
25. 0.31	27. $4 > 1$	29. $2 \leq 3$	31. $3 \geq 0$
33. $<$	35. $>$	37. $<$	39. $<$
41. $>$	43. $<$	45. $>$	47. $<$
49. $<$	51. $<$	53. $>$	55. $<$
57. $x < -2$	59. $1 < x < 4$		

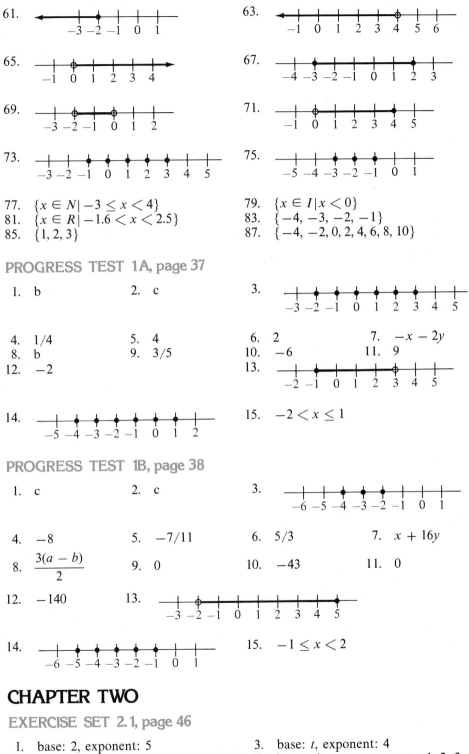

61.

63.

65.

67.

69.

71.

73.

75.

77. $\{x \in N \mid -3 \le x < 4\}$
81. $\{x \in R \mid -1.6 < x < 2.5\}$
85. $\{1, 2, 3\}$

79. $\{x \in I \mid x < 0\}$
83. $\{-4, -3, -2, -1\}$
87. $\{-4, -2, 0, 2, 4, 6, 8, 10\}$

PROGRESS TEST 1A, page 37

1. b

2. c

3.

4. 1/4
8. b
12. −2

5. 4
9. 3/5

6. 2
10. −6
13.

7. $-x - 2y$
11. 9

14.

15. $-2 < x \le 1$

PROGRESS TEST 1B, page 38

1. c

2. c

3.

4. −8

5. −7/11

6. 5/3

7. $x + 16y$

8. $\dfrac{3(a - b)}{2}$

9. 0

10. −43

11. 0

12. −140

13.

14.

15. $-1 \le x < 2$

CHAPTER TWO

EXERCISE SET 2.1, page 46

1. base: 2, exponent: 5
5. bases: 3, y; exponents: 1, 5

3. base: t, exponent: 4
7. bases: 3, x, y; exponents: 1, 2, 3

9. 3^3

11. $\left(\dfrac{1}{3}\right)^4$

13. $3y^4$

15. d

17. b^7 19. $6x^6$ 21. $-20y^9$ 23. $\dfrac{20}{21}v^8$

25. $-3x^4$ 27. c, d   29. $4x^4, -2x^2, x, -3; 4, -2, 1, -3$

31. $\dfrac{2}{3}x^3y, \dfrac{1}{2}xy, -y, 2; \dfrac{2}{3}, \dfrac{1}{2}, -1, 2$

33. $\dfrac{1}{3}x^3, \dfrac{1}{2}x^2y, -2x, y, 7; \dfrac{1}{3}, \dfrac{1}{2}, -2, 1, 7$

35.

Term	$3x^3$	$-2x^2$	3
Degree	3	2	0

37.

Term	$4x^4$	$-5x^3$	$2x^2$	$-5x$	1
Degree	4	3	2	1	0

39.

Term	$\frac{3}{2}x^4$	$2xy^2$	y^3	$-y$	2
Degree	4	3	3	1	0

41. 3   43. 3   45. 4   47. b
49. 13   51. 11   53. 176.2   55. πr^2
57. (a) area of entire field   (b) perimeter of rectangle on right
    (c) perimeter of square on left   (d) total amount of fencing

EXERCISE SET 2.2, page 50

1. $7x$
5. $4x^2 - x + 4$

3. $3x^3 - 6x^2$
7. $7x^2 + 3$

9. $-3rs$

11. $\dfrac{6}{5}rs^3 - 2r^2s^2 - r^2s + 2r^2 + 7$

13. $-x^2 - 4x + 14$
17. $24y$

15. $y^2 - 2x^2 - xy + 7y$
19. $6x^2 - 3$

21. $x^2y^2 + 3xy - 3y - x - 3$

23. $-\dfrac{19}{10}x^3 + 3x^2 - x - 2$

25. $-3rs^3 + 2rs - r + s - 3$
29. $4xy + 2y + 3$
31. $5r^2s^2 + rs^2 - r^2s - rs + r + s + 1$
33. $2s^2t^3 - 3s^2t^2 + 2s^2t + 3st^2 + st - s + 2t - 3$
35. $-2a^2bc + ab^2c - 2ab^3 + 3$

27. $2x^2 - 6x + 9$

37. $-260x + 13y + 17z$

EXERCISE SET 2.3, page 55

1. $6x^5$ 3. $6a^2b^3$ 5. $2x^3 + 6x^2 - 10x$
7. $-4s^4t^2 + 4s^4t - 12s^3$
11. $x^2 - x - 6$
15. $x^2 + 6x + 9$
19. $3x^2 - x - 2$
23. $6y^2 + 13y + 6$
27. $4y^2 - 25$
31. $x^4 + 4x^2 + 4$
35. $x^3 + 3x^2 - x - 3$
39. $3a^3 + 5a^2 + 3a + 10$

9. $8a^4b^2 + 4a^3b^3 - 4a^2b^4$
13. $y^2 + 7y + 10$
17. $s^2 - 9$
21. $2a^2 + a - 10$
25. $4a^2 + 12a + 9$
29. $9x^2 - 16$
33. $x^4 - 4$
37. $2s^4 - 3s^3 - 2s^2 + 7s - 6$
41. $2x^4 - x^3 + 8x^2 - 3x + 6$

43. $2x^4 + x^3 - 6x^2 + 7x - 2$ 45. $6x^5 - 4x^4 - 8x^3 + 14x^2 - 12x + 4$
47. $6a^3 + 2a^2 + 3a^2b - 2a^2b^2 + 3ab^2 - ab^3 + ab + b^2 - b^4$
49. $20x^2 - 60x + 45$ 51. $2x^3 + 3x^2 - 2x$
53. $x^3 + 4x^2 + x - 6$ 55. (a) -3 (b) 5
57. (a) 1 (b) 6 59. $4x^3y + 2x^2y$
61. $2x - 3$ 63. $-2x^2 - 6x + 16$
65. $4x - 20$ 67. c
69. $1.56x^2 - 9.18x + 13.5$ 71. $22.73y^4 - 4.57y^2 - 3.24$
73. $21.68x^3 - 2.90x^2 - 6.88x + 0.28$

EXERCISE SET 2.4, page 62

1. $2(x + 3)$ 3. $3(x - 3y)$ 5. $-2(x + 4y)$
7. $2(2x^2 + 4y - 3)$ 9. $5b(c + 5)$ 11. $y(1 - 3y^2)$
13. $-y^2(3 + 4y^3)$ 15. $3bc(a + 4)$ 17. $5r^3s^3(s - 8rt)$
19. $4(2a^3b^5 - 3a^5b^2 + 4)$ 21. $(x + 1)(x + 3)$
23. $(y - 5)(y - 3)$ 25. $(a - 3b)(a - 4b)$
27. $(y + 3)(y + 3)$ 29. $(5 - x)(5 - x)$
31. $(x - 7)(x + 2)$ 33. $(2 + y)(2 - y)$
35. $(x - 3)^2$ 37. $(x - 10)(x - 2)$
39. $(x + 8)(x + 3)$ 41. $(2x + 1)(x - 2)$
43. $(3a - 2)(a - 3)$ 45. $(3x + 2)(2x + 3)$
47. $(2m - 3)(4m + 3)$ 49. $(2x - 3)(5x + 1)$
51. $(2a - 3b)(3a + 2b)$ 53. $(5rs + 2t)(2rs + t)$
55. $(3 + 4x)(2 - x)$ 57. $25r^2 + 4s^2$
59. $2(x - 3)(x + 2)$ 61. $2(5x - 2)(3x + 4)$
63. $2b^2(3x - 4)(2x + 3)$ 65. $3m(3x + 1)(2x + 3)$

67. $5m^2n(5n^2 - 1)$ 69. $xy\left(1 + \frac{1}{4}x^2y^2\right)$

71. $(x^2 + y^2)(x^2 + y^2)$ 73. $(b^2 + 4)(b^2 - 2)$
75. $(3b^2 - 1)(2b^2 + 3)$ 77. $(x - 1)(2x + 7)$
79. $(2x - 1)^2(x + 2)^2[4(x + 2)(x + 1) - 3(2x - 1)^3(x + 3)]$
81. $50x(7 - 2x)^2(7 - 5x)$

EXERCISE SET 2.5, page 65

1. $(2x + y)^3$ 3. $(x - 2y)^3$ 5. $(3r + 2s)^3$ 7. $(2m - 5n)^3$

9. $\left(\frac{x}{2} - 2y\right)^3$ 11. $(x + 3y)(x^2 - 3xy + 9y^2)$

13. $(3x - y)(9x^2 + 3xy + y^2)$ 15. $(a + 2)(a^2 - 2a + 4)$

17. $\left(\frac{1}{2}m - 2n\right)\left(\frac{1}{4}m^2 + mn + 4n^2\right)$

19. $(x + y - 2)(x^2 + 2xy + y^2 + 2x + 2y + 4)$
21. $(2x^2 - 5y^2)(4x^4 + 10x^2y^2 + 25y^4)$ 23. $(3r^2 + 2s^2)(9r^4 - 6r^2s^2 + 4s^4)$

25. $\left(\frac{1}{4}x^2 + 5y^2\right)\left(\frac{1}{16}x^4 - \frac{5}{4}x^2y^2 + 25y^4\right)$

PROGRESS TEST 2A, page 66

1. base: $-1/5$, exponent: 4   2. 5
3. 31   4. s^3 5. $-3x^3 + 3x^2 + x - 2$

6. $2x^2y - x + 6y - 2$
7. $-x^3 + x^2 - 6x + 3$
8. $3x^2y + 3xy^2 - x^2 + 4y - 2$
9. $4x^2 - 20xy + 25y^2$
10. $3x^4 - 2x^3 + 11x^2 - 4x + 10$
11. $(x + 6)(x - 2)$
12. $2y(x - 3)(x - 1)$
13. $(2a + 7)(2a - 7)$

14. $(x - 5y)(3x + 2y)$
15. $\left(\dfrac{x}{5} - 5y\right)\left(\dfrac{x^2}{25} + xy + 25y^2\right)$

PROGRESS TEST 2B, page 67

1. base: $-\dfrac{2}{3}$, exponent: 5
2. 5

3. 2
4. $\dfrac{1}{2}s^2$

5. $6x^5 + 4x^4 + 2x^3 - x^2 + 1$
6. $x^2y + 6x^2 - 3y^2 - 10$
7. $-5x^3 + 2x^2 + 3x - 8$
8. $4x^2y^2 - xy^2 + xy + 2$
9. $9x^2 - 24x + 16$
10. $2x^2 - 3y^2 - xy - 2x - 2y$
11. $(r + 7)(r + 2)$
12. $3(x + 1)(x + 1)$
13. $16(y + 2x)(y - 2x)$
14. $(3x - 2)(x - 5)$

15. $\left(2a + \dfrac{b}{2}\right)\left(4a^2 - ab + \dfrac{b^2}{4}\right)$

CHAPTER THREE

EXERCISE SET 3.1, page 72

1. T
3. F
5. T
7. T
9. 4
11. $-5/4$
13. -2
15. 1
17. $-4/3$
19. 2
21. 6
23. 3
25. 4
27. $3/2$
29. $-10/3$
31. -2
33. 3
35. 5
37. $-7/2$
39. -7

41. 1
43. $\dfrac{8}{5 - k}$
45. $\dfrac{6 + k}{5}$

EXERCISE SET 3.2, page 77

1. $250
3. 50
5. 8.5 cents
7. 260
9. 80
11. no
13. $1.20
15. 8
17. 10, 38
19. 6
21. 9 and 12
23. 11
25. 68°F
27. $12.30
29. 4 meters and 8 meters
31. 9/2 meters by 27/2 meters
33. 9 centimeters

35. 65°, 65°, and 50°
37. $\dfrac{A - c}{B}$
39. $-\dfrac{2kt}{5A + 3bt}$

EXERCISE SET 3.3, page 83

1. c, d, e
3. a, c, d, e

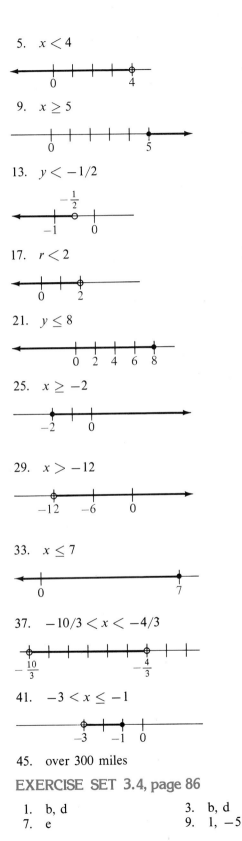

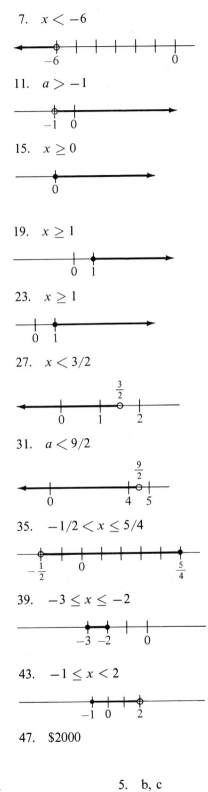

5. $x < 4$

7. $x < -6$

9. $x \geq 5$

11. $a > -1$

13. $y < -1/2$

15. $x \geq 0$

17. $r < 2$

19. $x \geq 1$

21. $y \leq 8$

23. $x \geq 1$

25. $x \geq -2$

27. $x < 3/2$

29. $x > -12$

31. $a < 9/2$

33. $x \leq 7$

35. $-1/2 < x \leq 5/4$

37. $-10/3 < x < -4/3$

39. $-3 \leq x \leq -2$

41. $-3 < x \leq -1$

43. $-1 \leq x < 2$

45. over 300 miles

47. $2000

EXERCISE SET 3.4, page 86

1. b, d   3. b, d   5. b, c
7. e   9. 1, −5   11. 11/2, 9/2

13. $1, -2$

15. $2, -2/3$

17. $2, -4/3$

19. $-5/2, 1/2$

21. $-5 < x < 5$

23. $x < -4$ or $x > 4$

25. all real x

27. $-8 < x < 2$

29. $x < -4$ or $x > 2$

31. $x \leq -1$ or $x \geq 7$

33. $-6 \leq x \leq 2$

35. all real x

37. $-1 \leq x \leq 2$

39. $x > 5/3$ or $x < -1$

41. $x < -1$ or $x > 0$

43. $-5 < x < 7$

45. $-7/2 < x < 9/2$

47. $|x - 100| \leq 2$; $x = 98, 99, 100, 101, 102$

PROGRESS TEST 3A, page 88

1. $15/4$

2. -4

3. F

4. $h = V/\pi r^2$

5. $x = \dfrac{-1}{k + 2}$

6. 70

7. 3200

8. $L = 21/2, W = 15/2$

9. $14, 16, 18$

10. $3000 at 6%, $2000 at 7%

11. $x > -2$

12. $x \geq 1$

13. $7/2, -3/2$

14. $-10 \leq x \leq 14$

15. $x > 1$ or $x < -6$

PROGRESS TEST 3B, page 89

1. -2

2. 5

3. F

4. $\dfrac{1}{H}(2A - CH) = B$

5. $\dfrac{1}{2(k - 1)}$

6. 40

7. 80

8. $W = 7, L = 4$

9. 25, 21

10. $4000 at 5%, $4000 at 8%

11. $x < -5/2$

12. $x \leq -1$

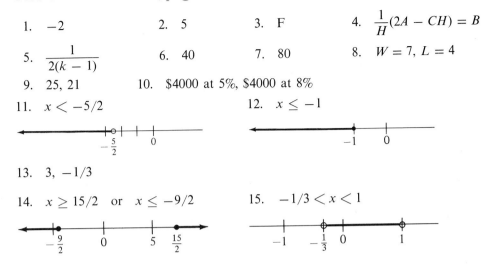

13. 3, $-1/3$

14. $x \geq 15/2$ or $x \leq -9/2$

15. $-1/3 < x < 1$

CHAPTER FOUR

EXERCISE SET 4.1, page 94

1. $J + M = 39$

3. $8n$

5. $T = 20 + C$

7. $B = 3 + 2R$

9. $x + 2x = 18$

11. 24, 27

13. 15, 45

15. 16, 28

17. 8

19. 5, 13

21. $W = 5, L = 15$

23. $W = 27, L = 43$

25. 5, 8, and 11 meters

EXERCISE SET 4.2, page 98

1. 10 nickels, 25 dimes
3. 14 ten-dollar bills, 8 twenty-dollar bills, 32 five-dollar bills
5. 300 children, 400 adults
7. 28 five-cent stamps, 26 ten-cent stamps, 18 fifteen-cent stamps
9. 61 three-dollar tickets, 40 five-dollar tickets, 20 six-dollar tickets

EXERCISE SET 4.3, page 101

1. $5000 at 7%, $3000 at 8%
3. $5806.45 at 8.5%, $4193.55 at 7%
5. $3000 in black and white, $1000 in color
7. $32,000
9. $3400 at 6%, $3700 at 8%, $13,600 at 10%

EXERCISE SET 4.4, page 105

1. 20 hours

3. 50 mph, 54 mph

5. 40 miles

7. 40 km/hr, 80 km/hr

9. 4 hours

EXERCISE SET 4.5, page 110

1. 20 pounds

3. 30 kilograms of 40%, 90 kilograms of 80%

5. 15 gallons

7. 5 quarts

9. 20 pounds

PROGRESS TEST 4A, page 111

1. $C = 4T - 3$
2. 17, 23
3. $L = 10/3$ cm, $W = 8/3$ cm
4. 3 quarters, 10 dimes, 17 nickels
5. six 30-pound, six 50-pound, eleven 60-pound crates
6. $8000
7. $6000 at 6.5%, $6200 at 7.5%, $12,300 at 9%
8. moped: 15 mph, car: 45 mph
9. 240 miles
10. 30 ounces of 60%, 90 ounces of 80%
11. 37.5 cubic centimeters

PROGRESS TEST 4B, page 111

1. $D = \dfrac{1}{3}R + 4$
2. 16, 32

3. side #1: 3.6 cm, side #2: 5.8 cm, side #3: 5.6 cm
4. fourteen $1-, four $5-, two $10-coupons
5. six 1-ounce, five 2-ounce, three 3-ounce samples
6. $3000 at 6%, $15,000 at 7.2%
7. $8000 at 6%, $6000 at 8%
8. 6 P.M.
9. 12 hours
10. 3.75 pounds
11. 20 gallons

CHAPTER FIVE

EXERCISE SET 5.1, page 118

1. T
3. F
5. T
7. F

9. $2x + 1$
11. $\dfrac{3x + 2}{3}$
13. $\dfrac{1}{2x^2 - 3}$
15. $a^2/5$

17. $\dfrac{1}{x - 4}$
19. $x - 4$
21. $\dfrac{3x + 1}{x + 2}$
23. $\dfrac{2x^2 + 5x + 3}{12}$

25. $\dfrac{3(a^2 - 16)}{b}$
27. $a/2$
29. $4/9$
31. $\dfrac{2(3x + 1)}{(x + 2)^2}$

33. $-2b(5 + a)$
35. $\dfrac{x + 3}{3x(x - 3)}$
37. $\dfrac{2(x + 2)(x - 2)^2}{(x + 1)(2x + 3)}$

39. $\dfrac{5y}{x - 4}$
41. $\dfrac{(2x + 1)(x - 2)}{(x - 1)(x + 1)}$
43. $\dfrac{(x - 2)^2}{(x + 3)(x - 3)}$

45. $\dfrac{(x + 2)(2x + 3)}{x + 4}$
47. $\dfrac{(x + 3)(x^2 + 1)}{x - 2}$
49. $\dfrac{x + 4}{(x + 1)(x - 5)}$

EXERCISE SET 5.2, page 123

1. xy
3. $2a$
5. $(b - 1)^2$
7. $(x - 2)(x + 3)$
9. $x(x^2 - 1)$
11. $7/x$
13. $3x/y$
15. 1

17. $x - 3$
19. $\dfrac{y - 14}{(y - 4)(y + 4)}$
21. $\dfrac{4}{a - 2}$

23. 2
25. $\dfrac{10 + x}{5x}$
27. $\dfrac{5x - 2}{x}$
29. $\dfrac{3x - 4}{(x - 1)(x - 2)}$

31. $\dfrac{3a^2 - 2b^2}{24ab}$ 33. $\dfrac{8x - 1}{6x^3}$ 35. $\dfrac{2(x + 1)}{3(x - 3)}$ 37. $\dfrac{x^2 + y^2}{x^2 - y^2}$

39. $\dfrac{x^2 + 2xy - y^2}{x^2 - y^2}$ 41. $\dfrac{r + 8}{r(r + 2)}$

43. $\dfrac{3x^2 - 4x - 1}{(x - 1)(x - 2)(x + 1)}$ 45. $-\dfrac{2a^3 - 3a^2 - 3a - 2}{a(a - 1)(a + 1)}$

47. $\dfrac{3x^2 - 4x - 12}{(x - 2)(x + 2)(x - 3)}$ 49. $-\dfrac{x^3 - 4x^2 + 3x - 1}{x(x - 2)(x + 2)(x - 1)}$

51. $\dfrac{17x + 26}{(x + 2)(x - 2)(x + 3)}$ 53. $\dfrac{2y^2 + y + 1}{y(y + 1)(y - 1)}$

EXERCISE SET 5.3, page 127

1. $\dfrac{x + 2}{x - 3}$ 3. $\dfrac{3x - 4}{5x^2}$ 5. $\dfrac{x(x + 1)}{x - 1}$

7. $4x(x + 4)$ 9. $\dfrac{a + 2}{a + 1}$ 11. $\dfrac{x + 3}{(3x - 7)(x + 2)}$

13. $a - b$ 15. $\dfrac{x - 2}{x}$ 17. $\dfrac{1 - x}{2}$

19. $\dfrac{a^3 - a^2 + 1}{a - 1}$ 21. $\dfrac{(y - 2)(y + 1)}{(y - 1)(y + 2)}$ 23. $\dfrac{x + 1}{2x + 1}$

EXERCISE SET 5.4, page 131

1. $\dfrac{5.08 \text{ cm}}{2 \text{ in}}$ or 5.08 cm:2 in   3. $\dfrac{16}{5}$ or 8:2.5

5. 3/4 or 3:4   7. 20, 16, 8   9. 18/5
11. 3/10   13. 5   15. 9, 21
17. $500   19. 60 ft   21. 55,000
23. 962   25. 24 cubic centimeters

EXERCISE SET 5.5, page 135

1. 10/3   3. 1   5. 9/2
7. 4   9. 4   11. no solution
13. 1/4   15. 12   17. 5/19
19. 2   21. 12/7   23. $x < 7$
25. $x \le -1$ 27. $x \ge 1$

EXERCISE SET 5.6, page 140

1. 18   3. 3/8, 3/4   5. 3/4
7. 4   9. 6/5 hours   11. 12/13 hours
13. 6 hours   15. 60/7 hours   17. 9 hours
19. 8 hours   21. 25 km/hr
23. 100 km/hr, 120 km/hr

PROGRESS TEST 5A, page 142

1. $\dfrac{12x^2(x + 2)}{y}$

2. $-\dfrac{(2x + 3)(3x - 1)}{(x + 4)^2}$

3. $-\dfrac{(2x - 1)(x - 2)(x - 1)}{3x}$

4. $5y^2(x - 1)^2$

5. $\dfrac{13}{x - 5}$

6. $\dfrac{7 - 5y^2}{2y(y + 1)}$

7. $3x$

8. $\dfrac{(x + 4)(x - 2)}{x - 3}$

9. 4

10. $291.67

11. -2

12. $x \le -33/5$

13. 2 hours

PROGRESS TEST 5B, page 143

1. $\dfrac{-6(y - 1)}{y^2(x - y)}$

2. $-\dfrac{(2x - 1)}{(3x - 1)(x + 1)}$

3. $-\dfrac{2x(x + 1)}{x + 2}$

4. $4x^2(y + 1)^2(y - 1)$

5. $\dfrac{2}{3 - x}$

6. $\dfrac{-2v^3 + 3v^2 + 6v + 8}{4v^2(v - 1)}$

7. $-2x$

8. $\dfrac{(x - 1)(x - 3)}{2(2 - x)}$

9. 7

10. 81

11. $11/12$

12. $x \le 24/5$

13. $7/3$ hours

CHAPTER SIX

EXERCISE SET 6.1, page 152

1. $A(2, 3)$, $B(-2, -1)$, $C(4, -1)$, $D(0, 5)$, $E(-4, 0)$, $F(-3, 4)$, $G(1, 1/2)$, $H(-1, 7/2)$

3.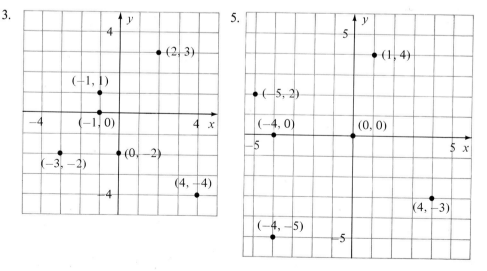

5.

7.

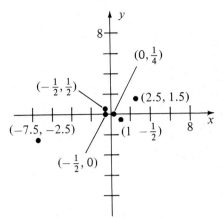

9. (a) $(-4, 0)$ (b) $(6, -2)$
11. I   13. IV
15. III   17. I
19. III   21. I
23. a, d   25. a, c

27.

x	1	$\dfrac{9}{2}$	0	3	-3	$\dfrac{3}{2}$
y	$\dfrac{8}{3}$	-2	4	0	8	2

29.

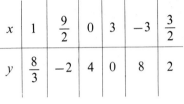

31.

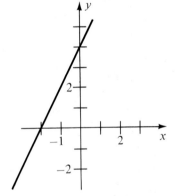

33.

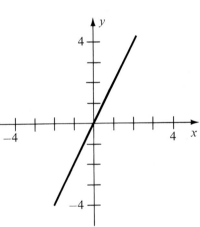

35.

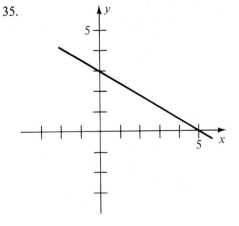

37.

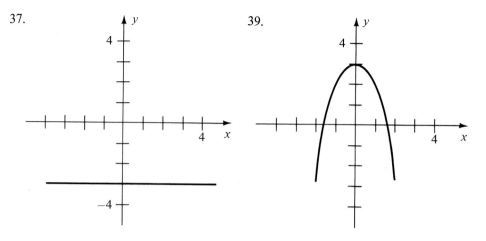

39.

41.

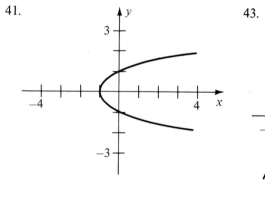

43.

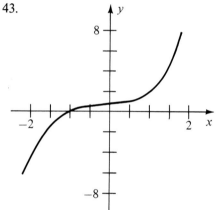

45.

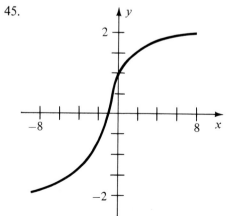

47.

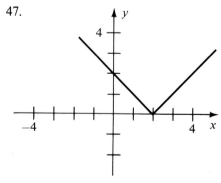

49.

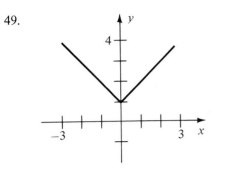

51.

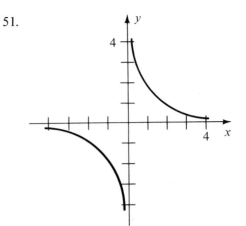

53.

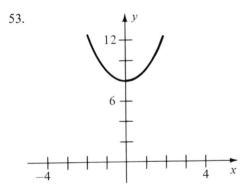

55.

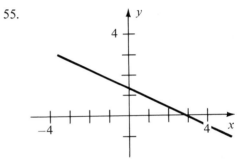

57.

59.

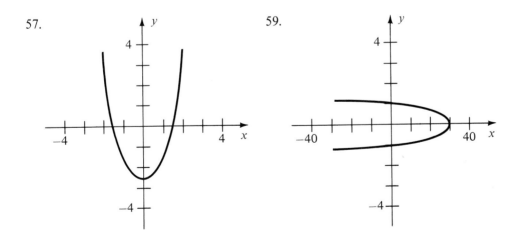

61. $x = 6$

63. $(2, 8)$

EXERCISE SET 6.2, page 162

1.
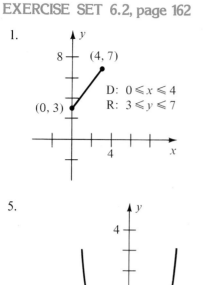

D: $0 \leqslant x \leqslant 4$
R: $3 \leqslant y \leqslant 7$

3.
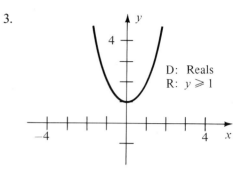

D: Reals
R: $y \geqslant 1$

5.
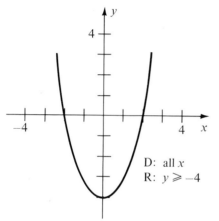

D: all x
R: $y \geqslant -4$

7.
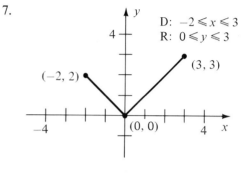

D: $-2 \leqslant x \leqslant 3$
R: $0 \leqslant y \leqslant 3$

9.
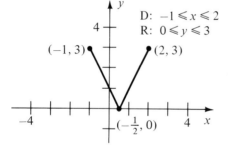

D: $-1 \leqslant x \leqslant 2$
R: $0 \leqslant y \leqslant 3$

11. all real numbers

13. $v \neq 3, -1$

15. $x \neq -1$

17. $x \neq 0$

19. function

21. not a function

23. function

25. function

27. not a function

29. not a function

31. 5

33. $2a^2 + 5$

35. $6x^2 + 15$

37. 3

39. $\dfrac{1}{x^2 + 2x}$

41. $a^2 + h^2 + 2ah + 2a + 2h$ 43. -0.92

45. $\dfrac{3x - 1}{x^2 + 1}$

47. $\dfrac{8x^2 + 2}{6x - 1}$

49. -0.21

51. $\dfrac{2(a - 1)}{4a^2 + 4a - 3}$

53. $\dfrac{a - 1}{a(a + 4)}$

55. $R(x) = \begin{cases} 300x, & 0 \leq x \leq 100 \\ 30{,}000 + 250(x - 100), & 100 < x < 150 \end{cases}$

57. $A(x) = 1.07x$

59. (a) 1001   (b) 16,004

EXERCISE SET 6.3, page 171

1.

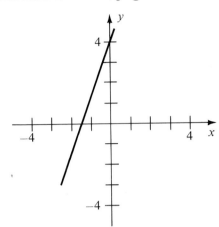

3.

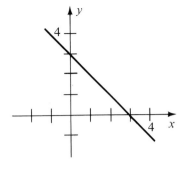

5.

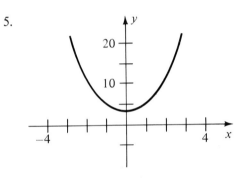

7.

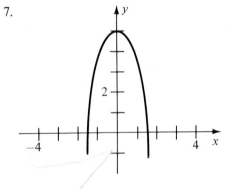

9.

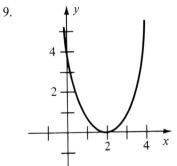

11.

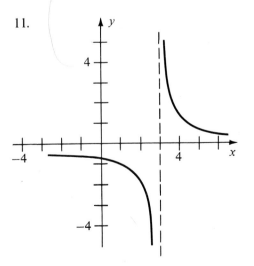

13.

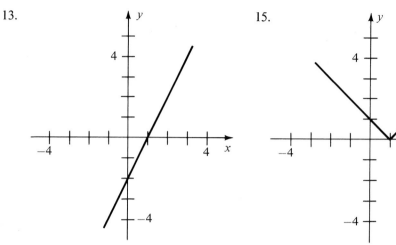

15.

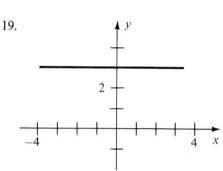

17.

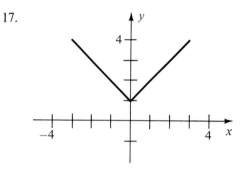

19.

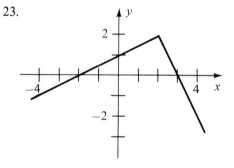

21.

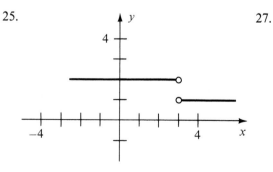

23.

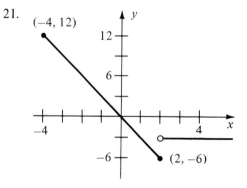

25.

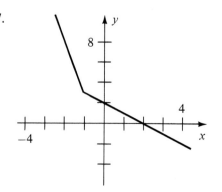

27.

29.

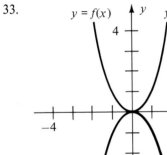

31.

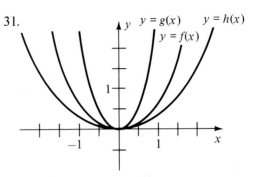

33.

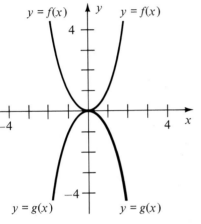

35.

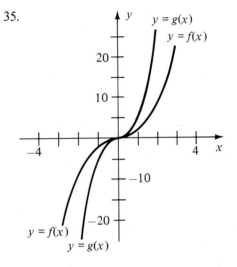

37.

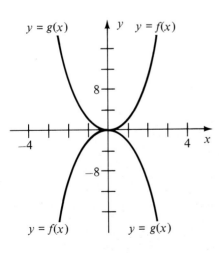

39.

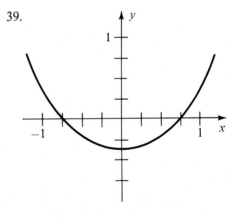

41.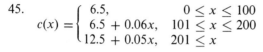

43.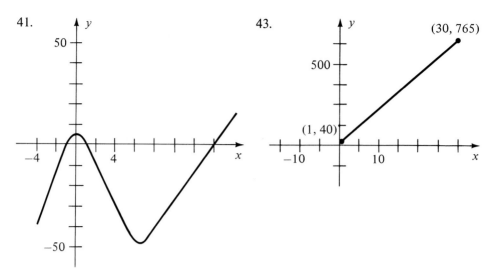

45.
$$c(x) = \begin{cases} 6.5, & 0 \le x \le 100 \\ 6.5 + 0.06x, & 101 \le x \le 200 \\ 12.5 + 0.05x, & 201 \le x \end{cases}$$

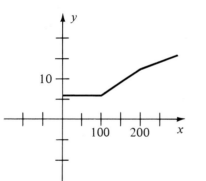

EXERCISE SET 6.4, page 175

1. always increasing

3. decreasing for $x \le 0$
   increasing for $x \ge 0$

5. decreasing for $x \ge 0$
   increasing for $x \le 0$

7. decreasing for $x \ge 2$
   increasing for $x \le 2$

9. decreasing for $x \le 0$
   increasing for $x \ge 0$

11. decreasing for $x \le 2$
    increasing for $x \ge 2$

13. decreasing for $x \le -1/2$
    increasing for $x \ge -1/2$

15. decreasing for $x \le -1$
    increasing for $x > -1$

17. always increasing

19. always increasing

21. decreasing for $-1 \le x < 0$ and $0 < x \le -1$
    increasing for $x \le -1$ and $x \ge 1$

23. decreasing for $x \le 0$
    increasing for $x \ge 0$

25. decreasing for $x \ge 40$
    increasing for $x \le 40$

27. decreasing for $2 \le x \le 5$
    increasing for $0 \le x \le 2$, $5 \le x \le 6$

EXERCISE SET 6.5, page 180

1. (a) 4   (b) $y = 4x$ (c)

x	8	12	20	30
y	32	48	80	120

3. (a) $-1/32$ (b) $-3/8$

5. (a) $1/10$ (b) $5/2$

7. (a) -3 (b) $-1/4$

9. (a) 512   (b) 512/125

11. (a) $M = r^2/s^2$ (b) $36/25$

13. (a) $T = 16pv^3/u^2$ (b) $2/3$

15. (a) 256 feet   (b) 5 seconds

17. 40/3 ohms

19. (a) 800/9 candlepower   (b) 8 feet

21. 6

23. 120 candlepower/ft^2

PROGRESS TEST 6A, page 183

1.

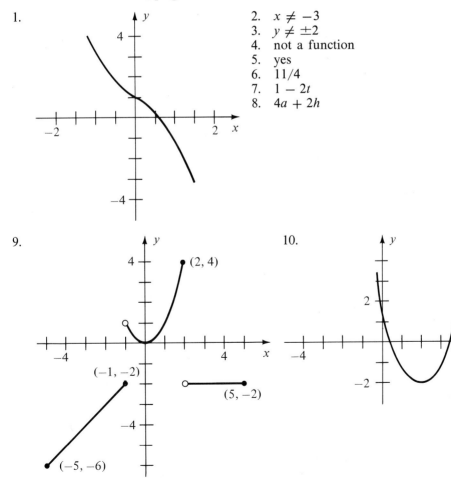

2. $x \neq -3$
3. $y \neq \pm 2$
4. not a function
5. yes
6. 11/4
7. $1 - 2t$
8. $4a + 2h$

9.

10.

11. 160

12. 1

13. $-1/4$

14. increasing for $x > 1$
    decreasing for $x < 1$

15. increasing for $1 < x < 3$
    decreasing for $x < 1$
    constant for $x \geq 3$

PROGRESS TEST 6B, page 184

1.
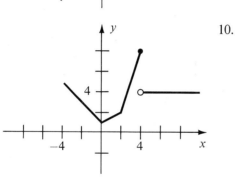

2. $t \neq 1/2$
3. $x \neq \pm 1$
4. function
5. yes
6. -3
7. $1 + a + \dfrac{a^2}{4}$
8. $-2a - h$

9.

10.
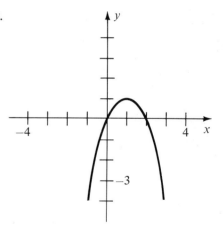

11. -1024 12. $-32/9$ 13. $65,536$
14. increasing for $x > 2$
    decreasing for $x < 2$
15. increasing for $0 < x < 3$
    decreasing for $-1 < x < 0$
    constant for $x < -1$ and $x > 3$

CHAPTER SEVEN

EXERCISE SET 7.1, page 191

1. 2, rising   3. -1, falling   5. $-3/2$, falling
7. 0   9. 1   11. no slope

EXERCISE SET 7.2, page 197

1. a, d   3. $2x - y = 3$; $A = 2$, $B = -1$, $C = 3$
5. $0x + y = 3$; $A = 0$, $B = 1$, $C = 3$
7. $3x - 2y = 7$; $A = 3$, $B = -2$, $C = 7$
9. $x + 0y = 1/2$; $A = 1$, $B = 0$, $C = 1/2$

11.

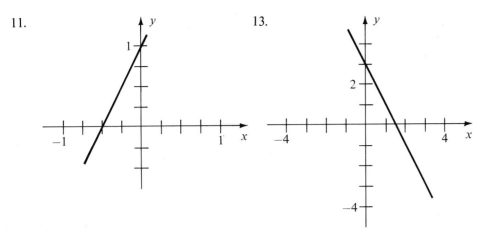

13.

15.

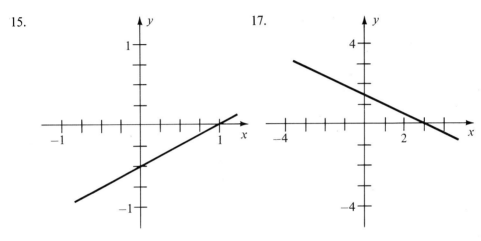

17.

19.

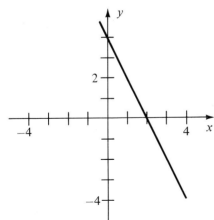

21.

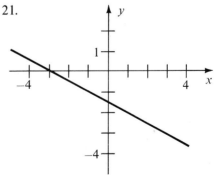

23. $y - 3 = 2(x + 1)$
27. $y - 4 = 2(x - 2)$
31. $y = 3x + 2$
35. slope $= 3$, y-intercept $= 2$
39. slope $= 0$, y-intercept $= 3$
43. slope $= 2/5$, y-intercept $= 3/5$
47. rises

25. $y = 3x$
29. $y = 2x/3$
33. $y = 2$
37. no slope or y-intercept
41. slope $= -3/4$, y-intercept $= 5/4$
45. slope $= 3/2$, y-intercept $= -3$
49. rises

51.   rises

53. $y = \dfrac{3}{4}x + \dfrac{11}{4}$

55. $y = -0.98x + 14.57$

57. $y = 0.83x - 7.04$

59. $y = -\dfrac{3}{4}x + \dfrac{5}{2}$

61. $c = 8 + 1.5h$, $c = $ charge in dollars, $h = $ hours
63. $c = 25 + 0.12s$, $c = $ charge in dollars, $s = $ number of shares

65.   (a) $F = \dfrac{9}{5}C + 32$ (b)   68°F

67. $1,000,000

EXERCISE SET 7.3, page 205

1. $y = 2$

3. $x = -2$

5. (a) $y = 3$ (b) $x = -6$

7. (a) $y = -5$ (b) $x = 4$

9. (a) $y = 0$ (b) $x = 0$

11. (a) $y = 0$ (b) $x = -7$

13. (a) $y = 5$ (b) $x = 0$

15. $y = 3$

17.

19.

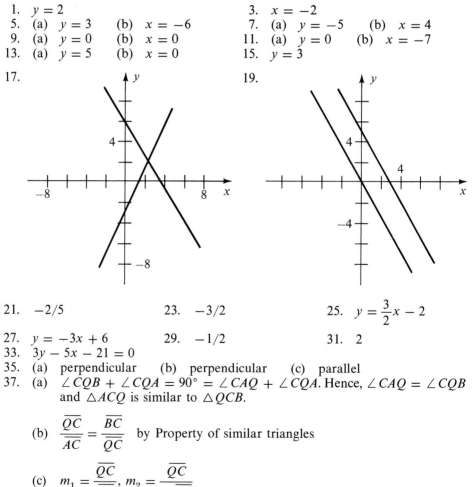

21. $-2/5$

23. $-3/2$

25. $y = \dfrac{3}{2}x - 2$

27. $y = -3x + 6$

29. $-1/2$

31. 2

33. $3y - 5x - 21 = 0$
35. (a) perpendicular   (b) perpendicular   (c) parallel
37. (a) $\angle CQB + \angle CQA = 90° = \angle CAQ + \angle CQA$. Hence, $\angle CAQ = \angle CQB$
     and $\triangle ACQ$ is similar to $\triangle QCB$.

     (b) $\dfrac{\overline{QC}}{\overline{AC}} = \dfrac{\overline{BC}}{\overline{QC}}$ by Property of similar triangles

     (c) $m_1 = \dfrac{\overline{QC}}{\overline{CA}}$, $m_2 = \dfrac{\overline{QC}}{-\overline{CB}}$

     (d) $m_1 = \dfrac{\overline{QC}}{\overline{CA}} = \dfrac{\overline{BC}}{\overline{QC}} = -\dfrac{1}{m_2}$ Hence, $m_1 m_2 = -1$ or $m_2 = -\dfrac{1}{m_1}$

EXERCISE SET 7.4, page 210

1.
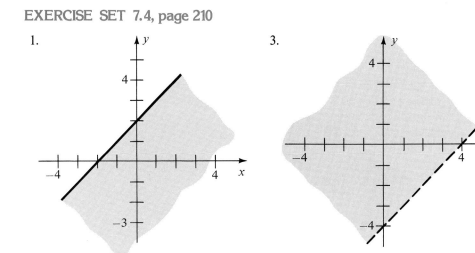

3.

5.
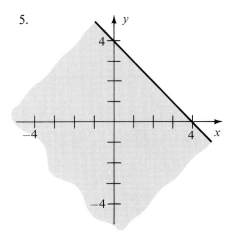

7.
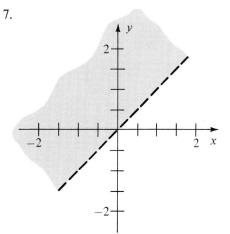

9.
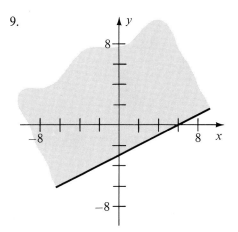

11.
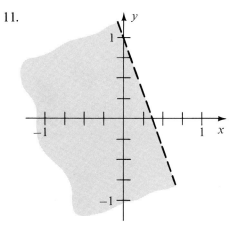

13.

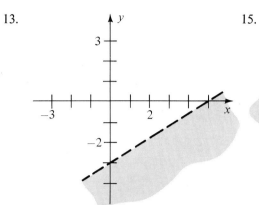

15.

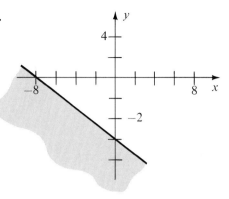

17.

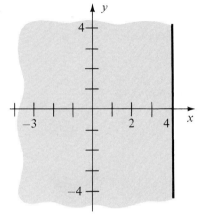

19.

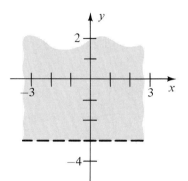

21.

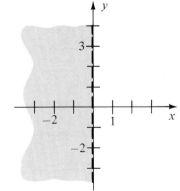

23.

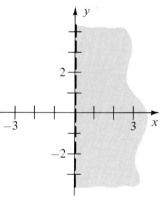

25.

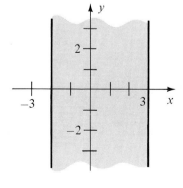

27.

29. $3x - 4y \leq 12$ 31. $y < -2$ 33. $-3 < x < 3$
35. $2x + 5y \leq 15; \ x \geq 0, \ y \geq 0$

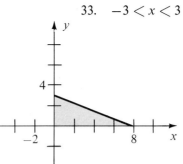

PROGRESS TEST 7A, page 213

1. 0 2. $-2/3$ 3. $7x + y + 5 = 0$
4. $x - 10y + 74 = 0$ 5. slope $= -5/2$, y-intercept $= 3$
6. slope $= 2/3$, rising   7. slope $= -1/4$, falling
8. $c = 20t + 5$ 9. $x = -11$
10. $y = -7/3$ 11. $y = 3x + 2$, same line
12. $2/7$ 13. $3x + 4y + 11 = 0$
14.   15.

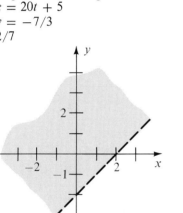

PROGRESS TEST 7B, page 214

1. no slope   2. -2 3. $y - 5x - 23 = 0$
4. $2x + 3y + 12 = 0$ 5. slope $= 4$, y-intercept $= -5/3$
6. slope $= 7/2$, rising   7. slope $= -5/4$, falling
8. $c = 13n + 2$ 9. $y = 5/4$
10. $x = 3$ 11. $y - 2x - 3 = 0$
12. $-5/3$ 13. $y = -3x$
14.   15.

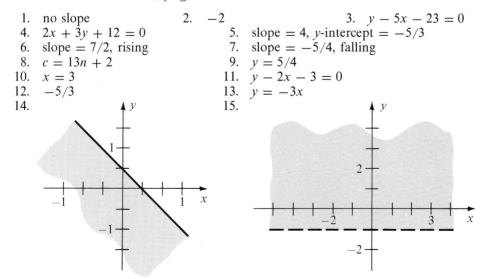

CHAPTER EIGHT

EXERCISE SET 8.1, page 218

1. $1/81$, base $= 1/3$, exponent $= 4$
3. -32, base $= -2$, exponent $= 5$
5. $(-5)^4$
7. $4^2 x^2 y^3$
9. x^6
11. b^4
13. $16x^4$
15. $-1/128$
17. x^{3m^2}
19. 81
21. $-x^3/y^3$
23. y^8
25. x^{19}
27. x^4
29. $-32x^{10}$
31. x^{4n}
33. 3^8
35. $1/x^2$
37. $3^5 x^{15}/y^{10}$
39. $30x^8$
41. $x^{27} y^8$
43. 1
45. $(3b + 1)^{25}$
47. $1/(2a + b)^2$
49. $27x^9 y^3$
51. $-27x^3/8$
53. $(2x + 1)^{10}$
55. $1/y^9$
57. $2^{2n} a^{4n} b^{6n}$
59. $-8a^6 b^9/c^6$
61. 21.49
63. -2.57
65. 302.90
67. 888.73

EXERCISE SET 8.2, page 223

1. 1
3. 1
5. 3
7. $-1/3^3$
9. $2 \cdot 4^3$ or 2^7
11. $-x^3$
13. y^6
15. $1/2^6 a^6$
17. $-4(5a - 3b)^2$
19. 25
21. $4y^3$
23. $1/x^7$
25. $1/3^6$
27. -2^9
29. x^9
31. 32
33. x^{18}
35. $4/x$
37. 2
39. $b^9/8a^6$
41. $a^4 b^6/9$
43. $a^9/b^9 c^6$
45. $a^2/b^4 c$
47. $a^9/3b^4$
49. x
51. $4a^{10} c^6/b^8$
53. $2y^5$
55. $1/y$
57. $a^4/4b^6 c^4$
59. $(a - b)^2/a + b$
61. 0.69
63. 0.00000499
65. 1.9
67. 3.20

EXERCISE SET 8.3, page 229

1. 8
3. $-1/5$
5. $4/25$
7. $1/c^{5/12}$
9. $2x^{13/12}$
11. $1/x^{17}$
13. 25
15. $x^{2/3} y$
17. $64a^6 b^3$
19. x^9/y^6
21. $x^{12} y^4$
23. $1/x^4 y^8$
25. $\sqrt[5]{\dfrac{1}{16}}$
27. $\sqrt[3]{x^2}$
29. $\sqrt[3]{\dfrac{1}{12^4 y^4}}$
31. $\sqrt[3]{\dfrac{144x^6}{y^4}}$
33. $\sqrt{\dfrac{27x^6 y^{12}}{8^3}}$
35. $\sqrt[4]{(2x^2 y^2 + z^3)^3}$
37. $8^{3/4}$
39. $(1/8)^{3/4}$
41. $x^{-7/3}$
43. $\left(\dfrac{4}{9} a^3\right)^{-1/4}$
45. $(3a^4 b^6)^{5/6}$
47. $(3x^3 + y^5)^{2/4}$
49. c
51. c
53. $2/3$
55. not real
57. $1/3$
59. 5
61. $1/2$
63. $7/2$
65. 54.82
67. 4.78
69. 0.15
71. 27.4

EXERCISE SET 8.4, page 234

1. $4\sqrt{3}$
3. $3\sqrt[3]{2}$
5. $2\sqrt[3]{5}$
7. $2/3$
9. $2/5$
11. x^4
13. $a^3\sqrt[3]{a^2}$
15. x^3
17. $7\sqrt{2}b^5$
19. $3y^5\sqrt[3]{4y}$
21. $2x^2\sqrt{5}$
23. $a^5b^3\sqrt{b}$
25. $x^2y^2\sqrt[3]{y^2}$
27. $ab^2\sqrt[4]{ab^2}$
29. $6x^3y^5\sqrt{2xy}$
31. $2b^3c^4\sqrt[3]{3bc^2}$
33. $2b^2c^3\sqrt[4]{3b^2}$
35. $2a^2bc\sqrt[3]{5a^2bc^2}$
37. $\sqrt{5}/5$
39. $4\sqrt{11}/33$
41. $\sqrt{3}/3$
43. $\sqrt{3y}/3y$
45. $2x\sqrt{2x}$
47. $3y\sqrt{2x}$
49. $-5x^3y^4\sqrt[3]{xy}$
51. $\frac{5}{2}\sqrt[3]{4a^2b}$
53. $2x^2\sqrt{5xy}/y^2$
55. $x\sqrt[3]{x^2y}/y$
57. $\sqrt[4]{216x^2y^3}/3y$
59. $x^3y^2\sqrt{xy}$
61. $2x^2y\sqrt[4]{2y^2}$
63. $2x^2y\sqrt[4]{3y^2z^2}$
65. $7y\sqrt[4]{27xy}/6x$
67. $\sqrt[4]{216x^2y^3}/12xy^3$
69. $\frac{4}{5}a\sqrt[3]{b^2}$

EXERCISE SET 8.5, page 237

1. $7\sqrt{3}$
3. $-2\sqrt[3]{11}$
5. $7\sqrt{x}$
7. $6\sqrt{2}$
9. $\frac{9}{2}\sqrt{y}$
11. $5\sqrt{6}$
13. $4\sqrt{3}$
15. $11\sqrt{5} - \sqrt[3]{5}$
17. $5\sqrt[3]{xy^2} - 4\sqrt[3]{x^2y}$
19. $\frac{3}{2}a\sqrt[3]{ab} + \frac{5}{2}\sqrt{ab}$
21. $2\sqrt[5]{2x^3y^2}$
23. $-3y\sqrt[3]{xy} - \frac{1}{2}y^2\sqrt{x}$
25. $-5\sqrt{5}$
27. $x(x\sqrt[3]{xy} - \sqrt[3]{x^2y} + 2x^2\sqrt{y})$
29. $3 + 4\sqrt{3}$
31. $2\sqrt{3} + 6\sqrt{2}$
33. $3xy$
35. $-4xy\sqrt[5]{x}$
37. $\sqrt{2} - 4$
39. $6\sqrt{6} + 4\sqrt{3} - 9\sqrt{2} - 6$
41. $5 - 2\sqrt{6}$
43. 0
45. $3x - 4y - \sqrt{6xy}$
47. $\sqrt[3]{4x^2} - 9$
49. $a^2 - 2b^2 - 2ab\sqrt[3]{ab} + \sqrt[3]{a^2b^2}$
51. $\frac{3}{7}(3 - \sqrt{2})$
53. $\frac{3(9 + \sqrt{7})}{74}$
55. $\frac{2\sqrt{x} - 6}{x - 9}$
57. $\frac{3(1 - 3\sqrt{a})}{9a - 1}$
59. $\frac{2(2 + \sqrt{2}y)}{2 - y}$
61. $2(\sqrt{2} - 1)$
63. $3 + 2\sqrt{2}$
65. $4 + \sqrt{15}$
67. $\frac{x - \sqrt{xy}}{x - y}$
69. $\frac{2a + \sqrt{ab} + 2\sqrt{a} + \sqrt{b}}{4a - b}$
71. $\frac{2\sqrt{2x} + (2 + \sqrt{2})\sqrt{xy} + y}{2x - y}$

EXERCISE SET 8.6, page 244

1. 1
3. $-i$
5. -1
7. $-i$
9. i
11. 1
13. i
15. $2 + 0i$
17. $-\frac{1}{2} + 0i$
19. $0 + 4i$
21. $0 - \sqrt{5}\,i$
23. $0 - 6i$

25. $2 + 4i$

27. $-\dfrac{3}{2} - 6\sqrt{2}\,i$

29. $0.3 - 7\sqrt{2}\,i$

31. $3 + i$

33. $8 - i$

35. $5 + i$

37. $-5 - 4i$

39. $-8 + 0i$

41. $2 - 6i$

43. $-1 - \dfrac{1}{2}i$

45. $-5 + 12i$

47. $5 + 0i$

49. $2 + 14i$

51. $4 - 7i$

53. 5

55. 25

57. 20

59. $-\dfrac{13}{10} + \dfrac{11}{10}i$

61. $-\dfrac{7}{25} - \dfrac{24}{25}i$

63. $\dfrac{8}{5} - \dfrac{1}{5}i$

65. $\dfrac{5}{3} - \dfrac{2}{3}i$

67. $\dfrac{4}{5} + \dfrac{8}{5}i$

69. $\dfrac{3}{13} - \dfrac{2}{13}i$

71. $\dfrac{2}{5} + \dfrac{4}{5}i$

73. $0 + \dfrac{1}{7}i$

75. $\dfrac{\sqrt{2}}{3} + \dfrac{1}{3}i$

77. $\dfrac{1}{a+bi} = \dfrac{1}{a+bi} \cdot \dfrac{a-bi}{a-bi} = \dfrac{a-bi}{a^2+b^2} = \dfrac{a}{a^2+b^2} - \dfrac{b}{a^2+b^2}i$

PROGRESS TEST 8A, page 247

1. $(x + 1)^{n-2}$

2. $\dfrac{1}{16}x^{12}y^8$

3. $-\dfrac{8y^6}{125x^3}$

4. $5x^6$

5. $\dfrac{1}{64x^{8/5}}$

6. $x^{4/3}y^2$

7. $25x^4/9y$

8. $\sqrt{(2y-1)^5}$

9. $(6y^5)^{1/3}$

10. $-4x^2y^4\sqrt{2y}$

11. $\sqrt{5x}/5$

12. $-4a^4b^3$

13. $-7\sqrt[3]{3}$

14. $5y\sqrt{x} - 8x\sqrt{y}$

15. $3(4x - 9y)$

16. $4 - 2i$

17. 2

18. i

19. $\dfrac{-4}{25} - \dfrac{22}{25}i$

20. $\dfrac{4}{25} - \dfrac{3}{25}i$

PROGRESS TEST 8B, page 248

1. x^{6n-2}

2. $-a^6b^3/27$

3. $-y^6/27x^9$

4. $\dfrac{4}{x+1}$

5. $-\dfrac{1}{216y^{3/2}}$

6. $\dfrac{x^{5/2}}{y^{4/5}}$

7. $\dfrac{x^{8/5}}{27^{2/5}y^{6/25}}$

8. $\sqrt[3]{(3x+1)^2}$

9. $(4x^3)^{1/5}$

10. $-6x\sqrt{xz}$

11. $x\sqrt{2xy}/y$

12. $-2x^5y^4\sqrt{y}$

13. $-2\sqrt[3]{2}$

14. $-3b\sqrt{a-1} - (a-1)\sqrt{b(a-1)}$

15. $-2(6a + 3b - 11\sqrt{ab})$

16. $2 + 2\sqrt{2}\,i$

17. -3

18. $-3i$

19. $\dfrac{7}{13} + \dfrac{4}{13}i$

20. $\dfrac{3}{34} + \dfrac{5}{34}i$

CHAPTER NINE

EXERCISE SET 9.1, page 254

1. ± 3
3. $\pm 5/2$
5. $\pm \sqrt{5}$
7. $3 \pm i\sqrt{2}$
9. $-5/2 \pm \sqrt{2}$
11. $-2 \pm \dfrac{\sqrt{3}}{2}i$
13. $\dfrac{5 \pm 2\sqrt{2}}{3}$
15. $\pm 2i$
17. $\pm \dfrac{8}{3}i$
19. $\pm i\sqrt{6}$
21. 1, 2
23. -2, 1
25. -4, -2
27. 0, 4
29. 1/2, 2
31. ± 2
33. 1/3, 1/2
35. 4, -2
37. $-1/2$, 4
39. 1/3, -3
41. $-\dfrac{1}{2} \pm \dfrac{1}{2}i$
43. 1, $-3/4$
45. $-\dfrac{1}{3} \pm \dfrac{\sqrt{2}}{3}i$
47. 3, -4
49. 0, $-1/3$
51. $\dfrac{-1 \pm \sqrt{11}}{2}$
53. -2, 2/3
55. $-\dfrac{5}{4} \pm \dfrac{\sqrt{7}}{4}i$
57. $\pm 1/2$
59. $-\dfrac{1}{4} \pm \dfrac{\sqrt{11}}{4}i$

EXERCISE SET 9.2, page 258

1. d
3. c
5. $a = 3, b = -2, c = 5$
7. $a = 2, b = -1, c = 5$
9. $a = 3, b = -1/3, c = 0$
11. -2, -3
13. 3, 5
15. 0, $-3/2$
17. $\dfrac{2}{5} \pm \dfrac{\sqrt{11}}{5}i$
19. $\dfrac{2}{5} \pm \dfrac{\sqrt{21}}{5}i$
21. 3, -4
23. 2/3, -1
25. ± 3
27. $\pm \dfrac{2\sqrt{3}}{3}$
29. 0, $-3/4$

EXERCISE SET 9.3, page 261

1. e
3. b
5. d
7. d
9. c
11. a
13. a
15. two complex roots
17. a double root
19. two real roots
21. two real roots
23. two complex roots
25. two complex roots
27. two real roots
29. a double root
31. 2
33. 0
35. 1
37. 0

EXERCISE SET 9.4, page 265

1. A: 3 hours, B: 6 hours
3. roofer: 6 hours, assistant: 12 hours
5. $L = 12$ feet, $W = 4$ feet
7. $L = 8$ centimeters, $W = 6$ centimeters
9. 10 feet
11. 5 or 1/5
13. 3, 7 or -3, -7
15. 6, 8
17. 150
19. 8

EXERCISE SET 9.5, page 269

1. 4   3. 2   5. 3
7. 0, 4   9. 5   11. 6
13. 2, 10   15. $u = x^2$ 17. $u = x^{2/3}$
19. $u = 1/y^2$ 21. $u = 1/x^{2/3}$ 23. $u = 2 + 3/x$
25. $u = x^2$; $\pm\sqrt{3}/3$, $\pm i\sqrt{2}$ 27. $u = 1/x$; $-3/2, 2$
29. $u = x^{1/5}$; $-1/32, -32$ 31. $u = 1 + 1/x$; $-2/7, 1/3$

EXERCISE SET 9.6, page 273

1. b, d, e   3. a, c, d, e   5. c, e
7. $-2, -3$ 9. $-1/2, 1$ 11. 0, 2
13. $-5, -3$ 15. $-1/2, 3$ 17. $-2/3, 1/2$
19. $x < -3, x > 2$ 21. $-1 < x < 5/2$

23. $x < -2, x > -3/2$ 25. $-3/2 < x < 1/2$

27. $x < -1, x > 1$ 29. $x \le -1, x \ge -1/3$

31. $x \ne -1$ 33. $2/3 \le x \le 1$

35. $2/5 \le x < 2/3$ 37. $x \ne 1$

39. $x \le -1/2, x > 3/2$ 41. $-2 < x < 2/3, x > 1$

43. $x \le -2, 2 \le x \le 3$ 45. $-5/3 < x < -1/2, x > 3$

47. (a) at least 101   (b) $0 \le x < 100$

PROGRESS TEST 9A, page 276

1. $\pm\dfrac{\sqrt{21}}{3}i$ 2. 4   3. $7/2, -1/2$ 4. $2 \pm 2\sqrt{15}\,i$

5. $4, -1/2$ 6. $2, -1/3$ 7. $\dfrac{3 \pm \sqrt{33}}{3}$

8. two real roots   9. two complex roots   10. $\dfrac{-3 \pm \sqrt{13}}{4}$

11. $0, -2/3$ 12. $-3 \le x \le 1/2$ 13. $-1 \le x < 1$

14. $L = 26$ meters, $W = 21$ meters   15. $3\sqrt{2}, 3\sqrt{2} + 6$ hours

PROGRESS TEST 9B, page 277

1. $\pm 3/2$ 2. 5 3. $-9, 21$

4. $-\dfrac{1}{6} \pm \dfrac{\sqrt{10}}{3}$ 5. $4, -3$ 6. $-1/2, -1$

7. $\dfrac{3}{2} \pm \dfrac{1}{2}i$ 8. a double root   9. two complex roots

10. $\dfrac{5 \pm i\sqrt{7}}{4}$ 11. $\pm \dfrac{\sqrt{15}}{3}i$ 12. $x \le -5/2, x \ge 2$

13. $x < -1/2, x \ge 1/2$ 14. 9 meters, 12 meters   15. 25

CHAPTER TEN

EXERCISE SET 10.1, page 283

1. $Q(x) = x - 2, R(x) = 2/(x - 5)$
3. $Q(x) = 2x - 4, R(x) = (8x - 4)/(x^2 + 2x - 1)$
5. $Q(x) = 3x^3 - 9x^2 + 25x - 75, R(x) = 226/(x + 3)$
7. $Q(x) = 2x - 3, R(x) = (-4x + 6)/(x^2 + 2)$
9. $Q(x) = x^2 - x + 1, R(x) = 0$
11. $Q(x) = x^2 - 3x, R(x) = 5/(x + 2)$
13. $Q(x) = x^3 + 3x^2 + 9x + 27, R(x) = 0$
15. $Q(x) = 3x^2 - 4x + 4, R(x) = 4/(x + 1)$
17. $Q(x) = x^4 - 2x^3 + 4x^2 - 8x + 16, R(x) = 0$
19. $Q(x) = 6x^3 + 18x^2 + 53x + 159, R(x) = 481/(x - 3)$

EXERCISE SET 10.2, page 287

1. -7 3. -34 5. 0 7. -1
9. 0 11. -62

13.

15.

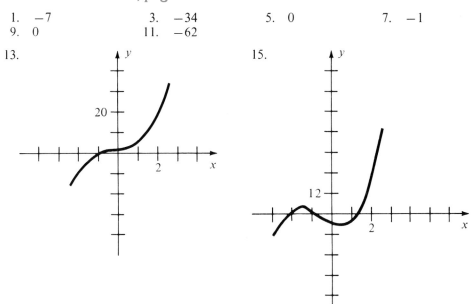

17.

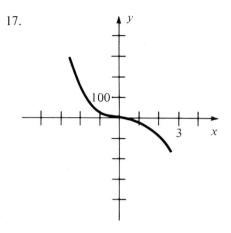

19. yes   21. no
23. yes   25. yes
27. $r = 3, -1$ 29. $5/2$

EXERCISE SET 10.3, page 295

1. 5
3. 25
5. 20
7. $-13/10 + 11i/10$
9. $-7/25 - 24i/25$
11. $8/5 - i/5$
13. $5/3 - 2i/3$
15. $4/5 + 8i/5$
17. $4/25 - 3i/25$
19. $9/10 + 3i/10$
21. $0 + i/5$
25. $x^3 - 2x^2 - 16x + 32$
27. $x^3 + 6x^2 + 11x + 6$
29. $x^3 - 6x^2 + 6x + 8$
31. $x^3/3 + x^2/3 - 7x/12 + 1/6$
33. $x^3 - 4x^2 - 2x + 8$
35. $3, -1, 2$
37. $-2, 4, -4$
39. $-2, -1, 0, -1/2$
41. $5, 5, 5, -5, -5$
43. $x^3 + 6x^2 + 12x + 8$
45. $4x^4 + 4x^3 - 3x^2 - 2x + 1$
47. $2, -1$
49. $(3 \pm i\sqrt{3})/2$
51. $-1, -2, 4$
53. $x^2 + (1 - 3i)x - (2 + 6i)$
55. $x^2 - 3x + (3 + i)$
57. $x^3 + (1 + 2i)x^2 + (-8 + 8i)x + (-12 + 8i)$
59. $(x^2 - 6x + 10)(x - 1)$
61. $(x^2 + 2x + 5)(x^2 + 2x + 4)$
63. $(x - 2)(x + 2)(x - 3)(x^2 + 6x + 10)$
65. $x - (a + bi)$

EXERCISE SET 10.4, page 301

	positive roots	negative roots	complex roots
1.	3	1	0
	1	1	2
3.	0	0	6
5.	3	2	0
	1	2	2
	3	0	2
	1	0	4
7.	1	2	0
	1	0	2
9.	2	0	2
	0	0	4
11.	1	1	6

13. $1, -2, 3$
15. $2, -1, -1/2, 2/3$
17. $1, -1, -1, 1/5$
19. $1, -3/4$
21. $3, 3, 1/2$
23. $-1, 3/4, \pm i$
25. $3/5, \pm 2, \pm i\sqrt{2}$
27. $0, 1/2, 2/3, -1$
29. $1/2, -4, 2 \pm \sqrt{2}$
31. $k = 3, r = -2$
33. $k = 7, r = 1$

PROGRESS TEST 10A, page 304

1. $Q(x) = 2x^2 - 5$, $R(x) = 11/(x^2 + 2)$
2. $Q(x) = 3x^3 - 7x^2 + 14x - 28$, $R(x) = 54/(x + 2)$
3. -25 4. -165
6. $x^3 - 2x^2 - 5x + 6$ 7. $x^4 - 6x^3 + 6x^2 + 6x - 7$
8. $2, \pm i$ 9. $-1, -1, (3 \pm \sqrt{17})/2$
10. $x^5 + 3x^4 - 6x^3 - 10x^2 + 21x - 9$
11. $16x^5 - 8x^4 + 9x^3 - 9x^2 - 7x - 1$
12. $x^2 - (1 + 2i)x + (-1 + i)$
13. $1/2, 1/2$
14. $1, 1 \pm i$ 15. $(x^2 - 4x + 5)(x - 2)$
16. 2 17. 1
18. none   19. $1, 1, -1, -1, 1/2$
20. $2/3, -3, \pm i$

PROGRESS TEST 10B, page 304

1. $Q(x) = 3x^3/2 + 3x^2/4 + 17x/8 + 15/16$, $R(x) = (49x - 17)/(2x^2 - x - 1)$
2. $Q(x) = -2x^2 + x + 1$, $R(x) = 0$ 3. -1
4. 24 6. $2x^4 - x^3 - 3x^2 + x + 1$
7. $x^3 - 4x^2 + 2x + 4$ 8. $1, 2, 2, 2$
9. $(-3 \pm \sqrt{13})/2, -3, -3$ 10. $8x^4 + 4x^3 - 18x^2 + 11x - 2$
11. $x^4 + 4x^3 - x^2 - 6x + 18$ 12. $x^4 - 4x^3 - x^2 + 14x + 10$
13. $(3 \pm \sqrt{17}/2)$ 14. $-2 \pm 2\sqrt{2}$
15. $(x^2 - 2x + 2)(2x^2 + 3x - 2)$ 16. 1
17. 2 18. $-1, 2/3, -2$
19. $-1/2, 3/2, \pm i$ 20. $0, 1/2, \pm \sqrt{2}$

CHAPTER ELEVEN

EXERCISE SET 11.1, page 317

1. (a) 5 (b) $x^2 + x - 1$ (c) 9
3. (a) $-5/4$ (b) $\dfrac{x^2 + 1}{x - 2}$ (c) 0
5. (a) $x \neq 2$ (b) all real numbers
7. (a) 1 (b) $2x^2 + x$ (c) $2x^3 + 2x^2 - x - 1$
9. (a) $-17/2$ (b) -34 (c) 2
11. (a) 21 (b) $4x^2 + 2x + 1$ (c) 105
13. (a) -5 (b) $4x + 3$ (c) 210
15. (a) 8 (b) $x + 6$ (c) $\sqrt{15}$
17. (a) $x \geq -2$ (b) all real numbers
19. (a) $x + 1$ (b) $x + 1$ 21. (a) $\dfrac{x - 1}{x}, -\dfrac{x + 1}{x}$
23. $f(x) = x^8$; $g(x) = 3x + 2$ 25. $f(x) = x^{1/3}$; $g(x) = x^3 - 2x^2$
27. $f(x) = x^{20}$; $g(x) = 3x^2 + 1$ 29. $f(x) = \sqrt{x}$; $g(x) = 4 - x$

31. $f(x) = x^{-10}$; $g(x) = 2 - 5x^2$ 33. $f(x) = \sqrt{x}$; $g(x) = \dfrac{x - 2}{x + 5}$

41. $f^{-1}(x) = \dfrac{x - 3}{2}$ 43. $f^{-1}(x) = -\dfrac{x - 3}{2}$

45. $f^{-1}(x) = 3x + 15$ 47. $f^{-1}(x) = \sqrt[3]{x - 1}$

49. $f^{-1}(x) = \sqrt{x}$ 51. (a) $f^{-1}(x) = 3x - 6$

                                                     (b) 0   (c) 2   (d) 3

53. (a) $h^{-1}(x) = -\dfrac{x - 2}{3}$ (b) 4/3   (c) 3   (d) -3

55. (a) $f^{-1}(x) = \sqrt[3]{x + 2}$ (b) $\sqrt[3]{5}$ (c) 1   (d) -2

57. yes   59. no   61. yes   63. no

EXERCISE SET 11.2, page 326

1.

3.

5.

7.

9.

11.

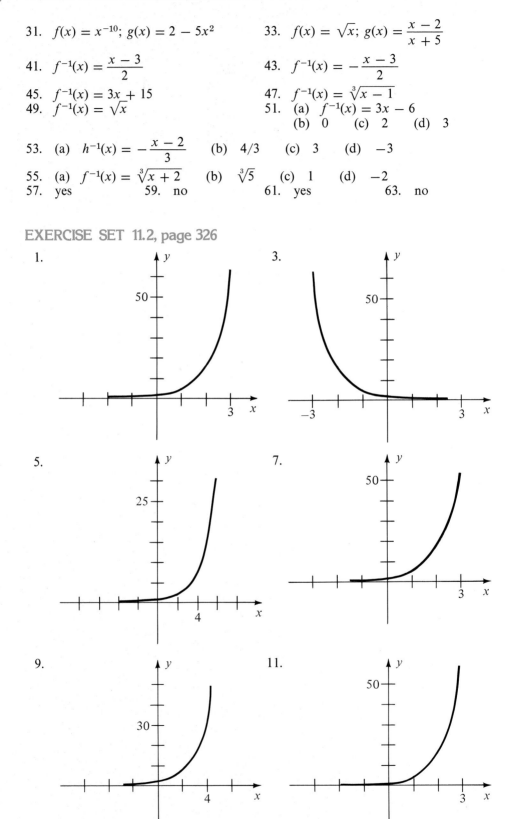

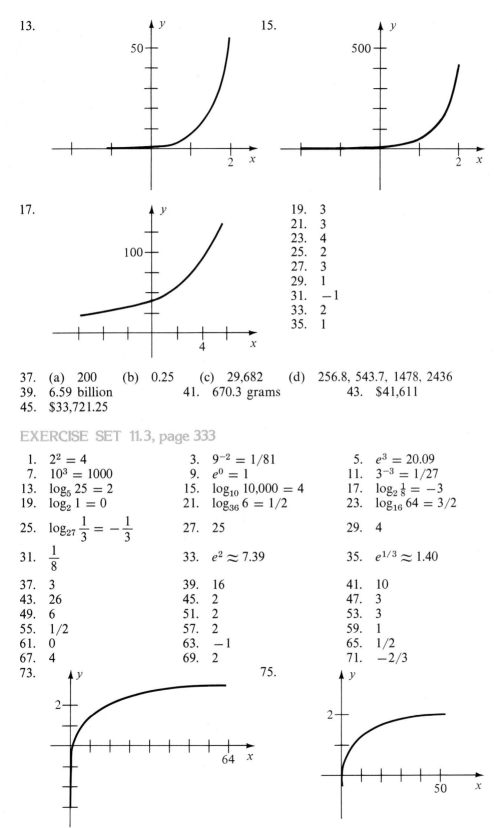

13.

15.

17.

19. 3
21. 3
23. 4
25. 2
27. 3
29. 1
31. −1
33. 2
35. 1

37. (a) 200   (b) 0.25   (c) 29,682   (d) 256.8, 543.7, 1478, 2436
39. 6.59 billion   41. 670.3 grams   43. $41,611
45. $33,721.25

EXERCISE SET 11.3, page 333

1. $2^2 = 4$

3. $9^{-2} = 1/81$

5. $e^3 = 20.09$

7. $10^3 = 1000$

9. $e^0 = 1$

11. $3^{-3} = 1/27$

13. $\log_5 25 = 2$

15. $\log_{10} 10,000 = 4$

17. $\log_2 \frac{1}{8} = -3$

19. $\log_2 1 = 0$

21. $\log_{36} 6 = 1/2$

23. $\log_{16} 64 = 3/2$

25. $\log_{27} \dfrac{1}{3} = -\dfrac{1}{3}$

27. 25

29. 4

31. $\dfrac{1}{8}$

33. $e^2 \approx 7.39$

35. $e^{1/3} \approx 1.40$

37. 3

39. 16

41. 10

43. 26

45. 2

47. 3

49. 6

51. 2

53. 3

55. 1/2

57. 2

59. 1

61. 0

63. −1

65. 1/2

67. 4

69. 2

71. −2/3

73.

75.

77. 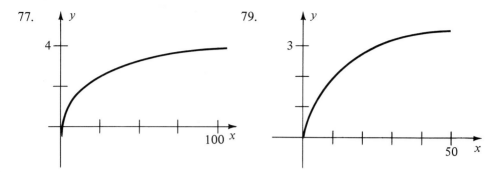   79.

EXERCISE SET 11.4, page 338

1. $\log_2 12 = \log_2 3 + \log_2 4$
3. $\log_4(8.4 \cdot 1.5) = \log_4 8.4 + \log_4 1.5$
5. $\log_2 5^3 = 3\log_2 5$
7. $\ln 15 = \ln 5 + \ln 3$
9. $\ln(4 \cdot 7) = \ln 4 + \ln 7$
11. $\ln 4^4 = 4 \ln 4$
13. $\log 120 + \log 36$
15. 4
17. $\log_a 2 + \log_a x + \log_a y$
19. $\log_a x - \log_a y - \log_a z$
21. $5 \ln x$
23. $2 \log_a x + 3 \log_a y$

25. $\dfrac{1}{2}(\log_a x + \log_a y)$
27. $2 \ln x + 3 \ln y + 4 \ln z$

29. $\dfrac{1}{2} \ln x + \dfrac{1}{3} \ln y$
31. $2 \log_a x + 3 \log_a y - 4 \log_a z$

33. $-2 \ln a$
35. 0.77
37. 0.94
39. 1.07
41. 0.07
43. -0.53
45. 0.56
47. $\log(x^2 \sqrt{y})$

49. $\ln \sqrt[3]{xy}$
51. $\log_a \dfrac{x^{1/3}y^2}{z^{3/2}}$
53. $\log_a \sqrt{xy}$

55. $\ln \dfrac{\sqrt[3]{x^2 y^4}}{z^3}$
57. $\log_a \dfrac{\sqrt{x-1}}{(x+1)^2}$
59. $\log_a \dfrac{x^3(x+1)^{1/6}}{(x-1)^2}$

EXERCISE SET 11.5, page 344

1. 2.725×10^3
3. 8.4×10^{-3}
5. 7.16×10^5
7. 2.962×10^2
9. 9.6501×10^2
11. 3.45×10^0
13. 0.5514
15. 1.1632
17. 1.5740
19. 1.5476
21. 1.8692
23. 4.6830
25. -0.4660
27. -2.2741
29. -0.0969
31. 2.520
33. 0.6000
35. 541.0
37. 7.900
39. 0.01550
41. 0.0504
43. 0.08551
45. 0.4878
47. 0.001096
49. 652.2
51. 1.524
53. 805.4
55. 0.05453
57. 0.4786
59. 0.003324
61. 1.028
63. 671.4
65. 2.115
67. 1883
69. 103.5
71. 3.837×10^8
73. 230.3
75. 0.02066
77. 21.36
79. 2.404
81. $12,910
83. 8%
85. 87.0

EXERCISE SET 11.6, page 348

1. $\log_5 18$

3. $(\log_2 7) + 1$

5. $\dfrac{\log_3 46}{2}$

7. $\dfrac{5 + \log_5 564}{2}$

9. $\dfrac{\log 2 + \log 3}{\log 3 - 2 \log 2}$

11. $-\log_2 15$

13. $-\dfrac{(\log_4 12) - 1}{2}$

15. $\ln 18$

17. $\dfrac{(\ln 20) - 3}{2}$

19. 500

21. $x = 1/2$

23. 5

25. 3

27. 8

29. $-1 + \sqrt{17}$

31. 36.62 years

33. 12.6 hours

35. 8.8 years

37. 27.47 days

39. 1.386 days

PROGRESS TEST 11A, page 351

1. (a) $-1/2$ (b) -42

2. $g[f(x)] = \dfrac{(2x + 4)}{2} - 2 = (x + 2) - 2 = x$

   $f[g(x)] = 2\left(\dfrac{x}{2} - 2\right) + 4 = (x - 4) + 4 = x$

3. 3

4. 2

5. $64^{1/2} = 8$

6. $\log_9 27 = 3/2$

7. $e^{-1/4} - 1$

8. 2

9. -2

10. 5

11. $-1/3$

12.

13. $\dfrac{1}{5}\left[3 \log(x - 1) + 2 \log y - \dfrac{1}{2}\log z\right]$

14. $\log_a \dfrac{\sqrt[3]{x}}{\sqrt{y}}$

16. $\dfrac{\log_2 14 + 1}{3}$

17. $\dfrac{199}{98}$

PROGRESS TEST 11B, page 352

1. (a) $\dfrac{2x - 1}{x^2 + 3}$ (b) 19

2. $f[g(x)] = -3\left(-\dfrac{1}{3}x + \dfrac{1}{3}\right) + 1 = (x - 1) + 1 = x$

   $g[f(x)] = -\dfrac{1}{3}(-3x + 1) + \dfrac{1}{3} = \left(x - \dfrac{1}{3}\right) + \dfrac{1}{3} = x$

3. $-2/3$

4. 8

5. $2^{-3} = 1/8$

6. $\log_{64} 16 = 2/3$

7. $e^{1/12}$

8. 27^3

9. -1

10. 5

11. $5/2$

12.

13. $2 \log x + \dfrac{1}{2} \log(2y - 1) - 3 \log y - \log z$

14. $\log \dfrac{(x - 1)x^3}{\sqrt[3]{x + 1}}$

15. 0.5

16. $\log_3(3/10)$

17. $-9/19$

CHAPTER TWELVE

EXERCISE SET 12.1, page 358

1. $3\sqrt{2}$

3. $4\sqrt{2}$

5. $\sqrt{65}$

7. $2\sqrt{10}$

9. $\sqrt{229}/4$

11. $\sqrt{13}$

13. $(5/2, 5)$

15. $(1, 5/2)$

17. $(-7/2, -1)$

19. $(0, -1/2)$

21. $(-1, 9/2)$

23. $(0, 0)$

EXERCISE SET 12.2, page 361

1. none

3. origin

5. y-axis

7. x-axis

9. y-axis

11. none

13. none

15. all

17. origin

19. y-axis

21. y-axis

23. y-axis

25. all

27. all

29. all

31. none

33. origin

EXERCISE SET 12.3, page 365

1. $(x - 2)^2 + (y - 3)^2 = 4$

3. $(x + 2)^2 + (y + 3)^2 = 5$

5. $x^2 + y^2 = 9$

7. $(x + 1)^2 + (y - 4)^2 = 8$

9. $(h, k) = (2, 3); r = 4$

11. $(h, k) = (2, -2); r = 2$

13. $(h, k) = \left(-4, -\dfrac{3}{2}\right); r = 3\sqrt{2}$

15. no graph

17. $(x + 2)^2 + (y - 4)^2 = 16; (h, k) = (-2, 4); r = 4$

19. $\left(x - \dfrac{3}{2}\right)^2 + \left(y - \dfrac{5}{2}\right)^2 = \dfrac{11}{2}; (h, k) = \left(\dfrac{3}{2}, \dfrac{5}{2}\right); r = \dfrac{\sqrt{22}}{2}$

21. $(x - 1)^2 + y^2 = \dfrac{7}{2}; (h, k) = (1, 0); r = \dfrac{\sqrt{14}}{2}$

23. $(x - 2)^2 + (y + 3)^2 = 8; (h, k) = (2, -3); r = 2\sqrt{2}$

25. $(x - 3)^2 + (y + 4)^2 = 18; (h, k) = (3, -4); r = 3\sqrt{2}$

27. $\left(x + \dfrac{3}{2}\right)^2 + \left(y - \dfrac{5}{2}\right)^2 = \dfrac{3}{2}$; $(h, k) = \left(-\dfrac{3}{2}, \dfrac{5}{2}\right)$; $r = \dfrac{\sqrt{6}}{2}$

29. $(x - 3)^2 + y^2 = 11$; $(h, k) = (3, 0)$; $r = \sqrt{11}$

31. $\left(x - \dfrac{3}{2}\right)^2 + (y - 1)^2 = \dfrac{17}{4}$; $(h, k) = \left(\dfrac{3}{2}, 1\right)$; $r = \dfrac{\sqrt{17}}{2}$

33. $(x + 2)^2 + \left(y - \dfrac{2}{3}\right)^2 = \dfrac{100}{9}$; $(h, k) = \left(-2, \dfrac{2}{3}\right)$; $r = \dfrac{10}{3}$

35. not a circle

EXERCISE SET 12.4, page 371

1.
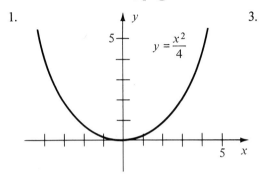

3.
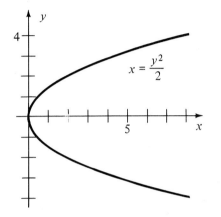

5.
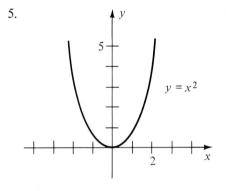

7.
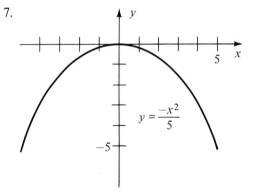

9.
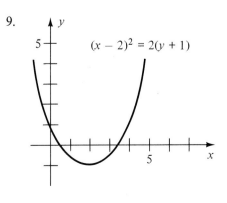

11.
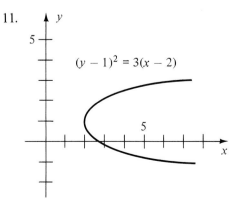

13.

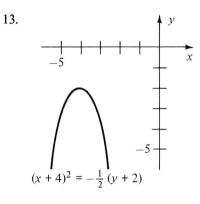

$$(x + 4)^2 = -\frac{1}{2}(y + 2)$$

15.

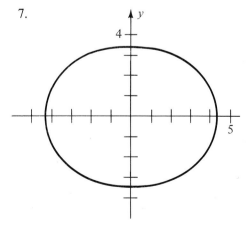

17. $V = (1, 2)$, $x = 1$, upward
19. $V = (2, 4)$, $y = 4$, opens left
21. $V = (1/2, -1/4)$, $x = 1/2$, downward
23. $V = (-1/3, 5)$, $y = 5$, opens right
25. $V = (3/2, -5/12)$, $x = 3/2$, upward
27. $V = (4, -3)$, $y = -3$, opens left
29. $V = (-1, -1)$, $x = -1$, downward

EXERCISE SET 12.5, page 379

1.

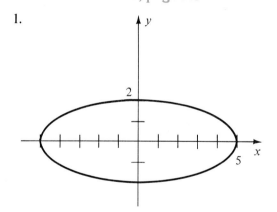

3.

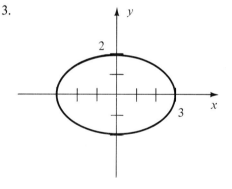

5.

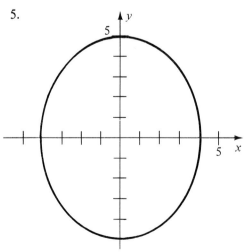

7.

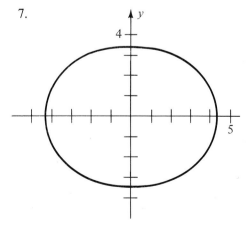

9. $\dfrac{x^2}{9} + \dfrac{y^2}{4} = 1$; $(0, \pm 2), (\pm 3, 0)$

11. $\dfrac{x^2}{4} + \dfrac{y^2}{1} = 1$; $(0, \pm 1), (\pm 2, 0)$

13. $\dfrac{x^2}{1} + \dfrac{y^2}{\frac{1}{4}} = 1$; $\left(0, \pm \dfrac{1}{2}\right), (\pm 1, 0)$

15. $\dfrac{x^2}{3} + \dfrac{y^2}{4} = 1$; $(0, \pm 2), (\pm \sqrt{3}, 0)$

17. $\dfrac{x^2}{\frac{1}{4}} + \dfrac{y^2}{\frac{9}{8}} = 1$; $\left(0, \pm \dfrac{3\sqrt{2}}{4}\right), \left(\pm \dfrac{1}{2}, 0\right)$

19.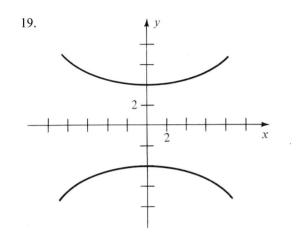

21.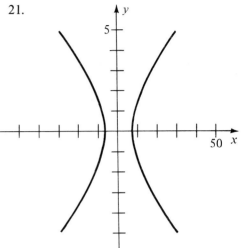

23.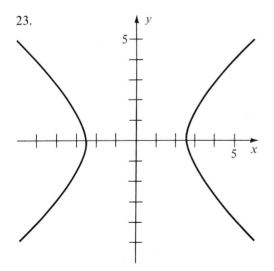

25.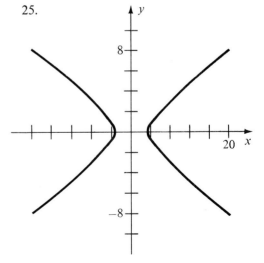

27. $\dfrac{x^2}{4} - \dfrac{y^2}{64} = 1$; $(\pm 2, 0)$

29. $\dfrac{y^2}{\frac{1}{4}} - \dfrac{x^2}{\frac{1}{4}} = 1$; $\left(0, \pm \dfrac{1}{2}\right)$

31. $\dfrac{x^2}{5} - \dfrac{y^2}{4} = 1$; $(\pm \sqrt{5}, 0)$

33. $\dfrac{y^2}{16} - \dfrac{x_2}{4} = 1$; $(0, \pm 4)$

35. $\dfrac{x^2}{4} - \dfrac{y^2}{8} = 1$; $(\pm 2, 0)$

37.

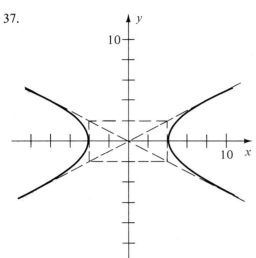

39.

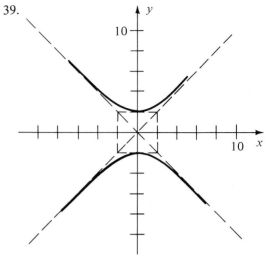

41.

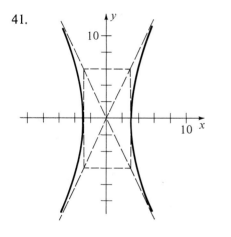

43.

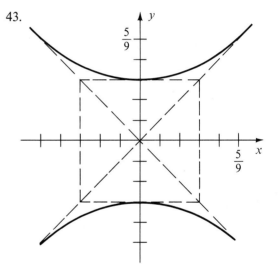

45.

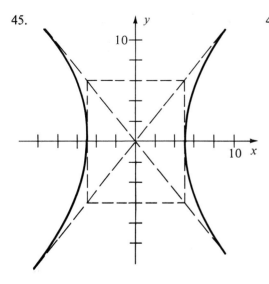

47.

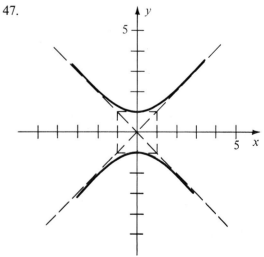

EXERCISE SET 12.6, page 382

1. parabola
3. circle
5. hyperbola
7. no graph
9. no graph
11. no graph
13. hyperbola
15. point
17. no graph
19. circle
21. parabola
23. no graph
25. circle
27. parabola
29. ellipse

PROGRESS TEST 12A, page 385

1. $7\sqrt{2}$
2. $(5/4, -1/8)$
5. x-axis
6. x-axis
7. $(h, k) = (4, -3);\ r = \sqrt{10}$
8. $V = (-1, 1/2);\ y = 1/2$

9. intercepts: $\left(0, \pm\dfrac{1}{2}\right)$; asymptotes: $y = \pm\dfrac{1}{6}x$

10. $\left(x - \dfrac{2}{3}\right)^{2} + (y + 3)^2 = 3$

11.
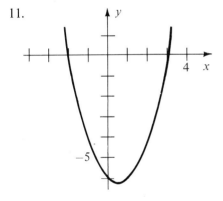

12.
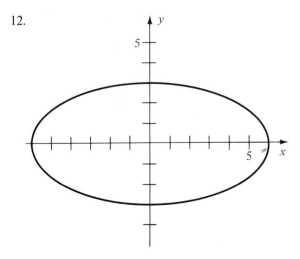

13. hyperbola
14. parabola
15. circle

PROGRESS TEST 12B, page 385

1. $2\sqrt{5}$ 2. $(-7/4, 1/4)$ 3. $(2, 5)$
5. origin   6. y-axis
7. $(h, k) = (-1/2, 3);\ r = 1/2$ 8. $V = (1/4, -2);\ x = 1/4$

9. intercepts: $\left(\pm\dfrac{\sqrt{6}}{3}, 0\right)$; asymptotes: $y = \pm\dfrac{\sqrt{6}}{4}x$

10. $(x + 1)^2 + \left(y + \dfrac{1}{2}\right)^2 = 5$

11.

12.

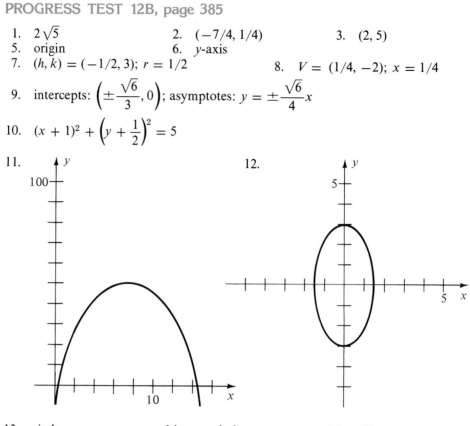

13. circle   14. parabola   15. ellipse

CHAPTER THIRTEEN

EXERCISE SET 13.1, page 393

1. e

3.

5.

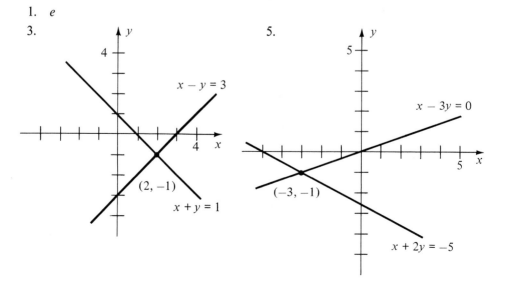

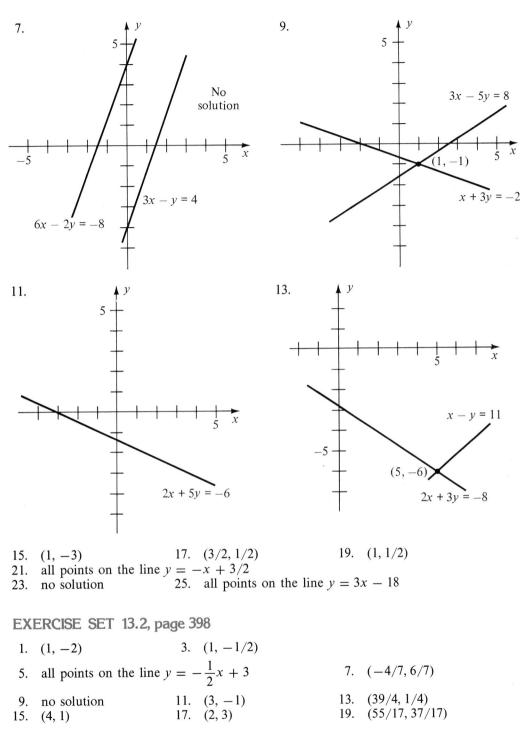

7.

No solution

$3x - y = 4$

$6x - 2y = -8$

9.

$3x - 5y = 8$

$(1, -1)$

$x + 3y = -2$

11.

$2x + 5y = -6$

13.

$x - y = 11$

$(5, -6)$

$2x + 3y = -8$

15. $(1, -3)$ 17. $(3/2, 1/2)$ 19. $(1, 1/2)$
21. all points on the line $y = -x + 3/2$
23. no solution   25. all points on the line $y = 3x - 18$

EXERCISE SET 13.2, page 398

1. $(1, -2)$ 3. $(1, -1/2)$

5. all points on the line $y = -\dfrac{1}{2}x + 3$ 7. $(-4/7, 6/7)$

9. no solution   11. $(3, -1)$ 13. $(39/4, 1/4)$
15. $(4, 1)$ 17. $(2, 3)$ 19. $(55/17, 37/17)$

EXERCISE SET 13.3, page 406

1. 25 nickels, 15 dimes   3. color: $2.50, black and white: $1.50
5. $4000 in bond A, $2000 in bond B
7. 10 rolls of 12″, 4 rolls of 15″

9.   8 pounds of $1.20 coffee, 16 pounds of $1.80 coffee
11. speed of bicycle $= 105/8$ mph, wind speed $= 15/8$ mph
13. 34
15. 30 pounds of nuts, 20 pounds of raisins
17. $6000 in type A, $12,000 in type B
19. 5 mg. Epiline I, 4 mg. Epiline II
21. (a) $R = 95x$

(b)

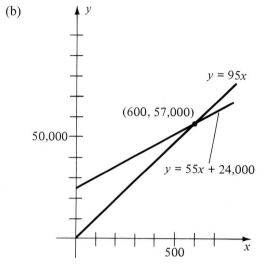

$y = 95x$

$(600, 57,000)$

$50,000$

$y = 55x + 24,000$

500

(c)   57,000
23. (a) $p = 3$ (b)   16

EXERCISE SET 13.4, page 410

1. $x = 2, y = -1, z = -2$
5.   no solution
9. $x = 1, y = 1, z = 0$
13. no solution
15. $x = 102/19, y = 86/19, z = -36/19$
17. no solution
21. 2 units of A, 3 units of B, 3 units of C
23. three 12″-sets, eight 16″-sets, five 19″-sets

3. $x = 1, y = 2/3, z = -2/3$
7. $x = 1, y = 2, z = 2$
11. $x = 1, y = 27/2, z = -5/2$

19. $x = 8, y = -12, z = -11$

EXERCISE SET 13.5, page 414

1. $x = 3, y = 2; x = 1/5, y = -18/5$
3. $x = 1, y = 1; x = 9/16, y = -3/4$
5. $x = 1, y = 2; x = 13/5, y = -6/5$
7. $x = \dfrac{-1 + \sqrt{5}}{2}, y = \dfrac{1 + \sqrt{5}}{2}; x = \dfrac{-1 - \sqrt{5}}{2}, y = \dfrac{1 - \sqrt{5}}{2}$
9. $x = 3, y = 2; x = 3, y = -2$
11. $x = 3, y = 2; x = -3, y = 2; x = 3, y = -2; x = -3, y = -2$
13. no solution
15. $x = \sqrt{2}, y = 5; x = -\sqrt{2}, y = 5; x = \sqrt{2}, y = -5; x = -\sqrt{2}, y = -5$
17. $x = 1, y = -1; x = 5/2, y = 1/2$
19. 6 and 8   21.   4 and 5

1.

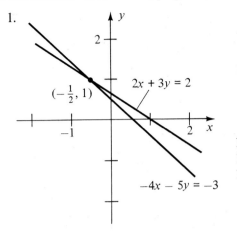

$(-\frac{1}{2}, 1)$

$2x + 3y = 2$

$-4x - 5y = -3$

2. $x = 5, y = -1$
3. $x = -4, v = 3/2$
4. $x = 1/3, y = 2/3$
5. no solution
6. $x = -1/4, y = -1/3$
7. $x = -66/13, y = 61/39, z = -5/3$
8. 45
9. plane: 600 kph, wind: 100 kph
10. A: 50 cents, B: 60 cents
11. $x = 575, R = \$9200$
12. $p = 3, S = 11$

1.

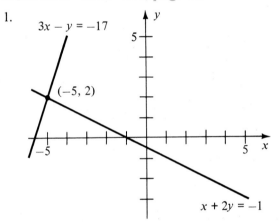

$3x - y = -17$

$(-5, 2)$

$x + 2y = -1$

2. $x = -1, y = 6$

3. $x = 1/3, y = -1/3$

4. $x = -1/2, y = 1/4$

5. all points on the line $y = \dfrac{7}{2}x - \dfrac{3}{2}$

6. $x = -3, y = -2$

7. $x = -5, y = 16/7, z = 11/21$

8. 37

9. boat: $\dfrac{35}{2}$ kph, current: $\dfrac{5}{2}$ kph

10. pencil: 3 cents, pen: 12 cents

11. 1100

12. $p = 6, S = 31$

CHAPTER FOURTEEN

1. 2×2 3. 4×3 5. 3×3
7. (a) -4 (b) 7   (c) 6   (d) -3
9. $\begin{bmatrix} 3 & -2 \\ 5 & 1 \end{bmatrix}, \begin{bmatrix} 3 & -2 & 12 \\ 5 & 1 & -8 \end{bmatrix}$

11. $\begin{bmatrix} \frac{1}{2} & 1 & 1 \\ 2 & -1 & -4 \\ 4 & 2 & -3 \end{bmatrix}, \begin{bmatrix} \frac{1}{2} & 1 & 1 & 4 \\ 2 & -1 & -4 & 6 \\ 4 & 2 & -3 & 8 \end{bmatrix}$

13. $\frac{3}{2}x + 6y = -1$
    $4x + 5y = 3$

15. $\begin{aligned} x + y + 3z &= -4 \\ -3x + 4y &= 8 \\ 2x + 7z &= 6 \end{aligned}$

17. yes
19. yes
21. $x = -13, y = 8, z = 2$
23. $x = 35, y = 14, z = -4$
25. The answer is not unique.
   A possible answer is
   $\begin{bmatrix} 1 & -2 & -1 & \frac{1}{2} \\ 0 & 1 & -1 & \frac{3}{2} \\ 0 & 0 & 12 & -11 \end{bmatrix}$
27. The answer is not unique.
   A possible answer is
   $\begin{bmatrix} 1 & 2 & 1 & 0 \\ 0 & 1 & -\frac{1}{5} & -\frac{1}{5} \\ 0 & 0 & 1 & 1 \end{bmatrix}$

29. $x = 2, y = 3$
31. $x = 2, y = -1, z = 3$
33. $x = 3, y = 2, z = -1$
35. $x = -5, y = 2, z = 3$
37. $x = -5/7, y = -2/7, z = -3/7, w = 2/7$

EXERCISE SET 14.2, page 432

1. (a) -6 (b) -1 (c) 1 (d) 7
3. (a) -6 (b) 1 (c) 1 (d) 7
5. (a) 11 (b) 12 (c) 4 (d) -12
7. (a) -11 (b) -12 (c) -4 (d) -12
9. F
11. F
13. T
15. 22
17. -8
19. 0
21. 52
23. -3
25. 0
27. -12
29. 0

EXERCISE SET 14.3, page 436

1. $x = 2, y = -2, z = 3$
3. $x = 4, y = 2, z = 0$
5. no solution
7. $x = 6, y = -3, z = -1$
9. $x = 1/3, y = -2/3, z = -1$

PROGRESS TEST 14A, page 438

1. (a) 3×2 (b) -3
2. $3x - 5y = 4, 4x + 2y = -1$
3. $x = 1/2, y = -1, z = -2$
4. (a) $\begin{bmatrix} 2 & 0 & 4 & -3 \\ 0 & 3 & -10 & -2 \\ -1 & 1 & 5 & 4 \end{bmatrix}$ (b) $\begin{bmatrix} 2 & 0 & 4 & -3 \\ 4 & 3 & -2 & -8 \\ 0 & 1 & 7 & \frac{5}{2} \end{bmatrix}$
5. $x = -5, y = 2$
6. $x = 3, y = 1/3, z = -2$
7. -10
8. $15, -7, -6$
9. -34
10. 256
11. $x = 2/3, y = -1/3$
12. $x = 3, y = -2, z = 5$

PROGRESS TEST 14B, page 439

1. (a) 2×3 (b) 0.6
2. $2.6x + 1.5y = -13, 0.2x - 3.7y = 7$
3. $x = 4, y = -1, z = -3/2$

4. (a) $\begin{bmatrix} -1 & 2 & 4 & 0 \\ 0 & 6 & 8 & -3 \\ -\frac{1}{2} & -5 & -2 & 7 \end{bmatrix}$ (b) $\begin{bmatrix} -1 & 2 & 4 & 0 \\ 3 & 0 & -4 & -3 \\ 0 & -6 & -4 & 7 \end{bmatrix}$

5. $x = 20/3,\ y = -10/3$ 6. $x = 3,\ y = -2,\ z = -5$
7. 4   8. 6, 0, -12
9. 40   10. -38
11. $x = 1,\ y = 1$ 12. $x = 3,\ y = -1,\ z = 1/2$

CHAPTER FIFTEEN

EXERCISE SET 15.1, page 447

1. 2, 4, 6, 8   3. 5, 7, 9, 11   5. 2, 5, 8, 11
7. 4/3, 5/3, 2, 7/3   9. 1, 2, 3, 4   11. 1/3, 4/5, 9/7, 16/9
13. yes   15. no   17. yes
19. yes   21. no   23. yes
25. 2, 6, 10, 14   27. 3, 5/2, 2, 3/2   29. -4, 0, 4, 8
31. 21, 17, 13, 9   33. 1/3, 0, $-1/3$, $-2/3$ 35. 25
37. -8 39. -2 41. 19/3
43. 821/160   45. 440   47. -126
49. 1720   51. 30   53. $n = 30,\ d = 3$
55. $n = 6,\ d = 1/4$ 57. -2

EXERCISE SET 15.2, page 454

1. yes, $r = 2$ 3. no   5. yes, $r = -3/4$
7. yes, $r = 1/5$ 9. no   11. 3, 9, 27, 81
13. 4, 2, 1, 1/2   15. 1/2, 2, 8, 32   17. -3, -6, -12, -24
19. -384 21. 3/32   23. $-160/81$
25. 1/243   27. 27/8   29. ±2
31. 7   33. $6\sqrt[3]{4},\ 24\sqrt[3]{2}$ 35. 1/4, 1/16
37. 1093/243   39. $-1353/625$ 41. 1020
43. 55/8   45. \$10,235   47. 58,594
49. 2   51. 3/4   53. 8/3
55. 1   57. 1/5

EXERCISE SET 15.3, page 460

1. $243x^5 + 810x^4y + 1080x^3y^2 + 720x^2y^3 + 240xy^4 + 32y^5$
3. $256x^4 - 256x^3y + 96x^2y^2 - 16xy^3 + y^4$
5. $32 - 80xy + 80x^2y^2 - 40x^3y^3 + 10x^4y^4 - x^5y^5$
7. $a^8b^4 + 12a^6b^3 + 54a^4b^2 + 108a^2b + 81$
9. $a^8 - 16a^7b + 112a^6b^2 - 448a^5b^3 + 1120a^4b^4 - 1792a^3b^5 +$
    $1792a^2b^6 - 1024ab^7 + 256b^8$

11. $\dfrac{1}{27}x^3 + \dfrac{2}{3}x^2 + 4x + 8$

13. $1024 + 5120x + 11,520x^2 + 15,360x^3$
15. $19,683 - 118,098a + 314,928a^2 - 489,888a^3$
17. $16,384x^{14} - 344,064x^{13}y + 3,354,624x^{12}y^2 - 20,127,744x^{11}y^3$
19. $8192x^{13} - 53,248x^{12}yz + 159,744x^{11}y^2z^2 - 292,864x^{10}y^3z^3$

21. 120   23. 12   25. 990
27. 5040   29. 120   31. 210

33. $-35{,}840x^4$ 35. $\dfrac{495}{256}x^8y^4$ 37. $2016x^{-5}$

39. $-540x^3y^3$ 41. $181{,}440x^4y^3$ 43. $-144x^6$

45. $\dfrac{35}{8}x^{12}$ 47. 4.8268

EXERCISE SET 15.4, page 467

1. 120   3. 146,016   5. 256
7. 5040   9. 720   11. 336
13. 90   15. 84   17. 3
19. 210   21. 120   23. 60
25. 336   27. 120   29. 84
31. 45   33. 1   35. n

37. $\dfrac{n^2 + n}{2}$ 39. 3003

41. (a) 15,600   (b) 17,576   43. 12,271,512
45. 240   47. 59,400

49. $\dfrac{(26!)^2}{6!4!22!20!}$

PROGRESS TEST 15A, page 471

1. $-3, 2, 5/7, 3/7$ 2. $-1/2$ 3. 32
4. 21   5. $n = 11, d = 4$ 6. 1/4
7. 189/16   8. 10

9. $x^4 - 2x^3y + \dfrac{3}{2}x^2y^2 - \dfrac{1}{2}xy^3 + \dfrac{1}{16}y^4$ 10. 3/5

11. 2520   12. 4800   13. 126
14. 720

PROGRESS TEST 15B, page 471

1. $2, 9/2, 28/3, 65/4$ 2. 13/2   3. 49/2
4. 135/4   5. $n = 25, d = -3/2$
6. 243   7. 11/2   8. 9/5

9. $\dfrac{x^8}{16} - \dfrac{1}{2}x^6y + \dfrac{3}{2}x^4y^2 - 2x^2y^3 + y^4$ 10. $\dfrac{7}{n^2 - n}$

11. 120   12. 14,400   13. 84
14. 14,400

INDEX

A

Absolute value, 28, 84
    in inequalities, 85
Abscissa, 146
Addition
    of algebraic fractions, 120
    of complex numbers, 241
    of fractions, 11
    of polynomials, 48
Algebraic expression, 16
Algebraic fractions, 113
    addition of, 120
    division of, 114
    least common denominator
      of, 120
    multiplication of, 114
    simplification of, 116
    subtraction of, 120
Algebraic numbers, 300
Algebraic operations, 16
Analytic geometry, 353
Arithmetic progression, 443
    common difference in, 443

Arithmetic sequence, 443
Arithmetic series, 445
Associative laws, 24
Asymptotes, 378
Augmented matrix, 421
Axis of symmetry, 367

B

Back-substitution, 408
Base, 41, 319
Binomial expansion, 456
Binomial formula, 457
Binomial theorem, 457
Break-even analysis, 401
Break-even point, 402

C

Cancellation principle, 10, 116
Cartesian coordinate system, 146
Center of a circle, 362
Characteristic, 340

Circle, 362
    center of, 362
    equation of, 363
    general form of, 364
    radius of, 364
Coefficient, 43
Coefficient matrix, 421
Cofactors, 429
    expansion by, 430
Coin problems, 95
Columns of a matrix, 420
Combination, 465
Common difference, 443
Common factor, 10
Common ratio, 449
Commutative laws, 24
Completing the square, 252
Complex fractions, 125
Complex numbers, 239
    conjugate, 242, 288
    equality of, 240
    imaginary part of, 239
    multiplication of, 241
    quotient of, 242
    real part of, 239
    reciprocal of, 243, 288
    sum of, 241
Composite function, 310
Compound interest, 325
Conic sections, 362, 380
    axes of symmetry, 380
    characteristics of, 382
Conjugate complex numbers, 242,
    288
Conjugate Roots Theorem, 293
Constant, 16
    of variation, 177, 178
Constant functions, 174, 190
Constant term, 44
Conversion period, 325
Coordinate axes, 146
Coordinates of a point, 146
Counting principle, 462
Cramer's rule, 433
Critical value, 270
Cube root, 225

D

Decreasing functions, 172, 190
Degree of a polynomial, 44

Degree of a term, 48
Demand, 404
Denominator, 7
Dependent variable, 155
Descartes' Rule of Signs, 297
Determinant, 428
    cofactor of, 429
    minor of, 429
    of order 3, 428
    of second order, 428
Dimension of a matrix, 420
Direct variation, 176
Directrix, 366
Discriminant, 259
Distance formula, 354
Distance problems, 102
Distributive laws, 26
Dividend, 6
Division
    of algebraic fractions, 114
    of fractions, 8
    of polynomials, 280
Divisor, 6
Domain, 155
Double root, 259

E

e, 301, 322
Element of a set, 1
Elementary row operations, 422
Elements of a matrix, 420
Ellipse, 372
    equation of, 372
    foci of, 372
Entries of a matrix, 420
Equations, 69
    absolute value in, 84
    equivalent, 70
    first-degree, 72
    fractions in, 132
    left-hand side of, 70
    linear, 72, 387
    literal, 76
    nonlinear systems, 411
    in one variable, 72
    polynomial, 280
    quadratic, 249
    right-hand side of, 70
    roots of, 70
    second-degree, 249

solution of, 70
simplification of, 70
systems of, 387, 408
Equilibrium price, 404
Equivalent equations, 70, 395
Equivalent fraction, 9, 121
Exponential decay models, 324
Exponential equations, 346
Exponential functions, 319
Exponential growth models, 323
Exponents, 41
    integer, 215, 220
    rational, 224
Extraneous solution, 266

F

Factor, 6, 285
    common, 10
    special, 63
Factorial, 458
Factoring of polynomials, 57
Factor Theorem, 286
Finite sequence, 442
First-degree equations, 72
    general, 196
Foci of an ellipse, 372
Foci of a hyperbola, 374
Focus of a parabola, 366
Formulas, 76
    distance, 354
    midpoint, 356
Fractions, 7
    addition of, 11
    algebraic, 113
    cancellation principle of, 10
    complex, 125
    division of, 8
    equivalent, 9
    multiplication of, 8
    reduced form of, 10
    simplifying, 116
    subtraction of, 11
Function(s), 155, 308
    addition of, 308
    composite, 310
    constant, 174
    decreasing, 173
    division of, 308
    domain of, 155
    evaluation of, 160

    exponential, 319
    graphs of, 166
    increasing, 173
    inverse, 314
    linear, 169
    logarithmic, 327
    multiplication of, 308
    notation for, 159
    one-to-one, 312
    polynomial, 171, 279
    quadratic, 170, 249
    range of, 155
    subtraction of, 308
    zeros of, 249, 280
Fundamental Theorem, 290

G

Gaussian elimination, 408
General first-degree equation, 196
Geometric mean, 450
Geometric progression, 449
    common ratio in, 449
Geometric sequence, 449
Geometric series, 451
Graph
    of an equation, 149
    of a function, 166
    of an inequality, 31, 207
    of a quadratic fraction, 260
Growth constant, 323

H

Half planes, 207
Horizontal line test, 313
Horizontal lines, 200
Hyperbola, 374
    asymptotes of, 378
    equations of, 375
    foci of, 374

I

Image, 155
Imaginary number, 239
Imaginary part, 239
Increasing functions, 173, 190
Independent variable, 155
Inequalities, 29, 79
    absolute value in, 85
    fractions in, 134

Inequalities (continued)
    graphing of, 31
    linear, 79
    second-degree, 270
Infinite geometric series, 453
Infinite sequence, 442
Integer, 2
Inverse functions, 314
Inverse variation, 178
Investment problems, 99
Irrational number, 3

J

Joint variation, 179

L

Least common denominator, 11, 120
Level of production, 401
Like terms, 48
Linear equations, 72
    roots of, 72
Linear Factor Theorem, 291
Linear function, 169, 195
Linear inequalities, 79, 207
    graph of, 207
Linear systems, 388, 408
    equivalent, 395
    inconsistent, 390
    solving by Cramer's rule, 433
    solving by elimination, 395
    solving by Gaussian
      elimination, 408
    solving by graphing, 388
    solving by substitution, 391
    in triangular form, 408
Literal equations, 76
Logarithmic equations, 346
Logarithmic functions, 328
Logarithms, 328
    computing with, 340
    equations with, 346
    natural, 328, 340
    properties of, 334

M

Mantissa, 340
Matrix, 419
    augmented, 421
    coefficient, 421
    columns of, 418
    determinant of, 428
    dimension of, 420
    elements of, 420
    entries of, 420
    pivot row of, 423
    rows of, 420
    square, 420
Member of a set, 1
Midpoint formula, 356
Minors, 429
Mixture problems, 106
Multiplication
    of algebraic fractions, 114
    of fractions, 8
    of polynomials, 52

N

Natural logarithms, 328, 340
Natural number, 2
Negative direction, 4
Negative number, 4
Negative of a number, 19
Nonlinear systems, 411
Nonnegative number, 4
Nonpositive number, 4
nth root, 225
    principal, 227
Number of Roots Theorem, 291
Numerator, 7

O

One-to-one function, 312
Ordered pair, 146
Ordinate, 146
Origin, 4, 146

P

Parabola, 170, 260, 366
    axis of, 367
    directrix of, 366
    focus of, 366
    standard forms of, 368
    vertex of, 367
Parallel lines, 202

Pascal's triangle, 456
Percent, 13
    of a number, 15
Permutation, 462
Perpendicular lines, 202
Point–slope form, 193
Polynomial functions, 171
Polynomial(s), 43
    addition of, 48
    coefficients of, 43
    constant term of, 44
    degree of, 44
    division of, 280
    equations, 280
    factoring of, 57
    functions, 279
    multiplication of, 52
    subtraction of, 48
    term of, 43
Positive direction, 4
Positive number, 4
Power, 41
Prime number, 11
Principal nth root, 227
Principal square root, 226, 227
Product of real numbers, 6
Progression
    arithmetic, 443
    common difference of, 443
    common ratio of, 449
    geometric, 449
Proportion, 129

Q

Quadrant, 146
Quadratic equations, 249
    roots of, 259
Quadratic formula, 255
Quadratic functions, 170, 249, 260
Quotient
    in division, 63
    of polynomials, 281
    of real numbers, 6

R

Radical equation, 266
Radical form, 227
Radical sign, 227

Radicals, 227, 230, 235
    simplified, 231
Radius, 362
Range, 155
Ratio, 128
Rational number, 2
Rational Root Theorem, 298
Rationalizing denominators, 232
Real number line, 4
Real number system, 3
Real part, 239
Reciprocal, 8
Rectangular coordinate system, 146
Recursive formula, 443
Reduced form, 10
Remainder
    in division, 280
Remainder Theorem, 284
Repeated root, 259, 292
Roots
    of an equation, 70
    of a linear equation, 72
    of multiplicity k, 292
    of a polynomial, 280
Rows of a matrix, 420

S

Scientific notation, 340
Second-degree equations, 249
Second-degree inequalities, 270
Sequence
    arithmetic, 443
    finite, 442
    geometric, 449
    infinite, 442
    term of, 442
Series
    arithmetic, 445
    geometric, 451
    infinite geometric, 453
Set, 1
    element of, 1
    member of, 1
    solution, 30
Set-builder notation, 33
Simplifying fractions, 116
Slope, 188
Slope–intercept form, 195
Solution set, 30

Solutions
        of an equation, 70
        of an inequality, 79
        of linear systems, 388
Square matrix, 420
Square root, 225, 228
        principal, 227
Straight line(s),
        equations of, 193, 195
        horizontal, 200
        parallel, 202
        perpendicular, 202
        slope of, 188
        vertical, 200
Subset, 2
Substitution of variable, 268
Subtraction
        of algebraic fractions, 120
        of fractions, 11
        of polynomials, 48
Supply, 404
Symmetry, 359, 360
Synthetic division, 282
Systems of linear equations, 387
        see also Linear systems

T

Term of a polynomial, 43
Term of a sequence, 442
Transcendental numbers, 301
Triangular form, 408

V

Variable, 16
        dependent, 155
        independent, 155
        substitution of, 268
Variation
        constant of, 177, 178
        direct, 176
        inverse, 178
        joint, 179
Vertical line test, 157
Vertex, 367
Vertical lines, 200

W

Work problems, 138

X

x-axis, 146
x-coordinate, 146

Y

y-axis, 146
y-coordinate, 146

Z

Zeros of functions, 249, 280